Sustainable and Low Carbon Development in the Energy Sector

Sustainable and Low Carbon Development in the Energy Sector

Guest Editors

Maria Dzikuć
Aleksander Jakimowicz

MDPI

Basel • Beijing • Wuhan • Barcelona • Belgrade • Novi Sad • Cluj • Manchester

Guest Editors
Maria Dzikuć
Faculty of Economics and Management
University of Zielona Góra
Zielona Góra
Poland

Aleksander Jakimowicz
Institute of Economics
Polish Academy of Sciences
Warsaw
Poland

Editorial Office
MDPI AG
Grosspeteranlage 5
4052 Basel, Switzerland

This is a reprint of the Special Issue, published open access by the journal *Energies* (ISSN 1996-1073), freely accessible at: https://www.mdpi.com/journal/energies/special_issues/91790UKX26.

For citation purposes, cite each article independently as indicated on the article page online and as indicated below:

Lastname, A.A.; Lastname, B.B. Article Title. *Journal Name* **Year**, *Volume Number*, Page Range.

ISBN 978-3-7258-7036-3 (Hbk)
ISBN 978-3-7258-7037-0 (PDF)
https://doi.org/10.3390/books978-3-7258-7037-0

Cover image courtesy of Aleksander Jakimowicz

Contents

About the Editors vii

Preface ix

Aleksander Jakimowicz
The Energy Transition as a Super Wicked Problem: The Energy Sector in the Era of Prosumer Capitalism
Reprinted from: *Energies* **2022**, *15*, 9109, https://doi.org/10.3390/en15239109 1

Aleksander Jakimowicz
The Future of the Energy Sector and the Global Economy: Prosumer Capitalism and What Comes Next
Reprinted from: *Energies* **2022**, *15*, 9120, https://doi.org/10.3390/en15239120 32

Aleksander Jakimowicz and Daniel Rzeczkowski
Contact Zones in the Energy Transition: A Transdisciplinary Complex Problem
Reprinted from: *Energies* **2023**, *16*, 3560, https://doi.org/10.3390/en16083560 58

Aleksander Jakimowicz
The Material Entropy and the Fourth Law of Thermodynamics in the Evaluation of Energy Technologies of the Future
Reprinted from: *Energies* **2023**, *16*, 3861, https://doi.org/10.3390/en16093861 106

Tetiana Kurbatova, Iryna Sotnyk, Olha Prokopenko, Iryna Bashynska and Uliana Pysmenna
Improving the Feed-in Tariff Policy for Renewable Energy Promotion in Ukraine's Households
Reprinted from: *Energies* **2023**, *16*, 6773, https://doi.org/10.3390/en16196773 129

Second Bwanakare
Energy Efficiency Forecast as an Inverse Stochastic Problem: A Cross-Entropy Econometrics Approach
Reprinted from: *Energies* **2023**, *16*, 7715, https://doi.org/10.3390/en16237715 147

Łukasz Markowski and Kamil Kotliński
Renewables, Energy Mix and Inflation in the European Union Countries
Reprinted from: *Energies* **2023**, *16*, 7808, https://doi.org/10.3390/en16237808 163

Alicja Kolasa-Więcek, Agnieszka A. Pilarska, Małgorzata Wzorek, Dariusz Suszanowicz and Piotr Boniecki
Modeling the Consumption of Main Fossil Fuels in Greenhouse Gas Emissions in European Countries, Considering Gross Domestic Product and Population
Reprinted from: *Energies* **2023**, *16*, 7906, https://doi.org/10.3390/en16237906 178

Eulalia Skawińska and Romuald I. Zalewski
New Foods as a Factor in Enhancing Energy Security
Reprinted from: *Energies* **2024**, *17*, 192, https://doi.org/10.3390/en17010192 196

David Machacek, Nazim Yasar, Fabio Widmer, Thomas Huber and Christopher Onder
Energy Management of Hydrogen Hybrid Electric Vehicles—Online-Capable Control
Reprinted from: *Energies* **2024**, *17*, 2369, https://doi.org/10.3390/en17102369 215

Evangelia Karasmanaki, Spyros Galatsidas and Georgios Tsantopoulos
Socioeconomic Factors Driving the Transition to a Low-Carbon Energy System
Reprinted from: *Energies* **2024**, *17*, 3576, https://doi.org/10.3390/en17143576 240

Christiana I. Kostaki, Pantelis A. Dratsas, Georgios N. Psarros, Evangelos S. Chatzistylianos and Stavros A. Papathanassiou
A Novel Method to Integrate Hydropower Plants into Resource Adequacy Assessment Studies
Reprinted from: *Energies* **2024**, *17*, 4237, https://doi.org/10.3390/en17174237 . . . . **255**

Athanasios T. Balafoutis and Bas Paris
How can Fossil-Energy-Free Technologies and Strategies (FEFTS) be adopted in European Farming?
Reprinted from: *Energies* **2024**, *17*, 4857, https://doi.org/10.3390/en17194857 . . . . **277**

Nilton Bispo Amado, Erick Del Bianco Pelegia, Ildo Luís Sauer, Welson Bassi, Julieta Andrea Puerto Rico and Carlos Germán Meza González
Who Should Own the Residual Rights over Distributed Resources?
Reprinted from: *Energies* **2024**, *17*, 5286, https://doi.org/10.3390/en17215286 . . . . **296**

Michał Kopacz, Dominika Matuszewska and Piotr Olczak
Carbon Capture and Storage (CCS) Implementation as a Method of Reducing Emissions from Coal Thermal Power Plants in Poland
Reprinted from: *Energies* **2024**, *17*, 6342, https://doi.org/10.3390/en17246342 . . . . **313**

Seigo Takahashi, Yuki Kinoshita, Nora Schelte, Semih Severengiz and Tetsuo Yamada
2-Stage Design of E-Moped-Sharing Service for Accessibility, Greenhouse Gas Emissions, and Cost Through Station and Supplier Selections
Reprinted from: *Energies* **2025**, *18*, 1644, https://doi.org/10.3390/en18071644 . . . . **333**

Muhammed Cavus, Huseyin Ayan, Dilum Dissanayake, Anurag Sharma, Sanchari Deb and Margaret Bell
Forecasting Electric Vehicle Charging Demand in Smart Cities Using Hybrid Deep Learning of Regional Spatial Behaviours
Reprinted from: *Energies* **2025**, *18*, 3425, https://doi.org/10.3390/en18133425 . . . . **355**

About the Editors

Maria Dzikuć

Maria Dzikuć is an associate professor at the University of Zielona Góra. Her research interests focus on the economic aspects of electricity and heat production from renewable and conventional sources. Dr. Dzikuć is the author of approximately 120 scientific publications and has led and worked as a research team member on international research projects funded by the National Science Centre (Poland).

Aleksander Jakimowicz

Aleksander Jakimowicz is a professor at the Institute of Economics of the Polish Academy of Sciences. In his research, he has demonstrated that, due to the simultaneous operation of the law of progressive complexity and the principle of emergence, markets and economies spontaneously converge on a distinct state known as the edge of chaos. His current research focuses on transdisciplinary issues related to renewable energy. In 2004, he was awarded the esteemed Bank Handlowy Prize for exceptional contributions to theoretical economic and financial thought, often referred to as the Polish Nobel Prize in Economics. In addition, he was honored with the distinction of Most Famous Pole of the Year 2004. In 2022, the Multidisciplinary Digital Publishing Institute (Basel, Switzerland) presented him with an award for the most outstanding article published in the scientific journal, *Entropy*.

Preface

The transition to a sustainable and low-carbon economy represents one of the most profound challenges facing modern civilization, due to its immense complexity, time constraints, and the absence of a central authority. This Reprint, titled "Sustainable and Low Carbon Development in the Energy Sector", compiles seventeen peer-reviewed articles that collectively address this multifaceted issue by bridging the gap between theoretical frameworks and practical implementation.

The scope of this volume is intentionally transdisciplinary. It moves beyond a purely technical perspective to encompass the economic, sociological, and thermodynamic dimensions of the energy transition. The aim is to provide a holistic analysis of decarbonization, ranging from the rigorous application of material entropy and thermodynamic laws in evaluating future technologies to the sociological examination of contact zones between citizens and local administrations. Simultaneously, the collection addresses critical operational challenges, including the forecasting of electric vehicle charging demand, the integration of hydropower into resource adequacy assessments, and the adoption of Fossil-Energy-Free Technologies (FEFTS) in European agriculture.

The motivation for this work stems from the urgent need to redefine current development paradigms. As the global economy shifts toward prosumer capitalism, standard economic models and isolated engineering solutions are no longer sufficient. This Reprint seeks to illuminate the complex interdependencies between grid stability, policy optimization, and social acceptance that define the modern energy landscape.

This volume is addressed to a diverse audience of researchers, policymakers, energy economists, and engineers. It serves as a vital resource for those seeking to understand not only the specific technologies driving the zero-emission shift but also the broader systemic mechanisms necessary to avert complex crises in the global energy sector.

Maria Dzikuć and Aleksander Jakimowicz
Guest Editors

energies

MDPI

Review

The Energy Transition as a Super Wicked Problem: The Energy Sector in the Era of Prosumer Capitalism

Aleksander Jakimowicz

Department of World Economy, Institute of Economics, Polish Academy of Sciences, Palace of Culture and Science, 1 Defilad Sq., 00-901 Warsaw, Poland; ajakimowicz@inepan.waw.pl

Abstract: The main objective of this paper is to demonstrate that the energy transition as part of prosumer capitalism is a socio-economic process whose complexity increases over time, which makes it an example of a super wicked problem. It comprises many new phenomena emerging spontaneously, and often unpredictably, in the energy markets. The main contemporary challenge involves such an energy sector transformation which will prevent climate change and will ensure the sustainable development of the global economy. However, this requires solving a large number of sub-problems in areas such as legislation, energy distribution, democracy, and cybersecurity. Therefore, this is a multidisciplinary issue. Moreover, the situation is complicated by the frequently omitted fact that energy transition is not part of the standard capitalism model, extensively described in handbooks and scientific literature, but it is conducted as part of a new economic system—prosumer capitalism, which has not been properly explored yet. However, a solution to this super wicked problem has to be found soon, as the energy system may be threatened with complexity catastrophe, which denotes exceeding the upper complexity limit associated with the breakdown of its adaptability. Therefore, developing effective techniques for alleviating the complexity catastrophe, including redefining the change management and complexity management methods to the global scale, becomes the top priority among the tasks faced by science.

Keywords: energy transition; super wicked problem; prosumer capitalism; energy policy; climate change; renewable energy; energy prosumerism; complexity catastrophe; change management; complexity management

1. Introduction

The prosumer model is highly useful in the energy transition, and its role and importance are greater than is commonly thought. It leads to a gradual transformation of the whole global economy and to the emergence of a new form of capitalism, called prosumer capitalism [1–3]. This stems from the fact that energy is at the base of each economic process. The initial parts of the paper present a review of the most popular definitions of prosumer and prosumption, beginning with the first concepts proposed by Alvin Toffler to the contemporary meanings of the words. The development of data communication technologies and widespread digitalization of the economy has unleashed the prosumer potential of societies and has led to the third great economic revolution in the history of mankind—the ICT revolution, which is the aftermath of the two earlier revolutions: the agricultural and the industrial [4]. Thus, traditional prosumption has been transformed into digital, which has accelerated the development of the renewable energy markets and their integration with traditional, fossil fuel-based markets. Moreover, digitalization enabled the decentralization of the green energy sector and, in consequence, a gradual departure from monopolies in the energy sector. At the same time, digitalization favors the electrification of many important sectors of the economy, especially transport and heating, which can contribute significantly to carbon dioxide emission reduction [5]. These changes have made it necessary to replace the old management principles and business models with new ones.

Therefore, the development of Toffler's research approach, called wikinomics, based on four principles and seven business models, has been discussed [6,7]. The principles include openness, peering, sharing, and acting globally, whereas the models include peer pioneers, ideagoras, prosumers, new Alexandrians, platforms for participation, global plant floor, as well as the wiki workplace. Wikinomics is based on digital prosumption, as everything else depends on it.

The next part of the paper focuses on prosumers as chief players in the energy market, and it describes the motives that encourage them to engage in economic activity. It presents their relations with the incumbent energy companies and the methods of participation in the markets and the potential for affecting them. It was necessary to define the new challenges faced by the prosumer energy sector in such fields as jurisdiction, distribution, democracy, and cybersecurity. Issues related to the development of smart grids and their effectiveness in the context of the Jevons paradox were also presented. Moreover, attention was drawn to energy prosumerism, which creates a framework for studying many new issues related to distributed generation, such as energy justice, energy citizenship, energy democracy, and energy poverty. Furthermore, the macroeconomic effects of prosumer activities and the development of new ways of organizing the economy were explained, with the gig economy and sharing economy as the best examples. Subsequently, the energy sector regulation methods were considered, given the differences between the modern and traditional electricity policy, stimuli that encourage people to become prosumers and democratic experimentalism.

According to scientific literature, the energy transition, which is humanity's response to climate change, is a process whose complexity increases over time. This is caused by a growing number of interdependencies between the major systems: social, legal, economic, and climatic. The study has shown that the energy transition creates an extremely complex, interrelated system of phenomena, described by various scientific disciplines, which is why it should be regarded as a super wicked problem. This hinders the reconciliation of two mutually exclusive socio-economic challenges, i.e., sustainable development of the global economy and elimination of the anthropogenic greenhouse effect. There is an additional difficulty in that the challenges appear as part of the new economic system, called prosumer capitalism, which has not yet been fully explored. When defining the system, Ritzer points out that its characteristic feature is a phenomenon called synergistically double exploitation [3] (pp. 82–86). In this case, the economic development is based on the doubly unpaid labor of people, who usually exist in two roles in the contemporary world: as workers and as prosumers. This results in an uncontrolled transfer of surplus value from societies to large enterprises and to international corporations. Today, this is particularly true for users of social media platforms, whose spontaneous and unpaid activity not only generates powerful profits for multinational corporations but also strengthens their economic and political power. Moreover, an energy system includes a similar phenomenon, called cross-subsidization, which involves the uneven distribution of the utility's fixed costs between prosumers and traditional consumers, to the disadvantage of the latter. It is probable that in future, prosumption processes can change from an economic development stimulus to a barrier. Incidentally, prosumer capitalism, in Toffler's terminology, forms the Fourth Wave of civilization development, which is also called the Connected Society [8,9].

A fundamental research gap in the energy transition literature is the overlooking of its complexity arising as a result of interactions between elements of the entire system. Focusing on individual subsystems, such as society, the environment, legislation, energy and environmental policy, distribution processes, the development of energy democracy, and technological progress, causes one to lose sight of the holistic picture of the phenomena. Meanwhile, we are dealing with a complex adaptive system that is inherently open and in which there are dynamic networks of interaction. This makes the energy transition as a whole evolve to the extent determined by the emergence and the edge of chaos. Emergence is the ability of a system to create orderly collective phenomena, the description of which is only possible at a level higher than the description applied to the constituent elements. The edge of chaos, on the other hand, is the distinguished state of the system located between periodic behavior and chaotic behavior, where its computational power is at its maximum. Near the edge of chaos, the system's complexity is nearly optimal, and its adaptability is at its highest. This article proposes a holistic view of energy transition, while maintaining a detailed description of its key components and the environment in which it occurs.

The research used the method of a systematic review of scientific literature on energy transition, which involved using the entire internet space. This means that it was not limited to specific literature databases, but the information was collected from a wide variety of sources. However, the scientific regime was strictly adhered to, which was to select only those research works that dealt with the most important elements of the modern energy system and the interdependencies between them. Special attention was paid to those subsystems that play an important role in generating the dynamic complexity of the entire system.

A combination of all the issues discussed in the paper shows that the global socio-economic system may soon find itself at a critical point associated with a complexity catastrophe. This is a certain system complexity limit, and crossing it breaks down the system's adaptability. Hence, there is a need to develop solutions which will restrict the complexity of processes in the energy sector. Therefore, developing proper methods of avoiding a complexity catastrophe should be one of the research directions which may be helped by redefining two approaches in the organization science to a global scale: change management and complexity management.

2. Prosumption as the Basic Market Force in the Modern Economy

2.1. Definition of Prosumption

The term "prosumption" is a portmanteau of production and consumption. It refers to a process where a consumer takes over some of the activities previously performed by a producer, usually by directly involving themselves in the design and production of various goods or services. This phenomenon was identified as early as in 1972 by Marshall McLuhan and Barrington Nevitt, who noted that the development of electric-information technology had made it possible to combine the roles of consumers and producers [10] (p. 4). As such, the prosumer is a new form of economic entity, who not only consumes but also designs, produces, and markets a given good or service. In other words, a prosumer leverages its innovation and creativity potential to produce goods/services.

The subject of prosumption was explored further in 1980 by Alvin Toffler, who posited that the economy is divided into two sectors [4] (pp. 282–305). Sector A comprises unpaid work performed by people to fulfill the needs of themselves, their families, or even their communities, whereas Sector B relates to the production of goods or services for sale or exchange. This classification shows that the officially acknowledged and valued economy—sector B—is complemented by another, invisible economy—sector A. Furthermore, the production of goods and services for private use lays the groundwork for the economic activity carried out by society, and thus, sector B could not exist without sector A. According to Toffler, prosumption at a time of information technology revolution—designated "the third wave" by Toffler himself—is, at its core, the shift of production from sector B to sector A by the people. This is directly coupled with a reduced role of the market in the economy—de-marketization. Prosumption is mainly driven by its benefits to the prosumers, stemming from the law of relative inefficiency. This law holds that as the production of goods becomes automated, the per-unit cost of these goods falls, thus increasing the relative cost of handcrafts and non-automated services. This makes production for one's own, private use a more cost-efficient proposition, thus spurring the growth of sector A. Among the many other factors that drive people to prosume are: rising inflation, the dissolution of second-wave bureaucratic service-provision systems characteristic of industrial civilization, the appearance of new, third-wave information technologies, and the rise in structural unemployment.

Prosumption leads to the externalization of the labor cost, meaning that people who produce goods/services for themselves take up part of the labor that had been originally provided by producers. This labor is mostly unpaid, and thus increases profits for some businesses and precipitates economic exploitation of prosumers.

2.2. Prosumption as the Basic Business Model of Wikinomics

Wikinomics is a field of research that discusses the impact of information technology on modern economic processes. As part of it, new principles for conducting business in the modern world and new models of global cooperation were identified. Digital technology can now be considered a top technique in the sense of Hicks, i.e., one that provides the highest rate of return and enables the introduction of the economy on a balanced growth path with the maximum rate of growth [11]. This explains the high, unprecedented economic growth and development that has occurred in recent decades in many countries of the world.

There are four principles of wikinomics: openness, peering, sharing, and acting globally. These principles are the most important factors of economic growth and development in the digital economy. Openness involves companies sharing part of their resources with the business environment in order to obtain external sources of creativity and innovation. Partnership is associated with the spontaneous transformation of some economic organizations from hierarchical to horizontal. Sharing means a less restrictive policy of companies regarding intellectual property, which applies, in particular, to patents and copyright. Information technology has removed existing barriers between business entities and enabled them to operate on a global scale [6,7].

An important achievement of wikinomics is to identify seven models of mass collaboration that have radically changed traditional business strategies. These include [6,7]:

1. Peer pioneers—volunteers creating innovative ventures outside the market sector;
2. Ideagoras—modern search systems for business partners and original scientific research experts;
3. Prosumers—people who are both producers and consumers;
4. New Alexandrians—people's cooperation to multiply, accumulate, systematize, and share all knowledge of humanity;
5. Platforms for participation—websites with relevant products and information technology that encourage large communities of partners to collaborate in order to create new products;
6. Global plant floor—production of goods and services as part of global cooperation;
7. Wiki workplace—meritocracy abolishing the hierarchy in the enterprise and connecting internal teams with external networks.

The principles and business models of wikinomics, which developed as a result of mass and spontaneous collaboration of people, led to the rapid growth and economic development of the entire global economy. However, one should not forget that this progress also has a dark side. This is due to the fact that the well-known phenomenon in classical economics, namely, the exploitation of people in the processes of production and distribution of goods and services, has been forgotten.

2.3. Types of Prosumption

2.3.1. Alvin Toffler's Three Waves of Prosumption

Prosumption is a complex and multi-dimensional topic, and has thus been relatively poorly explored in economics. The first classification of prosumption was put forward by Toffler himself. His taxonomy is based on characteristics of the three economic waves of humanity's economic history, which he analyzed. Using this criterion, he distinguished three forms of prosumption [4]:

1. First-wave prosumption of agricultural societies, where goods and services were mostly produced for private use, with only a small proportion exchanged between

people. Obviously, this impeded the growth of goods and service markets. In this period, sector A was relatively large, whereas sector B was far less prominent;

2. Second-wave prosumption was characteristic of the industrial society largely geared towards producing for trade, with minimal production for self-use. This was the reverse of the first-wave trend: consumption functions and production functions became largely separated. Sector B grew to gigantic proportions, which meant that Sector A became imperceptible to many economists. The progressing industrialization gave rise to an international trade network, a trend which Toffler named "the marketization of the world";
3. Third-wave prosumption appeared hand in hand with the informational technology revolution and restored the balance between Sector A and Sector B through the emergence of a new lifestyle, one based on production for exchange and production for self-use in roughly equal measures. Toffler argued that the market could no longer grow by absorbing new countries—that, on the contrary, it would start shrinking. This would bring about the aforementioned de-marketization—a reverse of the trend present during the second wave. Nevertheless, qualitative changes and development will continue, and the (necessarily reduced) market will give rise to a trans-market civilization which will deal with solving entirely new problems. The fourth wave is, therefore, to be expected.

By making this distinction, Toffler suggests that the renewed significance of sector A—brought on by the information technology revolution—will not only change the very underpinnings of our economic system, but also our belief system. A new, more holistic conception of an economy would be needed, one which would describe the phenomena of sector A and how they relate to sector B. The entire economic terminology would have to be redefined, as explaining the relationships between unmeasured production/productivity in sector A and measured production/productivity in sector B would require brand new models, metrics, and criteria [12]. Traditional measures of production, such as the gross national product, would slowly become obsolete if they fail to incorporate data on sector A economic activity. A fresh perspective on the problem of prices and costs would also have to be adopted, since the effectiveness of prosumption in sector A would affect the costs of private and public enterprises in sector B. Efficiency is currently determined by simply comparing different methods of producing a given good or service and is not compared across sectors, i.e., production in sector B is not measured against the corresponding efficiency in sector A. The rising importance of prosumption was also destined to cause inevitable changes in the belief system of a trans-market civilization. The development of the markets during the second wave gave rise to vulgar materialism and the preconception of economic stimuli as the primary force shaping people's lives, as expressed in marriage contracts and various social contracts. However, the market is more than just the economic structure—the psychosocial structure is a part of it as well. Therefore, in a trans-market civilization, we should see—according to Toffler—wider shifts in the worldview, or even new religions [4] (pp. 300–305). Many of these predictions have proven accurate, especially with regard to energy prosumption, as discussed further in the paper.

2.3.2. The Six Archetypes of Prosumers

Prosumption is a multi-faceted phenomenon and a focus of decades of research, with a wealth of studies detailing its taxonomy. Even so, the subject has yet to be fully explored. The most recent classifications distinguish between six archetypes of prosumers [13]:

1. DIY (do it yourself) prosumers—individuals who create goods and services exclusively for private use, without paying an external entity for them;
2. Self-service prosumers—prosumers who can perform partial self-service using technological tools, for example by repairing a bike or a computer by themselves, and thus co-creating value;
3. Customizing prosumers—prosumers who personalize products to better serve their own needs, especially with regard to entertainment, travel, or clothing;

4. Collaborative prosumers—individuals who provide for their own needs or the needs of others and thus create value without profit for any intermediary, e.g., by developing open-source software;
5. Monetized prosumers—prosumers who create value accessible to others via a commercial entity without being formally rewarded for their activity, e.g., through their activity on social media platforms;
6. Economic prosumers—individuals who receive formal incentives from commercial entities for the value they create for others, as best exemplified by prosumers who produce electricity for their own use and for sale.

2.3.3. Cooperation of Prosumers with Other Market Participants

Classifications of prosumption are increasingly being adapted to the changing markets and economies. In many cases, prosumers may well be categorized according to how they cooperate with other market entities [14,15]. Using this criterion, three types of prosumption can be identified [16] (pp. 51–57):

1. Individual prosumption, which is when a given prosumer does not engage in any cooperation whatsoever with other market entities, being completely independent from them and self-sufficient in their own activity;
2. Intra-prosumption refers to a collective form of activity where the actor is part of an organized group of prosumers and engages in participatory design or co-production (exclusively within the given group of prosumers);
3. Inter-prosumption occurs when a group of prosumers and a producer work together to develop new products/services or improve existing ones, for example, via social media. Inter-prosumption is further divided into its simple and complex varieties.
    - Simple inter-prosumption may take forms such as: lead users [17], the individualization (de-massification) of products and services [4] (pp. 271–274), production of a commodity by a producer, using information and starting material supplied by the prosumer, self-assembly [4] (pp. 294–295), and sharing a product with a business to perform a service (bike-sharing, car-sharing);
    - In contrast, complex inter-prosumption occurs both within a prosumer group and a producer group, with the two collectives also working together. The field of renewable energy production offers an example of public prosumers in the form of autonomous energy regions—demarcated areas of rural and agricultural land [18–21].

Notably, these two above taxonomies are largely in line with Toffler's views of the subject. Essentially, all these types of prosumption can be associated with one of the three waves of civilization development. It is, therefore, clear that the different types of prosumption can overlap, further speaking to the challenges faced by modern researchers attempting to define prosumption precisely. Producers of electricity who self-consume a portion of it and feed the excess to the power grid could be classified either as monetized prosumers or as entities engaging in intra-prosumption/inter-prosumption.

2.4. The Exemplification: The Interdependencies between Sector A and Sector B in the Polish Economy

Polish academia has seen a number of studies on prosumption in recent years, but virtually all of them focus on the market sector. Most of the analyses agree that society is showing a strong propensity toward prosumption, particularly in the banking services [22], food [23], and energy sectors [24]. Rather than a phenomenon exclusive to the youth (Generation C)—i.e., people born after 1990, reaching maturity after 2000, and entering the labor market after graduation—prosumption has been shown to also extend to the elderly, who are interested in pursuing energy prosumption [25,26]. It has also been demonstrated that prosumption is an important element of innovation marketing [27]. Of the literature analyzed for this paper, only one study indicated low societal interest in commercial prosumption [28].

The work by Szymusiak corroborates our earlier assumption that prosumption is more pervasive in Poland than in more developed countries such as Germany, with the added finding that individual prosumption is more popular in Poland [29], whereas Germans gravitate towards its collective forms (intra-prosumption and inter-prosumption) [14] (pp. 263–264). However, these results are from 2012, when the public administration sector was inelastic, poorly digitalized, and ill-suited to harnessing the prosumption capacity of the society, resulting in its sluggish response to changes in the private sector. Furthermore, there were very few incentives to energy prosumption. Though the Polish government has since been more successful in tapping into societal appetite for prosumption, especially with regard to the energy sector, its hand was forced by external factors, such as its obligation to uphold international agreements pertaining to the energy transition. While these changes promoted a new focus on developing collective forms of prosumption, the current situation is far from satisfactory. There persists a disparity between the private and public administration sectors in terms of harnessing the prosumption capacity of the society, and this asymmetry hinders the growth and development of both individual markets and the economy at large [30]. There is still not enough research on prosumption in public administration.

3. Prosumption in the Energy Sector

3.1. The Energy Prosumer as an Active Market Participant

3.1.1. Definition of the Energy Prosumer

Until recently, the supply side of the energy market was highly centralized, as it was dominated by large companies, i.e., generators and service providers. As such, consumers tended to be the passive segment of the market. As renewable energy technologies developed, this situation began to gradually change, primarily due to the emergence of relatively cheap photovoltaic solar panels, allowing many consumers to become energy prosumers—a term used to describe market actors who combine the characteristics of energy consumers and energy producers. When people are driven to prosume, it is usually due to economic factors—they hope to reduce utility bills, while also making a net profit. Energy prosumers may conduct commercial activity in one of three ways: engage in distributed generation (self-production and sale of electricity), provide energy storage services, or participate in energy efficiency and demand response programs. Energy efficiency is achieved when the same service is performed using less energy, whereas demand response programs aim to change the normal electricity consumption patterns of end-use customers on a voluntary basis, as energy prices change over time. This is accomplished by utilities charging higher prices for electricity at peak times or paying customers for reducing their electricity usage at certain times of day (under appropriate contracts). Demand response programs can function both on retail markets and wholesale markets [31].

In the European Union, energy prosumers are referred to as active consumers, though the terms are more or less interchangeable in meaning [32]. An "active consumer" is defined as [33] (p. 139):

> *a final customer, or a group of jointly acting final customers, who consumes or stores electricity generated within its premises located within confined boundaries or, where permitted by a Member State, within other premises, or who sells self-generated electricity or participates in flexibility or energy efficiency schemes, provided that those activities do not constitute its primary commercial or professional activity.*

Active consumers play a key role in the European energy transition—the switch from centralized, fossil fuel-based energy generation to the production of electricity from distributed renewable sources. The idea is that renewable energy prosumers boost competition on the electricity market by increasing the number of energy suppliers and curbing the economic power of large energy companies [34].

3.1.2. Barriers to Demand Response

The member states of the European Union are incentivized to promote and facilitate participation in demand response programs. Nevertheless, deploying such programs in prosumer markets may not be without problems. There are two types of barriers to demand response [35,36]:

1. Barriers to implicit demand response emerge when the consumer response to price change is disrupted by certain factors, e.g., lack of access to a dynamic pricing contract or lack of a smart meter (preventing two-way real-time communication);
2. Barriers to explicit demand response may be legal or logistical. Legal barriers arise when demand response is not allowed in certain markets, whereas logistical barriers emerge when technical demand response and aggregation is permitted by the applicable regulations but cannot be used due to particular market requirements, such as product definitions, excessive minimum bid sizes, or aggregation limits.

3.1.3. Motivators That Drive Customers to Be Active in the Energy Market

Recent years have seen substantial rise in the global number of prosumers who use photovoltaic panels to produce electricity for their own use and for sale. The main reason for households installing solar panels is to increase self-consumption, since the self-generated electricity may be cheaper than energy from the grid, which is subject to various taxes and fees. In other words, energy prosumers aim to reach a grid parity. The structure of the electricity price is fairly complex because, in addition to the charge for the energy consumed, it also includes a number of other components that are divided into fixed and variable ones [37] (p. 14). In general, the energy production costs make up only 36% of the on-grid electricity charge paid by the final consumer, an additional 26% being the network costs, and the remaining 38% consisting of taxes and levies [32]. In Poland, the primary motivating factors driving customers to engage in prosumption are the potential savings and lower electricity bills [38]. Utilization of battery storage systems may prove to be the next step towards energy self-sufficiency of households by reducing their dependence on the electricity grid. Simultaneous production, consumption, and storage of electricity is referred to as prosumage [39,40]. It has been shown that the achievable self-consumption rates for a 4 kWp PV system can be boosted by installing a battery, though the number of persons in the household and the installed storage capacity are also a factor. The cost-efficiency of prosumage is maximized when the ratio of battery capacity to peak power capability of a solar panel is 1:1 (kWh/kWp) [41]. Even an electric car battery can serve as a domestic backup power source, forming a vehicle-to-home (V2H) system when connected to the grid [42]. This solution can certainly support self-consumption.

3.1.4. Types of Energy Prosumers

Prosumers active in the energy markets form a highly diverse group, which is why it is difficult to neatly categorize them. Energy prosumers are most commonly classified according to the role they play in the market. With this criterion in mind, prosumers are divided into residential, commercial, and industrial prosumers [42]:

1. Residential prosumers mostly comprise those who live in dwellings supplied with power for regular domestic use, such as lighting, heating, cooling, and cooking. These include households, apartment buildings, housing associations, cooperatives, and collectives;
2. Commercial prosumers use energy to meet both their own needs and the needs of the community. This group includes businesses of different sizes, universities, department stores, shopping malls, hospitals, offices, and sport facilities;
3. Industrial prosumers include industrial operators which use on-site renewable energy systems to self-generate power for own use and supply the excess to the national grid or sell it to the local community [43]. This group mostly encompasses manufacturers and producers such as factories, mines, mills, industrial plants, and farms.

These three groups of consumers can be easily distinguished by the size of their energy system. Residential prosumers have an installed capacity limit of no more than 10 kW, commercial prosumers encompass installations over 10 kW and below 250 kW, while industrial prosumers employ installations of 250 kW and more [44,45]. There are also other ways of classifying energy prosumers—for example, by the type of energy they generate, the source of energy they use, or by their relationship with the power grid [42,44]. Importantly, prosumers may organize themselves into groups or communities, which frustrates attempts to classify them. Such organizations are also considered prosumers, as they often operate as single entities in the energy market. Prosumers may also establish virtual power plants that integrate multiple types of distributed energy resources to ensure reliable overall power supply [46,47].

3.1.5. Relations between Energy Prosumers and Incumbent Energy Companies

The emergence of prosumers in electricity markets has forced energy companies to rethink their traditional business models. The relations between these two groups of entities should be considered to be a unique type of stakeholder relations—co-producing prosumers behave fundamentally differently than consumers. Power companies initially enjoy a substantial advantage over the prosumer, owing to their ability to manage the power grids. In contrast, prosumers need grid services to purchase and sell electricity. Studies have demonstrated that the relations between prosumers (in their capacity as co-producers) and energy companies should be based on reciprocity, with the two groups working closely together towards the decentralized production of renewable energy [48]. Today, this relationship may even be the inverse of the original one, as energy companies have increasingly realized that competitiveness in renewable electricity generation is predicated on their ability to attract prosumers and develop constructive partnerships with them [15].

3.1.6. The Potential for Prosumers to Manipulate Energy Markets

Prosumers play a growing role in modern economic life, raising the question of whether they would be able to manipulate the power markets, whether as sellers or buyers. Modeling a market with a strategic prosumer has demonstrated that the prosumer can leverage its market power without impacting its net position in equilibrium. A prosumer in the short (long) position as a price-taker, i.e., a buyer of power from the main grid, is thus also in the short (long) position as a strategic entity, and vice-versa. Furthermore, a prosumer has influence over prices, but in terms of strategy, it is better served by assuming the role of a price-taker, rather than a strategic entity employing the Cournot strategy, as long as there is competition between other entities. Under certain conditions, the activity of the prosumer can lead to lower power prices for consumers, but at the expense of the producers [49].

3.2. *Challenges Facing the Prosumer Energy Sector*

The emergence of energy prosumers in the economy brought with it a number of new problems with regard to energy law and the energy regulatory system, which Jacobs describes as jurisdictional, distributional, and democracy puzzles [31,35]. In addition, there are issues related to cybersecurity.

3.2.1. Challenges in the Jurisdictional Sphere

The need to clarify legal regulations stems from the fact that energy prosumers combine the attributes of consumers and producers. This raises the question of whether they sell energy on a retail or wholesale basis, an important distinction in the context of net metering and demand response. Net metering policy is a way to encourage investment in on-site generation by paying prosumers for the electricity they supply to the grid at full retail rates, which are much higher than wholesale rates. Such incentives are thought to diminish the role of prosumers in the market, and it remains unclear whether the retail rate is the proper price in such a situation. In the case of demand response, where the price signal from the grid is intended to reduce electricity consumption, there are similar difficulties in classifying a given sale as retail or wholesale. There are doubts on whether to categorize actors who provide such grid services as consumers or producers [31]. The majority of current legal problems related to the energy transition center around four issues: the concept of prosumer, the introduction of demand response, the evolving roles of distribution system operators, and the birth of peer-to-peer trading [36,50].

The legislative processes in Poland may generate significant transaction costs, potentially discouraging prosumers from investing in renewable energy production and delaying the development of the green energy sector [51]. Despite the many favorable jurisdictional changes introduced in recent years, 45.8% of experts still consider labyrinthine regulations to be a major barrier to the development of renewable energy production, with an even larger share (72.2%) believing that further legislative changes are necessary [52].

3.2.2. Challenges in the Area of Distribution

Distribution-related conundrums stem from the divergent interests between prosumers and traditional consumers, being connected with the problem of cross-subsidization and pricing demand response. Incentives for prosumers, aimed at encouraging them to provide grid services, should not be offered at the expense of traditional consumers, especially not by burdening the latter with the prosumers' portion of fixed grid costs. These costs remain fixed, regardless of the short-term energy consumption, and are covered by a specific combination of fixed and usage-based charges. The initial financial incentives for prosumers (in the form of net metering programs) were considered necessary to overcome the cost barriers to the development of distributed generation systems. However, as the number of energy prosumers grew rapidly, distributional inequities became a potential problem. The biggest concern is cross-subsidization, which occurs when the principle of equal division of utility's fixed costs between prosumers and traditional consumers is compromised, and the latter are burdened with the larger share of the costs. Furthermore, the issue of cross-subsidization is connected to the utility death spiral [42]. Passing the increasing costs of electricity on to traditional consumers may end up creating a scenario where some of them decide to disconnect from the grid and switch to self-generation. This will increase costs for the remaining consumers, thus further encouraging others to abandon the grid. As a consequence, the utility will not have enough customers to cover fixed costs. Conversely, however, it should be noted that traditional consumers may procure a number of direct and indirect benefits from distributed generation development, including reduced peak loads or ancillary services, i.e., voltage support and reactive power.

Problems in distribution can also emerge in relation to pricing demand response, as is well-exemplified by the locational marginal price. This is the price that prosumers may receive for selling demand response into organized wholesale markets, set at a level corresponding to generation resources. This means that prosumers are compensated for each megawatt of power they have foregone, with a value equivalent of one megawatt of power generated and supplied to the grid using traditional generation resources. There is controversy on how to distribute the costs of this compensation among other energy market actors [31].

3.2.3. Challenges in the Domain of Democracy

The third type of issue relates to democracy and speaks to the need for inclusiveness in administrative processes. There are many arguments for involving prosumers in the process of shaping energy policy [31]:

- Firstly, efficient and fair administrative governance is predicated on including all stakeholders. Excluding energy prosumers from the process would mean that their major interests would not be represented;
- Secondly, if the stakeholders are able to participate in legislative processes from the very start, the proposed legal acts are more likely to be well-received and less likely to be questioned ex-post. Negotiated rule-making procedures can facilitate the exchange

of information between regulators and stakeholders and, consequently, help settle disputes and reach a consensus on proposed legal rules, which can help avoid many delays and associated costs;

- Thirdly, including prosumers in administrative procedure results in higher-quality decision-making.

In Poland, the issues of democracy related to the development of distributed power generation center around the social aspects of the energy transition, social dialogue and effective communication, solidarity and co-responsibility, as well as intensification of cooperation and synergy of actions [53].

3.2.4. Cybersecurity Challenges

Digitalization plays a strongly transformative role in the energy system, as it includes not only digitization, i.e., the conversion of analog data to digital, but also new business models of prosumer capitalism [42]. Digitalization has changed the traditional business models of the energy sector by enabling the development and integration of renewable and non-renewable energy production. This has produced new power systems boasting unprecedented complexity, with smart grids becoming their most crucial components [54]. Initially, the energy landscape had mainly evolved along the lines of an isolated and protected sector of the economy, gradually transforming into an open network, based on technologies that make use of the internet and other business networks. Furthermore, the diversification of energy sources (some of which are intermittent) and the resultant change in energy generation's structure have led to the creation of a power system that requires constant and real-time management. Balancing electricity demand and supply is only possible if the key system components have been digitalized, which is why digitalization is considered to be a prerequisite to energy transition. The rapid development of information and communication technologies has made electricity a key factor of production in modern economies. There is a certain paradox with regard to cybersecurity. Digitalization is a driver of electricity sector development and, at the same time, makes it more vulnerable. Smart grid solutions certainly bolster system efficiency and reliability while reducing costs of energy supply, but they also open up new and serious cybersecurity challenges [55]. The rapidly growing circulation of energy and information is being mediated by an increasing number of people using a wide variety of technical devices and smart software, making it much more difficult to address data security issues such as integrity, availability, and confidentiality [56]. These complex systems are represented, for example, by the U.S. power grid, often called the world's largest interconnected machine. In 2017, this grid consisted of over 7000 power plants, 55,000 substations, 160,000 miles of high-voltage transmission lines, and millions of miles of low-voltage distribution lines [57]. It should not come as a surprise that such complexity can open up a lot of potential vulnerabilities.

The electro-energetic system is a key part of the state's critical infrastructure, which includes systems composed of functionally related components in the form of buildings, equipment, installations, and services crucial for the security of the state and its citizens, ensuring the efficient functioning of public administration, institutions, and businesses. This makes the system a popular target of cyberattacks [58]. Its destabilization would be all the more dangerous due to the knock-on disruption of numerous other key sectors, including telecommunications, transportation, health care, or the water and sewage system, creating a threat to human life and health as well as a risk of significant material losses. Cybercriminals are usually motivated to act by politics, ideology, money, or moral sentiments [59]. One of the most notorious cyberattacks was launched against Ukraine on 23 December 2015 and resulted in the disruption of the energy grid, causing 6-hour blackouts which affected approximately 225,000 customers across multiple agglomerations [60]. Russia was accused of taking part in the attack, though, as of yet, no evidence has emerged that would categorically prove such involvement [61].

The progressing energy transition will require further digitalization of the energy industry, meaning that it will unavoidably keep growing more complex. As such, the risk of cybersecurity incidents cannot be fully eliminated. The combat these issues, a suitable cyberattack prevention policy will need to be developed, despite the huge legislative challenge it poses. The accommodate this need, the European Commission presented its cybersecurity strategy on 16 December 2020, with the general protection of the global and open internet as one of its primary goals [62]. Cyber insurance policies—a type of disaster insurance—may serve as a useful tool for mitigating the effects of cyberattacks "after the fact". However, this solution will not come cheap for policyholders, as insurance companies tend to set caps on cybersecurity policies, forcing utilities to increase these caps by combining policies from different insurers [63]. Reducing the complexity of energy systems still seems to be one of the best ways to protect against cyberattacks [64].

3.3. Smart Grids as a Fundamental Element of a Decentralized Energy System

Inherent to a decentralized prosumer model is a technological infrastructure consisting of smart grids equipped with innovative digital applications and smart metering systems. In the European Union, a smart grid is defined as [65]:

> *an electricity network that can integrate in a cost efficient manner the behaviour and actions of all users connected to it, including generators, consumers and those that both generate and consume, in order to ensure an economically efficient and sustainable power system with low losses and high levels of quality, security of supply and safety.*

In other words, smart grids are integrated systems that encompass not only technology, but also information, human and social influences, organizational and managerial supporting arrangements, political constraints, and appropriate legal solutions [66,67]. They are primarily intended to increase consumer participation in the energy market and increase energy efficiency through real-time balancing of this market, using tools such as demand response management. Smart grids are dual in nature, being a fusion of hardware (which stimulates the growth of prosumer energy and the development of the renewable resource market) and software (which provides relevant regulation). Smart grids also contribute to the expansion of the gig economy by facilitating the emergence of new market actors, blurring the line between producers and consumers, the employee and the self-employed, which is what Toffler refers to as the de-marketization of old structures by shifting economic activity from sector B (production for swap) to sector A (production for own use). One of the ways to optimize the operation of smart grids is to supplement them with energy storage capacity in the form of batteries or by using electric vehicles for this purpose. Smart grids must also conform to stringent security requirements, given the need to collect and process detailed consumption data as well as data relating to geo-location, tracking and profiling on the internet, video surveillance systems, and radio frequency identification systems [67] (p. 142). Consequently, they may help facilitate the achievement of the fundamental objectives of the European Union's energy policy, such as efficiency, security, and sustainability. In the near future, smart-grid energy systems will use Internet-of-Things (IoT) networks, connecting billions of smart objects such as smart meters, smart appliances, and a wide range of sensors [68–74].

Some of the crucial benefits of implementing smart grids are: an 80% reduction in costs of integrating renewable energy sources with the energy system, a 40–52% increase in the share of solar (PV) energy in the public grid, and a 15% reduction in global carbon dioxide emissions [75] (pp. 80–81). Forecasts regarding the development of smart grids around the world to 2030 envisage three scenarios with different volumes of investments [76] (pp. 304–308):

1. The best-case scenario supposes an increasing pace of smart grid development at about 30–40% annually, with total investments at around 60 billion USD;
2. The realistic scenario forecasts a growth of about 2% annually and an estimated 20 billion USD spent on investment;
3. The worst-case scenario sees a decrease of about 2% annually and a reduction in investment into smart grid technologies, falling below 10 billion USD.

When balancing traditional energy grids, utilities know what the supply is and only have to balance it with the demand, which usually fluctuates within a pre-established range. Preserving the right relationship between supply and demand required them to maintain a large reserve capacity, which did help prevent market imbalances, but also increased the overall costs and carbon dioxide emissions. In contrast, smart grids offer sophisticated means of controlling energy use, accounting for the fluctuations in the net load. Thus, the aim is to minimize the gap between total demand (load) and variable generation, demand and supply both being stochastic variables. Switching from a supply-driven to a demand-side energy policy requires that traditional risk metrics and risk management methods evolve, as well [67] (pp. 145–146).

3.4. Participation of Prosumers in Energy Markets

Recent years have seen the emergence of numerous prosumers on the energy markets, who simultaneously produce electricity from intermittent renewable resources, store it, and use home energy management and electric vehicle-to-grid (V2G) systems [42]. This necessitates the use of smart prosumer grids which support dynamic pricing, are adapted to distributed generation, and incorporate large-scale digital networking. This is not without its share of technical, institutional, economic, and social challenges, but overcoming them offers many potential benefits. Market structures can be divided into three types, according to how modern prosumers interact with the market [77]:

1. Peer-to-peer models. Such systems connect all prosumers with each other via decentralized, autonomous, and flexible peer-to-peer networks, allowing electricity producers and consumers to bid and sell/purchase energy services directly. The distribution grid usually charges a management fee as well as a tariff for distribution, which varies depending on the type and amount of the service and the distance between the supplier and the recipient;
2. Prosumer-to-grid models. These models are based on brokerage systems for microgrid-connected prosumers. These may be either prosumer-to-interconnected microgrids (where the market actors are connected to a microgrid, which in turn is connected to a larger network) or prosumer-to-islanded microgrids (where services are provided in autonomous microgrids). Each of these modes offers different incentives for prosumers. In the former, it is in the interest of prosumers to generate as much electricity as possible, as any excesses may be sold to the main grid. Conversely, for the island mode, energy services must be balanced at the microgrid level and surplus electricity must usually be limited to what can be stored;
3. Organized prosumer groups. In these models, groups of prosumers pool their resources and/or set up virtual power plants. This way, local prosumer markets can be created to support the functioning of smart cities. Such structures could give local organizations and communities a means to enhance their electricity demand management [78].

There has been a number of specific solutions developed recently to harness peer-to-peer platforms based on smart grids to trade electricity. Apart from maximizing prosumer profits, these approaches emphasize aspects such as optimality and fairness among prosumers [79,80], mitigation of power losses and prevention of over utilization of the power lines [81], complications of local demand and supply balancing [82], maintenance of prosumers' privacy [83], factors affecting the behavior patterns of individual prosumers [84], or a motivational psychology framework to help reduce carbon dioxide emissions and lower the cost of energy [85]. Peer-to-peer platforms may be adapted to work in the energy ecosystem of different countries, but there are various specific regulatory barriers that have to be overcome first [86,87]. There have also been proposals to introduce market models that integrate prosumer communities using peer-to-peer trade and residential storage with wholesale electricity markets [88].

In Poland, organized civic participation in the energy sector is facilitated by energy clusters and energy cooperatives, which exemplify renewable energy communities and citizen energy communities—the two forms of distributed-generation associations grouping market actors, as proposed by the European Union [89]. The term "cluster" has been clarified and popularized by Porter and refers to a geographic grouping of interrelated enterprises and institutions who engage in specific economic activity [90]. An energy cluster is a civil law agreement on the production, demand balancing, distribution, and/or trade of energy from renewable sources or other sources/fuels. The cluster can be joined by natural persons, legal persons, universities, research institutions, and local government bodies, which must operate in a distribution network with rated voltage below 110 kV. A single cluster (which is represented by a coordinator) cannot extend beyond the limits of a single district or five municipalities. The territorial extent of the cluster depends on the locations of grid connections for the participant generators and consumers of energy. An energy cooperative is a voluntary association of persons which generates electricity, biogas, and/or heat from a renewable energy source for the benefit of its members, while also balancing the demand for the electricity/biogas/heat. It must be connected to a local electricity distribution network with rated voltage below 110 kV, a gas distribution network, and/or a heat distribution network. Cooperatives are subject to a number of limitations. A cooperative must operate within a rural municipality, an urban–rural municipality, or with a territory of no more than three adjacent municipalities of either type. A cooperative must also not exceed one thousand members. The total installed electrical capacity must be sufficient to cover at least 70% of the cooperative's own demand and not exceed 10 MW, the total thermal power output should be less than 30 MW, and the combined capacity of biogas installations must not exceed 40 million m^3 [91–94].

3.5. The Jevons Paradox and the Energy Efficiency

One of the biggest challenges for prosumer markets is finding a way to meet the coming electricity demand, which is forecasted to increase. In other words, the question is whether smart grids designed for energy efficiency can avoid the well-known Jevons paradox, also known as the rebound effect [95,96]. In 1865, the English economist William Stanley Jevons noted that the same technological advancement which improved the efficiency of coal use in steam engines also increased the demand for coal as a relatively cheap power source. This meant that the steam engine proliferated across various industries, resulting in an increase in total coal consumption. He summarized his observation thus [95] (p. 123):

> *It is wholly a confusion of ideas to suppose that the economical use of fuel is equivalent to a diminished consumption. The very contrary is the truth. As a rule, new modes of economy will lead to an increase of consumption according to a principle recognised in many parallel instances.*

This paradox, when applied to smart grids, would mean that the increased energy use efficiency will also increase energy consumption, without a reduction in greenhouse gas emissions. As such, efficiency policies may be counterproductive to the objective pursued. In the context of climate change and sustainability goals, it is argued that the only way to avoid the Jevons paradox would be to introduce conservation policies that raise the fees for access to a given power supply. Such proposals include physical caps, such as quotas or rationing [97], environmental taxation [98], and emissions trading schemes [99–102].

Energy efficiency can also be viewed through a different lens—as a new type of fuel, which is sometimes referred to as the fourth fuel (after hard/brown coal, crude oil, and natural gas) or the invisible fuel. Its reserves can be increased in two ways. One is to mitigate energy waste in global and regional economic processes—for example, by eschewing the production of disposable, perishable, low-quality, useless, non-repairable products, as well as products too costly to repair (so that buying a new product becomes more economical). This would require a shift in the current trends and consumption patterns, which promote constant replacement of still functional products and equipment with new ones, as exemplified by cars and electronics. The second way is to reduce collective and individual energy consumption across all facets of society, in accordance with the maxim "the biggest innovation in energy is to go without" [103]. The fourth fuel is the cheapest of all, even compared to renewables [104] (p. 47). In the United States, the estimated cost of saving 1 kWh is 2.8 cents, whereas the typical retail fee for 1 kWh is 10 cents. In an electricity-consuming sector, saving 1 kWh can cost a mere one-sixth of a cent, meaning that the payback period for such an investment is measured in months, rather than years [105]. Energy efficiency potential in Poland is between 30 and 60 TWh/year, the equivalent of six to twelve 200 MW power units in conventional power plants that could be shut down. Increasing energy efficiency provides multiple benefits, such as energy conservation, carbon dioxide emission mitigation, and reduced energy costs. Eliminating energy waste and increasing energy efficiency of economic processes could greatly reduce global energy demand and thus reduce carbon emissions, far beyond what could be achieved by shutting down fossil fuel power plants alone [103].

In the context of energy waste, it should be noted that the embodied energy in some products is enormous. An estimated 27,780 kWh goes into producing a medium-sized car. The global production in 2015 was 72 million vehicles, meaning that the 1944 TWh of energy was embodied in these products—more than 12 times the electricity production of Poland. Doubling the service life of a fully operational car could create energy savings of 972 TWh, with a corresponding reduction in carbon dioxide emissions of 900 million tons per year. Slightly lower savings are achievable by extending the life of electronic equipment such as laptops, phones, and tablets, the global production of which consumed 278 TWh energy in 2015 [103,106].

3.6. Energy Prosumerism as a Framework for Studying Phenomena Related to Distributed Generation

3.6.1. Definition of Energy Prosumerism

The growing importance of energy prosumers in the electricity market has not only led to major changes in the energy sector but also to the emergence of completely novel socio-economic trends, necessitating a broader view of prosumption processes. The most important of these trends is prosumerism, defined as a set of social behaviors and prosumer attitudes striving to change the traditional criteria for evaluating products, standard of living, and quality of life. These new criteria take into consideration all benefits of prosumption, the material and the non-material, both in the course of production and consumption [16] (pp. 306–310). Energy is a prerequisite to launching any production process and, therefore, lies at the forefront of prosumerism. Modern energy prosumerism is a social movement facilitated by the growing participation of prosumers in renewable energy generation, which translates to numerous economic, social, and environmental benefits for the society [42]. The movement aims to move away from the centralized and monopolized energy system to a decentralized democratic energy system. Prosumerism is constantly growing in importance, which is why it has framed the efforts to study energy justice, energy citizenship, energy poverty, energy democracy, and the sharing economy [107]. These terms are related and overlapping, but still worth discussing in greater detail below.

3.6.2. Energy Justice

Of the issues related to energy justice, the availability and affordability of energy services, intragenerational equity, intergenerational equity, and responsibility are the ones most frequently cited [32]. Intragenerational equity relates to the distribution of burdens within separate generations, including the current generation and future generations [108]. Conversely, intergenerational equity concerns the distribution of burdens between the current generation and future generations—a major problem in a situation where the current generation pays the costs of climate change mitigation while future generations benefit [109]. Studies have demonstrated that energy prosumers could help achieve energy justice goals, such as access to basic energy services for everyone and responsible use of energy sources. However, this cannot happen without relevant legislation that ensures intragenerational equality [32].

3.6.3. Energy Citizenship

Energy citizenship concerns the active participation of individuals in energy generation from renewable sources, including by material engagement with the use of new technologies. Solar energy and the use of devices such as electric cars, smart meters, and photovoltaic panels play a major part in this process [110]. This also shows that pursuers of energy projects are aware of climate change risks and strive to mitigate them through various means. One of them is political activism, such as taking part in protest and support movements. Such individuals feel the need to pursue energy justice and to prevent energy poverty as part of their value system [107]. The different approaches to energy citizenship include those which call for going beyond the *homo oeconomicus* model, i.e., beyond using financial incentives to encourage consumer involvement in the energy transition through investment in innovative energy technologies, as this may lead to inequalities and exclusion of poorer citizens. It is, therefore, necessary to transcend individualistic approaches and start treating energy as an ecological resource to be managed via collective decision-making, rather than a commodity. In an energy citizenship model, electricity should not be a problem of individual consumers, but should drive well-being on the level of neighborhoods and districts [111]. In light of the above, energy citizenship should be understood to be renewable energy prosumerism supported by political prosumerism, with the latter laying the groundwork for a new kind of socioeconomic and political order [112].

3.6.4. Energy Poverty

Energy poverty is related to the idea of energy justice in the sense that one is the reverse of the other. The United Kingdom uses a fairly strict definition of energy poverty as a situation where a household is forced to spend more than 10% of its income on maintaining an adequate level of warmth [113]. This refers to the energy costs required to achieve a minimum or standard level of heating to preserve one's health and maintain thermal comfort, while also being able to satisfy basic needs such as adequate lighting, cooking, or refrigeration [114]. This definition is also unofficially used in other countries, but is considered insufficient.

Energy poverty is caused by three factors: high energy costs, low home energy efficiency, and relatively low household income [115–117]. Some have also identified a link between energy poverty and gender, pointing to the underrepresentation of women in the energy sector and the increased prevalence of energy poverty among older women (compared to men of the same age). Accordingly, it has been posited that some issues of energy in relation to gender should be integrated into energy justice. That is why it is a worthy cause for the prosumer movement to gradually eliminate this issue by uplifting the socio-economic conditions to give citizens access to affordable and clean energy [107].

In 2016, energy poverty affected around 4.6 million people in Poland, who make up 1.3 million (12%) of Polish households. Energy poverty has been one of the major contributors to air pollution—low-income residents use coal stoves to heat their homes, especially in rural areas, where most houses were built with large usable areas and low energy efficiency. Of the people affected by energy poverty, 2.5 million (i.e., more than half) were also poor in economic terms. However, the other 2.1 million have experienced energy poverty, but not economic poverty. It follows that energy poverty is not always caused by economic poverty, but may stem from other factors [53].

Energy poverty is strictly linked to the problem of low emission, i.e., releasing the by-products from burning solid, liquid, and gaseous fuels by emitters located at a height of no more than 40 m [118]. A survey encompassing three provinces of southern Poland showed that coal is the most popular fuel for heating buildings and preparing domestic hot water, used by 41% of the respondents. It is also one of the cheapest energy sources. Nevertheless, of the respondents not using heat from the heating network (68% total respondents), 27% admitted to burning waste in heating stoves. This is a concern, as such waste often contains high amounts of damaged building materials containing asbestos and other hazardous substances. Additionally, 30% of the respondents reported being aware that burning low-quality fuels emits toxic gases, highly hazardous to the human body [119]. An analysis of annual household energy costs shows that, taking into account the installations needed to heat the house and prepare domestic hot water, the cheapest solution (compared to hard coal) is to use a ground source heat pump, which has an investment payback period of 15 years. Gas and biomass are slightly less economical options [120].

3.6.5. Energy Democracy

At its core, energy democracy is built around the idea that renewable energy generation is more than just an environmentally-friendly way to supply electricity—the rapid take-up of renewable energy may also end up creating more equitable societies by reshaping wealth distribution and political power [121,122]. Energy democracy is an emerging social movement, aiming to spark a transition from a centralized energy system based on fossil fuels to a model of dispersed and locally produced renewable energy, with a particular focus on cooperative and public ownership of the energy infrastructure [123,124]. In other words, energy democracy leverages the relationship between energy ownership and political, economic, institutional, and cultural foundations of society to redistribute political and economic power. The social change potential of renewable energy lies in three factors: the dispersed nature of renewable energy, which is a catalyst for changes in politics and ownership structures in the economy; the use of steady, perpetual, and free resources such as sunlight, wind, and water; and lastly, the abundance and widespread accessibility of renewable resources, removing the need to compete for them. Energy democracy incorporates three types of social activism: resisting fossil-fuel-dominated energy systems, restructuring energy systems, and reclaiming energy infrastructure [122]. The movement also draws on values such as energy justice, environmentalism, and racial equity. The literature includes three different approaches to energy democracy: as a process driven forward by a popular movement, an outcome of decarbonization, and a goal or ideal to which stakeholders aspire [125]. Energy democracy operationalization (i.e., specific innovatory initiatives intended to put its principles into practice) is pursued along three lines: popular sovereignty, participatory governance, and civic ownership [126]. The social movement strives to destabilize power relations of a fossil-fuel-based energy system and replace them with democratic relations to establish a community-based decision-making authority [127]. It bears emphasizing that, in practice, the main goal of energy democracy is social justice and economic development rather than climate change [128]. The movement strives for the ideal of completely replacing fossil fuels with renewable energy, the prime example of which is the city of Burlington, Vermont—the first US city 100% powered by clean energy [129].

3.7. *Macroeconomic Effects of the Energy Prosumer Activities and Digitalization of the Power Industry*

3.7.1. Impact of Prosumer Households on Macroeconomic Indicators

The current state of knowledge on the macroeconomic impact of energy prosumer activity is still relatively limited. Nevertheless, the rising number of prosumers should be expected to eventually produce marked changes across the entire economy. The current body of research suggests that household prosumer activity in the energy market may have some effect on the most important macroeconomic indicators, increasing private consumption, GDP, and employment. Prosumer households enjoy higher disposable incomes thanks to their self-production of electricity and the guaranteed remuneration for feed-in. This, in turn, enables them to increase consumption and invest in power generation technologies, both of which stimulate the economy. Prosumer activity also helps curtail carbon dioxide emissions into the atmosphere [130]. Digitalization of the power industry, which is transforming the traditional operating and business models, has had even more of an impact on the economy as a whole. These trends have led to a reshaping of the entire macroeconomic structure and, thus, to the emergence of such systems as the gig economy and the sharing economy.

3.7.2. Gig Economy

The shift towards a decentralized energy system in the United States and the European Union has occurred within the framework of a new market system termed the "gig economy" (characterized by a growing number of temporary jobs). Development of information technology is one of the leading factors driving organizations and enterprises to increasingly employ independent workers for specific tasks [131–136]. This change in the economic order has led to the emergence of new market actors such as energy prosumers, due to the related decentralization of the energy sector necessitated by climate change. This is attributable to fundamental, top-down changes across the entire energy value chain, which gradually push potential, competences, and leverage away from EU institutions, states, and corporations into the hands of prosumers. This shift is accompanied by bottom-up action to empower prosumers and grant them a larger role in the energy transition. This results in energy democratization through the decreased economic power of energy companies (which, until recently, had a monopoly on energy markets) and the higher share of prosumer-generated electricity in the total energy balance [35].

3.7.3. Sharing Economy

The sharing economy is a novel socio-economic trend of peer-to-peer sharing of goods and services via online social media. The sharing economy is closely related to—and sometimes even conflated with—collaborative consumption [137]. Though some forms of collaborative consumption (such as eating meals socially with a group of friends, travelling to visit a loved one, or a family washing its clothes in a single washing machine) have been a subject of study for a long time [138], it is only now that information technology has uplifted collaborative consumption to a completely different level [139]. As one of its underlying principles, this model of consuming and using goods/services calls for expanding traditional property rights and the possession of things by moving away from rigid materialism and popularizing renting or sharing [140,141]. The idea is to counteract the ubiquitous overconsumption of resources driven by short product lifecycles, planned obsolescence, and the consumers' constant search for ever-new products/services. With this in mind, collaborative consumption can be viewed as a potential catalyst of sustainability [142]. Waste can be reduced by embracing sharing, renting, gifting, bartering, swapping, lending, and borrowing. Extensive interaction between different communities, made possible by network technologies, can help avoid unnecessary purchases, find ways to use idle assets, and reuse unwanted possessions [143]. Collaborative consumption is based on the triad of: a platform provider, a peer service provider, and a customer [144]. The platform provider offers matchmaking, making it easier for the customer to procure assets of a peer service provider.

The success of the sharing economy lies in the capacity of digital platforms to generate network effects [137]. Ongoing debates on the sharing economy seem to paint it as a niche of innovation, a force of decentralization and transformation of existing technical and socio-economic structures (regimes). Within this framework, it is defined in its six aspects as [145]: an economic opportunity; a more sustainable form of consumption; a pathway to a decentralized, equitable, and sustainable economy; creating unregulated marketplaces; reinforcing the neoliberal paradigm; and an incoherent field of innovation. The combination of the first two aspects may cause the sharing economy to become dominated by corporate co-option, which would likely complicate the transition to sustainability. This phenomenon is clearly complex and has, therefore, yet to be precisely delineated, but the extensive body of research does point to the following definition [146]:

> *The sharing economy is an IT-facilitated peer-to-peer model for commercial or non-commercial sharing of underutilized goods and service capacity through an intermediary without a transfer of ownership.*

This conception is consolidated and systemic in that it draws on the majority of research to date and covers most aspects of the phenomenon. The sharing economy may be defined as a socioeconomic system supported by digital technology and characterized by five attributes [147]: temporary access, transfer of economic value, platform mediation, expanded consumer role, and crowdsourced supply. For the sharing economy to grow, one-sided users of digital platforms (users who have been exclusively providers or consumers) would have to assume the other role. In other words, the idea is to create as many prosumers (actors who both provide and consume) as possible. Research indicates that one-sided users are driven to become prosumers not only by economic benefits, but also by factors such as trust, gratitude, enjoyment of the activity, and sustainability of collaborative consumption [137,148]. In this light, the sharing economy has much in common with the gig economy and wikinomics. As a prosumer-based phenomenon, its importance for the energy sector cannot be overestimated.

3.8. Methods of Regulating the Energy Sector in a Prosumer Economy

3.8.1. Traditional versus Contemporary Electricity Policy

The evidence indicates that prosumption does not conflict with traditional electricity law norms. One such norm—one of the more crucial ones—is the reliability of electricity service, i.e., maintaining a constant balance between electricity supply and demand while ensuring a stable grid frequency. This merits further exploration, as the renewable energy sources most commonly used by prosumers (such as sunlight or wind) create completely new complications for energy system reliability due to their intermittent nature. The wind does not always blow and the sun does not always shine. A rapid upsurge in the total amount of distributed generation may easily cause an overload of distribution grids [42]. These problems may be alleviated with specific types of distributed resources, such as demand response programs and energy storage, and there are high hopes for combining photovoltaic panels with energy storage systems. Another major norm of a traditional energy policy requires that consumers have access to electricity at reasonable rates while ensuring stable viability utilities—a prerequisite to stable and reliable electricity access. Prosumption may prove beneficial in this regard as well, by helping achieve a better pricing policy, even with the costs of incorporating distributed energy resources into the system and implementing demand-side management. One factor that could keep such costs down would be the reduced demand for new transmission grids, as without the need to transmit energy over long distances, line losses would be reduced [31].

The modern electricity law norms, which emphasize eliminating environmental externalities and boosting industry competition, are highly convergent with prosumption processes [31,42]. Climate change has resulted in the relationship between energy and environmental law being closer now than ever before. Environmental considerations must be increasingly taken into account when designing energy systems. Prosumption mediated by distributed renewable energy resources causes environmental benefits in the form of reduced pollution. Another energy regulatory norm relates to competition. It is assumed that competition provides a major improvement over the old energy market model (which was based on strictly regulated monopolies) by reproducing its benefits without the regulatory inefficiencies. Prosumption based on distributed generation obviously increases market competition by ensuring a diversified supply of electricity. Demand response resources provide the same advantage, as they allow operators to balance energy markets not only by increasing supply, as in the traditional model, but also through contraction of demand.

3.8.2. Economic Stimuli Shaping Prosumption

Measures used to regulate prosumption within the framework of current economic policy include mandates, incentives, environmental policies, and structural reforms [31,42].

1. Mandates are specific renewable power generation targets for municipalities, provinces, or entire countries, often alternatively referred to as renewable portfolio standards. These mandates require utilities operating within a given area to achieve a specified share of renewables in the generated electricity by a certain date. Mandates tend

to support prosumption by incentivizing the utilities covered by the mandate to purchase renewable customer-generated electricity [149].

2. Economic incentives encompass a wide variety of economic stimuli intended to boost prosumption. Examples include tax credits, rebates, or savings designed to reduce the cost of deploying renewables technologies. Many incentives encourage the creation of advanced energy storage systems, such as the installation of behind-the-meter batteries. Pricing programs incentivize capital investment, for example, in net energy metering, to ensure customers are remunerated at a generous rate for the energy sold back to the grid. Another option is the feed-in tariff, which allows homeowners with photovoltaic panels to conclude long-term contracts utilities, guaranteeing a specific rate for the excess energy fed to the grid. The rates usually vary depending on the source of renewable energy, which is done to support selected technologies to promote an economically optimal and sustainable energy mix [150]. Some incentives may be designed to eliminate barriers that discourage prosumption. These include granting prospective prosumers the right to sunlight by building solar fences around protected buildings to ensure they are not shaded by new constructions, obligating utilities to publish distribution grid information, so that individuals are able to see where the best prosumption opportunities are, and finally, ensuring that local governments reduce installation costs.
3. Environmental regulations are a form of indirect incentives to engage in prosumption. They make it less cost-efficient to operate large, centralized fossil fuel power plants, thus encouraging utilities (at least, in theory) to generate electricity from renewables. This results in increased competitiveness of renewable distributed generation, demand response, and energy storage measures. Such policies tend to include greenhouse gas caps and/or carbon dioxide pollution standards imposed on existing coal-fired power plants operating within the given area. Meeting these requirements encourages prosumption by making it easier to involve consumers in distributed generation and demand response programs.
4. Structural reforms mainly relate to the transition from centralized, hub-and-spoke power production systems to a model based on on-site generation and demand-side management (mediated by distributed system platforms and intelligent network platforms, which coordinate the market). Another form of structural reforms is the municipalization of the electricity system—transferring the regulatory responsibility to a new entity, such as a commune, by means of public and localized regulation of the electricity sector. This promotes supply-side democratization, decentralization, and decarbonization of the energy system on the local level, in particular by granting the commune control over its electrical distribution, purchasing, and generation. Such measures may drive replacement of the local investor-owned utility with an electric cooperative or municipal utility, perhaps owing to the common perception that new entities are better for the development of distributed generation and will be able to handle the growing number of solar prosumers. Though municipalization does entail certain costs, it can also help democratize the energy market and empower prosumers to shape energy policy [31].

As a type of economic incentive, subsidies are of great importance for the development of household solar power plants—whether within or outside the grid. As shown by a study conducted in the Czech Republic, public support reduces initial investment costs and, consequently, cuts the payback period down by 3–4 years. This bolsters the economic efficiency of prosumer photovoltaic systems, which in turn translates to encouraging economic indicators such as cash flow, cumulative cash flow, discounted cash flow, net present value, and internal rate of return [151]. According to a case study in Poland, the best economic performance is achieved when the net subsidy value accounts for 19.10% to 27.20% of the total investment costs and when the photovoltaic installation has an installed capacity of 10 kWp [152].

3.8.3. Democratic Experimentalism

Modern-day prosumption is determined by electricity policy that is incredibly complex and falls under the trend of democratic experimentalism, which represents something of a counterpoint to the completely rational approach to policymaking. The experimentalist approach may be warranted, given the inexperience of policy-makers and utilities in regulating renewable energy prosumption, which makes it difficult to effectively design suitable solutions. Polish legislation offers several examples of decision-makers learning from practical experience [153–155]. This mode of thinking can be supported by the incrementalist paradigm, which states that some problems are too complex to predict all the effects of a given policy choice. This issue can be approached from two standpoints. On the one hand, experimentation does seem to be a justified approach—after all, prosumption is based on complicated and rapidly developing technologies. On the other hand, fully leveraging the fruits of the incrementalist approach would require some measure of centralized control over electricity generation and/or coordination of electricity-related efforts. It would also be prudent for the central government to establish a set of objectives for prosumption policy and metrics for measuring the effectiveness (and thus, usefulness) of different policy types [31]. As it is, energy policymaking remains a super wicked problem.

4. Discussion

Thus far, the information about renewable generation prosumers and which energy ecosystems they should operate in is still unreliable and scattered across different sources—so much so that a novel smart knowledge network would be needed to collate the data into one coherent and logical whole (which could have the added benefit of reducing the uncertainty of the collated knowledge and eliminating the current information redundancy). This speaks to the necessity of literature review initiatives to collect dispersed knowledge [78,156] and identify the research hotspots [157] and prospective lines of prosumer energy development such as the Energy Internet [158–162], uses of blockchain technology [68,163,164], or development of innovative energy business model designs incorporating a deep reinforcement learning technique [165].

As it is, formulating a reliable and universal network of knowledge on social policy to prevent environmental degradation worldwide is an extremely challenging and complex proposition—a textbook example of a wicked problem [166,167]. Tackling social policy problems usually ends in failure due to the highly complex nature of the considered issues. Even formulating the issue itself may pose a challenge. The pluralistic society is completely unable to agree on how to define "undisputable public good" or "equity". In addition, there are no reliable criteria for evaluating policies aimed at dealing with social issues—the success or failure of such measures can only be speculated upon. Optimal solutions can only be achieved by applying stringent constraints. The resultant solutions can be neither final nor objective. This is what separates wicked problems from tame problems often encountered in science [168]. Examples of wicked problems include homelessness, social injustice, drug trafficking, natural hazard, and healthcare. Wicked problems can also emerge in socio-technical settings, as exemplified by the problem of energy efficiency [169] and low carbon energy transition [170].

Some issues are further classified as super wicked problems, which include, first and foremost, the problem of global climate change. Super wicked problems have four key characteristics [171]:

1. There is a time limit on finding the solution,
2. Those seeking to solve the problem are also causing it,
3. There is no central authority appointed to find a solution,
4. Certain policies designed to prevent problems are irrational and impede future progress.

This confluence of factors begets a policy-making "tragedy" impervious to traditional analytical methods, preventing action that could mitigate disastrous consequences in the future. It should be possible to overcome this tragedy by applying path-dependent policies designed to impose certain limits on our future collective selves. Sticky interventions in the form of well-chosen trajectories may be a useful tool to achieve this, progressively and incrementally entrenching positive trends across the entire expanding population [171]. Issues of environmental lawmaking are part of what makes climate change a super wicked problem. Environmental law is susceptible to being gradually undermined due to the extent of financial resources that need to be deployed in a short timeframe to achieve goals many decades or even centuries away. To ensure long-term effectiveness, climate change legislation must be designed according to institutional design features which would uncouple its programmatic implementation from powerful political and economic interests mired in short-term thinking. One proposal is to introduce some measure of planned flexibility to environmental law by incorporating asymmetric precommitment strategies. On the one hand, this would complicate changes to legislation; on the other, it would open up ways to introduce certain amendments to deal with other long-term problems, as long as they are in line with greenhouse gas emissions reductions [172].

Several methods have been put forward to tackle wicked problems, including authoritative, competitive, and collaborative strategies [173], as well as an approach based on managed networks [174]. Some make use of issue mapping and visual language [175–177]. While these proposals are interesting and help find solutions to wicked problems, none of them constitute a fundamental breakthrough in social sciences, which only proves that human skill and intuition is very difficult to replace. Visualization of wicked problems can be considered to be a step in the right direction—after all, according to René Thom, the creator of the theory of morphogenesis, the dilemma of all scientific explanation lies in the choice between magic and geometry [178] (p. 5).

5. Conclusions

The survey of the scientific literature on the energy transition shows that it raises the complexity of the interrelated subsystems of the global economy, which include society, economy, legislation, and the environment. At the same time, societies cannot give up on either developing the green energy sector (as it is a remedy for global warming) or striving for sustainable development of the global economy (as this is a condition for the survival of humanity). These two mutually exclusive goals are the quintessence of the present situation in which the whole world has found itself. Thus, the energy transition meets all four conditions of being regarded as a super wicked problem. It is obvious that the time available for finding a solution to the problem is limited, as the climate system can reach the point-of-no-return at any moment, the crossing of which is particularly dangerous as the hazards cannot then be overcome [179]. Some legislative solutions create new problems, such as the contradictions between the short-term and the long-term prospects of the energy policy, the cross-subsidization linked with the transfer of utility's fixed costs between energy prosumers and traditional consumers, or an uncertain experimentalist approach. Moreover, there is no central body to coordinate the energy transition on a global scale. Many local attempts at a concentration of efforts in this regard may be made, such as the European Union's climate policy, but, naturally, their range is limited. There are also plans for solarization of the global energy infrastructure, but this requires thorough changes in policy and outlook on the world for mankind's transition to a new socio-economic system, called solar communism [180,181]. Finally, there are many examples of irrational policies with an adverse impact on the future, as the complexity of the problem makes it difficult to predict all of the consequences of adopting a specific solution.

The situation is made more complicated by the need for the energy transition in prosumer capitalism, which is a relatively new and not fully known economic system. The basic problem, already noted by Toffler, is the externalization of the labor cost linked closely to synergistically double exploitation, extensively discussed by Ritzer. Unpaid human labor results in a transfer of the surplus value from prosumers and consumers to large enterprises and international corporations, thereby boosting their profits. Until now, economic exploitation did not damage the existing economic relations. However, one cannot be certain that it will not happen in the future. If this resulted in disruptions in the energy markets, it would be dangerous for the whole world.

Complexity theory has a large cognitive potential, which can be used in studying organizations and organizational change [182,183]. There is no doubt that the energy transition creates a highly complicated system of interdependencies between society, the economy, legislation, and the environment. It is a huge challenge to organization science. According to the general systems theory, the development of such complex structures is governed by certain laws. The number of states possible in a simple system, comprising few components and interdependencies between them, can be much smaller than the number of challenges originating in the environment. Its adaptability is then low, and the system's increase in complexity is a positive process. It must be enriched with new elements and proper connections between them. However, development can have not only lower but also upper limits. If a system has many components which are interrelated in many ways, an increase in its complexity above a certain level may reduce its adaptability. Kauffman called it the complexity catastrophe [184] (pp. 52–54), [185–187]. It can occur in both biological and socio-economic systems, and the complexity can be reduced to a safe level in both [188,189]. Complexity catastrophe is a well-known phenomenon in organization science, where it has well-established applications [190–194]. Development of the complexity reduction methods in this science can be useful in raising the effectiveness of the socio-economic policy by preventing the energy transition from becoming a super wicked problem.

Making use of the accomplishments of the general systems theory to reduce the complexity of the issues of the energy transition can be supported by a related method called change management [195–199]. This is usually understood to denote a systematic approach, within which organizational strategies, structures, procedures, and technologies aimed at adapting the organization to a change in the external conditions and the business environment are defined, developed, and implemented. Many methods of an organization adapting to changes have been developed, for example, the three-stage change model by Kurt Lewin, which includes unfreezing, moving, and freezing group standards [200], or a more developed model by John Kotter, which comprises eight steps: establishing a sense of urgency, creating the guiding coalition, developing a vision and strategy, communicating the change vision, empowering employees for broad-based action, generating short-term wins, consolidating gains and producing more change, and anchoring new approaches in the culture [201].

In recent years, both system control methods, avoiding or alleviating complexity catastrophes, based on complexity science and change management based on management and organization science were combined to make a new approach called complexity management [202–209]. The main concepts in change and complexity management are usually microeconomic, which may be necessary and useful for exemplification, as this is confirmed by numerous experiments. However, when it comes to the energy transition, change management and complexity management must be redefined to make those approaches global in nature. However, this may not be enough to solve the major issues related to energy transition. All attempts at alleviating complexity catastrophes by developing and implementing proper methods of change management and complexity management directed towards reducing the energy transition complexity may be doomed to failure. This is caused by the lack of a central organization in the form of a global super-organization responsible for solving the problem on a global scale. Since the problem is global, its solution must also be global.

Funding: This research received no external funding.

Data Availability Statement: Not applicable.

Conflicts of Interest: The author declares no conflict of interest.

References

1. Ritzer, G.; Jurgenson, N. Production, consumption, prosumption: The nature of capitalism in the age of the digital 'prosumer'. *J. Consum. Cult.* **2010**, *10*, 13–36. [CrossRef]
2. Ritzer, G. Prosumer capitalism. *Sociol. Q.* **2015**, *56*, 413–445. [CrossRef]
3. Ritzer, G. Prosumption: Contemporary capitalism and the "new" prosumer. In *The Oxford Handbook of Consumption*; Wherry, F.F., Woodward, I., Eds.; Oxford University Press: New York, NY, USA, 2019; pp. 75–93. ISBN 9780190695583.
4. Toffler, A. *The Third Wave*; William Morrow and Company: New York, NY, USA, 1980. ISBN 0-688-03597-3.
5. Astarloa, B.; Kaakeh, A.; Lombardi, M.; Scalise, J. *The Future of Electricity: New Technologies Transforming the Grid Edge*; World Economic Forum: Cologny/Geneva, Switzerland, 2017. Available online: https://www3.weforum.org/docs/WEF_Future_of_Electricity_2017.pdf (accessed on 5 January 2022).
6. Tapscott, D.; Williams, A.D. *Wikinomics: How Mass Collaboration Changes Everything*; Portfolio/Penguin: New York, NY, USA, 2006. ISBN 978-1-59184-138-8.
7. Tapscott, D.; Williams, A.D. *Macrowikinomics: Rebooting Business and the World*; Portfolio/Penguin: New York, NY, USA, 2012. ISBN 978-1-59184-356-6.
8. Montgomery, B. The 5th Wave. *The IOT Magazine*. 26 October 2016. Available online: https://theiotmagazine.com/the-5th-wave-4ee1ad8b3e9e (accessed on 10 June 2022).
9. Ranjith, P.V.; Varma, A.J.; Ashwini, J. Connected societies through digital transformation. In *Emerging Challenges, Solutions, and Best Practices for Digital Enterprise Transformation*; Sandhu, K., Ed.; IGI Global: Hershey, PA, USA, 2021; pp. 199–220. ISBN 978-1-7998-8587-0. [CrossRef]
10. McLuhan, M.; Nevitt, B. *Take Today: The Executive as Dropout*; Harcourt Brace Jovanovich: New York, NY, USA, 1972. ISBN 0-15-187830-7.
11. Hicks, J.R. *Capital and Growth*; Oxford University Press: London, UK, 1965.
12. Toffler, A.; Toffler, H. *Revolutionary Wealth: How It Will Be Created and How It Will Change Our Lives*; Currency Doubleday: New York, NY, USA, 2006. ISBN 978-0-385-52207-6.
13. Lang, B.; Dolan, R.; Kemper, J.; Northey, G. Prosumers in times of crisis: Definition, archetypes and implications. *J. Serv. Manag.* **2021**, *32*, 176–189. [CrossRef]
14. Szymusiak, T. Prosumpcja—Wyzwanie dla marketingu oraz zarządzania. Charakterystyka oraz klasyfikacja współczesnego prosumenta. Studium przypadku: Polacy a Niemcy. In *Finanse, Rachunkowość i Zarządzanie. Polska, Europa, Świat 2020*; Adrianowski, D., Patora, K., Sikorski, J., Eds.; Wydawnictwo Uniwersytetu Łódzkiego: Łódź, Poland, 2013; pp. 253–266. ISBN 978-83-7525-827-1. [CrossRef]
15. Łaskawiec, K. Rola prosumenta na rynku energii elektrycznej. *Zeszyty Naukowe Politechniki Śląskiej. Seria Organizacja Zarządzanie* **2014**, 361–371. Available online: https://bibliotekanauki.pl/articles/321730 (accessed on 24 September 2022).
16. Szymusiak, T. *Prosument—Prosumpcja—Prosumeryzm: Ekonomiczne Oraz Społeczne Korzyści Prosumpcji na Przykładzie Polski Oraz Niemiec (Podejście Naukowe)*; Wydawnictwo Bezkresy Wiedzy: Saarbrücken, Germany, 2015. ISBN 978-3-639-89210-9.
17. von Hippel, E. Lead users: A source of novel product concepts. *Manag. Sci.* **1986**, *32*, 791–805. Available online: http://web.mit.edu/evhippel/www-old/papers/evh-01.htm (accessed on 12 December 2021). [CrossRef]
18. Ślusarz, G.; Gołębiewska, B.; Cierpiał-Wolan, M.; Twaróg, D.; Gołębiewski, J.; Wójcik, S. The role of agriculture and rural areas in the development of Autonomous Energy Regions in Poland. *Energies* **2021**, *14*, 4033. [CrossRef]
19. Maśloch, P.; Maśloch, G.; Kuźmiński, Ł.; Wojtaszek, H.; Miciuła, I. Autonomous Energy Regions as a proposed choice of selecting selected EU regions—Aspects of their creation and management. *Energies* **2020**, *13*, 6444. [CrossRef]
20. Maśloch, G. Budowa autonomicznych regionów energetycznych w Polsce—Utopia czy konieczność? *Stud. Prawno-Ekon.* **2018**, *106*, 251–264. [CrossRef]
21. Gostomczyk, W. Możliwości tworzenia autonomicznych regionów energetycznych wykorzystujących odnawialne źródła energii. *Roczniki Naukowe Stowarzyszenia Ekonomistów Rolnictwa Agrobiznesu* **2018**, *20*, 47–52. [CrossRef]
22. Lebiejko, A. Prosumer—A new trend of active consumption on the example of banking services. *J. Interdiscip. Res.* **2011**, *1*, 65–69. Available online: http://www.magnanimitas.cz/01-02 (accessed on 12 September 2021).
23. Matysik-Pejas, R.; Szafrańska, M.; Krasnodębski, A. Prosumer attitudes as a new component of consumer behaviour on the food market. In *International Scientific Days 2016. The Agri-Food Value Chain: Challenges for Natural Resources Management and Society. Conference Proceedings*; Horská, E., Kapsdorferová, Z., Hallová, M., Eds.; Slovak University of Agriculture in Nitra: Nitra, Slovak Republic, 2016; pp. 812–819. ISBN 978-80-552-1503-7. [CrossRef]
24. Żabińska, I. Rozwój energetyki prosumenckiej opartej o OZE w Polsce. *Systemy Wspomagania Inżynierii Produkcji Problemy Zarządzaniu Środowiskiem* **2017**, *6*, 83–95. Available online: https://bibliotekanauki.pl/articles/113177 (accessed on 20 September 2022).
25. de Pourbaix, P. Prosumer of the XXI century—New challenges to commerce and marketing. *Acta Sci. Pol. Oeconomia* **2016**, *15*, 89–97. Available online: https://bibliotekanauki.pl/articles/38096 (accessed on 24 September 2022).
26. Zalega, T. Wykorzystanie odnawialnych źródeł energii w gospodarstwach domowych seniorów w Polsce w świetle wyników badań własnych. In *Rola Odnawialnych Źródeł Energii w Rozwoju Społeczno-Ekonomicznym Kraju i Regionu*; Nowak, A.Z., Szałański, M., Zborowska, W., Eds.; Wydawnictwo Naukowe Wydziału Zarządzania Uniwersytetu Warszawskiego: Warsaw, Poland, 2016; pp. 48–68. ISBN 978-83-65402-43-1.

27. Koszembar-Wiklik, M. Komunikacja marketingowa a komercjalizacja nowych technologii i produktów. *Zeszyty Naukowe Politechniki Śląskiej. Seria Organizacja Zarządzanie* **2016**, 251–260. Available online: https://bibliotekanauki.pl/articles/323613 (accessed on 24 September 2022).
28. Iwanicka, A. Formy działalności prosumpcyjnej wybranych grup nabywców finalnych. *Handel Wewnętrzny* **2016**, *363*, 56–68. Available online: https://bibliotekanauki.pl/articles/561860 (accessed on 22 September 2022).
29. Jakimowicz, A.; Rzeczkowski, D. The impact of public administration digitalization on the decarbonization of the economy. *Energies* **2021**, *14*, 5739. [CrossRef]
30. Jakimowicz, A.; Rzeczkowski, D. Prosumption in the public administration sector. *Acta Phys. Pol. A* **2016**, *129*, 1011–1017. [CrossRef]
31. Jacobs, S.B. The Energy Prosumer. *Ecol. Law Q.* **2016**, *43*, 519–579. [CrossRef]
32. Milčiuvienė, S.; Kiršienė, J.; Doheijo, E.; Urbonas, R.; Milčius, D. The role of renewable energy prosumers in implementing energy justice theory. *Sustainability* **2019**, *11*, 5286. [CrossRef]
33. Directive (EU) 2019/944 of the European Parliament and of the Council of 5 June 2019 on common rules for the internal market for electricity and amending Directive 2012/27/EU (Text with EEA relevance). *Off. J. Eur. Union* **2019**, *L 158*, 125–199. Available online: https://eur-lex.europa.eu/eli/dir/2019/944/oj (accessed on 18 November 2021).
34. Popczyk, J. *Energetyka Rozproszona. Od Dominacji Energetyki w Gospodarce Do Zrównoważonego Rozwoju, od Paliw Kopalnych Do Energii Odnawialnej i Efektywności Energetycznej*; Polski Klub Ekologiczny Okręg Mazowiecki: Warsaw, Poland, 2011. ISBN 978-83-915094-1-8.
35. Leal-Arcas, R.; Lesniewska, F.; Proedrou, F. Prosumers: New actors in EU energy security. In *Netherlands Yearbook of International Law 2017. Shifting Forms and Levels of Cooperation in International Economic Law: Structural Developments in Trade, Investment and Financial Regulation*; Amtenbrink, F., Prévost, D., Wessel, R.A., Eds.; T.M.C. Asser Press: The Hague, The Netherlands, 2018; Volume 48, pp. 139–172. ISBN 978-94-6265-242-2. [CrossRef]
36. Lavrijssen, S.; Carrillo Parra, A. Radical prosumer innovations in the electricity sector and the impact on prosumer regulation. *Sustainability* **2017**, *9*, 1207. [CrossRef]
37. Bukowski, M.; Pankowiec, A.; Szczerba, P.; Śniegocki, A. *Przełomowa Energetyka Prosumencka. Dlaczego Źródła Rozproszone Mogą Doprowadzić Do Przewrotu Na Rynku Energii*; Warszawski Instytut Studiów Ekonomicznych: Warsaw, Poland, 2014. Available online: https://wise-europa.eu/2014/03/26/przelomowa-energetyka-prosumencka/ (accessed on 18 September 2021).
38. Ropuszyńska-Surma, E.; Węglarz, M.; Szwabiński, J. Energy prosumers: Profiling the energy microgeneration market in Lower Silesia, Poland. *Oper. Res. Decis.* **2018**, *28*, 75–94. [CrossRef]
39. Green, R.; Staffell, I. "Prosumage" and the British electricity market. *Econ. Energy Environ. Policy* **2017**, *6*, 33–49. [CrossRef]
40. Schill, W.-P.; Zerrahn, A.; Kunz, F. Prosumage of solar electricity: Pros, cons, and the system perspective. *Econ. Energy Environ. Policy* **2017**, *6*, 7–31. [CrossRef]
41. Becker, A.; Loges, H.; Kippelt, S.; Gitis, A.; Merei, G.; Echternacht, D.; Müller, M.; Zeh, A.; Kleimaier, M.; Leuthold, M.; et al. Electricity Storage Systems in Medium- and Low-Voltage Networks. In *ETG-Fb. 147: International ETG Congress 2015: Die Energiewende—Blueprints for the New Energy Age. Proceedings, November 17–18, 2015, World Conference Center, Bonn (ETG-Fachbericht)*; Power Engineering Society in the VDE (ETG)—VDE Verlag GmbH: Berlin, Germany, 2015; pp. 1–8. ISBN 978-3-8007-4121-2.
42. Kotilainen, K. Energy prosumers' role in the sustainable energy system. In *Affordable and Clean Energy. Encyclopedia of the UN Sustainable Development Goals*; Leal Filho, W., Azul, A.M., Brandli, L., Salvia, A.L., Wall, T., Eds.; Springer Nature Switzerland AG: Cham, Switzerland, 2021; pp. 1–14. ISBN 978-3-319-95863-7. [CrossRef]
43. Masera, D.; Couture, T. *Industrial Prosumers of Renewable Energy. Contribution to Inclusive and Sustainable Industrial Development*; United Nations Industrial Development Organization, Vienna International Centre: Vienna, Austria, 2015. Available online: https://www.unido.org/sites/default/files/2015-04/PROSUMERS_Energy_0.pdf (accessed on 7 January 2022).
44. Rickerson, W.; Couture, T.; Barbose, G.; Jacobs, D.; Parkinson, G.; Chessin, E.; Belden, A.; Wilson, H.; Barrett, H. *Residential Prosumers: Drivers and Policy Options (Re-Prosumers)*; Technical Report; IEA—Renewable Energy Technology Deployment: Berkeley, CA, USA, 20 September 2014. Available online: https://www.osti.gov/servlets/purl/1163237 (accessed on 8 January 2022).
45. Study on "Residential Prosumers in the European Energy Union". JUST/2015/CONS/FW/C006/0127, Framework Contract EAHC/2013/CP/04. European Commission. Prepared by: GfK Belgium Consortium. 2 May 2017. Available online: https://ec.europa.eu/info/sites/default/files/study-residential-prosumers-energy-union_en.pdf (accessed on 9 January 2022).
46. Popławski, T.; Dudzik, S.; Szeląg, P.; Baran, J. A Case study of a virtual power plant (VPP) as a data acquisition tool for PV energy forecasting. *Energies* **2021**, *14*, 6200. [CrossRef]
47. Subramanya, R.; Yli-Ojanperä, M.; Sierla, S.; Hölttä, T.; Valtakari, J.; Vyatkin, V. A virtual power plant solution for aggregating photovoltaic systems and other distributed energy resources for Northern European primary frequency reserves. *Energies* **2021**, *14*, 1242. [CrossRef]
48. Olkkonen, L.; Korjonen-Kuusipuro, K.; Grönberg, I. Redefining a stakeholder relation: Finnish energy "prosumers" as co-producers. *Environ. Innov. Soc. Transit.* **2017**, *24*, 57–66. [CrossRef]
49. Ramyar, S.; Liu, A.L.; Chen, Y. A power market model in presence of strategic prosumers. *IEEE Trans. Power Syst.* **2020**, *35*, 898–908. [CrossRef]
50. Pieńkowski, D. Rethinking the concept of prosuming: A critical and integrative perspective. *Energy Res. Soc. Sci.* **2021**, *74*, 101967. [CrossRef]

51. Burchard-Dziubińska, M. Prosument na rynku energii w Polsce—Próba oceny w świetle teorii kosztów transakcyjnych. *Gospodarka Praktyce Teorii* **2015**, *41*, 5–19. [CrossRef]
52. Pietrzak, M.B.; Igliński, B.; Kujawski, W.; Iwański, P. Energy transition in Poland—Assessment of the renewable energy sector. *Energies* **2021**, *14*, 2046. [CrossRef]
53. Worek, B.; Kocór, M.; Micek, D.; Lisek, K.; Szczucka, A. Społeczny wymiar rozwoju energetyki rozproszonej w Polsce—Kluczowe czynniki i wyzwania. *Energetyka Rozproszona* **2021**, 105–118. [CrossRef]
54. Kucharska, A. Cybersecurity challenges in Poland in the face of energy transition. *Rocznik Instytutu Europy Środkowo-Wschodniej* **2020**, *18*, 141–159. [CrossRef]
55. Barichella, A. *Cybersecurity in the Energy Sector. A Comparative Analysis between Europe and the United States. Études de l'Ifri*; Institut Français des Relations Internationales: Paris, France, 2018. ISBN 978-2-36567-830-8. Available online: https://www.ifri.org/en/publications/etudes-de-lifri/cybersecurity-energy-sector-comparative-analysis-between-europe-and (accessed on 17 January 2022).
56. Rice, E.B.; AlMajali, A. Mitigating the risk of cyber attack on smart grid systems. *Procedia Comput. Sci.* **2014**, *28*, 575–582. [CrossRef]
57. Govindarasu, M.; Hahn, A. Cybersecurity of the Power Grid: A Growing Challenge. *The Conversation*. 24 February 2017. Available online: https://theconversation.com/cybersecurity-of-the-power-grid-a-growing-challenge-73102 (accessed on 16 January 2022).
58. Malko, J.; Wojciechowski, H. Sektor energetyczny i cyberbezpieczeństwo. *Nowa Energia* **2015**, *1*, 10–15. Available online: https://nowa-energia.com.pl/2015/02/18/ne-01-2015/ (accessed on 18 January 2022).
59. Skokowski, D. Cyberbezpieczeństwo sektora elektroenergetycznego. *Sektor Elektroenergetyczny* **2018**, *1*, 141–146. Available online: http://pliki.wydawnictwo.murator.pl/ksiazeczki/SektorElektroenergetyczny2018/files/assets/basic-html/index.html#8 (accessed on 17 January 2022).
60. Lee, R.M.; Assante, M.J.; Conway, T. *Analysis of the Cyber Attack on the Ukrainian Power Grid. Defense Use Case*; TLP: White; Industrial Control Systems, Electricity-Information Sharing and Analysis Center: Washington, DC, USA, 18 March 2016. Available online: https://nsarchive.gwu.edu/sites/default/files/documents/3891751/SANS-and-Electricity-Information-Sharing-and.pdf (accessed on 18 January 2022).
61. Styczynski, J.; Beach-Westmoreland, N. *When the Lights Went Out. A Comprehensive Review of the 2015 Attacks on Ukrainian Critical Infrastructure*; Booz Allen Hamilton Inc.: McLean, VA, USA, 2019. Available online: https://www.boozallen.com/content/dam/boozallen/documents/2016/09/ukraine-report-when-the-lights-went-out.pdf (accessed on 19 January 2022).
62. The EU's Cybersecurity Strategy for the Digital Decade. Joint Communication to the European Parliament and the Council; European Commission, High Representative of the Union for Foreign Affairs and Security Policy: Brussels, Belgium, 16 December 2020. Available online: https://digital-strategy.ec.europa.eu/en/library/eus-cybersecurity-strategy-digital-decade-0 (accessed on 20 January 2022).
63. Smith, D.C. Enhancing cybersecurity in the energy sector: A critical priority. *J. Energy Nat. Resour. Law* **2018**, *36*, 373–380. [CrossRef]
64. Govindarasu, M.; Hahn, A. As Russians Hack the US Grid, a Look at What's Needed to Protect It. *The Conversation*. 7 August 2018. Available online: https://theconversation.com/as-russians-hack-the-us-grid-a-look-at-whats-needed-to-protect-it-100489 (accessed on 20 January 2022).
65. Regulation (EU) No 347/2013 of the European Parliament and of the Council of 17 April 2013 on guidelines for trans-European energy infrastructure and repealing Decision No 1364/2006/EC and amending Regulations (EC) No 713/2009, (EC) No 714/2009 and (EC) No 715/2009 (Text with EEA relevance). Available online: http://data.europa.eu/eli/reg/2013/347/oj (accessed on 2 December 2021).
66. Leal-Arcas, R.; Lesniewska, F.; Proedrou, F. Smart grids in the European Union: Assessing energy security, regulation & social and ethical considerations. *Columbia J. Eur. Law* **2018**, *24*, 293–391. Available online: https://ssrn.com/abstract=3062957 (accessed on 3 December 2021).
67. Leal-Arcas, R.; Lesniewska, F.; Proedrou, F. Prosumers as new energy actors. In *Africa-EU Renewable Energy Research and Innovation Symposium 2018 (RERIS 2018), 23–26 January 2018, National University of Lesotho: On occasion of NULISTICE 2018*; Mpholo, M., Steuerwald, D., Kukeera, T., Eds.; Springer Nature Switzerland AG: Cham, Switzerland, 2018; pp. 139–151. ISBN 978-3-319-93438-9. [CrossRef]
68. Casquiço, M.; Mataloto, B.; Ferreira, J.C.; Monteiro, V.; Afonso, J.L.; Afonso, J.A. Blockchain and Internet of Things for electrical energy decentralization: A review and system architecture. *Energies* **2021**, *14*, 8043. [CrossRef]
69. Hossein Motlagh, N.; Khatibi, A.; Aslani, A. Toward sustainable energy-independent buildings using Internet of Things. *Energies* **2020**, *13*, 5954. [CrossRef]
70. Dobrowolski, Z. Internet of Things and other e-solutions in supply chain management may generate threats in the energy sector—The quest for preventive measures. *Energies* **2021**, *14*, 5381. [CrossRef]
71. Shakerighadi, B.; Anvari-Moghaddam, A.; Vasquez, J.C.; Guerrero, J.M. Internet of Things for modern energy systems: State-of-the-art, challenges, and open issues. *Energies* **2018**, *11*, 1252. [CrossRef]
72. Hossein Motlagh, N.; Mohammadrezaei, M.; Hunt, J.; Zakeri, B. Internet of Things (IoT) and the energy sector. *Energies* **2020**, *13*, 494. [CrossRef]
73. Strielkowski, W.; Streimikiene, D.; Fomina, A.; Semenova, E. Internet of Energy (IoE) and high-renewables electricity system market design. *Energies* **2019**, *12*, 4790. [CrossRef]

74. Alam, S.E.; Shorten, R.; Wirth, F.; Yu, J.Y. Distributed algorithms for Internet-of-Things-Enabled prosumer markets: A control theoretic perspective. In *Analytics for the Sharing Economy: Mathematics, Engineering and Business Perspectives*; Crisostomi, E., Ghaddar, B., Häusler, F., Naoum-Sawaya, J., Russo, G., Shorten, R., Eds.; Springer Nature Switzerland AG: Cham, Switzerland, 2020; pp. 125–149. ISBN 978-3-030-35031-4. [CrossRef]
75. Kucharska, A. Społeczny wymiar "Energiewende"—Rozwój energetyki prosumenckiej w Austrii. *Studia Humanistyczne AGH* **2018**, *17*, 67–84. [CrossRef]
76. Strielkowski, W. *Social Impacts of Smart Grids: The Future of Smart Grids and Energy Market Design*; Elsevier: Amsterdam, The Netherlands, 2020. ISBN 978-0-12-817770-9.
77. Parag, Y.; Sovacool, B.K. Electricity market design for the prosumer era. *Nat. Energy* **2016**, *1*, 16032. [CrossRef]
78. Espe, E.; Potdar, V.; Chang, E. Prosumer communities and relationships in smart grids: A literature review, evolution and future directions. *Energies* **2018**, *11*, 2528. [CrossRef]
79. Leong, C.H.; Gu, C.; Li, F. Auction mechanism for P2P local energy trading considering physical constraints. *Energy Procedia* **2019**, *158*, 6613–6618. [CrossRef]
80. Long, C.; Zhou, Y.; Wu, J. A game theoretic approach for peer to peer energy trading. *Energy Procedia* **2019**, *159*, 454–459. [CrossRef]
81. PankiRaj, J.S.; Yassine, A.; Choudhury, S. An auction mechanism for profit maximization of peer-to-peer energy trading in smart grids. *Procedia Comput. Sci.* **2019**, *151*, 361–368. [CrossRef]
82. Long, C.; Wu, J.; Zhang, C.; Cheng, M.; Al-Wakeel, A. Feasibility of peer-to-peer energy trading in low voltage electrical distribution networks. *Energy Procedia* **2017**, *105*, 2227–2232. [CrossRef]
83. El-Baz, W.; Tzscheutschler, P.; Wagner, U. Evaluation of energy market platforms potential in microgrids: Scenario analysis based on a double-sided auction. *Front. Energy Res.* **2019**, *7*, 41. [CrossRef]
84. Spasova, B.; Kawamoto, D.; Takefuji, Y. Evaluation of the effects of bidding strategy with customized pricing on the individual prosumer in a local energy market. *Adv. Sci. Technol. Eng. Syst. J.* **2019**, *4*, 366–379. [CrossRef]
85. Tushar, W.; Saha, T.K.; Yuen, C.; Morstyn, T.; McCulloch, M.D.; Poor, H.V.; Wood, K.L. A motivational game-theoretic approach for peer-to-peer energy trading in the smart grid. *Appl. Energy* **2019**, *243*, 10–20. [CrossRef]
86. Kokchang, P.; Junlakarn, S.; Audomvongseree, K. Business model and market designs for solar prosumer on peer to peer energy trading in Thailand. *IOP Conf. Ser. Earth Environ. Sci.* **2020**, *463*, 012127. [CrossRef]
87. Shrestha, A.; Bishwokarma, R.; Chapagain, A.; Banjara, S.; Aryal, S.; Mali, B.; Thapa, R.; Bista, D.; Hayes, B.P.; Papadakis, A.; et al. Peer-to-peer energy trading in micro/mini-grids for local energy communities: A review and case study of Nepal. *IEEE Access* **2019**, *7*, 131911–131928. [CrossRef]
88. Zepter, J.M.; Lüth, A.; del Granado, P.C.; Egging, R. Prosumer integration in wholesale electricity markets: Synergies of peer-to-peer trade and residential storage. *Energy Build.* **2019**, *184*, 163–176. [CrossRef]
89. Kupiec, B. Analiza prawno-porównawcza klastrów energii i spółdzielni energetycznych. *Przegląd Prawniczy Uniwersytetu Warszawskiego* **2020**, *19*, 64–94. [CrossRef]
90. Porter, M.E. Clusters and the new economics of competition. *Harv. Bus. Rev.* **1998**, *76*, 77–90. Available online: https://hbr.org/1998/11/clusters-and-the-new-economics-of-competition (accessed on 15 January 2022).
91. Ustawa z dnia 20 lutego 2015 r. o odnawialnych źródłach energii. *Dziennik Ustaw Rzeczypospolitej Polskiej*. 2015; poz. 478 z późn. zm. Available online: http://isap.sejm.gov.pl/isap.nsf/DocDetails.xsp?id=wdu20150000478 (accessed on 16 January 2022).
92. Mataczyńska, E. Społeczności energetyczne—Od regulacji unijnych do polskich. In *Klastry Energii. Regulacje, Teoria i Praktyka*; Mataczyńska, E., Kucharska, A., Eds.; Wydawnictwo Naukowe Instytutu Polityki Energetycznej im. I. Łukasiewicza: Rzeszów, Poland, 2020; pp. 31–49. ISBN 978-83-946727-8-2. Available online: https://www.instytutpe.pl/wp-content/uploads/2019/09/Klastry-energii.-Regulacje-teoria-i-praktyka.pdf (accessed on 24 September 2021).
93. Marzec, T. Prawne perspektywy rozwoju spółdzielni energetycznych w Polsce. *Internetowy Kwartalnik Antymonopolowy Regulacyjny* **2021**, *2*, 24–40. [CrossRef]
94. Mucha-Kuś, K.; Sołtysik, M.; Zamasz, K.; Szczepańska-Woszczyna, K. Coopetitive nature of energy communities—The energy transition context. *Energies* **2021**, *14*, 931. [CrossRef]
95. Jevons, W.S. *The Coal Question; An Inquiry Concerning the Progress of the Nation, and the Probable Exhaustion of Our Coal-Mines*, 2nd ed.; Macmillan & Co.: London, UK, 1866.
96. Polimeni, J.M.; Mayumi, K.; Giampietro, M.; Alcott, B. *The Jevons Paradox and the Myth of Resource Efficiency Improvements*; Earthscan: London, UK, 2008. ISBN 978-1-84407-462-4.
97. Alcott, B. Jevons' paradox. *Ecol. Econ.* **2005**, *54*, 9–21. [CrossRef]
98. Freire-González, J.; Puig-Ventosa, I. Energy efficiency policies and the Jevons paradox. *Int. J. Energy Econ. Policy* **2015**, *5*, 69–79. Available online: http://www.econjournals.com/index.php/ijeep/article/view/965/565 (accessed on 4 December 2021).
99. Bayer, P.; Aklin, M. The European Union Emissions Trading System reduced CO_2 emissions despite low prices. *Proc. Natl. Acad. Sci. USA* **2020**, *117*, 8804–8812. [CrossRef]
100. Convery, F.J.; Redmond, L. The European Union Emissions Trading Scheme: Issues in allowance price support and linkage. *Annu. Rev. Resour. Econ.* **2013**, *5*, 301–324. [CrossRef]
101. Nield, K. Fraud on the European Union Emissions Trading Scheme: Effects, vulnerabilities and regulatory reform. *Eur. Energy Environ. Law Rev.* **2011**, *20*, 255–289. [CrossRef]

102. Mandaroux, R.; Dong, C.; Li, G. A European Emissions Trading System powered by distributed ledger technology: An evaluation framework. *Sustainability* **2021**, *13*, 2106. [CrossRef]
103. Jędral, W. OZE i efektywność energetyczna w kontekście wyzwań dla wytwarzania i użytkowania energii elektrycznej w Polsce. *Rynek Energii* **2019**, 3–8. Available online: https://www.cire.pl/pliki/2/2019/01__jedral.pdf (accessed on 7 October 2021).
104. Ahn, D.P. *Principles of Commodity Economics and Finance*; The MIT Press: Cambridge, MA, USA, 2018. ISBN 9780262038379.
105. Invisible Fuel. *The Economist*. 17 January 2015. Available online: https://www.economist.com/special-report/2015/01/15/invisible-fuel (accessed on 15 January 2022).
106. Smil, V. Numbers don't lie. Embodied energy: Mobile devices and cars. *IEEE Spectr.* **2016**, *53*, 26. [CrossRef]
107. Campos, I.; Marín-González, E. People in transitions: Energy citizenship, prosumerism and social movements in Europe. *Energy Res. Soc. Sci.* **2020**, *69*, 101718. [CrossRef]
108. Kverndokk, S.; Rose, A. Equity and justice in global warming policy. *Int. Rev. Environ. Resour. Econ.* **2008**, *2*, 135–176. [CrossRef]
109. Kverndokk, S.; Nævdal, E.; Nøstbakken, L. The trade-off between intra- and intergenerational equity in climate policy. *Eur. Econ. Rev.* **2014**, *69*, 40–58. [CrossRef]
110. Ryghaug, M.; Skjølsvold, T.M.; Heidenreich, S. Creating energy citizenship through material participation. *Soc. Stud. Sci.* **2018**, *48*, 283–303. [CrossRef]
111. Olivadese, R.; Alpagut, B.; Revilla, B.P.; Brouwer, J.; Georgiadou, V.; Woestenburg, A.; van Wees, M. Towards energy citizenship for a just and inclusive transition: Lessons learned on collaborative approach of Positive Energy Districts from the EU Horizon 2020 Smart Cities and Communities Projects. *Proceedings* **2020**, *65*, 20. [CrossRef]
112. Gierszewska, G.; Seretny, M. Sustainable behavior—The need of change in consumer and business attitudes and behavior. *Found. Manag.* **2019**, *11*, 197–208. [CrossRef]
113. Sovacool, B.K. Fuel poverty, affordability, and energy justice in England: Policy insights from the Warm Front Program. *Energy* **2015**, *93 Part 1*, 361–371. [CrossRef]
114. Moore, R. Definitions of fuel poverty: Implications for policy. *Energy Policy* **2012**, *49*, 19–26. [CrossRef]
115. Day, R.; Walker, G.; Simcock, N. Conceptualising energy use and energy poverty using a capabilities framework. *Energy Policy* **2016**, *93*, 255–264. [CrossRef]
116. González-Eguino, M. Energy poverty: An overview. *Renew. Sustain. Energy Rev.* **2015**, *47*, 377–385. [CrossRef]
117. Bouzarovski, S. Energy poverty in the European Union: Landscapes of vulnerability. *Wiley Interdiscip. Rev. Energy Environ.* **2014**, *3*, 276–289. [CrossRef]
118. Kaczmarczyk, M. (Ed.) *Niska Emisja: Od Przyczyn Występowania do Sposobów Eliminacji*; Redakcja Globenergia—Geosystem Burek, KOTYZA, S.C.: Kraków, Poland, 2015. ISBN 978-83-64339-02-8.
119. Pudło, M.; Wrzalik, A. Managing prosumer energy in the aspect of low emission reduction. *Zeszyty Naukowe Politechniki Częstochowskiej Zarządzanie* **2017**, 51–63. Available online: https://wz.pcz.pl/znwz/zeszyt-naukowy-numer-26.html (accessed on 28 September 2021). [CrossRef]
120. Biadacz, R.; Wysłocka, E. Wykorzystanie odnawialnych źródeł energii przez gospodarstwa domowe. *Przegląd Organizacji* **2014**, *896*, 30–36. [CrossRef]
121. Szulecki, K.; Szwed, D. Społeczne aspekty OZE: Którędy do energetycznej demokracji? In *Odnawialne źródła energii w Polsce. Wybrane Problemy Bezpieczeństwa, Polityki i Administracji*; Księżopolski, K.M., Pronińska, K.M., Sulowska, A.E., Eds.; Dom Wydawniczy Elipsa: Warsaw, Poland, 2013; pp. 184–211. ISBN 978-83-7151-992-5.
122. Stephens, J.C. Energy democracy: Redistributing power to the people through renewable transformation. *Environ. Sci. Policy Sustain. Dev.* **2019**, *61*, 4–13. [CrossRef]
123. Benander, L.; Horowitz, D.A.; Baker, I. New economy energy cooperatives bring power to the people. In *Energy Democracy: Advancing Equity in Clean Energy Solutions*; Fairchild, D., Weinrub, A., Eds.; Island Press: Washington, DC, USA, 2017; pp. 195–217. ISBN 978-1-61091-929-6. [CrossRef]
124. Sweeney, S. Conflicting agendas: Energy democracy and the labor movement. In *Energy Democracy: Advancing Equity in Clean Energy Solutions*; Fairchild, D., Weinrub, A., Eds.; Island Press: Washington, DC, USA, 2017; pp. 113–138. ISBN 978-1-61091-929-6. [CrossRef]
125. Szulecki, K.; Overland, I. Energy democracy as a process, an outcome and a goal: A conceptual review. *Energy Res. Soc. Sci.* **2020**, *69*, 101768. [CrossRef]
126. Szulecki, K. Conceptualizing energy democracy. *Environ. Politics* **2018**, *27*, 21–41. [CrossRef]
127. Schoolman, A.; Delman, B. Building power through community-based project development. In *Energy Democracy: Advancing Equity in Clean Energy Solutions*; Fairchild, D., Weinrub, A., Eds.; Island Press: Washington, DC, USA, 2017; pp. 219–238. ISBN 978-1-61091-929-6. [CrossRef]
128. Islar, M.; Busch, H. "We are not in this to save the polar bears!"—The link between community renewable energy development and ecological citizenship. *Innov. Eur. J. Soc. Sci. Res.* **2016**, *29*, 303–319. [CrossRef]
129. Stephens, J.C.; Burke, M.J.; Gibian, B.; Jordi, E.; Watts, R. Operationalizing energy democracy: Challenges and opportunities in Vermont's renewable energy transformation. *Front. Commun.* **2018**, *3*, 43. [CrossRef]
130. Flaute, M.; Großmann, A.; Lutz, C.; Nieters, A. Macroeconomic effects of prosumer households in Germany. *Int. J. Energy Econ. Policy* **2017**, *7*, 146–155. Available online: https://www.econjournals.com/index.php/ijeep/article/view/3386 (accessed on 27 September 2021).

131. Dolber, B.; Rodino-Colocino, M.; Kumanyika, C.; Wolfson, T. (Eds.) *The Gig Economy: Workers and Media in the Age of Convergence*; Routledge—Taylor & Francis Group: New York, NY, USA, 2021. ISBN 978-0-367-69021-2.
132. Mulcahy, D. *The Gig Economy: The Complete Guide to Getting Better Work, Taking More Time Off, and Financing the Life You Want*; AMACOM—American Management Association: New York, NY, USA, 2016. ISBN 9780814437339.
133. Sinicki, A. *Thriving in the Gig Economy: Freelancing Online for Tech Professionals and Entrepreneurs*; Apress Media LLC: Bicester, UK, 2019. ISBN 978-1-4842-4089-2.
134. Prassl, J. *Humans as a Service: The Promise and Perils of Work in the Gig Economy*; Oxford University Press: New York, NY, USA, 2018. ISBN 978-0-19-879701-2.
135. Kessler, S. *Gigged: The End of the Job and the Future of Work*; St. Martin's Press: New York, NY, USA, 2018. ISBN 9781250097897.
136. Duggan, J.; McDonnell, A.; Sherman, U.; Carbery, R. *Work in the Gig Economy: A Research Overview*; Routledge—Taylor & Francis Group: New York, NY, USA, 2022. ISBN 978-0-367-36792-3.
137. Lang, B.; Botha, E.; Robertson, J.; Kemper, J.A.; Dolan, R.; Kietzmann, J. How to grow the sharing economy? Create prosumers! *Australas. Mark. J.* **2020**, *28*, 58–66. [CrossRef]
138. Felson, M.; Spaeth, J.L. Community structure and collaborative consumption: A routine activity approach. *Am. Behav. Sci.* **1978**, *21*, 614–624. [CrossRef]
139. Sutherland, W.; Jarrahi, M.H. The sharing economy and digital platforms: A review and research agenda. *Int. J. Inf. Manag.* **2018**, *43*, 328–341. [CrossRef]
140. Belk, R. Sharing. *J. Consum. Res.* **2010**, *36*, 715–734. [CrossRef]
141. Belk, R. You are what you can access: Sharing and collaborative consumption online. *J. Bus. Res.* **2014**, *67*, 1595–1600. [CrossRef]
142. Heinrichs, H. Sharing economy: A potential new pathway to sustainability. *GAIA Ecol. Perspect. Sci. Soc.* **2013**, *22*, 228–231. [CrossRef]
143. Piscicelli, L.; Cooper, T.; Fisher, T. The role of values in collaborative consumption: Insights from a product-service system for lending and borrowing in the UK. *J. Clean. Prod.* **2015**, *97*, 21–29. [CrossRef]
144. Benoit, S.; Baker, T.L.; Bolton, R.N.; Gruber, T.; Kandampully, J. A triadic framework for collaborative consumption (CC): Motives, activities and resources & capabilities of actors. *J. Bus. Res.* **2017**, *79*, 219–227. [CrossRef]
145. Martin, C.J. The sharing economy: A pathway to sustainability or a nightmarish form of neoliberal capitalism? *Ecol. Econ.* **2016**, *121*, 149–159. [CrossRef]
146. Schlagwein, D.; Schoder, D.; Spindeldreher, K. Consolidated, systemic conceptualization, and definition of the "sharing economy". *J. Assoc. Inf. Sci. Technol.* **2020**, *71*, 817–838. [CrossRef]
147. Eckhardt, G.M.; Houston, M.B.; Jiang, B.; Lamberton, C.; Rindfleisch, A.; Zervas, G. Marketing in the sharing economy. *J. Mark.* **2019**, *83*, 5–27. [CrossRef]
148. Hamari, J.; Sjöklint, M.; Ukkonen, A. The sharing economy: Why people participate in collaborative consumption. *J. Assoc. Inf. Sci. Technol.* **2016**, *67*, 2047–2059. [CrossRef]
149. Zajączkowska, M. Prospects for the development of prosumer energy in Poland. *Oecon. Copernic.* **2016**, *7*, 439–449. [CrossRef]
150. Błażejowska, M. Rozwój energetyki prosumenckiej w świetle nowej ustawy o odnawialnych źródłach energii. *Roczniki Naukowe Stowarzyszenia Ekonomistów Rolnictwa Agrobiznesu* **2015**, *17*, 38–43. Available online: https://bibliotekanauki.pl/articles/868862 (accessed on 20 September 2022).
151. Zimmermannová, J.; Pawliczek, A.; Čermák, P. Public support of solar electricity and its impact on households—Prosumers. *Organizacija* **2018**, *51*, 4–19. [CrossRef]
152. Szeląg-Sikora, A.; Sikora, J.; Niemiec, M.; Gródek-Szostak, Z.; Suder, M.; Kuboń, M.; Borkowski, T.; Malik, G. Solar power: Stellar profit or astronomic cost? A case study of photovoltaic installations under Poland's national prosumer policy in 2016–2020. *Energies* **2021**, *14*, 4233. [CrossRef]
153. Kuchmacz, J.; Mika, Ł. Description of development of prosumer energy sector in Poland. *Polityka Energetyczna Energy Policy J.* **2018**, *21*, 5–20. [CrossRef]
154. Słupik, S. Energetyka prosumencka i jej wpływ na rynek energii elektrycznej. *Zeszyty Naukowe Uniwersytetu Szczecińskiego. Studia Prace Wydziału Nauk Ekonomicznych Zarządzania* **2014**, *2*, 127–136. Available online: http://yadda.icm.edu.pl/yadda/element/bwmeta1.element.ekon-element-000171363365 (accessed on 14 September 2021).
155. Ginalski, Z. *Energetyka Prosumencka na Obszarach Wiejskich*; Centrum Doradztwa Rolniczego w Brwinowie, Oddział w Radomiu: Radom, Poland, 2016. ISBN 978-83-63411-57-2.
156. Chen, T.; Alsafasfeh, Q.; Pourbabak, H.; Su, W. The next-generation U.S. retail electricity market with customers and prosumers—A bibliographical survey. *Energies* **2018**, *11*, 8. [CrossRef]
157. Dai, C.; Cheng, K.; Lei, Y.; Yang, Y. Research hotspots and evolution of energy prosumer: A literature review and bibliometric analysis. *Math. Probl. Eng.* **2020**, *2020*, 5703101. [CrossRef]
158. Joseph, A.; Balachandra, P. Energy Internet, the future electricity system: Overview, concept, model structure, and mechanism. *Energies* **2020**, *13*, 4242. [CrossRef]
159. Joseph, A.; Balachandra, P. Smart grid to Energy Internet: A systematic review of transitioning electricity systems. *IEEE Access* **2020**, *8*, 215787–215805. [CrossRef]
160. Hua, H.; Hao, C.; Qin, Y.; Cao, J. A class of control strategies for Energy Internet considering system robustness and operation cost optimization. *Energies* **2018**, *11*, 1593. [CrossRef]

161. Liu, Q.; Wang, R.; Zhang, Y.; Wu, G.; Shi, J. An optimal and distributed demand response strategy for Energy Internet management. *Energies* **2018**, *11*, 215. [CrossRef]
162. Zhou, K.; Yang, S.; Shao, Z. Energy Internet: The business perspective. *Appl. Energy* **2016**, *178*, 212–222. [CrossRef]
163. Zeng, Z.; Li, Y.; Cao, Y.; Zhao, Y.; Zhong, J.; Sidorov, D.; Zeng, X. Blockchain technology for information security of the Energy Internet: Fundamentals, features, strategy and application. *Energies* **2020**, *13*, 881. [CrossRef]
164. Siano, P.; De Marco, G.; Rolán, A.; Loia, V. A survey and evaluation of the potentials of distributed ledger technology for peer-to-peer transactive energy exchanges in local energy markets. *IEEE Syst. J.* **2019**, *13*, 3454–3466. [CrossRef]
165. Chen, T.; Huang, C.; Hu, F.; Zhao, J.; Su, W. Self-learning prosumer in competitive local energy market. In Proceedings of the 2019 IEEE Innovative Smart Grid Technologies—Asia (ISGT Asia), Chengdu, China, 21–24 May 2019; pp. 3085–3090. [CrossRef]
166. Churchman, C.W. Wicked problems. *Manag. Sci.* **1967**, *14*, B-141–B-142. [CrossRef]
167. Lönngren, J.; van Poeck, K. Wicked problems: A mapping review of the literature. *Int. J. Sustain. Dev. World Ecol.* **2021**, *28*, 481–502. [CrossRef]
168. Rittel, H.W.J.; Webber, M.M. Dilemmas in a general theory of planning. *Policy Sci.* **1973**, *4*, 155–169. [CrossRef]
169. Thollander, P.; Palm, J.; Hedbrant, J. Energy efficiency as a wicked problem. *Sustainability* **2019**, *11*, 1569. [CrossRef]
170. Haukkala, T. *The Wicked Problem of a Low Carbon Energy Transition. Structure, Agency and Framing in the Multi-Actor Process of Solar PV Deployment in Finland*; Aalto University publication series: Doctoral Dissertations 65/2019; Aalto University, School of Business, Department of Management Studies: Helsinki, Finland, 2019. ISBN 978-952-60-8501-2. Available online: https://aaltodoc.aalto.fi/handle/123456789/37503 (accessed on 10 January 2022).
171. Levin, K.; Cashore, B.; Bernstein, S.; Auld, G. Overcoming the tragedy of super wicked problems: Constraining our future selves to ameliorate global climate change. *Policy Sci.* **2012**, *45*, 123–152. [CrossRef]
172. Lazarus, R.J. Super wicked problems and climate change: Restraining the present to liberate the future. *Cornell Law Rev.* **2009**, *94*, 1153–1233. Available online: http://scholarship.law.cornell.edu/clr/vol94/iss5/8 (accessed on 11 January 2022).
173. Roberts, N. Wicked problems and network approaches to resolution. *Int. Public Manag. Rev.* **2000**, *1*, 1–19. Available online: https://journals.sfu.ca/ipmr/index.php/ipmr/article/view/175 (accessed on 11 January 2022).
174. Ferlie, E.; Fitzgerald, L.; McGivern, G.; Dopson, S.; Bennett, C. *Making Wicked Problems Governable? The Case of Managed Networks in Health Care*; Oxford University Press: Oxford, UK, 2013. ISBN 9780199603015. [CrossRef]
175. Kunz, W.; Rittel, H.W.J. Issues as Elements of Information Systems. Working Paper No. 131, July 1970 (reprinted May 1979). Available online: https://citeseerx.ist.psu.edu/viewdoc/citations?doi=10.1.1.134.1741 (accessed on 12 January 2022).
176. Ritchey, T. Wicked problems: Modelling social messes with morphological analysis. *Acta Morphol. Gen.* **2013**, *2*, 1–8. Available online: https://www.swemorph.com/pdf/wp.pdf (accessed on 14 January 2022).
177. Horn, R.E.; Weber, R.P. *New Tools for Resolving Wicked Problems: Mess Mapping and Resolution Mapping Processes*; MacroVU(r), Inc.; Strategy Kinetics, LLC: San Francisco, CA, USA, 2007. Available online: https://www.strategykinetics.com/new_tools_for_resolving_wicked_problems.pdf (accessed on 14 January 2022).
178. Thom, R. *Structural Stability and Morphogenesis: An Outline of a General Theory of Models*; W.A. Benjamin, Inc.: Reading, MA, USA, 1975. ISBN 0-8053-9277-7.
179. Randers, J.; Goluke, U. An earth system model shows self-sustained thawing of permafrost even if all man-made GHG emissions stop in 2020. *Sci. Rep.* **2020**, *10*, 18456. [CrossRef] [PubMed]
180. Schwartzman, D. Solar communism. *Sci. Soc.* **1996**, *60*, 307–331. Available online: https://www.jstor.org/stable/40403574 (accessed on 10 November 2021).
181. Schwartzman, P.; Schwartzman, D. *The Earth Is Not for Sale: A Path out of Fossil Capitalism to the Other World That is Still Possible*; World Scientific Publishing Co., Pte. Ltd.: Singapore, 2019. ISBN 978-981-3234-24-6.
182. Burnes, B. Complexity theories and organizational change. *Int. J. Manag. Rev.* **2005**, *7*, 73–90. [CrossRef]
183. McKelvey, B. Emergent order in firms: Complexity science vs. the entanglement trap. In *Complex Systems and Evolutionary Perspectives on Organisations: The Application of Complexity Theory to Organisations*; Mitleton-Kelly, E., Ed.; Pergamon—Elsevier Science: Amsterdam, The Netherlands, 2003; pp. 99–125. ISBN 0-08-043957-8.
184. Kauffman, S.A. *The Origins of Order: Self-Organization and Selection in Evolution*; Oxford University Press: New York, NY, USA, 1993. ISBN 0-19-505811-9.
185. Kauffman, S.A.; Levin, S. Towards a general theory of adaptive walks on rugged landscapes. *J. Theor. Biol.* **1987**, *128*, 11–45. [CrossRef] [PubMed]
186. Kauffman, S.A.; Weinberger, E.D. The NK model of rugged fitness landscapes and its application to maturation of the immune response. In *Molecular Evolution on Rugged Landscapes: Proteins, RNA and the Immune System. The Proceedings of the Workshop on Applied Molecular Evolution and the Maturation of the Immune Response, Held March, 1989 in Santa Fe, New Mexico*; Perelson, A.S., Kauffman, S.A., Eds.; CRC Press—Taylor & Francis Group: Boca Raton, FL, USA, 2018; pp. 135–175. ISBN 978-0-201-52150-4.
187. Weinberger, E.D. Local properties of Kauffman's N-K model: A tunably rugged energy landscape. *Phys. Rev. A* **1991**, *44*, 6399–6413. [CrossRef] [PubMed]
188. Solow, D.; Burnetas, A.; Tsai, M.; Greenspan, N.S. Understanding and attenuating the complexity catastrophe in Kauffman's *NK* model of genome evolution. *Complexity* **1999**, *5*, 53–66. [CrossRef]
189. McKelvey, B. Avoiding complexity catastrophe in coevolutionary pockets: Strategies for rugged landscapes. *Organ. Sci.* **1999**, *10*, 294–321. [CrossRef]

190. Westhoff, F.H.; Yarbrough, B.V.; Yarbrough, R.M. Complexity, organization, and Stuart Kauffman's The Origins of Order. *J. Econ. Behav. Organ.* **1996**, *29*, 1–25. [CrossRef]
191. Solow, D.; Burnetas, A.; Tsai, M.; Greenspan, N.S. On the expected performance of systems with complex interactions among components. *Complex Syst.* **2000**, *12*, 423–456. Available online: https://www.complex-systems.com/abstracts/v12_i04_a05/ (accessed on 20 September 2022).
192. Vidgen, R.; Bull, L. Applications of Kauffman's coevolutionary NKCS model to management and organization studies. In *The SAGE Handbook of Complexity and Management*; Allen, P., Maguire, S., McKelvey, B., Eds.; SAGE Publications: London, UK, 2011; pp. 201–219. ISBN 978-1-84787-569-3.
193. McKelvey, B. Self-organization, complexity catastrophe, and microstate models at the edge of chaos. In *Variations in Organization Science. In Honor of Donald T. Campbell*; Baum, J.A.C., McKelvey, B., Eds.; Sage Publications: Thousand Oaks, CA, USA, 1999; pp. 279–307. ISBN 0-7619-1125-1.
194. Levinthal, D.A. Adaptation on rugged landscapes. *Manag. Sci.* **1997**, *43*, 934–950. Available online: https://www.jstor.org/stable/2634336 (accessed on 19 September 2022). [CrossRef]
195. Hayes, J. *The Theory and Practice of Change Management*, 4th ed.; Palgrave Macmillan: Hampshire, UK, 2014. ISBN 978-1-137-27534-9.
196. Cook, S.; Macaulay, S.; Coldicott, H. *Change Management Excellence: Using the Four Intelligences for Successful Organizational Change*; Kogan Page: London, UK, 2004. ISBN 0-7494-4033-3.
197. Voehl, F.; Harrington, H.J. *Change Management: Manage the Change or It Will Manage You*; CRC Press—Taylor & Francis Group: Boca Raton, FL, USA, 2016. ISBN 978-1-4822-1419-2.
198. Harrington, H.J. *Innovative Change Management (ICM): Preparing Your Organization for the New Innovative Culture*; CRC Press—Taylor & Francis Group: Boca Raton, FL, USA, 2018. ISBN 978-1-351-24855-6.
199. Cameron, E.; Green, M. *Making Sense of Change Management: A Complete Guide to the Models, Tools and Techniques of Organizational Change*, 5th ed.; Kogan Page: London, UK, 2020. ISBN 978-1-78966-045-6.
200. Lewin, K. Frontiers in group dynamics: Concept, method and reality in social science; Social equilibria and social change. *Hum. Relat.* **1947**, *1*, 5–41. [CrossRef]
201. Kotter, J.P. *Leading Change*; Harvard Business School Press: Boston, MA, USA, 1996. ISBN 0-87584-747-1.
202. Marti, M. *Complexity Management: Optimizing Product Architecture of Industrial Products*; Deutscher Universitäts-Verlag: Wiesbaden, Germany, 2007. ISBN 978-3-8350-0866-3.
203. Boulton, J.G.; Allen, P.M.; Bowman, C. *Embracing Complexity: Strategic Perspectives for an Age of Turbulence*; Oxford University Press: New York, NY, USA, 2015. ISBN 978-0-19-956525-2.
204. Jackson, M.C. *Critical Systems Thinking and the Management of Complexity*; John Wiley & Sons: Chichester, UK, 2019. ISBN 978-1119118374.
205. Gorzeń-Mitka, I.; Okręglicka, M. Managing complexity: A discussion of current strategies and approaches. *Procedia Econ. Financ.* **2015**, *27*, 438–444. [CrossRef]
206. Grösser, S.N. Complexity management and system dynamics thinking. In *Dynamics of Long-Life Assets: From Technology Adaptation to Upgrading the Business Model, Grösser, S.N., Reyes-Lecuona, A., Granholm, G., Eds.*; Springer Open: Cham, Switzerland, 2017; pp. 69–92. ISBN 978-3-319-45437-5. [CrossRef]
207. Stacey, R.D.; Griffin, D.; Shaw, P. *Complexity and Management: Fad or Radical Challenge to Systems Thinking?* Routledge—Taylor & Francis Group: New York, NY, USA, 2002. ISBN 0-415-24760-8.
208. McMillan, E. *Complexity, Management and the Dynamics of Change: Challenges for Practice*; Routledge—Taylor & Francis Group: New York, NY, USA, 2008. ISBN 978-0-415-41722-8.
209. Burguillo, J.C. *Self-Organizing Coalitions for Managing Complexity: Agent-Based Simulation of Evolutionary Game Theory Models Using Dynamic Social Networks for Interdisciplinary Applications*; Springer International Publishing AG: Cham, Switzerland, 2018. ISBN 978-3-319-69896-0. [CrossRef]

energies

MDPI

Review

The Future of the Energy Sector and the Global Economy: Prosumer Capitalism and What Comes Next

Aleksander Jakimowicz

Department of World Economy, Institute of Economics, Polish Academy of Sciences, Palace of Culture and Science, 1 Defilad Square, 00-901 Warsaw, Poland; ajakimowicz@inepan.waw.pl

Abstract: This paper describes the present and the future of the energy sector in relation to the dominant and constantly evolving form of the global economic system. These considerations have their starting point in transformations of the energy sector in prosumer capitalism, which has dramatically changed the picture of the global economy in recent years. Subsequently, a futuristic approach is applied to determine the role and importance of energy from renewable sources for further human development. The main objective of the paper is to explain the current situation of the energy sector in prosumer capitalism and to extrapolate these relationships for the future, considering the need to enter the path of sustainable development to eliminate the global warming processes and climate changes. A review of the existing scientific literature was applied as the research method. The historical wave concept, proposed by Toffler, was found to be highly useful because of its high potential in futurology, where it enables one to study megatrends. The Fourth Wave was linked to prosumer capitalism, and it provided the base for defining the next ones: the Fifth Wave of Computing (ecosocialism) and the Sixth Wave in the form of technological and energy communism (solar communism). It also turned out that the key to solving mankind's energy problems lies in the global mean entropy budget. The literature review shows that founding the global energy system on solar radiation is the only known method for eliminating the anthropogenic greenhouse effect, which is the source of global warming and, consequently, of climate change. Therefore, the second law of thermodynamics provides a physical, economic, and logical justification for introducing a new and ultimate management form—solar communism—by 2050.

Keywords: energy transition; prosumer capitalism; synergistically double exploitation; futurology; dot-communism; revised diagram of the Toffler's civilization revolutions; second law of thermodynamics; global mean entropy budget; solar energy; solar communism

1. Introduction

The future of the energy sector is an urgent topic nowadays, as the survival of humanity and each of us depends on it. However, this sector is part of a larger whole, i.e., of the national or global economy, and it determines its growth. This became particularly obvious recently as it has become necessary to increase the share of renewable energy in the total energy balance to stop global warming, which is the source of climate change. Among the characteristic features of the present energy system, there is an increase in the share of prosumers in energy generation, a feature unknown before. Prosumption is a consequence of the ICT revolution and widespread digitalization of nearly all sectors of the economy and spheres of human life. It has led to the development of new business models, which have transformed previous economic systems. Traditional capitalism, known from economy handbooks and the scientific literature, has been replaced by a new system called prosumer capitalism.

Prosumer capitalism is regarded in this paper as the starting point for describing the future of the energy sector and the global economy. There is every indication that this system of management is now on the decline. Prosumer capitalism is not a stable form of economic process organization due to the externalization of the labor cost and synergistically double exploitation. Previous management systems—producer capitalism and consumer capitalism—also involved exploitation, although of a different kind. Producer capitalism should be classified as a singly exploitative economic system, as the exploitation of workers is its main feature. Consumer capitalism, in turn, is doubly exploitative, as it makes it possible to exploit both workers and consumers. On the other hand, a new and more intensive form of exploiting people emerged in prosumer capitalism, which permeates many areas of human activity at the same time [1,2]. Other flaws of this system include a self-propelled cycle of production and consumption, a constantly growing debt of work, microfascisms associated with power relations, and the robotization of prosumers and transforming them into a new type of commodity. Other threats may be associated with the emergence of huge multinational corporations, with considerable economic and political power given to them by social media users. All of this may lead to overproduction, overconsumption, social inequalities, ecological disaster, and climate change. Therefore, it becomes manifest that—on the one hand—prosumer capitalism facilitates the energy transition, as it accelerates the growth of the green energy sector and its integration with energy generation from fossil fuels, and—on the other—it can stimulate irrationality and excessive energy consumption. Thus, it is a temporary form of management, and it will have to be transformed into something different in future. Incidentally, proposed solutions to prosumer capitalism problems are, at the same time, strong stimuli for its transformation and improvement. From the futurology perspective, covering the period until the mid-21st century, one can list the successive systems of management, which include ecosocialism, dot-communism, and solar communism.

The visions of the future discussed in this paper are evaluated by means of a tool called terminology management, which is useful in knowledge organization and presentation and helps to imbue words with a sense. It helped to identify prosumer capitalism as the Fourth Wave in Toffler's civilization revolution pattern. It is also known as the Connected Society. The difference between individual waves arises from the importance of prosumption in the respective economic systems, i.e., switching cyclically between marketization and demarketization. This provided the basis for identifying the next two waves, which are consistent with the goals of the global energy transition and involve ensuring a compromise between the sustainable growth of the global economy and the need to eliminate global warming, which is responsible for climate change. The Fifth Wave of Computing is convergent with the socioeconomic agenda of ecosocialism, whereas the Sixth Wave of Technological and Energy Communism is associated with a new economic system called solar communism. The futuristic research horizon in this paper is restricted to the year 2050, as by this time—according to the optimistic scenario—mankind will have eliminated the anthropogenic greenhouse effect, and its whole energy needs will be satisfied with solar radiation.

All of the discussions on the future of the energy sector should start with the second law of thermodynamics and with defining the Earth's heat budget. The concept of thermodynamic entropy was introduced to economics by Nicholas Georgescu-Roegen, which contributed to the emergence of a new trend in economic studies called ecological economics. It involves the following reasoning. Human economic activity consists of transforming the limited natural resources of low entropy into high entropy waste, which shows that there are physical limits to global economic growth. A constant increase in the physical throughput is not possible due to the exhaustion of the sources of energy, raw materials and space, and biosphere destruction. Future knowledge will not be able to eliminate those limitations. Thus, entropy sets the ultimate limits for economic growth processes and becomes the physical basis for many socioeconomic forecasts. Apparently, all this appears sensible and logical, but this reasoning does not lead to a correct outcome due to incorrect assumptions. Nicholas Georgescu-Roegen claimed that the Earth is a closed system, so it cannot exchange matter or energy with its surroundings. However, this is a definition of an isolated system in thermodynamics, whereas closed systems are systems that do not exchange matter with the surroundings, but they are open to energy exchange.

The Earth meets these conditions. Correcting Georgescu-Roegen's error proves that the solar radiation which reaches the Earth can be a source of work whose recycling is infinite. Therefore, not only qualitative but also quantitative human growth is possible. Thus, we are reaching the economic system called solar communism, with dot-communism as its initial phase, as announced by wikinomics.

This paper outlines the genesis of scientific communism and the way it is understood by classical economists. Solar communism, about which little is known so far, does not have to be equivalent to scientific communism. They can be two distinct socioeconomic systems, although some similarities between them can be expected. The main idea of solar communism lies in tapping the whole global economy into the huge stream of solar radiation that is reaching the Earth continuously, which should allow for the gradual elimination of the anthropogenic greenhouse effect and, consequently, of global warming and climate change. Thus, what is left is the natural greenhouse effect, which is beneficial to the biosphere and on which its existence hinges. The issues that were touched on here are elaborated on further in this paper.

The high originality and innovation of studies presented in this paper arise from the following:

1. It is one of the first and one of the few comprehensive futuristic studies concerning energy transition. It presents a global perspective of catastrophic climate change based on the laws of thermodynamics and the global mean entropy budget. The type of the economic system, which is the starting point for the transition, is also taken into account, which is very important and which is often omitted.
2. This paper identifies a rather important mistake made by the creator of ecological economics, Nicholas Georgescu-Roegen, concerning the thinking of the Earth as an isolated system in a thermodynamic sense, i.e., one closed for both the flow of energy and of matter. This misunderstanding became a source of erroneous visions of human civilization's future and ecological catastrophism, as well as various dystopias, such as a global entropic apocalypse. According to these views, mankind is likely to be annihilated in the near future because of the impossibility of continuous economic growth as a consequence of natural resource exhaustion and biosphere destruction. In fact, the Earth is—thermodynamically—a closed system, and it does not exchange matter with space, but it is open to the influx of solar energy. This means that mankind can grow in an unobstructed manner, both qualitatively and quantitatively. It is noteworthy that the Earth—as referred to by Nicholas Georgescu-Roegen—is also a closed system, but understood differently—in fact, his definition of a closed system is the same as what is called an isolated system in thermodynamics.
3. Solar energy is the only one of all renewable resources that is practically unlimited and easily accessible. The Sun is an inexhaustible source of energy, which distinguishes it from other renewable resources. The solar energy received by the Earth's surface during an hour could power the global economy for a year. Moreover, mankind's ability to acquire and store it is increasing.
4. For these reasons, the fourth law of thermodynamics, formulated by Georgescu-Roegen, is redundant and can be reduced to the second law of thermodynamics.
5. Moreover, Georgescu-Roegen's pessimism about mankind's future also included his view that there is a small chance of ever making effective use of solar energy. He considered all the contemporary technologies for harvesting this energy to be parasites of energy technologies based on fossil fuel combustion. This slowed down research aimed at the improvement of solar energy conversion methods and decreased economists' and politicians' interest in its practical use.
6. The literature review, a comparative analysis, and terminology management showed that the three waves concept formulated by Alvin Toffler can be expanded to include another three waves: Fourth—prosumer capitalism, Fifth—ecosocialism, and Sixth—solar communism, which gives it enormous futurological potential, allowing for studying mankind's future until 2050.

7. Solar communism may seem similar to a utopian system today, and it was initially perceived as such, even by the author of this paper. However, there was something in it which made me abandon these prejudices and analyze all the pros and cons of the concept from a logical perspective. Ultimately, it turned out that solar communism is not based on political or social assumptions but on one of the most basic laws of physics and economics—the second law of thermodynamics. The natural entropy budget of the Earth and its imbalance caused by the anthropogenic greenhouse effect, as well as the catastrophic climate change which threatens mankind, do not leave us with any choice of a different path for survival and growth. Pointing out that energy is one of the most valuable goods and—as such—requires socialization on a global scale is another element of originality and innovativeness of this study.

2. Prosumer Capitalism as the Fourth Wave in Alvin Toffler's Terminology

2.1. Definition of Prosumer Capitalism in a World of Networked Technologies

There can be no doubt whatsoever that prosumption is one of the most important economic processes in modern economies. To appreciate how potent that force is, one only needs to realize that even the emergence of the World Wide Web—one of the most powerful knowledge tools of all time—was the result of prosumption [3] (pp. 178–179). Currently, the approach to the concept is twofold, as it may be optimistic or pessimistic. In the former perspective, prosumption implies an expansion of individual freedom, as consumers thus gain greater control over production, while the latter perspective emphasizes facilitated exploitation on the part of the producers since parts of the labor process are transferred to the consumers. In other words, prosumption alleviates alienation, on the one hand, thanks to which consumers become active participants in production processes, yet it exacerbates exploitation, on the other, as customers perform unpaid labor, resulting in the transfer of surplus value and reduced production costs. Fisher argues that these two dimensions of prosumption are mutually complementary and that there exists a dialectical relationship between them. In the era of the dominance of network technologies, prosumption is heralded as a new social compact between capital and labor; therefore, it is hardly surprising that it has become the basis for a new classification of various forms of capitalism [4].

Nowadays, the most developed countries in the world are dominated by the form of economies' development, which has been called prosumer capitalism [5–8]. This is due to the proliferation of prosumption processes, which are gaining on an unprecedented scale through the use of ICTs. This kind of prosumption is referred to as the digital prosumption [9]. This led to the need to reinterpret the economic history of the world and to redefine forms of development of the capitalist economy. Prosumer capitalism is the successor of the two previous forms—producer capitalism and consumer capitalism. All three forms of capitalism differ from each other in their attitude to the exploitation of workers and prosumers. Producer capitalism is a singly exploitative economic system because it is based on the exploitation of workers. Consumer capitalism is a doubly exploitative economic system in which exploitation of both employees and consumers is a significant source of profit for enterprises. In the case of prosumer capitalism, we are facing a completely new form of exploitation named synergistically double exploitation [2] (pp. 82–83). In order to clarify this issue, two types of prosumption should be distinguished: prosumption-as-production (p-a-p) and prosumption-as-consumption (p-a-c). The first occurs in organizations focused on production and work, such as factories and offices, while the second appears in organizations related to consumption, e.g., in shopping malls. Therefore, we have two types of prosumers, prosumers-as-producers (p-a-ps) and prosumers-as-consumers (p-a-cs), who have been exploited at various places and times. Until recently, the exploitation of p-a-ps occurred in factories and offices and happened during the work day, and the exploitation of p-a-cs was observed mostly in shopping venues after work hours and on weekends. In prosumer capitalism, the exploitation of both types of prosumers is more and more common in the same conditions and occurs at the same time. Thus, the exploitation interpenetrates two forms of prosumption, p-a-p and p-a-c, which leads to synergy resulting in a new, enhanced form of human exploitation [1,2].

Empirical research on prosumer capitalism is only in the initial stage. Among the few exceptions in this area are studies on the pop culture industry [10], and public administration [11]. Diagnosing the threats of a new form of economy–prosumer capitalism seems to be extremely crucial. For now, the free-of-charge work of prosumers does not appear to be a major barrier to economic growth, but it may soon change. It seems that societies should now develop appropriate remedies to prevent possible future economic crises. The paper formulates the thesis that the first line of defense against prosumer capitalism risks should be the government sector.

2.2. The Socioeconomic Effects of Prosumer Capitalism

2.2.1. Main Threats Related to Prosumer Capitalism

According to Ryder, numerous warnings against the prosumer economy are to be found in the science fiction literature. However, such threats are often dismissed by researchers due to the limitations imposed by the language and the requirements of a purely scientific academic discourse. The dystopias analyzed by that author establish a research framework that not only enables one to determine the impact of mass prosumption on people's lives but also to take a critical look at the emergence of a new economic system known as surveillance capitalism. It turns out that in such an economic system, the activities of the prosumer resemble the mindless behavior characteristic of a robot which obeys and does not even realize that it is being controlled. The principal dangers of prosumer capitalism discussed in dystopias include the alienation of the individual caused by the involvement of robots in the self-perpetuating cycle of production and consumption, as well as the entrapment of citizens in an increasing debt of work. One also stresses the significance of computer-like, internalized microfascisms, which are rooted in a complex network of power relations and shape people's desires, prompting them to intensify and continue prosumption processes indefinitely. This includes such adverse phenomena as overproduction and overconsumption, social inequalities, ecological disasters, and climate change [12].

Moreover, science fiction has long conveyed warnings that prosumption might become a social institution. These fears came true with the development of social media, which resulted in the disempowerment of numerous users who were thus trapped in a cycle of repetitive, meaningless free work. People in this system act in the manner of robots as they perform free labor for multinational corporations, which reap substantial profits as a result, at the simultaneous expense of users who lose personal freedoms and privacy. This endless cycle of prosumption consists of the production and consumption of unnecessary content, while at the same time, social media companies sell users' data and target them with advertising that reinforces and perpetuates this vicious cycle. Increasingly often, the absence of prosumption and manifestations of prosumptive attitudes may be associated with the loss of social status. The progressive robotization of prosumers should be attributed to the fact that advertising agencies and marketers are able to program and manipulate them. The ultimate prosumer is unlikely to be an individual who can autonomously choose and produce a product, but one who becomes a commodity they want to consume and, as a result, consumes a small part of themselves, as exemplified by the users of social media. The emergence of multinational corporations, whose economic and political power matches that of certain countries, is another noteworthy issue [12].

The emergence of prosumer capitalism is associated with two types of economic effects. First of all, prosumption processes certainly accelerate economic growth because, owing to them, enterprises can benefit from additional, external sources of innovation and creativity without incurring the associated fixed costs. Nevertheless, the growing profits of enterprises result from the unpaid labor of workers and prosumers. The basic research question focuses on how society can be saved from the adverse effects of synergistically double exploitation. Although the threats from the prosumer capitalism do not appear to be too noticeable today, this situation may change in the near future.

2.2.2. Outline of Socioeconomic Policy Counteracting the Harmful Effects of Prosumer Capitalism

The remedy proposed by the author, which aims to avoid the effect of synergistically double exploitation, is based on the method of extinguishing fires known in firefighting as controlled fires. According to this method, combustible materials must be removed by burning them before they are reached by the firefront [13]. Similarly in the economy, there is a large but presumably finite prosumption resource of the society that is mainly used in the private sector in prosumer capitalism. The government sector does not sufficiently use this resource, which is the reason for significant asymmetry between the two sectors. Meanwhile, public administration is a serious economic force that can prevent prosumer exploitation in and outside of the workplace. If this sector had obtained the still free part of the prosumption resource, then, assuming its limited quantity in society, this could constitute a certain barrier in the further acquisition of this resource by the market sector. In the context, the increase in private companies' profits resulting from the use of the prosumption resource of society would be inhibited, which would also mean a limitation of the synergistically double exploitation. A prerequisite for this to happen is lighting a fire in the public sector, and therefore introducing all models of wikinomics, with prosumption in the first place, into this sector [14,15]. In this way, the natural creativity and ingenuity of people would be redirected from the unpaid aid in profit increasing of private enterprises to the realization of socially useful goals, which would also be a factor of economic growth and development. However, for this to happen, public administrations need to adapt many information technologies so far only used in the market sector, which is a prerequisite for establishing a balance between both sectors in the use of society's prosumption resources. The public sector is not currently ready to compete with the market sector for the prosumption resources of society due to the lack of an appropriate technological and organizational base that would enable the implementation of all of the digital prosumption principles.

Harmonization of the public administration sector with the market sector encounters a number of barriers that can be divided into technological and mental ones. The first ones concern the technical requirements of the platforms for participation. Therefore, they can easily be overcome with the right inputs, while the latter ones are not. Breaking the mental barrier is only possible through the transformation of clients into prosumers, which requires simultaneous bottom-up actions of citizens and the implementation of the principle of openness in the public administration itself.

2.3. *A New Era of Change in World History Is upon Us*

Towards the end of the first decade of the 21st century, societies began to realize that the teleinformatics revolution, which may be dated from 1970 to 2005, had reached its declining phase and that a new age of change was beginning in the history of the world. This era has already begun and though it has received no standard appellation so far, various authors still attempt to identify its key features. The following offers a brief overview of a number of futuristic approaches in order to formulate a common conceptual framework for analyzing changes in the energy sector in the future and, while doing so, subject them to expert validation. These include the following:

1. Maynard and Mehrtens take advantage of Toffler's terminology and refer to the era in question as the Fourth Wave, which sees an integration of the core business principles, environmental concern, personal integrity, and spiritual values [16].
2. Daniel Pink elaborates on the idea of the Conceptual Age based on creators and empathizers as well as the six senses that will prove crucial to success in the new economy: design, story, symphony, empathy, play, and meaning [17].

3. David Houle also draws on Toffler and employs the term "the Shift Age" to describe the 2007–2030 period, where the primary emphasis is on consciousness and where change is driven by three forces: the shift towards globalization of culture and politics, the flow of power and influence to individuals, and the acceleration of electronic connectedness [18,19].
4. Maria Azua explores the vision of a revolution called the Social Age, whose prevalent approach, informed by communication and collaboration, is actuated by three forces as well: information overload created during the Information Age, standardization of technology that has commoditized key conduits of communication, and low-cost two-way Internet communication, including wikis and other social networking tools [20].
5. Jemielniak and Przegalinska define the concept of the collaborative society as a formation of enduring cooperative groups whose members have utilized network technologies to develop specific relationship patterns. Moreover, emerging technologies are greatly important in this case, as they become supermultipliers of cooperative outcomes insofar as they are used by an increasing number of people [21].
6. Hoffman and Ehrenfeld highlight the challenges facing management systems in the Anthropocene, a new geological epoch characterized by a significant impact of human economic activity on the Earth's ecosystems. Their model, based on punctuated equilibrium, distinguishes three waves of environmental management in the period from 1970 to the present: regulatory compliance (1970), strategic environmentalism (1990), and sustainability (2010). Occurring today, the fourth wave is referred to as Managing in the Anthropocene. To prevent adverse changes to the ecosystem on a global scale, it is necessary to adjust the functioning of the autonomous invisible hand by changing our culture and business management models accordingly. Systems thinking plays a paramount role in the pursuit of that goal in that it engenders new forms of the following six elements: partnerships, materials use and supply chains, domains of corporate activity, organizations, economic models, and metrics [22].
7. Klaus Schwab formulated the well-known and much-discussed concept of the Fourth Industrial Revolution, which denotes the fusion of cutting-edge technologies that will gradually eliminate the boundaries between the physical, digital, and biological realms. It is expected that technology breakthroughs in fields such as artificial intelligence, robotics, the Internet of Things, autonomous vehicles, 3D printing, nanotechnology, biotechnology, materials science, energy storage, and quantum computing will multiply humanity's creative capabilities, with virtually all areas of life changing as a result [23,24]. This revolution will not only open up numerous opportunities but also pose challenges [25]. In particular, profound changes are to be expected in the labor markets [26]. In this context, one also speaks of Industry 4.0 as a revolutionary reconstruction of the global value chains by smart factories, which will provide flexible cooperation between virtual and physical production systems. An interesting feature within the concept of the Fourth Revolution is its emphasis on the remodeling of legacy energy systems through departure from fossil fuels to prosumer energy based on renewable energy sources [27]. Naturally, prosumption will not be confined to just one sector but will assume a variety of forms in many domains.

As may be seen, all concepts are similar in that they apply to a knowledge-based society, where network technologies have penetrated almost all areas of life, where social networking tools enable balanced and integrated use of information to enhance creativity and cooperation of agents, as well as strengthen their ties with the environment. The future is not entirely predictable, so there are, and must be, terminological differences between the various futuristic approaches, which appear similar in their general formulation. For this reason, some attention is due to a field of knowledge designated as terminology management, which today serves as a community tool for organizing and representing knowledge, as well as aligning concepts with reality. The essential features of the new era of change include meaning-making and understanding. In today's world, terminology management has shifted from the highly specialized science known as taxonomy to social network users [28].

At this juncture, it is necessary to confront some of the challenges of terminology management and argue that the concept of historical waves proposed by Toffler is a useful tool for studying socioeconomic development of humanity and affords insights far into the future, going well beyond the revolutionary description of the post-industrial society of the Third Wave [29]. In addition, the proposition has been subjected to expert validation consisting of the above seven opinions. Consequently, it may be legitimately stated that prosumption is one of the foremost elements in each of the futuristic scenarios discussed, not to mention the modern energy sector, where it operates as a key transformative factor. This is supported by Ritzer's view that we are now dealing with a form of socioeconomic and political system based on digital prosumption [30]. Hence, the most appropriate name for the fourth industrial revolution in Toffler's nomenclature is prosumer capitalism.

2.4. Prosumer Capitalism as a Herald of the Communist Economy

Wikinomics is a discipline that studies the impact of information technology on economic growth. Teleinformatics has changed the principles of conducting business and encouraged the emergence of new models of global collaboration. The rapid development of the world economy in recent decades is the effect of these changes [14,15]. It resulted in the emergence of a new form of capitalism called prosumer capitalism. However, these successes are supported by exploitation of prosumers both in the workplace and outside of it. In this way, prosumption from the economic growth factor gradually becomes its barrier. However, there seems to be a solution to this problem. It stems from the asymmetry between the market and the government sectors, which is a consequence of poorer adaptation of digital prosumption in the latter. A gradual increase in prosumption in the public sector, especially in public administration, should reduce the extent of prosumer exploitation in the market sector, as long as the services in the public sector coproduced by them are available to them free of charge. This issue is open for the time being, and it may be a promising area of future research. However, one should point out that this leads one towards the reasoning presented by classical economists, who developed a vision of scientific communism.

As Van Dijck and Nieborg point out, the wikinomics program is in many respects a business manifesto to persuade the public to opt for a different socioeconomic paradigm by introducing a revolutionary way of thinking with respect to the processes of production and consumption. The models of cocreation that it advocates are not subject to choice by economic actors since, according to Tapscott and Williams, they are inevitable owing to historical necessity. In fact, the task of this manifesto of mass creativity and peer production is to develop—relying on network technologies—a concept of public collectivism, which is to fuse the two former opposites into an integral whole: producers and consumers; collective (nonmarket, public) mode of production and commercial (market, private) mode of production; the foundations of traditional business economics and the sociopolitical ideals of the 1960s counterculture; communalist thinking and the rational foundations of business; grassroots values of commonality and fundamental capitalist values; profit-driven platforms and nonprofit virtual collectives; commerce and commons; creativity and consumerism; capital-intensive, profit-oriented industrial production; and labor-intensive, nonprofit peer production. All of this is supposed to take place within a commodity culture. Emerging from this marriage, one sees a picture of the hegemonic ideology of cultural collectivism based on mass collaboration and communal creativity to be supported by all. The capitalist model of enterprise must therefore be supplanted by a modality of value creation relying on a shared public–private system [31].

Van Dijck and Nieborg demonstrated that the above mindset is grounded in several assumptions which are not always true: all content producers are equally creative, while their motivations to participate stem from similar desires; all users display the same spirit of collectivism that inspires them to create communities; users affiliated with commercial online communities are equated with people who use nonprofit social exchange sites. Simultaneously, they emphasize that, in fact, this is not about user-provided content but about the enormous economic value of metadata, connections, and profiled actions, new commodities of the digital age which enable Web 2.0 platforms to connect people to advertisers and products [31].

The ultimate goal of the wikinomics manifesto is to declare the victory of dot-communism over the final stages of traditional capitalism. Dot-communism stands for an economic system which includes such elements as a workforce composed only of free agents, a decentralized gift or barter economy without property, and a technological architecture which determines the political space [32,33]. Arguably, the first step in this direction is prosumer capitalism, in which the hierarchical model of production and consumption processes has been replaced by the cocreation model. Hence, value-creation processes are shifting from products and enterprises toward active networks of cocreators.

3. Is Solar Communism the Future of the Energy Sector?

3.1. The Origins of Communism as a Socioeconomic System

Schwartzman claims that the roots of communism lie in the Bible [34] (p. 16). It is also asserted that communism is synonymous with practical Christianity and that Jesus Christ was the first communist [35] (p. 593). As the well-known socialist theorist Karl Kautsky notes, Christian communism resulted from the desire of the poor to abolish all private property, and joint ownership was supposed to guarantee that misery and poverty would be eliminated. Their source is the same as in other forms of communism—the disenfranchisement of the popular masses. In early Christianity, the communist sphere encompassed articles of consumption, their distribution, and joint consumption. Furthermore, rural areas witnessed communism in production since jointly organized labor was possible there. In agriculture, consumption was directly linked to production, given that almost all production was intended for private consumption [36]. In other words, Christian communism essentially consisted of agricultural activity in the circumstances of Toffler's First Wave, which is why it occurred mainly in rural regions. Production was mostly undertaken to meet one's own consumption needs rather than to be sold. Using contemporary language, prosumption prevailed there. The factors which united Christian communities were communal suppers and the organization of mutual aid. As soon as wealthy individuals emerged among them, they gradually lost their proletarian character.

In the Middle Ages, monasteries were the primary form of technical and economic improvement since the cooperative mode of production, which was characteristic of those entities, dovetailed perfectly with the existing conditions of rural production. Unsurprisingly, monastic communism was the main form of Christian communism at the time. Within its framework, the Christian ideal of communal consumption and production, i.e., prosumption, could once again be reified. This structure of Christian monasteries has survived to the present day. However, monastic communism could not become the basis of the social economy because it required celibacy and conflicted with individual marriage. As such, it was limited to a minority. On the other hand, the conditions of production in the large urban centers resulted in the dispersion of proletarians [36] (pp. 345, 452, 455). The development of communism in the cities faced enormous obstacles from the very outset due to the separation of production and consumption, which, as Toffler noted, led to the gradual marketization of the world, and that trend deepened even further during the Second Wave associated with the Industrial Revolution. The shape of the economic system is not determined by how consumption is arranged but by the organization of production. The discrepancy between production and consumption was exacerbated by the division of labor, which, in turn, resulted from expanding industrial production. Hence, the cities saw only communism of consumption, confined to joint consumption of foodstuffs and other goods, so it operated within common households and families. In Kautsky's contemporary utopian communism, which functioned in the circumstances of the Second Wave, attempts were made to couple communism of consumption with the communism of production, but this led to its failure due to disputes and misunderstandings [36] (p. 454). The failure of utopian communism should be attributed to the separation of the consumption and production sectors for the sake of exchange, which resulted in the collapse of prosumption. It may therefore be concluded that communism and prosumption go hand in hand: what supports the former also sustains the latter.

Kautsky also underlines that religion was the primary factor in the development of Christian communism. As the author himself admits, there is no communism without religion. The strength of the Christian community lay in the propagation of messianic ideas, belief in the Christ Crucified and personal resurrection. Not only did it enable the group to survive as a secret organization inside the Roman state, but it also enabled it to become an unassailable power. Originating from the proletariat, the Crucified Messiah accomplished what no one else before had, or has since been able to, achieve: he conquered Rome, coerced Caesars into obedience, and gained dominion over the entire world [36] (pp. 380–381, 451).

According to Kautsky, modern communism is altogether different in its nature from Christian communism. By definition, the former is a macroeconomic system encompassing the entire society whereby the proletariat must acquire state power in order for it to become a reality, whereas the latter is inherently a microeconomic system. Modern communism should be the communism of concentration of wealth and concentration of production [36] (pp. 467–468).

An elaborate paradigm of modern communism was envisioned by Lenin, who emphasized that the basic trait of this socioeconomic system is the complete withering away of the state. In his view, the state is an organization of violence which is brought to bear for one social class to exercise oppression over another. Social classes are in evidence when the members of society differ in their relation to the social means of production. Since under communism, society will be classless, the state as an instrument of oppression will also prove dispensable. A similar fate is destined to befall democracy since, to Lenin, the term denotes a state that presupposes subordination of the minority to the majority, which is tantamount to the systematic oppression of one part of the population by another. It is only in a communist society that democracy may become complete and thus unnecessary, withering away along with the state in consequence. Only then will it be possible to realize the principle of "from each according to his ability, to each according to his needs". The state will wither away during the transition period from the proletarian revolution to the consummation of communism. Under this new system, everyone will be able to partake in the management of socioeconomic life and take over functions previously held by the state [37]. As Engels notes, as social classes disappear, society "will put the whole machinery of state where it will then belong: into the museum of antiquities, by the side of the spinning wheel and the bronze axe" [38] (p. 160). These notions turn out to coincide with the ideas of political prosumerism and correspond to what one understands today as digital prosumer capitalism [30]. In addition, Lenin would have certainly approved of the present-day energy transition based on prosumer electricity markets, given that he fully appreciated the importance of electricity for the socioeconomic development of humanity. In his opinion, "communism is Soviet power plus the electrification of the whole country since industry cannot be developed without electrification" [39] (p. 419).

Priddat and Jansen elaborate on the idea of virtualization of the state, which consists of enhanced political prosumerism, a goal achieved by increasing citizen participation in the political planning process. This translates into delegating responsibility through society to society so that citizens may take control, assume greater responsibility for their own affairs, and focus on local problems. It is the citizens who should run the state, not the other way around. The development of eGovernment will make it possible to tackle the three challenges of civil society: the integration of the citizenry, co-opetition, and policy networks, as well as the virtual production of public goods. As a result, the size of the state will be reduced by half while its influence will grow twofold. The government's use of citizens' advice may allow it to solve problems to better effect. The paradox of the withering state is also present here: as globalization advances, local politics (glocalization) gains increasing significance. The production of multiple public goods (with the exception of law, infrastructure, and, in part, education) can be outsourced to civic organizations or private enterprises, which can simultaneously cooperate and compete with one another. Another matter of interest is the concept of the zero-gravity state, which utilizes virtual production structures and value-adding networks to optimize its state quota [40].

3.2. The Essence of Global Solar Communism

It is not economic, social, and political ideas which lie at the core of solar communism, but the laws of physics. Prior to discussing this issue, the basic concepts involved have to be defined. According to the second law of thermodynamics, the entropy of an isolated system, i.e., one which does not exchange energy and matter with its surroundings, must increase insofar as any changes occur in it. An increase in entropy is tantamount to decreased ability to perform work, which results from the transformation of low-entropy energy into waste heat. Consequently, thermodynamic entropy can be defined as a randomized state of energy that cannot be utilized to perform work (being unavailable). Thermodynamics also defines closed systems which do not exchange matter with the environment but are open due to the energy flow [41].

The introduction of thermodynamic entropy into economics is credited to Nicholas Georgescu-Roegen, who made it a measure to determine the ultimate limits of the development of economic processes and a physical foundation employed in social predictions [42]. The processes of production and consumption consist of transforming low entropy into high entropy, resulting in waste [42] (p. 18). Since the resources of low entropy in the human surroundings are limited and become gradually depleted, as is the case with fossil fuels, there must exist physical limits to further growth of the world economy. The continued increase of the physical throughput will sooner or later become impossible due to the depletion of energy, materials, and space, with concurrent destruction of the biosphere. Future knowledge will not be capable of removing the strictures imposed on the world economy by finitude, entropy, and ecological dependence [43] (p. 199). The concept of entropy generated by economic processes was expanded by Jeremy Rifkin, who approached entropy as pollution, a marker of cosmic disorder, and a primary cause of environmental destruction [44,45]. Járosi and Kovács attribute the shape of the modern energy transition in Europe to the catastrophic visions and intellectual influence of Jeremy Rifkin, who continued the work of Georgescu-Roegen. The Hungarian authors are skeptical of the European Union's energy policy and emphasize that its efforts have minor significance for global climate protection and cannot prevent the rise of anthropogenic greenhouse gas emissions in emerging economies [46].

To Georgescu-Roegen, a closed system does not exchange matter and energy with the environment [47] (pp. 7–8). He further observes that a closed system cannot ceaselessly perform work at a constant rate [48] (p. 304). However, this is how thermodynamics describes an isolated system as opposed to a closed one [49] (pp. 3–4). If the Earth is an isolated system, then there must indeed be physical limits to the development of the world economy since, under the second law of thermodynamics, matter and energy are irrevocably dissipated [47] (p. 8). However, this is not the case, since the Earth is a closed system and thus does not exchange matter with space (except for space vehicles and meteorites), but remains open to solar energy transfer. David Schwartzman notes the mistake Georgescu-Roegen made and demonstrates that the conversion of low-entropy, high-temperature energy into high-entropy, low-temperature heat can be a source of work whose recycling is infinite. An elucidation of the differences between isolated systems and closed systems helps one understand why Georgescu-Roegen took a pessimistic view of the use of solar energy in the economy—he considered all solar systems known in 1981 to be parasites of fossil-fuel-based energy technologies [50] (pp. 70–71, 198). This confusion of concepts is also crucial in solving the problem of optimizing humanity's relationship with nature, since it supplies the physical foundation for the development of solar communism [41]. Thus, the vision of an entropic apocalypse culminating in the annihilation of civilization proved erroneous.

Given the energy budget at the Earth's surface, the solar energy flux reaching the surface/atmosphere is equal to the flux radiating back into space. The natural greenhouse effect is due to the absorption of heat radiation by water and carbon dioxide molecules in the atmosphere which reradiate heat to the surface. Any economy which does not rely on direct solar flux but uses fossil fuels, as well as nuclear and geothermal energy, affects the heat budget of the Earth because the emission of heat radiation exceeds the natural flux from the surface. This also involves the enhanced greenhouse effect, whose principal source is anthropogenic carbon dioxide and methane. The global economy would not have an impact on the Earth's surface heat budget only if the use of solar energy were not associated with net transfers of these greenhouse gases to the atmosphere and ocean systems [41].

The physical underpinning for the development of solar communism is staggering. The world economy could increase its energy consumption tenfold using only 1% of the solar radiation reaching the Earth's surface without changing its current heat budget. This means that anthropogenic greenhouse gas emissions and direct waste heat production can be avoided. The annual flux of solar radiation is ten times the total energy stored in the global coal resource or in one million billion (10^{15}) barrels of crude oil. The energy payback time, or the number of years it takes for photovoltaic panels to generate the amount of electricity needed to produce them, is 1–4 years (as of 1994). The U.S. demand for electricity could be fully met by installing silicon-based photovoltaic panels in an area equal to a square with a side measuring 140 km, which is much smaller than the area currently occupied by U.S. military installations. The introduction of solar communism will not be possible without optimal relations between the technosphere and the ecosphere and, therefore, adherence to the principle of maximizing the containment of the technosphere [41].

3.3. *Conditions Necessary for the Implementation of Solar Communism*

It follows from the laws of thermodynamics that economic development on the Earth's surface may be continued by humankind even in the relatively distant future, provided that it relies on solar energy. Direct utilization of its minor proportion by humans would mean connecting the world economy to an immense flux of energy capable of performing inconceivable work. This portion of solar energy would eventually be converted into waste heat in any case, as evidenced by the natural heat budget. If the world economy were based solely on solar energy, it would increase the physical throughput (industrial processing) in the technosphere without a negative impact on the biosphere, provided that the production–consumption cycle is closed. This would enable the export of entropy flux into space in much the same way as it happens with the natural biosphere powered by solar energy. However, before these possibilities can be made a reality, the basic principle of communism—"from each according to his ability, to each according to his needs"—has to be reinterpreted in a manner which will render it applicable not only to humans but also to the Earth's ecosystem [51]. For this to be accomplished, the thermodynamic processes taking place in nature must be supported by appropriate socioeconomic policies on a global scale.

As Schwartzman observes, the transition to solar communism requires gradual application of that modified principle already in the near future, i.e., in a mixed social formation referred to as ecosocialism, which should be situated on the timeline immediately after prosumer capitalism. In his view, ecosocialism—a combination of ecological and socialist movements—is a prerequisite for the reification of the ultimate system: a global solar community. Since ecosocialism does not presuppose the occurrence of the phenomenon known as synergistically double exploitation, it must inevitably be clearly distinguished from prosumer capitalism. The end of the era of prosumer capitalism confronts humanity with the great bifurcation: the choice between relegating global capitalism to prehistory or a climate catastrophe of unimaginable consequences caused by unsustainable capital reproduction processes [52].

The notion that future civilization should rely on the solar radiation flux reaching the Earth's surface stems from the fact that it is the most abundant and most easily accessible source of energy, while its negative impact on health and the environment is negligible compared to fossil fuels. In addition, solar energy technologies are developing globally at a nearly exponential growth rate [34]. Today, the global anthropogenic energy flux (waste heat) equals only 0.03% of the solar flux on land surfaces. This is equivalent to saying that one hour of solar flux reaching the Earth's surface supplies an amount of energy equal to the annual consumption of energy by the global population. In this case, the second law of thermodynamics does not limit economic development. The impact of converting solar energy into useful work poses much less of a threat to ecosystems than the current hazards since the resulting waste heat is discharged into space and does not actually increase the natural flux [53]. In this way, humans may adapt their economic undertakings to the entropic flow from the Earth's surface and take advantage of the existence of a heat sink in outer space, which guarantees ecological balance and biospheric self-organization [54] (pp. 161–172).

The sufficient and necessary conditions for the attainment of solar communism include material and political requirements as well as the demilitarization of the world. The material requirements are as follows: the existence of a global and highly efficient solar energy infrastructure; containment and precautionary measures in environmental policy; demilitarization of technologies and propagation of state-of-the-art information technologies around the world; concentration of the population in green cities and the establishment of large biospheric reserves. Among the political requirements, one should list a transnational workforce organized bottom-up, accelerated democratization of the global society, a transnational ecosocialist political movement, and partial achievement of the material requirements cited above. Further indispensable elements include social governance of production and consumption at all levels (from local to global), universal equity among people, communal ownership of land, and termination of the production of value based on labor time [52] (pp. 483–489).

It has previously been argued that one of the preconditions for the realization of communism is increased labor productivity under capitalism. With respect to solar communism, the list may be supplemented with state-of-the-art information technologies, renewable energy, and organic agriculture based on agroecology [51].

The main goal of ecosocialism is the global class struggle for ecological sustainability [51]. In their quest to end the domination of the big capital, the working class, middle strata, and indigenous peoples represent a countervailing force to the national and transnational ruling classes [52]. Specifically, ecosocialists have targeted the nuclear military industrial fossil fuel complex as the guardian of the interests of the transnational capital class and thus one of the greatest threats to the transformation of current prosumer capitalism into solar communism. In 2015, total defense outlays amounted to USD 2.03 trillion. Ecosocialists consider this sector to be the reproduction center of global capital due to the fact that it exerts a powerful influence on the internal and external policies of the major capitalist powers and seeks to control the world's oil reserves. In addition, the fossil fuel and nuclear industries are closely associated with this sector, which translates into the ability to use the threat of nuclear attack to achieve imperial policy goals. The complex is also seen as a major obstacle to the emergence of a global multiclass alliance striving to reduce global greenhouse gas emissions and transition fully to solar energy [34].

Global prosumer capitalism prioritizes the reproduction of capital over the needs of humans and nature. This results in a reduction in global biodiversity and habitat destruction, which some believe may contribute to pandemics [34]. Therefore, the emergence of solar communism spells the ultimate end of the reproduction of the militarized fossil capital, and the consequent cessation of value creation. In such circumstances, the pursuit of class power will no longer be feasible [53]. A global solar communist society will be founded on a steady-state biophysical economy [34] (p. 22).

By consuming fossil fuels and using nuclear energy, humankind has incurred an entropic debt with the environment and, therefore, with space. Powering the global economy with solar energy will lead to the repayment of that debt in the form of nonincremental waste heat, which is impossible with unsustainable equivalents of such energy [34].

It is usually asserted that nuclear power does not affect climate change. However, environmentalists see it as a threat not only because of its ties to the military–industrial complex, but also because the existing fossil-fuel-based energy infrastructure powers all subsystems of the nuclear fuel cycle. Realignment of the global economy to solar energy may be delivered in sufficient time to avoid catastrophic climate change and prevent irreversible changes. Environmentalists emphasize that the transition to solar communism will be possible within the next 30 years if 1–2% of the current annual energy consumption, including the 85% involving fossil fuels, is allocated to solar and wind energy production each year [52] (pp. 491–492). The extant fossil fuel reserves will suffice to meet the needs of the global solar community, provided that they are not consumed as fuel. Moreover, solar energy tallies with the decentralized, democratic model of governance and control befitting a future global solar community [51]. Moreover, the current mean harmonized EROI (energy return on energy invested) values for solar photovoltaic systems range from 8.7 to 34.2, depending on the technology used, which is very promising, and these values could be even higher in the future [55].

3.4. The Exemplification: Possibilities of Developing Solar Power Plants in Poland

Solarization of the world—if it indeed takes place—will most likely begin with grassroots efforts by individual regions and countries. The use of solar radiation to produce electricity will be particularly important in this respect. The entire surface of the territory of Poland has similar solar energy resources, ranging from 980 to 1100 kWh/m^2 per year. Under standard test conditions (STC), an optimally located photovoltaic system can approximately yield 1000 kWh annually from each kilowatt of capacity installed. Depending on the technology used, this rate is usually in the 950–1025 kWh/kWp range [56] (pp. 132–133).

In recent years, the market for electricity involving direct solar radiation has grown rapidly. In late April 2022, the total installed capacity of the photovoltaic systems in Poland reached 9998.2 MW, in excess of double the figure for April of the previous year (4739.6 MW). At present, the prosumer share of installed capacity is almost 80%, and the number of devices they use exceeds 1 million. In the first four months of 2022, solar radiation generated 1782.6 GWh of electricity, which accounted for 2.8% of the national production. In the equivalent period last year, it amounted to 760.9 GWh and 1.3%, respectively [57].

In Poland, potential investors can estimate the overall cost of a photovoltaic system based on the following linear function [58]:

$$C(p) = 5845p + 3897, \tag{1}$$

where *C(p)* represents the cost in PLN, while *p* stands for the nominal power expressed in kWp. The calculations show that without external financing, the expected payback period of the system is 16 to 24 years.

At present, one of the most serious obstacles to the development of solar power in the country is the limited availability of connection capacities for new prosumers. Even so, forecasts concerning further development of solar power plants are very optimistic, but they do not appear likely to replace conventional power plants within a reasonable time horizon. Thus, the achievement of solar communism would indeed have to be global and involve the transmission of electrical power from the regions of the world with the highest solar exposure to the rest.

4. Discussion

Converting fossil-fuel-based energy systems to renewable energy is a super wicked problem, as many issues still need to be resolved in areas such as energy and environmental law, distribution, democracy, and cybersecurity. This is the most important challenge today, as previous unsustainable methods of energy extraction have led to unprecedented environmental destruction, impoverished biodiversity, and, ultimately, catastrophic climate change. Continued uncontrolled emissions of pollutants might cause tipping points of irreversible changes in the atmosphere to be crossed, possibly leading to the annihilation of humanity. This raises the question of the future of civilization, but in order to attempt an answer, it is necessary to return to the past.

A useful framework for economic and political futurism was established by Toffler, who developed the concept of three waves in the economic history of humankind, from the earliest times to the ICT revolution [29]. The concept is informed by the division of the economy into two sectors. Sector A includes the production of subsistence goods and services that do not reach the market, whereas sector B spans production intended for sale. Toffler noted that successive waves could be identified by determining the extent to which the share of these sectors in the economy changed. In this way, one can identify certain regularities in the long-term development of economies and even attempt to predict the future. In this concept, modes of energy production are essential, as it is the foundation of all human economic activity.

As shown earlier, the Fourth Wave in the development of civilization is prosumer capitalism. The appearance of prosumer capitalism is not only related to the acceleration of economic development, but also to certain risks. The biggest risk associated with it is that the development factor is the free-of-charge work of prosumers. There is a conviction that, as a result, prosumption processes can turn from a factor of economic development into its barrier. To prevent this, an appropriate economic policy of the state is necessary. The best way to achieve this is to use a fire extinguishing strategy by controlling its inciting in other places. In other words, it would be appropriate, in the light of the experience of the market sector, to introduce prosumption to the public sector and, in particular, to the public administration sphere, which may have key implications here. In this way, the public sector would be synchronized with the market sector. Preventing the negative effects of prosumer capitalism requires the improvement of the existing public services or even the introduction of new ones, which means the need for a thorough reorganization and development of the public administration. This would allow governments to improve their economic policies and prepare for new challenges by taking account of grassroots initiatives of the citizens. Introducing prosumption to the public administration would not be possible without adequate ICT support.

Prosumption in the market sector is rapidly developing, and a long time ago it took a form of the digital prosumption. However, the introduction of prosumption in the public administration is still delayed, which is a source of considerable asymmetry in the development of these sectors. Hence, in many economies worldwide, there is an uncontrolled phenomenon described by Ritzer, which he calls synergistically double exploitation [1,2]. The point is that the basis for economic growth is a doubly free-of-charge work of people. This is more and more common in the situations when the same people once act as employees and another time as prosumers. The public administration sector is predisposed to be the first line of defense against such situations.

Currently, the development of political prosumerism, which manifests itself as the virtualization of the state and an increase in the share of public goods produced by citizens, is the most important method of counteracting human exploitation. This will mean that the state is shrinking. Thus, assuming the simultaneous development of information and computing technologies, the dot-communism ideas will be implemented, which will create the basis for further transformations and may contribute to the emergence of solar communism. Therefore, one of the basic weaknesses of prosumer capitalism—founding economic growth on exploitation—becomes the main stimulus for the implementation of the global solar communist program. Making use of the power of prosumption in acquiring solar energy may make it possible to solve one of mankind's major problems: to successfully complete the energy transition, thereby liquidating the anthropogenic greenhouse effect.

Figure 1 shows the concept of economic and political development of humankind from the earliest times, which witnessed the agricultural revolution, until the near future (from the current point in time until the mid-21st century), including the Fifth Wave of Computing (ecosocialism) and the Sixth Wave—technological and energy communism (solar communism). In each of these waves, the share of A and B sectors in the economy fluctuated, which was indicative of either demarketization or marketization of the world's production and consumption systems. In other words, prosumption and production for exchange are the main driving forces of human economic activity. Considering the first four waves, i.e., the agricultural, industrial, and teleinformatics revolutions, as well as digital prosumer capitalism, one sees that the two forces act alternately: the growth of the prosumption sector is accompanied by diminution of the production-for-sale sector, and conversely. This brings to mind the steady, slow breathing of the gigantic economic and political structure created by societies called civilization, which not only encompasses the present but also extends into the past and future. Every aspiration must be followed by expiration. The agricultural revolution involved a strong prosumption sector, while the industrial revolution led to a massive increase in production for exchange. The ICT revolution restored the importance of prosumption and triggered processes of demarketization; the subsequent digital prosumer capitalism reversed the trend, mainly due to the rise of powerful global corporations linked to social media. Since the first four waves have already been quite thoroughly studied and belong to the past, it might be worthwhile to devote a certain amount of attention to futurology and focus on the last two waves, the germs of which can already be observed.

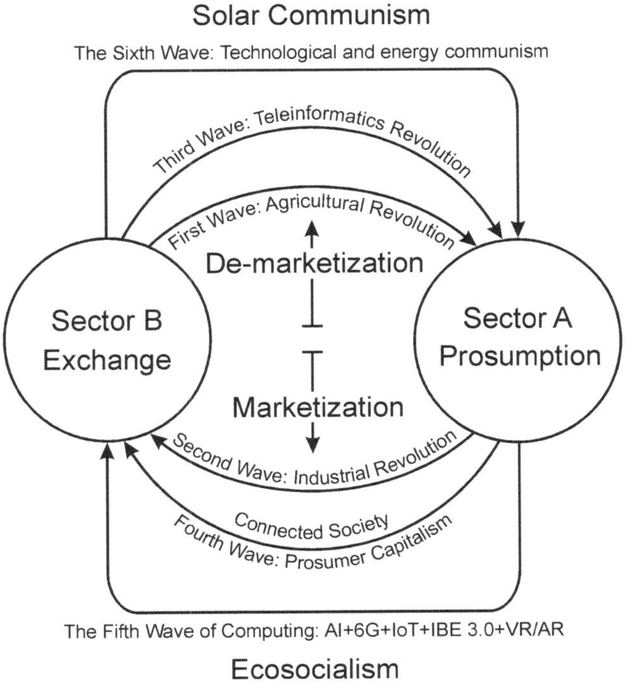

Figure 1. The revised diagram of Toffler's civilization revolutions.

The existence of all these waves would not have been possible without the energy systems which supported production processes in each era. The use of particular energy sources has always been contingent on the degree of technological advancement. However, the third wave introduced a new quality, i.e., digitalization, the potential of which was only fully appreciated during the Fourth Wave—digital prosumption capitalism. Digitalization should be distinguished from digitization [59]. Digitization denotes the formal conversion of data from analog to digital using information and communication technology. The meaning of digitalization is much broader, as it refers both to the digital transformation of technology as well as business models and processes at all levels of the economy [60,61]. Without digitalization, it would be impossible to decentralize the energy sector and gear it toward renewable energy sources. Decentralization causes electricity prices to drop, enabling the electrification of key sectors in the economy. Moreover, electricity is essential to digital technologies. Thus, digitalization, decentralization, and electrification together form a virtuous cycle in which all elements reinforce one another [62]. Moreover, without digitalization, humanity could not even attempt to prevent catastrophic climate change. Its absence in the government sector would hamper bottom-up energy transition [63]. Moreover, digitalization prompts the merging of the energy system into an integral whole, which not only fosters the development of energy production but also yields advantages in the heating sector, communications, and mobility [64].

Figure 1 shows that following the Fourth Wave, also known as the Connected Society, the cycle of the alternating rise of prosumption (demarketization) and production for exchange (marketization) as the driving forces of economic development came to an end. Immediately after the Fourth Wave, civilization will take another deep breath in, and the Fifth Wave will ensue, continuing the transformation which started with digitalization and consisting of the refinement of computational methods. Numerous innovative companies are expected to emerge in the future, which will temporarily strengthen the market sector even further. Only just beginning, the Fifth Wave is associated with the development of the Internet of Things (IoT), which will prospectively combine all manifestations of human life, including physical assets, into one large integrated network. Within such a system, people will interact with other people, but things will also interact with things. In particular, power grids will be fully interactive, with smart meters and other devices balancing energy production and consumption over time and space to an unprecedented extent. It is predicted that by 2025 the IoT market will span tens of billions of connected devices, reaching a value in excess of USD 11 trillion. The interoperability of encrypted data and authentication of users and devices will be critical in that system. The issues concerning privacy and security, as well as the management of tens of billions of digital certificates and standard keys, are likely to be resolved and aided thanks to the latest developments in Identity Based Encryption 3.0 (IBE 3.0) [65].

In the course of the Fifth Wave, one can expect the rapid spread of the fourth paradigm of science, a concept advanced by Jim Gray, which involves a shift in research methodology due to the increasing use of data. This means the emergence of data-intensive science, also known as eScience, which provides an integrating framework for the three previous paradigms: empirical evidence, scientific theory, and computational methods (large-scale computational simulation). The new paradigm gives rise to a scientific cycle in which the former three paradigms interact and augment one another. The main goal of the fourth paradigm is to find a way to integrate humans and computers so that researchers obtain increased capabilities of combining data from different sources with a view to discovering new patterns in that data [66,67].

It is likely that the Fifth Wave will see the spread of the sixth generation (6G) standard for wireless communications technologies supporting broadband cellular data networks, which will replace the 5G standard currently being deployed worldwide. It is presumed that 6G networks will support applications far beyond the current mobile experience through solutions such as virtual and augmented reality (VR/AR), instant communications anywhere, pervasive intelligence, and the IoT. It will be a combination of the physical world, software-based digital systems, artificial intelligence, and biological systems [68]. The 6G networks will use frequencies from 100 GHz to 3 THz, which substantially surpass the 5G range. In all probability, these bands harbor major potential since they are within the unused and unexplored spectrum [69]. The new system means higher speeds, lower latency, and masses of bandwidth. It is likely that 6G will deliver data at a rate of 1 TB per second, far exceeding the 10 to 20 gigabits per second capacities of current 5G networks [70].

Artificial intelligence (AI) is also expected to develop rapidly in the Fifth Wave, which may culminate in a technological singularity. The latter concept was developed and popularized by Vernor Vinge, who noted that technological advances in programming may lead to the creation of agents capable of self-improvement, resulting in a powerful superintelligence which is far superior to human intelligence in qualitative terms [71]. Ray Kurzweil predicts that this might happen as early as 2045 [72]. Given that the upper limit of intelligence and, therefore, the capabilities of superintelligent machines are unknown, all events beyond the time horizon of a technological singularity should be considered unpredictable and unfathomable. Should this come to pass, however, the future of solar communism would also be called into question unless that superintelligence finds the above thermodynamic evidence sufficient and accelerates the use of solar energy.

All of this is in line with the ecosocialist agenda since, as observed previously, one of the chief aspirations of ecosocialism is to promote the development of state-of-the-art information technologies. Additionally, taking social aspects into account, the Fifth Wave may be labelled as ecosocialism as it anticipates continued global class struggle to restore ecological balance and reverse catastrophic climate change, in which the middle class has a considerable role to play [73]. Here, the development of prosumer energy markets and the thermal retrofitting of the existing building stock constitute some of the major challenges.

Four future scenarios have been advanced concerning the transition from the present-day digital prosumer capitalism (the Connected Society), through the Fifth Wave of Computing (ecosocialism), to what is expected to emerge in the middle of this century (solar communism). At that point, having taken a double deep breath in, occasioned by the Fourth and Fifth Waves, civilization will exhale in the form of the Sixth Wave. Some of these predictions may resemble visions familiar from science fiction, but they are nevertheless worth quoting here since the future is unknown and nothing can be ruled out in advance. In the order of increasing probability, they are as follows [74]:

- Scenario 1: A 0.1% likelihood of occurrence. Current trends in global capitalism are upheld, and economic development continues to rely on fossil fuel energy. The implemented climate protection programs do not match the scale of the threat. Still, the projected effects of global warming are not in evidence due to the emergence of unforeseen negative feedback occurring in the climate system. Admittedly, an increase in atmospheric carbon dioxide levels will trigger acidification of the oceans, but this will be neutralized by the remarkable adaptive abilities of marine organisms.
- Scenario 2: A 10% chance of occurrence. The proliferation of high-efficiency thin-film photovoltaics and the use of ocean currents and high-altitude wind energy will lead to the rapid decarbonization of the global economy and the emergence of cheap green energy, thus undermining the position of the military–industrial complex. Massive grassroots civic initiatives will precipitate the rise of the ecosocialist movement, which, gaining political power as time goes by, will restore peace and justice, as well as accelerate solarization processes around the world.
- Scenario 3: A 30% likelihood of occurrence. A transnational ecosocialist movement proves to be the only remedy to resource wars and the increasing impact of global warming. High-efficiency solar power supplants fossil fuels and nuclear power, promoting the removal of atmospheric carbon dioxide to levels below 350 ppm (parts per million). This is the minimum level marking the tipping points of irreversible climate change. Once this value is exceeded, humanity may lose control of the climate altogether (at present, CO_2 concentration is at an alarming 410 ppm). In this scenario, per capita energy consumption is at a level which ensures an average life expectancy

for all citizens, along with state-of-the-art health and education. At the same time, biodiversity is preserved, and organic farming expands around the world. Unsustainable production and consumption processes gradually abate.

- Scenario 4: A 60% chance of occurrence. This is a genuine dystopia and displays the highest likelihood of occurrence. Crises in the world economy are suppressed as a result of fascist repression by the authorities and the introduction of inexpensive drugs. A new ruling caste emerges to gain the upper hand in the ongoing class struggle, while the military–industrial complex plays an ever-greater role in the reproduction and expansion of global capital. A quarter of humanity consumes 95% of all produced goods and services, which is taking place in the midst of a deepening global ecocatastrophe starting as early as 2025. Melting permafrost is releasing methane, which prompts rapid melting of the ice caps and flooding of coastal cities around the world. Ocean acidification destroys marine life, and the quantity of hydrogen-sulfide-producing bacteria increases, causing most of humanity to die from starvation and disease by 2050. A few thousand of the richest people survive along with their families; they live in air-conditioned rooms powered by nuclear energy and grow organic food in greenhouses.

Scenario 1 has very little chance of occurring on its own while being fraught with too great a risk of human passivity if the presumed negative feedback in the climate system does not exist. Scenarios 2 and 3 are considered the most desirable solutions, although they may be difficult to accomplish as they necessitate demilitarization, solarization, and the creation of organic agriculture [75].

In the contemporary world, there are visible germs of future change, and it seems that they increase the probability that scenarios 2 and 3 will come to pass. The post-pandemic period saw a distinct acceleration in the process of advanced technology use, mainly artificial intelligence and machine learning, which resulted in productivity increase in many economic sectors. The acceleration of digitalization of the global economy supports the decentralization of energy systems and the development of electrification. Moreover, the pandemic helped to make people aware of what the next climate crisis may look like, i.e., to show such of its features as its fast-moving, systemic nature, and global range. It would certainly affect many spheres of social and economic life. Making people aware of such threats may accelerate beneficial changes in such areas as power engineering, mobility, and agriculture, and so this happened. Currently, we see the acceleration of the measures aimed at reducing pollution and intensifying the struggle against climate change, as many countries' recovery plans take into account environmental policy priorities. Electric cars are becoming increasingly popular, and new methods of carbon capture from the atmosphere are being developed. The development of stakeholder capitalism, which creates a bridge between businesses and communities, is also noteworthy. In this system, businesses are legally required to take into account the interests of all stakeholders in the decision-making processes, even at the expense of changing their governance structures [76]. Other systemic changes in the form of cross-cultural solidarities, increased interest in the transnational dimension of prosumption, and the emergence of technology-empowered prosumer global citizenship are also consistent with ecosocialism and solar communism [77]. Moreover, energy prosumerism, which includes not only material changes in the infrastructure but also visions of social, cultural, political, economic, and institutional changes, is growing rapidly. It also came to embrace both the market paradigm and its supplements in the form of municipal and communal paradigms [78]. Furthermore, work aimed at the development of new methods for solving the optimal scheduling of combined heat and power units is of great importance [79].

Solar radiation is the largest renewable energy resource that exists, and the possibilities of using and storing it have increased with the advancement of technical progress. Examples include concentrated solar power plants, which not only generate electricity but can also store solar energy. The principle of the operation of such systems lies in using light-reflecting surfaces to concentrate the direct solar radiation. There are four main technologies used to focus beams of solar light: linear Fresnel reflectors, parabolic troughs, central receivers, and parabolic dishes. Photons are focused on a container with a special liquid, which is heated up to high temperatures. The heat obtained in this manner powers a steam turbine, which generates electricity [80]. Owing to such solutions, solar radiation becomes a dispatchable form of energy, as stored heat can be used to generate electricity when needed, both at night and during the day. As of the end of 2021, the cumulative installed global solar thermal power capacity amounted to 6800 MW, which is clear progress since 2014, when 4584 MW was reported [81] (pp. 7–8). It is noteworthy that the conversion of solar energy to heat and further to electricity does not increase the natural greenhouse effect, i.e., it is neutral from a climate change perspective.

Technical progress also helps to overcome the intermittent nature of solar radiation as a source of energy. It has been found that photovoltaic panels can generate electricity under very adverse conditions, even at night. Such weather conditions as a high level of cloudiness and rain considerably reduce the performance of solar cells. However, this can be helped by creating an energy harvesting system comprising not only a solar cell, but also a triboelectric nanogenerator, which converts mechanical energy into electricity [82–84]. This device helps to produce electricity both from solar radiation and from falling raindrops. Energy is also produced at night, as long as it rains. Another solution that can help to overcome the limitations of photovoltaic panel use is to equip them with a thermoelectric generator, which can produce electricity at night by making use of the temperature difference between a solar cell and the surrounding air. In this manner, with a clear night sky, one can achieve 50 mW/m^2 nighttime power generation. The generator also works during the day and provides additional power, which adds to the direct output of photovoltaic panels [85]. Another idea is to use a nighttime photovoltaic cell, also known as a thermoradiative cell, which generates electrical current while emitting infrared radiation (heat) towards extremely cold deep space. The Earth is used as a heat source in this device, and the night sky performs the function of a heat sink. Such cells can generate 50 W/m^2 under ideal conditions, i.e., a quarter of what a conventional solar panel can generate during a day. Moreover, it will operate during the day if direct sunlight is blocked [86]. In this manner, it may be easier to balance the power grid in the day–night cycle.

5. Conclusions

The second law of thermodynamics and the global mean entropy budget point to the precise direction for the energy transition. According to the revised diagram of Toffler's civilization revolutions, the starting point is prosumer capitalism, which—on the basis of the digitalization of economic processes—changed the traditional business models dramatically. This enabled the decentralization of the existing energy system and incorporating in it the energy from renewable sources. The Fifth Wave is linked to further digitalization of the energy system and the global economy, improvement of computation methods, and the development of the Internet of Things. Both waves, the Fourth and the Fifth, are supposed to pave the way for the Sixth Wave associated with solar communism, which should take place by 2050. However, this kind of development requires a radical reconstruction of socioeconomic systems existing around the world currently, and its success relies on common demilitarization, solarization of the global energy infrastructure, and efficient organic agriculture.

It is worth attempting to analyze solar communism and how it will be different from scientific communism known from the works of classical economists. There is every chance for Engels' and Lenin's vision of the withering away of the state to be fulfilled, which is testified to by the current trends in political prosumerism development, consisting of the virtual production of public goods becoming widespread and the growth of importance of glocalization processes. The same should be expected in terms of democracy. The emergence of modern technologies that enable one to use energy from renewable sources favors the development of prosumption in the energy sector, which leads to decentralization and democratization. If energy democracy becomes complete, it could turn out to be no longer needed, similar to the state itself. Moreover, energy democracy can be a bottom-upward force which initiates and enforces political democracy. Furthermore, electricity generation with highly efficient thin-film photovoltaic panels is in line with Lenin's idea, according to which communism is equivalent to the electrification of the whole country and its expansion to the whole world. Obviously, this new system will be based on sharing and participation, which will make it similar to Christian communism. Kautsky's predictions of the concentration of wealth and concentration of production in solar communism will probably not come true. A great role in it will be played by prosumption, not limited to the energy sector, and the negative processes associated with it will be alleviated or completely eliminated. The main features of solar communism also include a certain type of universality, which consists of the ability to function on all economic levels, starting with the microeconomics and mesoeconomics and ending with the macroeconomics of the world economy. This may come true provided mankind creates a worldwide social community, with generating energy from solar radiation being its major unifying factor. Energy will probably be owned collectively, whereas when it comes to the private ownership of the means of production, one would expect a considerable reduction of its share in the global economy. Thus, the categorization of people into social classes would tend to disappear. However, much less can be said about the moral values in such a community and its belief systems.

Today, two main methods for achieving solar communism are defined in the ecosocialist program. The first one is a multidimensional class struggle, whose aim is to eliminate the rule of transnational militarized fossil and financial capital, whereas the other consists of progress in the development of digital technologies. Close international cooperation will be required to pursue this goal. The road to the new system should be marked out by the developing social and political movement called ecosocialism. It proposes that a global entity should be established to lead and coordinate class struggle. It should unite the global working class, its allies, and indigenous communities, which requires constant reinforcing of transnational solidarity. When it comes to the material conditions necessary for solar communism to emerge, an important role can be played by such countries as Venezuela, as it has the largest proven oil reserves in the world, which can be used to accelerate the global solar energy transition, or China, which has the largest renewable energy capacity and is the leading producer of renewable energy technology in the world [87,88]. Looking from the perspective of current knowledge, this is the only possibility of solving the super wicked problem of the current energy transition. However, one cannot exclude the possibility that completely new solutions to problems associated with global warming and climate change will appear by 2050.

The conclusions from this study can be summarized as follows:

1. Prosumer capitalism, which facilitates the energy transition, is not a stable system because of its many adverse phenomena. Two of them are of the greatest importance: the externalization of the labor cost and synergistically double exploitation. This makes it a provisional economic system which has to be replaced with something else in future.
2. An in-depth review of many methods for examining the future showed Toffler's three waves concept to have the greatest futurological potential as it provides for predicting megatrends. Taking the studies conducted so far and using the terminology management, one can conclude that prosumer capitalism, also referred to as the Connected Society, can be identified as the Fourth Wave. With this point of reference, the next two waves could be determined: the Fifth Wave as ecosocialism, which will involve the improvement of computation methods, and the Sixth Wave as solar communism, i.e., the use of the global infrastructure of solar energy conversion by the whole of mankind.
3. Toffler's expanded concept of waves of civilization development covers the period until 2050, and it is based on the laws of physics, especially the second law of thermodynamics and the entropy budget of the Earth. For these reasons, its futurological

potential is the highest compared to the other available methods for predicting the future.

4. In the most probable scenario, the present and future socioeconomic systems can be placed on a time axis in the following sequence: prosumer capitalism as the present moment and the starting point, followed by ecosocialism, dot-communism, and solar communism.
5. Nicholas Georgescu-Roegen, the creator of ecological economics, wrongly regarded the Earth as an isolated system in a thermodynamic sense, i.e., one that does not exchange matter or energy with the environment. This had immense consequences for the environmental policy and became a source of particularly strong ecological pessimism, which resulted in delays and errors in the energy transition. These mainly included underestimating the role and importance of solar energy. It affected the whole of mankind, which was wrongly informed that there was an absolute limit to economic growth and development, whereas now we know that, basically, there is no such limit.
6. Because of Georgescu-Roegen's mistake, his fourth law of thermodynamics is redundant, and it can be reduced to the second law of thermodynamics.
7. In fact, the Earth is—thermodynamically—a closed system, which means that it does not exchange matter with space, but it is open to the influx of solar energy. In consequence, solar radiation of low entropy and high temperature can be a source of work which can be repeated in perpetuity. Therefore, there are no limits to mankind's economic growth.
8. Due to Georgescu-Roegen's pessimistic views of an entropic apocalypse, the potential of solar energy was not appreciated in time. He regarded all of the technologies for acquiring this type of energy as parasites feeding on fossil-fuel-based energy technologies.
9. The outlook for solar energy is astounding, and it is developing along with technological progress. Solar energy can be converted into heat and stored, and modern photovoltaic panels can generate electricity even under adverse conditions, e.g., in cloudy or rainy weather. There are also thermoradiative cells, which can work not only during the day but also at night.
10. The basic idea of solar communism involves the whole global economy being powered by inexhaustible solar energy. Only this method can prevent global warming and catastrophic climate change. The global nature of solar communism is the condition of its effectiveness, which means that the cooperation of all countries is required.
11. Currently, all methods of energy generation other than solar power engineering, for example, fossil fuel combustion or nuclear fission, disrupt the natural greenhouse effect and interfere with the environment by emitting excess entropy as waste heat. Likewise, solar geoengineering can disturb the natural entropy budget of the Earth.
12. Elimination of the anthropogenic greenhouse effect and restoring the balance of the natural entropy budget of the Earth is the main goal of the energy transition.
13. Mankind has to pay the entropic debt, i.e., a type of thermal debt, to the cosmos, which is constantly growing, as soon as possible. It disturbs the natural relationships, not only in the Earth's ecosystem but also in the universe.
14. Fundamental conditions for introducing solar communism include demilitarization, solarization of the global energy infrastructure, and the development of organic agriculture.
15. One can agree with Schwartzman's opinion that China is likely to play the leading role in the energy transition by 2050 because it is already the global leader in solar power engineering. This is consistent with the Polish Prophecy from Tęgoborze of 1893 on the future of the world. It claims that "the Dragon will renew its face" [89] (p. 6). This view is confirmed by contemporary studies of globalization and geopolitics, particularly considering the New Silk Road [90].

Funding: This research received no external funding.

Data Availability Statement: Not applicable.

Conflicts of Interest: The author declares no conflict of interest.

References

1. Ritzer, G. Prosumer capitalism. *Sociol. Q.* **2015**, *56*, 413–445. [CrossRef]
2. Ritzer, G. Prosumption: Contemporary capitalism and the "new" prosumer. In *The Oxford Handbook of Consumption*; Wherry, F.F., Woodward, I., Eds.; Oxford University Press: New York, NY, USA, 2019; pp. 75–93. ISBN 9780190695583.
3. Toffler, A.; Toffler, H. *Revolutionary Wealth: How It Will Be Created and How It Will Change Our Lives*; Doubleday: New York, NY, USA, 2007; ISBN 978-0-385-52207-6.
4. Fisher, E. The dialectics of prosumption in the digital age. In *Digital Labour and Prosumer Capitalism: The US Matrix*; Fraysse, O., O'Neil, M., Eds.; Palgrave Macmillan: Hampshire, UK, 2015; pp. 125–144. ISBN 978-1-137-47390-5. [CrossRef]
5. Ritzer, G.; Jurgenson, N. Production, consumption, prosumption: The nature of capitalism in the age of the digital 'prosumer'. *J. Consum. Cult.* **2010**, *10*, 13–36. [CrossRef]
6. Antonio, R.J. Is prosumer capitalism on the rise? *Sociol. Q.* **2015**, *56*, 472–483. [CrossRef]
7. Zwick, D. Defending the right lines of division: Ritzer's prosumer capitalism in the age of commercial customer surveillance and big data. *Sociol. Q.* **2015**, *56*, 484–498. [CrossRef]
8. Dusi, D. Beyond prosumer capitalism: Retaining the original understanding of prosumption. *Curr. Sociol.* **2017**, *66*, 663–681. [CrossRef]
9. Dusi, D. The perks and downsides of being a digital prosumer: Optimistic and pessimistic approaches to digital prosumption. *Int. J. Social. Sci. Humanit.* **2016**, *6*, 375–381. [CrossRef]
10. Siuda, P.; Troszyński, M. Natives and tourists of prosumer capitalism: On the varied pro-prosumer activities of producers exemplified in the Polish pop culture industry. *Int. J. Cult. Stud.* **2016**, *20*, 545–563. [CrossRef]
11. Jakimowicz, A.; Rzeczkowski, D. Prosumption in the public administration sector. *Acta Phys. Pol. A* **2016**, *129*, 1011–1017. [CrossRef]
12. Ryder, M.J. Lessons from science fiction: Frederik Pohl and the robot prosumer. *J. Consum. Cult.* **2020**, *22*, 246–263. [CrossRef]
13. Sutherland, C.R. Encountering the burn: Prescribed burns as contact zones. *Environ. Plan. E Nat. Space* **2019**, *2*, 781–798. [CrossRef]
14. Tapscott, D.; Williams, A.D. *Wikinomics: How Mass Collaboration Changes Everything*; Portfolio: New York, NY, USA, 2006; ISBN 978-1-59184-138-8.
15. Tapscott, D.; Williams, A.D. *Macrowikinomics: Rebooting Business and the World*; Portfolio/Penguin: New York, NY, USA, 2012; ISBN 978-1-59184-356-6.
16. Maynard, H.B., Jr.; Mehrtens, S.E. *The Fourth Wave: Business in the 21st Century*; Berrett-Koehler Publishers: San Francisco, CA, USA, 1996; ISBN 978-1-881052-15-9.
17. Pink, D.H. *A Whole New Mind: Why Right-Brainers Will Rule the Future*; Riverhead Books: New York, NY, USA, 2006; ISBN 1-59448-171-7.
18. Houle, D. *The Shift Age*; Sourcebooks: Naperville, IL, USA, 2011; ISBN 978-140-227-390-2.
19. Houle, D. *Entering the Shift Age: The End of the Information Age and the New Era of Transformation*; Sourcebooks: Naperville, IL, USA, 2012; ISBN 978-140-227-21-72.
20. Azua, M. *The Social Factor: Innovate, Ignite, and Win through Mass Collaboration and Social Networking*; IBM Press: Upper Saddle River, NJ, USA, 2010; ISBN 978-0-13-701890-1.
21. Jemielniak, D.; Przegalińska, A. *Collaborative Society*; The MIT Press: Cambridge, MA, USA, 2020; ISBN 978-026-253-791-9.
22. Hoffman, A.J.; Ehrenfeld, J.R. The fourth wave: Management science and practice in the age of the Anthropocene. In *Corporate Stewardship: Achieving Sustainable Effectiveness*; Mohrman, S.A., O'Toole, J., Lawler, E.E., III, Eds.; Routledge—Taylor & Francis Group: New York, NY, USA, 2017; pp. 228–246. ISBN 978-1-78353-259-9.
23. Schwab, K. *The Fourth Industrial Revolution*; World Economic Forum: Cologny/Geneva, Switzerland, 2016; ISBN 978-1-944835-01-9.
24. Wieczorek, P. Czwarta Rewolucja Przemysłowa. Wizja Przemysłu Nowej Generacji—Perspektywa Dla Polski. *Kontrola Państwowa* **2018**, *63*, 89–115. Available online: https://bibliotekanauki.pl/articles/416919 (accessed on 26 September 2021).
25. Xu, M.; David, J.M.; Kim, S.H. The Fourth Industrial Revolution: Opportunities and challenges. *Int. J. Financ. Res.* **2018**, *9*, 90–95. [CrossRef]
26. Hirschi, A. The Fourth Industrial Revolution: Issues and implications for career research and practice. *Career Dev. Q.* **2018**, *66*, 192–204. [CrossRef]
27. Schwab, K.; Davis, N. *Shaping the Future of the Fourth Industrial Revolution: A Guide to Building a Better World*; Currency: New Yorke, NY, USA, 2018; ISBN 978-198-482-261-1.
28. Albuquerque, A.; Costa, R. Terminology as a Sense Making Social Tool. In *ECSM 2015—The Proceedings of the 2nd European Conference on Social Media: School of Accounting and Administration at the Polytechnic Institute of Porto, Portugal, 9–10 July 2015*; Mesquita, A., Peres, P., Eds.; Academic Conferences Ltd.: Reading, UK, 2015; pp. 11–18. ISBN 978-1-910810-31-6. Available online: https://run.unl.pt/handle/10362/26665?mode=full (accessed on 30 October 2021).

29. Toffler, A. *The Third Wave*; William Morrow and Company: New York, NY, USA, 1980; ISBN 0-688-03597-3.
30. Ritzer, G.; Jandrić, P.; Hayes, S. The velvet cage of educational con(pro)sumption. *Open Rev. Educ. Res.* **2018**, *5*, 113–129. [CrossRef]
31. Van Dijck, J.; Nieborg, D. Wikinomics and its discontents: A critical analysis of Web 2.0 business manifestos. *New Media Soc.* **2009**, *11*, 855–874. [CrossRef]
32. Kelly, K. The New Socialism: Global Collectivist Society Is Coming Online, Wired—Culture, 22 May 2009. Available online: https://www.wired.com/2009/05/nep-newsocialism/?currentPage=1 (accessed on 3 June 2022).
33. Moglen, E. The dotCommunist Manifesto, January 2003. Available online: http://moglen.law.columbia.edu/publications/dcm.html (accessed on 4 June 2022).
34. Schwartzman, D. *The Global Solar Commons, the Future That Is Still Possible: A Guide for 21st Century Activists*; The Solar Utopia.org Press: Galesburg, IL, USA, 6 October 2020. Available online: https://www.theearthisnotforsale.org/solarcommons2021.pdf (accessed on 15 December 2021).
35. McLellan, D. Marxism. In *Jesus in History, Thought, and Culture: An Encyclopedia*; Houlden, L., Ed.; ABC-CLIO, Inc.: Santa Barbara, CA, USA, 2003; Volume 1, Entries A–J; pp. 593–597. ISBN 1-57607-857-4.
36. Kautsky, K. *Foundations of Christianity: A Study in Christian Origins*, 2nd ed.; International Publishers CO., Inc.: New York, NY, USA, 1925.
37. Lenin, V.I. *The State and Revolution: The Marxist Theory of the State and the Tasks of the Proletariat in the Revolution*; Haymarket Books: Chicago, IL, USA, 2014; ISBN 978-1-60846-498-2.
38. Engels, F. *The Origin of the Family, Private Property, and the State*; Resistance Books: Chippendale, Australia, 2004; ISBN 187-664-635-7.
39. Lenin, V.I. Our foreign and domestic position and the tasks of the party: Speech delivered to the Moscow Gubernia Conference of the R.C.P.(B.), 21 November 1920. In *Collected Works*, 2nd ed.; Lenin, V.I., Ed.; Progress Publishers: Moscow, Russia, 1974; Volume 31, April–December 1920; pp. 408–426.
40. Priddat, B.P.; Jansen, S.A. From civil society to eGovernment: The virtualisation of the state. In *Progressive Governance for the XXI Century: Contributions to the Berlin Conference*; Schröder, G., Kocka, J., Neidhardt, F., Eds.; Kluwer Law International: The Hague, The Netherlands, 2002; pp. 51–61. ISBN 978-90-41-11774-8.
41. Schwartzman, D. Solar communism. *Sci. Soc.* **1996**, *60*, 307–331. Available online: https://www.jstor.org/stable/40403574 (accessed on 10 November 2021).
42. Georgescu-Roegen, N. *The Entropy Law and the Economic Process*; Harvard University Press: Cambridge, MA, USA, 1971; ISBN 0-674-25781-2.
43. Daly, H.E.; Cobb, J.B., Jr.; Cobb, C.W. *For the Common Good: Redirecting the Economy Toward Community, the Environment, and a Sustainable Future*, 2nd ed.; Updated And Expanded; Beacon Press: Boston, MA, USA, 1994; ISBN 0-8070-4705-8.
44. Rifkin, J.; Howard, T. *Entropy: A New World View*; The Viking Press: New York, NY, USA, 1980; ISBN 0-670-29717-8.
45. Rifkin, J.; Howard, T. *Entropy: Into the Greenhouse World, Revised ed.*; Bantam Books: New York, NY, USA, 1989; ISBN 0-553-34717-9.
46. Járosi, M.; Kovács, P. Energy policy of Hungary. *Civ. Rev.* **2018**, *14*, 67–80. [CrossRef]
47. Georgescu-Roegen, N. *Energy and Economic Myths: Institutional and Analytical Economic Essays*; Pergamon Press Inc.: New York, NY, USA, 1976; ISBN 0-08-021027-9.
48. Georgescu-Roegen, N. Afterword. In *Entropy: Into the Greenhouse World, Revised ed.*; Rifkin, J., Howard, T., Eds.; Bantam Books: New York, NY, USA, 1989; pp. 299–307. ISBN 0-553-34717-9.
49. Kolesnikov, I.M.; Kolesnikov, S.I.; Vinokurov, V.A.; Zaikov, G.E. *Thermodynamics of Spontaneous and Non-Spontaneous Processes*; Nova Science Publishers, Inc.: Huntington, NY, USA, 2001; ISBN 1-56072-904-X.
50. Georgescu-Roegen, N. Energy, matter, and economic valuation: Where do we stand?/Reply. In *Energy, Economics, and the Environment: Conflicting Views of an Essential Interrelationship*; Daly, H.E., Umaña, A.F., Eds.; Routledge—Taylor & Francis Group: New York, NY, USA, 2018; pp. 43–79, 193–200. ISBN 978-0-367-01910-5. [CrossRef]
51. Schwartzman, D. Beyond eco-catastrophism: The conditions for solar communism. In *Socialist Register 2017: Rethinking Revolution*; Panitch, L., Albo, G., Eds.; The Merlin Press: London, UK, 2016; Volume 53, pp. 143–160. ISBN 978-0-85036-725-6.
52. Schwartzman, D. The great bifurcation and prospects for solar communism in the twenty-first century. *Int. Crit. Thought* **2013**, *3*, 480–495. [CrossRef]
53. Schwartzman, D. Ecosocialism or ecocatastrophe? *Capital. Nat. Social.* **2009**, *20*, 6–33. [CrossRef]
54. Schwartzman, D. *Life, Temperature, and the Earth: The Self-Organizing Biosphere*; Columbia University Press: New York, NY, USA, 1999; ISBN 0-231-10212-7.
55. Bhandari, K.P.; Collier, J.M.; Ellingson, R.J.; Apul, D.S. Energy payback time (EPBT) and energy return on energy invested (EROI) of solar photovoltaic systems: A systematic review and meta-analysis. *Renew. Sustain. Energy Rev.* **2015**, *47*, 133–141. [CrossRef]
56. Szymański, B. *Małe Instalacje Fotowoltaiczne. Teoria, Praktyka, Prawo, Ekonomika*; Globenergia—Geosystem Burek, Kotyza S.C.: Kraków, Poland, 2013.
57. Mikołajuk, H.; Zatorska, M.; Stępniak, E.; Wrońska, I. *Informacja Statystyczna o Energii Elektrycznej*; Biuletyn Miesięczny—Ministerstwo Klimatu i Środowiska—Agencja Rynku Energii S.A.: Warsaw, Poland, 2022; Volume 4. Available online: https://www.are.waw.pl/badania-statystyczne/wynikowe-informacje-statystyczne/publikacje-miesieczne (accessed on 20 June 2022).
58. Mikulik, J.; Jurasz, J. Determination of photovoltaic installation nominal power based on electrical energy consumption profile in the context of prosumer policy. *Przegląd Elektrotechniczny* **2015**, *91*, 99–101. [CrossRef]

59. Kotilainen, K. Energy prosumers' role in the sustainable energy system. In *Affordable and Clean Energy. Encyclopedia of the UN Sustainable Development Goals*; Leal Filho, W., Azul, A.M., Brandli, L., Salvia, A.L., Wall, T., Eds.; Springer Nature: Cham, Switzerland, 2021; pp. 1–14. ISBN 978-3-319-95863-7. [CrossRef]
60. Cont, W.; Romero, C.; Lleras, G.; Unda, R.; Celani, M.; Gartner, A.; Capelli, L.; Zipitría, L.; Besfamille, M.; Figueroa, N.; et al. *IDEAL 2021: Reducing Service Gaps: How Digitalization Can Improve the Use of Infrastructure*; Corporación Andina de Fomento—Development Bank of Latin America: Caracas, Venezuela, 2021; ISBN 978-980-422-223-8. Available online: http://scioteca.caf.com/handle/123456789/1763 (accessed on 3 January 2022).
61. Collin, J.; Hiekkanen, K.; Korhonen, J.J.; Halén, M.; Itälä, T.; Helenius, M. (Eds.) *IT Leadership in Transition—The Impact of Digitalization on Finnish Organizations; Aalto University Publication Series: Science + Technology, 7/2015*; Aalto University, School of Science, Department of Computer Science, ACIO Research Program: Helsinki, Finland, 2015; ISBN 978-952-60-6243-3. Available online: http://urn.fi/URN:ISBN:978-952-60-6243-3 (accessed on 4 January 2022).
62. Astarloa, B.; Kaakeh, A.; Lombardi, M.; Scalise, J. *The Future of Electricity: New Technologies Transforming the Grid Edge*; World Economic Forum: Cologny/Geneva, Switzerland, 2017. Available online: https://www3.weforum.org/docs/WEF_Future_of_Electricity_2017.pdf (accessed on 5 January 2022).
63. Jakimowicz, A.; Rzeczkowski, D. The impact of public administration digitalization on the decarbonization of the economy. *Energies* **2021**, *14*, 5739. [CrossRef]
64. Kucharska, A. Społeczny wymiar "Energiewende"—rozwój energetyki prosumenckiej w Austrii. *Studia Humanist. AGH* **2018**, *17*, 67–84. [CrossRef]
65. Montgomery, B. The 5th Wave. *The IOT Magazine*, 26 October 2016. Available online: https://theiotmagazine.com/the-5th-wave-4ee1ad8b3e9e (accessed on 10 June 2022).
66. Hill, C.A. Moving social science into the fourth paradigm: The data life cycle. In *Big Data Meets Survey Science: A Collection of Innovative Methods*; Hill, C.A., Biemer, P.P., Buskirk, T.D., Japec, L., Kirchner, A., Kolenikov, S., Lyberg, L.E., Eds.; John Wiley & Sons, Inc.: Hoboken, NJ, USA, 2021; pp. 713–731. ISBN 978-111-897-632-6. [CrossRef]
67. Hey, T.; Tansley, S.; Tolle, K. (Eds.) *The Fourth Paradigm: Data-Intensive Scientific Discovery*; Microsoft Research: Redmond, WA, USA, 2009; ISBN 978-0-9825442-0-4.
68. Kinney, S. Bell Labs President Says 6G Will Mix Physical, Digital and Biological Systems. *RCR Wireless News*, 24 April 2019. Available online: https://rcrwireless.com/20190424/5g/6g-physical-digital-biological (accessed on 15 June 2022).
69. Rappaport, T.S.; Xing, Y.; Kanhere, O.; Ju, S.; Madanayake, A.; Mandal, S.; Alkhateeb, A.; Trichopoulos, G.C. Wireless communications and applications above 100 GHz: Opportunities and challenges for 6G and beyond. *IEEE Access* **2019**, *7*, 78729–78757. [CrossRef]
70. Boxall, A.; Lacoma, T. What Is 6G, How Fast Will It Be, and When Is It Coming? *Digital Trends*, 3 June 2021. Available online: https://www.digitaltrends.com/mobile/what-is-6g/ (accessed on 17 June 2022).
71. Vinge, V. Technological singularity. *Whole Earth Rev.* **1993**, *81*, 88–95. Available online: https://frc.ri.cmu.edu/~{}hpm/book98/com.ch1/vinge.singularity.html (accessed on 18 June 2022).
72. Kurzweil, R. *The Singularity Is Near: When Humans Transcend Biology*; Duckworth Overlook: London, UK, 2008; ISBN 978-0-7156-4016-6.
73. Svobodová, I. Energy efficiency in buildings: Middle-class status as a trigger for the European fight against climate change. *Eur. View* **2019**, *18*, 163–170. [CrossRef]
74. Schwartzman, D. 4 Scenarios for 2050. *Capital. Nat. Social.* **2013**, *24*, 49–53. [CrossRef]
75. Schwartzman, P.; Schwartzman, D. *The Earth Is Not for Sale: A Path Out of Fossil Capitalism to the Other World That Is Still Possible*; World Scientific Publishing Co. Pte. Ltd.: Singapore, 2019; ISBN 978-981-3234-24-6.
76. Sneader, K.; Singhal, S. *The Next Normal Arrives: Trends That Will Define 2021—And Beyond*; McKinsey Global Publishing: New York, NY, USA, January 2021. Available online: https://www.mckinsey.com/featured-insights/leadership/the-next-normal-arrives-trends-that-will-define-2021-and-beyond (accessed on 5 November 2022).
77. Zhang, L. Fashioning the feminine self in "prosumer capitalism": Women's work and the transnational reselling of Western luxury online. *J. Consum. Cult.* **2017**, *17*, 184–204. [CrossRef]
78. Brown, D.; Hall, S.; Davis, M.E. What is prosumerism for? Exploring the normative dimensions of decentralised energy transitions. *Energy Res. Soc. Sci.* **2020**, *66*, 101475. [CrossRef]
79. Goudarzi, A.; Zhang, C.; Fahad, S.; Mahdi, A.J. A hybrid sequential approach for solving environmentally constrained optimal scheduling in co-generation systems. *Energy Rep.* **2021**, *7*, 3460–3479. [CrossRef]
80. Van Sark, W.; Corona, B. Concentrating solar power. In *Technological Learning in the Transition to a Low-Carbon Energy System: Conceptual Issues, Empirical Findings, and Use, in Energy Modeling*; Junginger, M., Louwen, A., Eds.; Academic Press: London, UK, 2020; pp. 221–231. ISBN 978-0-12-818762-3. [CrossRef]
81. China Solar Thermal Alliance. *Blue Book of China's Concentrating Solar Power Industry 2021*; China Solar Thermal Alliance: Beijing, China, 2021. Available online: https://www.solarpaces.org/wp-content/uploads/Blue-Book-on-Chinas-CSP-Industry-2021.pdf (accessed on 10 November 2022).
82. Liu, Y.; Sun, N.; Liu, J.; Wen, Z.; Sun, X.; Lee, S.-T.; Sun, B. Integrating a silicon solar cell with a triboelectric nanogenerator via a mutual electrode for harvesting energy from sunlight and raindrops. *ACS Nano* **2018**, *12*, 2893–2899. [CrossRef] [PubMed]
83. Wang, Y.; Yang, Y.; Wang, Z.L. Triboelectric nanogenerators as flexible power sources. *NPJ Flex. Electron.* **2017**, *1*, 10. [CrossRef]

84. Kim, D.W.; Lee, J.H.; Kim, J.K.; Jeong, U. Material aspects of triboelectric energy generation and sensors. *NPG Asia Mater.* **2020**, *12*, 6. [CrossRef]
85. Assawaworrarit, S.; Omair, Z.; Fan, S. Nighttime electric power generation at a density of 50 mW/m^2 via radiative cooling of a photovoltaic cell. *Appl. Phys. Lett.* **2022**, *120*, 143901. [CrossRef]
86. Deppe, T.; Munday, J.N. Nighttime photovoltaic cells: Electrical power generation by optically coupling with deep space. *ACS Photonics* **2020**, *7*, 1–9. [CrossRef]
87. Schwartzman, D.; Saul, Q. The Path to Climate Justice Passes through Caracas. *MRonline*, 25 June 2019. Available online: https://mronline.org/2019/06/25/the-path-to-climate-justice-passes-through-caracas/#_edn3 (accessed on 1 October 2022).
88. Schwartzman, D. Is Transition Possible? China and the Prospects for a Global Ecological Civilization. *Climate and Capitalism*, 17 September 2019. Available online: https://climateandcapitalism.com/2019/09/17/china-and-the-prospects-for-a-global-ecological-civilization/ (accessed on 9 November 2021).
89. "Gdy Czarny Orzeł Znak Krzyża Splugawi ... " Sensacyjna Przepowiednia z Roku 1893 o Przyszłych Losach Świata. *Ilustrowany Kuryer Codz.* **1939**, *30*, 6. Available online: http://mbc.malopolska.pl/dlibra/docmetadata?id=77196&from=publication (accessed on 11 November 2022).
90. Fereński, P.J.; Soboń, G. Why Is the New Silk Road Considered a Threat for the Western World, Including Poland? National-conservative Tendencies in North America and Europe. *Kult. Hist. Glob.* **2017**, *22*, 63–72. Available online: http://www.khg.uni.wroc.pl/files/7%20KHG_22%20Ferenski_Sobon.pdf (accessed on 11 November 2022).

energies

MDPI

Article

Contact Zones in the Energy Transition: A Transdisciplinary Complex Problem

Aleksander Jakimowicz [1,*] and Daniel Rzeczkowski [2]

[1] Department of Economic Policy, Institute of Economics, Polish Academy of Sciences, Palace of Culture and Science, 1 Defilad Sq., 00-901 Warsaw, Poland
[2] Department of Market and Consumption, Institute of Economics and Finance, Faculty of Economic Sciences, University of Warmia and Mazury in Olsztyn, 1/327 Cieszyński Sq., 10-720 Olsztyn, Poland
* Correspondence: ajakimowicz@inepan.waw.pl

Abstract: The success of energy transition relies on what happens in the contact zone, the area between citizens and municipality governments, which still awaits more thorough research. This article employs the concept of the contact zone both as a theory describing processes of developing energy prosumerism on a local level, and as a research method which enables one to uncover phenomena that are critical to attaining climate objectives. The research field was the Warmia and Mazury Province in Poland, which is the region with the lowest socio-economic potential both in Poland and in the European Union. The analyzed contact zone was divided into two parts: the human administrative legal contact zone and the more-than-human energy contact zone. To describe the relationships occurring in these subzones, the authors used empirical data originating from a survey addressed to citizens living in the above province. The aim was to explore the respondents' knowledge about current prosumption processes and the opportunities to implement them in the local government sector. Multiple correspondence analysis was used to analyze the data. The main findings were the low knowledge of citizens about prosumption, which was represented by a large number of the prosumption principles not indicated by the respondents, and—on the other hand—the evidence that local communities expect the implementation of digital prosumption, which they know from the market sector, in public administration. It was also demonstrated that the absence of citizens' involvement in the energy transition is a consequence of two historic colonialisms, German and Russian, which had a huge influence on the emergence of an autocratic management style in the analyzed region. Comparison of the analyzed contact zones with two reference zones showed that grassroots movements in the province are initiated mainly by external factors of a nationwide character. In the Warmia and Mazury contact zones, civic initiatives are in the early stage of development, although they display all features of developed zones, such as autoethnographic gestures, transculturation, struggle, violence, and anti-conquest.

Keywords: digitalization; energy transition; local government sector; virtuous cycle of digitalization; decentralization and electrification; more-than-human energy contact zone; human administrative legal contact zone; autoethnographic texts; transculturation; multiple correspondence analysis; principles of prosumption

1. Introduction

In this article, we take the reader on a journey along new trails leading to the knowledge of distant and unknown realms, as we consider a problem that has not yet been recognized in science despite the fact that it is extremely important for preventing global warming and catastrophic climate change. The issue is extremely transdisciplinary, cross-sectional, complex, and fascinating. Here, we define the concept of contact zones as complex systems containing elements belonging to seemingly separate fields of knowledge, such as humanities, natural sciences, and social and technical sciences. What is important is that the interactions of the components of contact zones can be decisive for the success or failure of the energy transition. That is why we consider this issue to be cutting-edge. It should be noted at the outset that this is a completely new and original problem, and it has not yet received its proper place in science. So far, there is no literature on the subject since contact zones have not been taken into account as a factor in the development of renewable energy. Aiming to rectify this, we have produced this article as a result. We address it to representatives of various scientific disciplines in order to arouse the interest of the scientific community in this topic and encourage other scientists to develop and supplement the ideas and concepts contained herein.

Contact zones are an issue studied mainly in the humanities, although one can also find applications of the concept to the study of human relations with animate and inanimate nature. However, these ideas are basically unknown in economics, social sciences, technical sciences, and natural sciences. In order to define contact zones for the energy transition, we have selected from the multidisciplinary literature on the subject all the components of these zones that may be relevant to the development of renewable energy. The most important components of contact zones are concepts that may come as a surprise to scholars involved in the transformation of current energy systems, which are still mostly based on fossil fuels. The list we have compiled, included in the last section of the article, is certainly not complete, but may provide a useful starting point for further research in this area. The features of the contact zones we have selected are not limited to a selected research context but can apply to all contact zones found anywhere in the world. They are universal in the sense that they can be applied locally, nationally, internationally and even globally.

Complementing the energy transformation issue with contact zones certainly increases the complexity of the problem itself as we add additional elements and variables, but the primary benefit is to bring it closer to reality, and this is crucial. Contact zones, like the transformation itself, are complex systems that cannot be studied by analytical methods alone. These zones contain many different quantitative, qualitative, tangible, and intangible elements, some of which have not yet been fully explored by the scientific fields to which they belong. These are primarily people, the expectations of individuals, local communities and entire nations, as well as the economy, energy law, technology, and many elements of animate and inanimate nature. Since contact zones are most often terra incognita in science, except, of course, in the humanities, this must lead to the conclusion that current knowledge of the energy transition has overlooked many important and critical elements for the development of renewable energy. By introducing contact zones into science, the humanities can contribute to achieving such important goals for humanity as limiting global warming and preventing climate change.

The structure of the article diverges somewhat from accepted standards due to the originality of the subject matter under study and the need for a broader explanation of some of the ideas and concepts applicable here. Section 2 includes a literature review, research questions, and a research hypothesis, and indicates that the concept of contact zones will be used not only as a theory but also as one of the primary research methods. Section 3 gives a definition of contact zones and characterizes their most important elements, providing model applications in humanities and environmental studies. The purpose of Section 4 is to demonstrate that digitalization is the driving force behind energy transition and that contact zones are its birthplace. Section 5 discusses the two most important types of contact zones found in energy transformation: the human administrative legal contact zone and the more-than-human energy contact zone. Section 6 presents a summary of our previous research on digitalization and distributed generation in the province of Warmia and Mazury, since—in view of the aforementioned lack of scientific literature on the subject—they provide, for the time being, the only, albeit imperfect, frame of reference for the research presented here. Section 7 verifies the research hypothesis using multiple correspondence analysis and presents the relevant calculations. Section 8 discusses the results of the calculations and presents the characteristic features of the Warmia and Mazury contact zones. The rationale for the need for further comparative research to clarify the conundrum of unselected prosumption principles by respondents is also presented. Section 9 describes the two reference contact zones of Żurawlów and Kraków, which were chosen for comparative analysis and to clarify the problem of citizens' unchosen prosumption principles. Section 10 is devoted to discussing the results of the empirical research and comparative analysis. It also discusses the basic characteristics of the contact zones studied in both national and international contexts. Section 11 provides a comprehensive summary of the results of the research and presents a list of those features of the contact zones that are most important from the perspective of energy transition.

2. Literature Review

It is widely believed that the main responsibility for the prevention of climate change rests with governments and international institutions. However, such an approach seems incomplete because, according to the conservative worldview, responsibility for the environment should also be based on private property rights and a strong emotional bond between citizens and their own household and their immediate environment. Free economies based on private property rights and the rule of law are less energy-intensive, pivot more quickly to renewable energy production and emit less carbon dioxide than economies without such features [1]. Solving environmental problems cannot be monopolized by the state, but should be treated as a permanent element of people's daily lives. The role of the government is to create the conditions for proper motivation, from which true civic responsibility for the climate can spontaneously arise and develop. The essence of this motivation is the citizens' identification with not only groups of people close to them, but also with their places of residence and the customs adopted by them. This concept is known as oikophilia, which should be understood as a love for the household and the desire to protect it, as well as the ability to sacrifice for your surroundings. This is related to people's desire to settle in a specific place in preference to another. Environmental protection in a particular region is most effective only if it is carried out by the people living there.

The motivation resulting from oikophilia is deeply rooted in the human psyche and has a long-term character, as it arouses responsibility not only towards the closest household members but also towards past and future generations, and thus encompasses all people associated with a given place. The sentiment of oikophilia includes people's most important moral, aesthetic and spiritual values, which makes it the basic causative factor of their actions. It has a lot in common with friendship, a sense of beauty, a respect for the sacred and allows one to distinguish prices on the market from what is priceless at home. It offers the hope of achieving climate goals at both local and global levels. A cogent environmental policy must therefore be based on the proper motivation of the electorate and at least a partial departure from homo economicus because, otherwise, it will be difficult to resist the consumption of what is most valuable by global entropy [2]. Oikophilia stands out among the resources beneficial for the environment by being the easiest to acquire and the easiest to renew. The only way to renew it is through appropriate education based on the values mentioned above.

Environmental law is important for climate protection, but its effectiveness may be limited by people's lack of proper motivation. The worst environmental problems were not solved by centralized projects resulting from state planning, but by civic initiatives, which, once successful, became a source of legislation. There is ample evidence that environmental quality is best where oikophilia is strong. In fact, it seems that environmental protection is almost impossible without civil associations driven by oikophilia. Thus, the importance of the bottom-up approach based on the local responsibility of citizens for the energy transition cannot be overstated. Local initiatives should be conducted under a regime of private property, in which oikophilia strengthens market signals and enables local environmental problems to be solved. Successes on a global scale can come later, when proven solutions are taken up in other places [1].

Many climate goals cannot be achieved without the involvement of people. In the European Union, efforts are being made to increase energy efficiency and decarbonize the building stock, which is expressed in legislation [3]. It turns out that the largest polluters include not only coal-fired power plants, airlines, or shipping companies, but buildings first and foremost, which account for 40% of energy consumption and 36% of greenhouse gas emissions. As the European Commission points out, about 75% of buildings currently do not meet the energy efficiency requirements. Their renovation is expected to reduce total energy consumption by 5–6% and reduce emissions of CO_2 by almost 5%. This is a very ambitious goal, given that only 0.4–1.2% of building stock is renovated annually depending on the country, and the implementation of this goal by 2050 would require at least doubling the current rates of renovations [4]. As reported by the Building Performance Institute Europe, only 2.5% of building stock is highly energy-efficient, i.e., complies with the Energy Performance Certificate (EPC) label A. Accordingly, as many as 97.5% of buildings do not currently belong to class A, which means that they should be upgraded in order to achieve the assumed climate objectives [5]. It will be very difficult to implement these plans without making room for civic initiatives and using the resources offered by oikophilia on a larger scale.

The implementation of EU legislation in Poland brought new energy efficiency standards for buildings into force in 2021. The changes concern the maximum value of a facility's annual demand indicator for non-renewable primary energy used for heating, ventilation, domestic hot water preparation, cooling, and lighting, as well as the heat transfer coefficient of walls, roofs, ceilings, and flat roofs for all types of buildings. For example, for single-family residential buildings, the partial index (including only heating, ventilation and hot water) decreased from 95 to 70 $\text{kWh}/\left(\text{m}^2{\cdot}\text{year}\right)$, while the second index was reduced from 0.23 to 0.2 $\text{W}/\left(\text{m}^2{\cdot}\text{K}\right)$. Meeting these standards is a prerequisite for recognizing a facility as a building with low energy consumption [6].

In Europe, more than 60% of buildings are residential, of which up to 70% are owner-occupied. Renovation of these buildings and their transformation into energy-saving and decarbonized stock requires involving the owners and identifying convenient sources of financing. These are the people who will determine Europe's energy consumption over the next few decades. Most owners of residential buildings belong to the middle class, so their financial situation is determined by their ownership of the property or apartment. In addition, the level of education in this group is higher than average, so they understand that renovation not only reduces energy costs, increases the value of the house, and extends its useful life, but also has a positive impact on their quality of life. They are also aware of the risks of environmental pollution. Therefore, they tend to take care of their buildings, try to keep them in good technical condition, and are willing to invest in increasing their energy efficiency. This is how homeowners strive to provide a sense of safety, health, and comfort for themselves and their loved ones. It should also be emphasized that buildings are the main financial asset that will be inherited by future generations. Ownership of the building also implies an interest in the immediate surroundings, encourages the establishment of social relations, and increases motivation for civic engagement. Hence, building owners should be treated not only as consumers but as active players in the energy transition as well. Economic strength, education, and oikophilia are the three features of the middle class that predestine it for the role of a key climate change stakeholder, making it the leading force in the decarbonization of Europe [7].

An important facet of the energy transition as a whole is the need to call upon local initiatives and involve citizens in this project. Just as the modernization and decarbonization of buildings requires the involvement of owners, the decentralization of the energy sector is similarly based on the transformation of existing consumers into energy prosumers, who will play an active role in the market by producing energy for their own needs, storing it, or diverting its surplus to the grid. All of these projects require a two-way flow of information between citizens and the local government sector, which takes place in an as-yet undefined common space known in the humanities as the contact zone. It is to some extent an imaginal sphere, having, to a greater or lesser extent, the attributes of ordinary physical space–time or cyberspace, which contains such elements as ideas, desires, and expectations, as well as material and intangible values.

This article attempts to determine the elements necessary to conduct the energy transition, which should operate in the contact zone between the citizens in Warmia and Mazury and local governments. The starting point for these considerations was the division of the contact zone under study into two sub-zones: the human administrative legal contact zone and the more-than-human energy contact zone. The main objective of this study is to determine the relations within two identified contact sub-zones between citizens and the municipal governments in Warmia and Mazury. In order to streamline the energy transition processes, it was necessary to identify the prosumer potential of society and its usability in local public administration. In particular, the aim was to learn society's attitude to digital prosumption and digitalization of the local government sector. The emphasis was placed on digitalization for a reason, as this is a key element in the development of the green energy sector. The ongoing energy transition process encompasses three mutually propelling components: digitalization, decentralization, and electrification. Of these components, the most important is digitalization, which refers to the digital transformation of technologies as well as business models and processes, and which plays a key role in controlling the flow of energy and information.

Digitalization is an intermediary medium between citizens and the local government in both contact zones under study. Moreover, it plays a strong transformative role in the development of energy generation from renewable sources. This paper seeks answers to three research questions:

1. How can digitalization transform the existing models of local government operation so that it can stimulate energy prosumerism development?
2. Do citizens expect the public administration to release the prosumer potential of the society and to incorporate it into a virtuous cycle comprising interrelations between digitalization, decentralization, and electrification?
3. What are the major similarities and differences between the contact zones in Warmia and Mazury and the two reference contact zones in Żurawlów and in Kraków?

The prosumption rules are the fundamental elements of a contact zone, without which the energy transition is not possible. In prosumer capitalism, they are a kind of understanding between citizens and the municipal governments, whose approval by both parties is a guarantee of the success of all joint enterprises. The authors formulated ten basic rules of prosumption and tried to see if they could be accepted by citizens as the basis for energy prosumerism. With the research questions listed above, the issue in hand was formulated in the following manner.

> *The research hypothesis: society is ready to participate in improving the local government sector by introducing digital prosumption to the contact zones. In other words, digitalization of the contact zones is based on the ten proposed rules of prosumption supported by digital technologies.*

The use of the contact zone as a theory and the research method in the energy transition is the most important added value of this paper, as there are still few studies of the subject matter. This helped to identify a research gap overlooked by scholars. According to the major discovery presented in this paper, events in the contact zone are crucial to the success or failure of the energy transition on the local level. In order to justify this claim, we refer to empirical studies which employ multiple correspondence analysis and comparative analysis. The empirical data were provided by a survey whose aim was to find out what members of the public think about the digital implementation of the prosumption rules in the municipal governments, as proposed in this paper. Thus created, the economic stimulus should lead to a change in the existing local government operation models. Interesting findings were obtained using multiple correspondence analysis. However, the authors were able to refer to these findings only by comparing them with their previous research because of the originality of the subject matter dealt with, resulting in the absence of any other studies in this field. Therefore, it was an additional aim to broaden the scope of research and to use the contact zone not only as a theory that would help to explain numerous phenomena but also as a research method. In order to achieve this, a comparative analysis was performed to determine the similarities and differences between the characteristic features of the contact zones in Warmia and Mazury and the features of two reference contact zones in Poland. In one zone, farmers in Żurawlów fought against the Chevron Corporation about shale gas extraction, whereas in the other, covering one of the major cities in Poland—Kraków—there was (and still is) a complex social conflict about the causes of air pollution and the choice of the optimum remedies.

3. Contact Zones in Science

3.1. Definition of the Contact Zone and Its Most Important Elements

The concept of the contact zone was introduced by Mary Louise Pratt to denote an area where different cultures can communicate, where linguistic and cultural meetings take place, where negotiations are conducted, or even where battles are fought to determine the common history and relations of power. Contact zones should be understood as

> *social spaces where cultures meet, clash, and grapple with each other, often in contexts of highly asymmetrical relations of power, such as colonialism, slavery, or their aftermaths as they are lived out in many parts of the world today [8].*

This concept has a lot to do with colonialism, so it can be treated as a synonym for the colonial frontier [9] (pp. 6–7). Explaining the emergence of contact zones, Pratt refers to a manuscript from 1613 written in two languages, Quechua and Spanish, by Felipe Guamán Poma de Ayala and titled *The First New Chronicle and Good Government*. In 1908, it was discovered by the German scholar Richard Pietschmann as he was viewing the collection of the Danish Royal Library. The manuscript was intended by the author, whose origin was Andean, as a letter addressed to King Philip III of Spain, in which, as an indigenous subject, he described the history of the conquest of the Inca Empire by the Spanish, waged harsh criticism of their colonial rule, and also presented principles of good governance of the country. The document is an extensive work, with a total of 1200 pages and containing 398 full-page illustrations, but it is not known whether the king ever received it. The Danish Royal Library has produced a high-quality digital facsimile of the original manuscript, along with a transcription that is now available online [10]. In addition, the English translation of the manuscript by Roland Hamilton is recommended to the reader [11]. An interesting two-volume edition of the work in Spanish with transcription, prologue, notes, and chronology developed by Franklin Pease García [12] also exists.

Nowadays, it is worth noting that Guamán Poma describes in his letter the basic principles of good governance, by which he understands the combination of Incan social and economic organization with European technical thought and Christian theology, to be adapted to the practical needs of the Andean nations [13]. His views on political issues were complex, but systematic, coherent, and unequivocal. He was not a supporter of colonialism and the direct rule of foreigners, instead supporting native rule and calling for the restoration of traditional Andean governance and the restitution of Andean lands and properties. He condemned the greed of both civil and ecclesiastical officials. For this reason, his views can be considered anticlerical, although, on the other hand, among his postulates was the institutionalization of the Christian religion. In his eyes, the Andeans were civilized Christians, while the Spaniards he regarded as lost sinners. He called for the creation of a sovereign Andean state, which would form part of a universal Christian empire ruled by the Spanish king. It is worth noting here that despite harsh criticism of the unjust and exploitative Spanish rule, he does not reject the Spanish king because, according to the New Testament view, he believes that all power comes from God. As Rolena Adorno aptly sums up, *Guamán Poma was anti-Inca but pro-Andean, anticlerical but pro-Catholic* [14].

Pratt cites this text as an example of a standard product of a contact zone, as Guamán Poma can be considered an intermediary and mediator in negotiations between native Peruvian and Spanish colonial societies and institutions. This illustrates the sociocultural and economic complexities resulting from the conquest of the Inca empire by the Spanish. The Andean author chose the title of his work with deliberation because the chronicle was the main writing apparatus for the Spaniards, which they used to present their American conquests. He decided to appropriate the official Spanish genre and use it for a new description of the history of the Christian world, in the center of which he placed the Andean peoples, not the European peoples, and where Cuzco took the place of Jerusalem. In this way, the chronicle of Guamán Poma became what Pratt calls an autoethnographic text, which means *a text in which people undertake to describe themselves in ways that engage with representations others have made of them* [8]. Ethnographic texts are characterized by descriptions of how European metropolitan subjects present themselves to their others, i.e., mainly conquered peoples. On the other hand, autoethnographic texts are responses to the former prepared by such defined others or intended to enter into dialogue with them. In other words, in the first type of text, the oppressed societies are presented in a certain way by the colonizers, while in the second, the tables are turned, and the colonized peoples and cultures selectively cooperate with the literary genres and idioms of the metropolis, thus appropriating them. These elements are combined with indigenous idioms, as a result of which self-representations are created, entering into metropolitan modes of understanding in order to change them. We are dealing here with something similar to the mirror dance, in which images can be reflected many times [9] (p. 143).

Another phenomenon existing in the contact zone is transculturation, meaning the process of selective and creative use by the conquered or marginalized peoples of dominant or metropolitan culture. Subjugated peoples do not have an influence on what reaches them from the metropolitan culture, but they can decide what elements they will transfer to their own culture and what purposes they will use them for [8,9]. This term was introduced by Cuban anthropologist Fernando Ortiz to denote cultural change during which something is always given in exchange for what is received. During this give and take, two sides of the equation are modified. Transculturation, within the meaning of Ortiz, does not mean the necessity of adapting one culture to another, but the cooperation of both active cultures, contributing to the creation of a new civilizational reality, which is exemplified by the Cuban counterpoint of two commodities, tobacco and sugar, the first of which is native and the second brought in from the outside [15].

The contact zone is a place where cultures that have so far developed independently, and thus had distinct geographical locations and histories, now meet, establish lasting relationships, and create a new society, often of a colonial nature, in which there may be fundamental inequalities, coercion and insurmountable conflicts. In the contact zones created by invasion and violence, there is a struggle for interpretive power regarding the perception of culture, which takes place in the conditions of radical heterogeneity, ambiguities, and enormous difficulties in determining the meanings of key elements of the experienced reality. It should be noted that contemporary Indigenous peoples, unlike the Western world, identify their culture with survival. Therefore, in such a system, being "different" means for them the need to play the role of a double agent and living in a bifurcated world. In order to survive, one should create oneself as a "self" for oneself, while for the colonizer, one should create oneself as an "other". In the contact zones, there is also the issue of self-determination of Indigenous peoples, meaning the possibility of controlling the conditions in which they will shape their relations with the nation-state or the world economy. For this reason, they are an arena of fierce negotiations concerning the mutual obligations of the conquerors and the conquered, and culture itself may also be the subject of negotiations. Relations between colonizers and the colonized are the medium from which culture, language, society, and consciousness are formed [16].

3.2. Contact Zones in the Humanities

It follows from the definition of the contact zone that negotiations are an important element of it, as only they allow for a compromise and coexistence of different cultures and societies. However, the existence of strong negotiating potential in contact zones is sometimes considered questionable since they are based on contact languages that, although ensuring survival, have lower symbolic and personal values for users than native languages. These zones do not give an accurate answer to the question of how cultural and political differences can be articulated and negotiated in order to avoid battle. It is also pointed out that Pratt does not explain how a public space is created in which people have a reason to negotiate with competing views or factions. Therefore, there is a danger of treating contact zones as a multicultural bazaar, where opposing views are presented in a harmless way and there is no conflict between them. The missing elements of the contact zone are, therefore, factors indicating a conflict, i.e., the mutual interpenetration of competing perspectives based on the multidirectional flow of information [17]. For this reason, it is proposed that the contact zone be reconceptualized and recognized as a forum whose functioning is maintained by continuous negotiations, interactions, and compromises between perspectives and individuals. Its essence is, therefore, a dispute over the differences regarding competing interests and views [18]. It is also proposed that the negotiations be looked at as a form of civil dialogue, which is possible when the will to compromise is shared by all parties. In such cases, the languages and protocols of negotiation are usually determined by the dominant groups and should therefore be placed in specific historical and political dimensions [19]. Civil dialogue must be preceded by the acquisition of real power relations, i.e., a strict definition of social roles and their immunity from routine questions about their legitimacy [20] (p. 236).

The concept of the contact zone has found many applications in the humanities, and it is particularly useful in various pedagogical spaces [21]. This idea can be used in research on multiculturalism [22,23], and giving it historical significance and organizing English studies around a concept thus defined allows for a significant improvement in teaching, which at the same time reduces the focus placed on literary or chronological periods and racial or gender categories [24]. It is also useful in museology, as it includes cultural actions, creations, and transformations of identities, all of which run along controlled and intercultural frontiers emphasizing the existence of nations, peoples, and locales. Museums should be understood as contact zones in which things and people move, not as centers or destinations [25]. There are also concerns that the neo-colonial nature of these contact zones has a negative impact on the empowerment of the source communities whose patrimonies are exhibited, which, after all, should be the fundamental purpose of museums. Some hold the opinion that the restoration of the colonial character of the museum as a contact zone results from the role of autoethnography as an instrument of appropriation [26]. Apart from the above, discursive contact zones are used to describe non-local and non-physical spaces in which differences and similarities between different environments of devotional fitness, both Christian and non-Christian, are negotiated [27].

3.3. *Contact Zones in Environmental Studies*

The original concept of the contact zone can be extended by including political ecologies, multispecies ethnography, and more-than-human geographies, which makes it both a method and a scientific theory extremely useful in environmental studies. Thus, key issues such as creating knowledge about the environment, nature management, and counteracting the exclusive governance of wildlife may be at the center of interest. In places where the environment is managed and explored, transformational processes take place, characterized by co-presence and unpredictability, and may also result in intimate violence. The more-than-human contact zones can help achieve three overlapping aims: finding a better explanation of non-human agency by taking multiple perspectives into account, preventing acts of violence and injustice, and facilitating the decolonization of the knowledge production process.

These interdisciplinary contact zones are dynamic, taking into account rhythmic patterns, phases, and ephemeral phenomena, by which they enable the study of variable human relations with the environment. Such an approach combines many important elements into a holistic unity, the most important of which are environmental and extended geographical research, postcolonial studies, and decolonization projects [28]. The inter-species contact zones involve interactions between people and other forms of life and, unfortunately, often asymmetrical encounters between capitalism and non-human nature, coercion, exploitation, neglect, destruction, violent oppression, wrongdoing, and other types of threats. Currently, these spaces serve to link together various phenomena, such as biodiversity conservation, climate change, and energy transition, which until recently have been the subject of separate policy discourses, now greatly increasing the ability to manage them. Due to the transformative nature of these zones, they should be places where environmental policy properly shapes and renegotiates existing relations and where effective sustainable development programs are implemented [29]. The contact zone, when extended to environmental, geographical, and biological elements, has many potential implementations.

Helen F. Wilson uses the concepts of the encounter and the contact zone as an analysis tool in research on multispecies scholarship and, for this purpose, refers to the documentary series on biological life in the oceans [30]. Multi-species contact zones located in the deep sea are the arena of struggles between filmmakers and technology, knowledge, various life forms, and natural phenomena in asymmetrical and changing power relations. It turns out that filmmaking of this type is ambivalent because although it shapes environmental awareness and consumer behavior, it also poses a threat to the oceans by initiating the development of maritime tourism and encouraging further exploration. Journeying across the seas in search of places beyond the human imagination and new underwater worlds waiting to be discovered has an undeniable imperial subtext, as it entails the inclusion of these marginal spaces into a realm of authority. Thus, it is a form of anti-conquest, i.e., a strategy of representation foregrounding gestures of innocence, whose task is to divert attention from the simultaneous pursuit of hegemony [9] (p. 7). Thus, oceanic contact zones are filled with extremely complex interactions involving humans, many species of animals, marine organisms and plants, and elements of inanimate nature. Technology imposes a means of contact by removing barriers between the benthic zone and the viewers [30].

Another more-than-human contact zone is discussed by Jenny R. Isaacs, focusing on people dealing with migratory shorebird conservation in the Delaware Bayshore in New Jersey's Cape May County [31]. On the beaches of this area, there are human–shorebird encounters whose asymmetry is a source of imperial domination and violence. During some stages of conservation research, such as bird banding, there is often aggression and brutality. This is interpreted as an expression of animality/coloniality, namely, reductive conceptions of the animal, common to colonialism and nature protection. The progressive globalization of nature conservation defines the ways in which this phenomenon manifests itself. Animality/coloniality may appear or disappear in a multispecies contact zone when face-to-face encounters occur and touch is practiced. This also illustrates the conservation paradox of interpreting harm as care, which is an example of conspicuous innocence in the field of bird and nature conservation. It occurs when there is a need to use force so that the bodies of individual birds can be marked and described in the name of the higher goal of survival of the entire species. In this approach, nature protection is presented as a neo-colonial form of anti-conquest, which is a confirmation of Western European authority and control using new definitions of care and concern. The phenomenon of anti-conquest can also be an expression of settler moves to innocence. Remote monitoring technologies are a new craft of the contact zone, as they allow an expansion of the encounter space and extend control over species, space, and time, even to the entire hemisphere. The contact zone concept is useful as an aid in developing harm reduction strategies that may arise during nature conservation activities [31].

Another development of the idea of the contact zone in environmental research is the intertidal contact zone studied by Aurora Fredriksen [32]. The interspecies meeting place in question is the Bay of Fundy's Minas Passage in the Canadian province of Nova Scotia, where extreme tides occur, having the largest range in the world. It is a very dynamic environment, as an average flood tide in the Bay of Fundy brings 14 billion tons of seawater, which crosses the Minas Passage, 5 km wide, and flows into the Minas Basin. The speed of tidal currents in the passage may exceed 10 knots, i.e., 5 m/s. The potential for renewable energy generation here was estimated at 7000 megawatts, so it is not surprising that it is attractive for the governments of Canada and Nova Scotia in their attempts to decarbonize their energy systems. Underwater turbines located in the path of tidal streams enable the transformation of tidal energy into electricity. However, there is a serious problem with the fact that, along with the massive flood tides flowing through the narrow Minas Passage, a huge number of fish and other marine animals are pushed in and out of the pool twice a day, being the primary source of livelihood for many local fishermen. In the newly created tidal energy industry, environmental assessments, which determine the impact of underwater turbine work on marine fauna and flora, are still incomplete despite the intensive work of scientists. Fredriksen studied the generation of environmental knowledge in two contact zones. The first zone is the fishing contact zone, which is variable and constantly recreated during direct and natural meetings of traditional fishers with marine wildlife. The second zone, the scientific contact zone, includes researchers' meetings with marine organisms, staged and remote, since they are mostly mediated by technology. In these two contact zones, two types of knowledge are produced, which overlap only to a small degree. Practical ecological knowledge acquired by anglers collides with scientific knowledge focusing on the pursuit of objectivity and detachment [32]. The usefulness of Pratt's concept is revealed here in two important ways related to the creation of knowledge in more-than-human contact zones. The first concerns embody cognition and thus take into account the ways in which bodies are involved in meetings, while the second indicates that cognition and action in contact zones are connected with unequal power relations. In the intertidal contact zone, it is therefore necessary to take into account the possibilities and limits of the two ways of learning, one of which assumes direct contact with marine animals and the other does not. It is to be expected that the importance of embodied knowing will increase with the intensification of ecological changes. This explains why, for many people working and living in the region, underwater turbines are an unwanted intrusion that can disrupt existing ecological relations in the bay.

A very interesting case is the fire contact zone, created as a result of prescribed burns in Canadian national parks. Colin R. Sutherland points out that a fire in a national park is not always associated with disasters and destruction but is skillfully used as an effective tool for active landscape management, as it helps to create healthier and more diverse ecosystems and reduces the risk of large fires [33]. The method of environmental protection using prescribed burns is increasingly popular in various countries, but in many places it is contrary to contemporary policies in forestry, municipal planning, and agriculture. Western fire management practices are the result of settler colonialism and are mainly based on a suppression approach, while all other possibilities are overlooked. Thus, the ways of learning about landscapes and living with them used by the Indigenous peoples of North America have been superseded. For hundreds or even thousands of years, burning practices have been a natural part of shaping the environment in which those peoples lived. With the development of settlements, Canada has transformed landscapes and deteriorated the condition of some ecosystems, particularly landscape biodiversity. Forests have become less resistant to insects and forest diseases and the threat of large and uncontrolled wildfires has increased. All of this is directly related to the disruption of the relations of Indigenous peoples with fire. According to Sutherland, the return of fire as a method of managing national parks should be understood not only as a departure from the suppression of combustion but also as a departure from the suppression of burning practices and knowledge [33]. In this way, two different fire contact zones can be observed. The first zone is a typical human contact zone, in which there are meetings between people in matters related to politics, environmental legislation, and landscape management plans. During the clash of different attitudes and views, an organizational principle was developed called "ecological integrity", which enabled the efficient management of national parks, making fire not only a natural phenomenon but also a method of achieving intended goals.

The second contact zone is essentially more-than-human, because it refers to the embodied contact of people with burning landscapes and thus dying fauna and flora. Interpreting these results, it can be stated that prescribed burns are certain form of decolonization because the knowledge of the First Nations has been noticed and institutionalized—but also a form of anti-conquest, because restoring them to power over the land and the environment is not an option.

4. Basic Elements of Energy Transition

4.1. Digitalization as a Driving Force of Energy Transition

This article is a continuation of research on the energy transition of Warmia and Mazury Province, Poland, with particular emphasis on the impact of the cooperation between the citizens and the municipal governments in developing the renewable energy sector. Our previous findings show that the digitalization of public administration is a critical factor in the energy transition in this region [34]. Digitalization can improve electricity services and optimize the use of energy infrastructure by eliminating gaps in connection to the power grid, the cost of provision, and power supply quality, among others [35] (p. 19). In addition, the digitalization of public administration can optimize control and regulatory activities throughout the economy, which also increases the efficiency of the energy sector. Three contemporary trends, i.e., electrification, decentralization, and digitalization, create complex chains of events in the form of a virtuous cycle and, based on the positive feedback loops existing there, reinforce each other's impact [36]. The share of electricity in the production of renewable energy is constantly increasing, as a result of which the electrification of large sectors of the economy, mainly transport and heating, is particularly important in terms of reducing carbon dioxide emissions. Decentralization is caused by a decrease in the cost of electricity production from renewable sources, which results in the conversion of consumers, until recently passive actors on the energy market, into prosumers, i.e., active participants in this market. In turn, digitalization strengthens the other two trends, contributing to the creation of a smart electricity system, of which an integral part is the smart grid, enabling the bi-directional flow of electricity and information as well as automated operation of the system in real-time.

A distinction should be made between digitization and digitalization. Digitization means the conversion of analogue data into digital data by information and communication technology, while digitalization has a much broader meaning and refers to the digital transformation of both technology and business models and processes [35,37]. For this reason, digitalization creates additional opportunities for creating value and, thus, is a new source of revenue for enterprises. In addition, and most importantly, it is a major factor in the development of the energy system, as it enables the introduction of energy from distributed generation into it [38]. Digitalization is also the source of the phenomenon of convergence, understood as the simultaneous development of such fields as electricity production (including the growing role of prosumer power generation), heating, communication, mobility, and intelligent buildings [39].

4.2. The Contact Zone as the Birthplace of Digitalization

Figure 1 shows the most important dependencies in the modern economy from the point of view of the energy transition. Citizens, mostly middle-class, play a central role here, as they can make a decisive contribution to achieving climate goals. The light purple circle in the middle of the figure denotes them. Citizens can influence the energy transition via the local government sector, which has been marked with a hatched ring. The elements representing these two groups of actors do not come into direct contact with each other due to their distinctiveness; hence, mutual communication requires the existence of a certain intermediary space. Encounters between citizens and representatives of the municipal governments take place in the contact zone represented by the white ring. It is now almost literally terra incognita due to the fact that there is basically no research on the role and importance of this zone in the context of renewable energy development and climate protection. This article is the first attempt so far to investigate this research gap and to show how important such a contact zone is for the development of the energy sector and what its characteristics are. The large blue circle symbolizes the entire national economy, and the three smaller circles placed around its perimeter indicate the three most important features of modern economic transformation: digitalization, decentralization, and electrification. The whole system constitutes a powerful economic machine whose proper functioning is necessary to achieve climate goals.

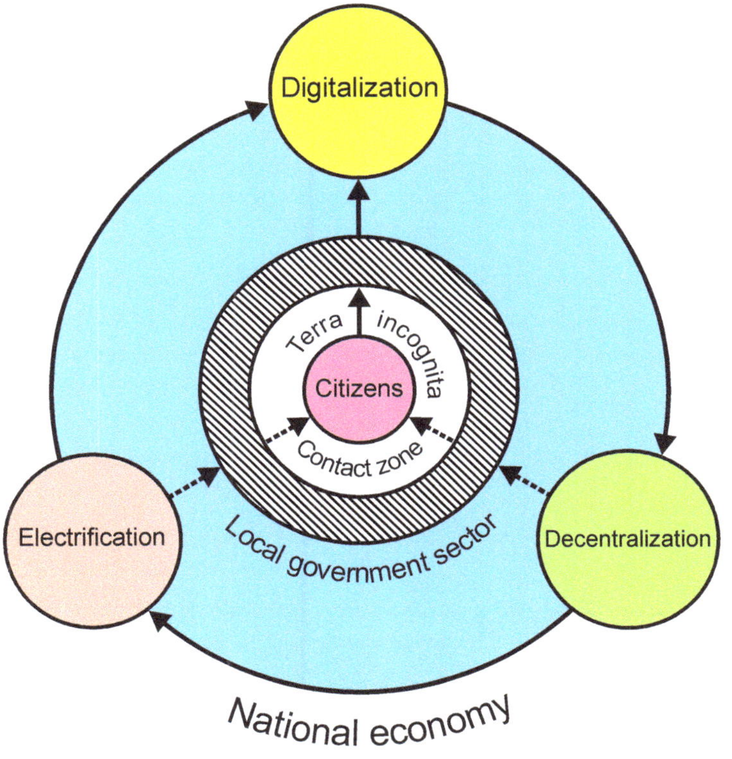

Figure 1. Forces driving the modern economy. The influence of citizens (middle class) on public administration causes the introduction of digital prosumption into this sector, which initiates a virtuous cycle involving digitalization, decentralization, and electrification.

As indicated by the arrows in Figure 1, the influence of citizens on energy transition takes place via the local government sector, and the meeting of both parties takes place in the contact zone. This is the place where critical changes take place; it is there that all this economic machinery is ignited and put into operation, but this will not be possible without civic initiatives and decisive civic action. Polish citizens have extensive experience in the field of digital prosumption, derived from the market sector, and expect the same from services provided by public administration. Unfortunately, to date, there remains a large asymmetry between the market sector and the local government sector in terms of the implementation and use of digital technologies. The market sector is already in a new phase of capitalism called prosumer capitalism, while the local government sector is still stuck in producer and consumer capitalism. Under these circumstances, citizens are taking steps towards the digitalization of the municipal public administration in order to change its business models and procedures to be more friendly and open to the needs of local communities. This is symbolized by a continuous upward arrow that runs through the white ring of the contact zone. Typically, these activities meet with the understanding of the authorities and, depending on the financial possibilities and other conditions, are implemented sooner or later. The support of digitalization by the local government is symbolized by another continuous arrow pointing upwards and running through the blue circle of the national economy. Digitalization is the most important element of change, as it enables the decentralization of the energy sector and its shift to renewable sources. This phenomenon is indicated by a semi-circular arrow running around the perimeter of the large blue circle, from the yellow digitalization circle to the green decentralization circle. As a result of decentralization, electricity generation becomes cheaper, which allows the electrification of the most important sectors of the economy. This is indicated by another semi-circular arrow running from the green decentralization circle to the pink electrification circle. The entire cycle is closed by the last arrow connecting the pink electrification circle with the yellow digitization circle. This is because electricity is the basis of digital technologies. In this way, a virtuous cycle is created, where individual elements reinforce each other's effect. Notably, decentralization and electrification have a return impact on citizens through the local government sector, which is symbolized by dashed arrows. The place where these critical relationships operate is the contact zone.

5. Types of Contact Zones in the Energy Transition

Filling out the area marked in Figure 1 as terra incognita, i.e., defining the contact zone between citizens and the municipal governments, requires describing the types of meetings that take place here. Furthermore, economic models and processes that change under the influence of digital technologies must be taken into account. It is also important to define the purpose of these meetings. This is a similar situation to the previously discussed case of using the contact zone concept in environmental studies concerning prescribed burns, where it functions as a dual method. On the one hand, we saw struggles between people in the field of environmental legislation and related issues, while on the other hand, there were encounters and confrontations of people with burning landscapes.

In the contact zone shown in Figure 1, there are encounters and struggles between citizens and representatives of the local government sector, which are supposed to lead to the optimal fulfilment of set climate goals on the local, national, and international scales. As shown in Figure 2, this zone has a dual character as it consists of two partially overlapping contact zones. One of these can be called the human contact zone, while the other is the more-than-human contact zone. The light green circle shows the administrative legal contact zone, in which encounters happen between citizens on the one hand and the legal system and municipal governments on the other. The light pink circle symbolizes the energy contact zone, which is the place where people meet with the energy system, the legal basis for its functioning, and the broadly understood ecology. The intersection of these circles, marked in purple, represents the energy law, which regulates the principles of energy production and their compliance with current climate objectives.

In the administrative legal contact zone, legal solutions are being worked on to facilitate the development of the renewable energy sector and climate protection through the renovation of the building stocks. Broadly, legal contact zones are such zones

> *in which rival normative ideas, knowledge, power forms, symbolic universes and agencies meet in unequal conditions and resist, reject, assimilate, imitate, subvert each other, giving rise to hybrid legal and political constellations in which the inequality of exchanges are traceable [40] (p. 449).*

In the course of interaction with administrative legal processes, citizens learn about the structures of state power, the prevailing logic of governance, and ways of improving it. They try to modify their exercise of political liberal rights by using legal and administrative tools and discourse, as appropriate. Past experience indicates that struggles in this contact zone may take place both within formal institutional channels and outside them [41].

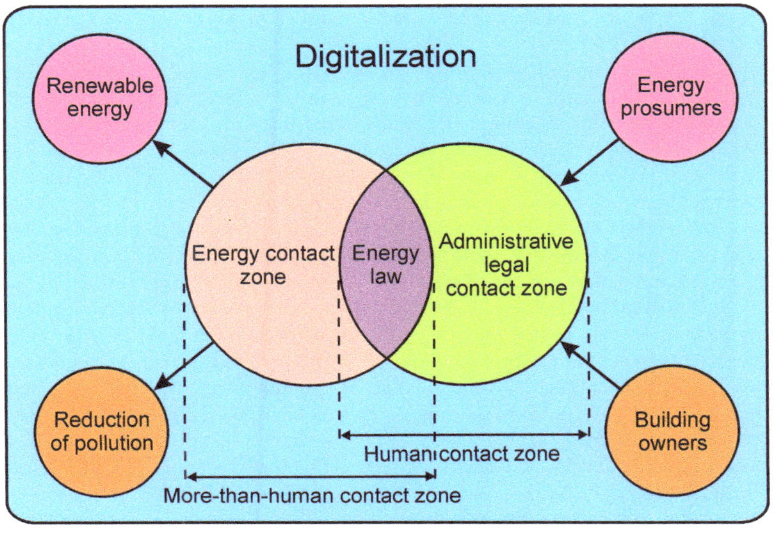

Figure 2. Types of contact zones in the energy transition.

In the energy contact zone, citizens encounter both what is human and what has a more-than-human character. An integral part of this zone is the energy law, which deals with problems related to energy production and its impact on the environment, relations of prosumers with energy companies, renewable energy sources, human–natural relations, environmental issues, preserving biodiversity, reducing pollutant emissions, meeting climate goals, and issues related to energy efficiency in buildings. As mentioned above, the middle class, particularly energy prosumers and building owners, have a big role to play here. The two groups have different goals. Energy prosumers specialize in the production of energy from renewable sources, while building owners focus on using the latest construction techniques to increase the energy efficiency of their buildings. Of course, the same people may find themselves in both groups. These groups are shown in the right part of Figure 2, while the left side shows their goals. Both groups and goals are represented by circles, with the colors of the circles emphasizing the relationships between them. For example, the energy prosumers are represented by a purple circle, and the same color is given to the circle corresponding to their objective, i.e., the production of renewable energy. Similarly, the two circles representing the building owners and their goal of reducing pollution are marked in orange. Achieving these objectives depends on the presence of citizens in two interconnected contact zones: the human administrative legal contact zone and the more-than-human energy contact zone.

Currently, almost all these phenomena and processes are shaped by digitalization, which in Figure 2 is marked as the blue background. It is the most important transforming factor, as it accelerates the processing of information and expedites the activities taking place in both contact zones. In a modern energy system, digitalization plays the role of a lever, i.e., a simple machine consisting of a rigid beam and a fulcrum, which allows obtaining a greater force by applying a lesser force. The law of the lever was proven by the Greek mathematician and inventor Archimedes of Syracuse in his work *On the Equilibrium of Planes or the Centres of Gravity of Planes* [42] (pp. 192–194). As noted by Pappus of Alexandria, during his research on levers, Archimedes made his famous statement: *give me a place to stand on, and I will move the Earth* [43] (p. 460). In this case, digital technologies play the role of the rigid beam, and the fulcrum is the interactions occurring in both contact zones, namely, the struggles of people with each other and with surrounding nature. These interactions are based on prosumption and will be formulated precisely later in the article.

6. Summary of Previous Research on Digitalization and Distributed Generation in the Warmia and Mazury Province

Figure 3 summarizes the authors' previous research on the impact of municipal governments on the development of the green energy sector in the Warmia and Mazury Province [34]. The need for an energy transition stems from the signing by Poland of international agreements aimed at limiting anthropogenic greenhouse gas emissions. The starting point was to determine whether the inhabitants of the region are interested in digital prosumption, which is a prerequisite for using public services related to distributed generation and decarbonization of the local economy. The answer came in the affirmative, as the public already had some experience in digital prosumption from the market sector. Historical determinants of prosumption were also of great importance because, in the socialist economy, prosumption was one of the basic methods of coping with the widespread shortage of goods and services [44]. Therefore, the local government sector, with societal support, decided to build a fiber-optic backbone distribution network and implement the 2025 Digital Plan for Warmia and Mazury. Next, the quality of municipal government websites was examined, which led to the conclusion that in terms of digitalization, public administration is significantly delayed in relation to the market sector. It was established that the cause of this is complex and has not only an economic basis, but also results from social, political, historical, geographical, and environmental issues. Furthermore, it turned out that municipalities with websites of the highest quality, which have the ability to jump-start a new business model in the form of social platforms for participation, are closely related to the transport infrastructure. This is due to the fact that the Internet infrastructure largely overlaps with communication routes [45]. This leads to the conclusion that the infrastructural growth poles and development axes contribute to the acceleration of the province's digitalization at the municipal level, and the participation platforms facilitate the use of roads, railway lines, and inland waterway transport. In this way, we obtain something like a virtuous cycle, in which both elements mutually reinforce [46,47].

The inclusion of electricity production from renewable sources in this cycle is an essential condition for removing the asymmetry between the public administration sector and the market sector, and thus for decarbonizing the economy of Warmia and Mazury. The region has good conditions for the development of small hydro- and wind power stations, which was the main focus of the research. Determining the best location for such objects is the province of molinology. According to the final conclusion of the research, the energy transition is one of the most important factors of social change, which may result in the shift of public administration from producer and consumer capitalism to prosumer capitalism, catching up with the market sector.

In Figure 3, the impact of similar factors, elements, variables, and actions is marked with the same color. Sustainable development of the national and local economy requires that the public administration sector achieves a similar level of digitalization as the market sector, which is to ensure an appropriate level of compatibility between them in terms of economic models and processes. For this reason, both sectors were marked with circles of the same size, placed at the same level and given the same light purple color. Similarly, there are groups of factors such as acceptance of digital prosumption by society and digitalization at the municipal level (orange), prosumption and platforms for participation (blue), and sources of asymmetry and elimination of the gap between public administration and the market (purple). Infrastructural growth poles and development axes are correlated with the quality of websites and Internet infrastructure (green). Moreover, three important groups of non-economic factors shaping the province in question include geographical, environmental, and social issues in the region, historical determinants of prosumption, and molinology, which is why all these were marked in yellow. The area occupied by producer and consumer capitalism was marked in pink, and the space of prosumer capitalism was marked in light green.

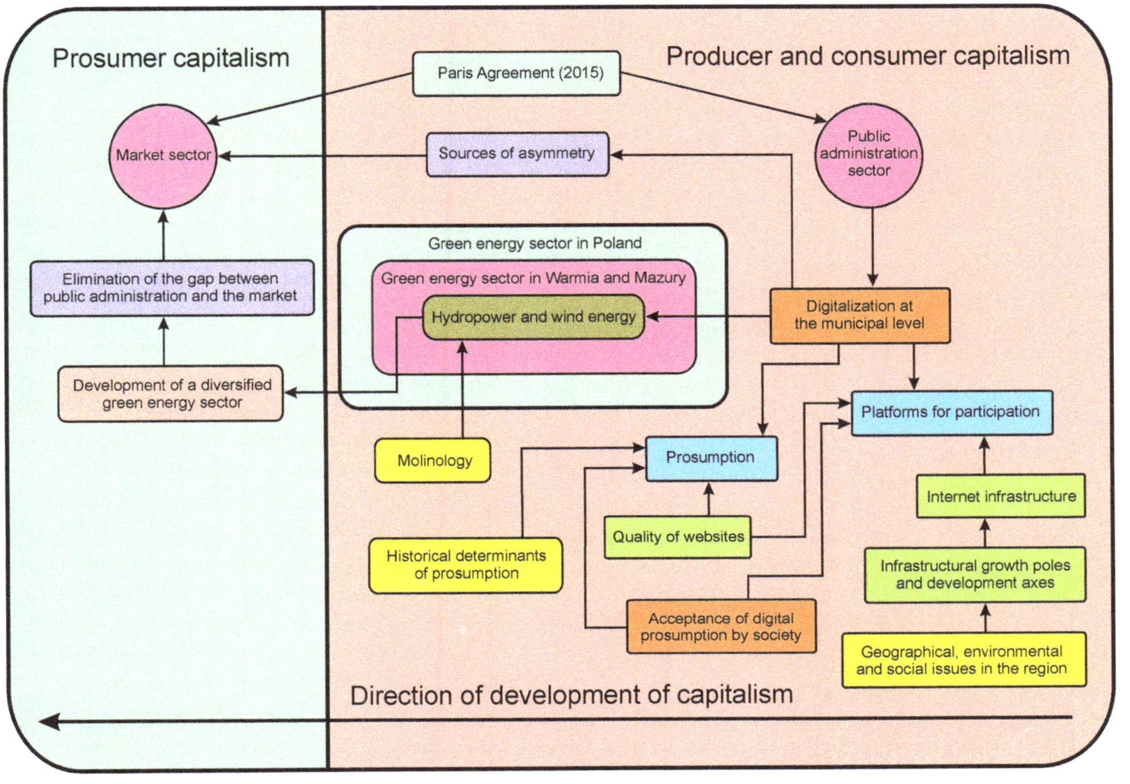

Figure 3. The first stage of research on the impact of citizens on the digitalization of municipal offices. Acceptance by the inhabitants of the Warmia and Mazury Province of digital prosumption in the local government sector was a prerequisite for the development of distributed generation.

In previous studies, the emphasis was placed on demonstrating that the inhabitants of Warmia and Mazury Province accept digital prosumption, which was to be a preliminary step towards a reform of public administration. The main issue was determining whether digital prosumption was acceptable to the local society. A positive answer was obtained, and at the same time, it was indicated that in order to accelerate the development of the prosumer power industry in the region, a change in the model of functioning of the local government sector was needed. Correspondence analysis was used to discover relationships between categorical variables in the database. The quality of municipal government websites was also studied using the binary method, and it was found that most of them do not play the role of social participation platforms.

This article presents the results obtained from an extended database on how digitalization can change operational and business models in local public administration for it to become the main driver in the development of energy prosumerism. In other words, this article seeks to determine whether the local government sector is able to release the prosumer potential of the society and incorporate it into a virtuous cycle involving feedback between electrification, decentralization, and digitalization (Figure 1). Previously, the focus was on determining the ability of the society to participate in the sphere of digital prosumption related to public administration, while in the current article the reverse dependence is examined, namely, whether municipal governments will be able to release the prosumption potential of citizens and direct it to the development of a regional prosumer power industry. A more advanced research method in the form of multiple correspondence analysis is another innovation compared to past research. This article found that, in Warmia and Mazury, citizens are the driving force of the energy transition, with the middle class having an outsized role [7]. The local government sector needs to be reformed in order to change from a barrier to the development of renewable energy into a catalyst for change.

7. Materials and Methods

7.1. The Research Hypothesis: Principles of Prosumption in the Contact Zones

The introduction of digitalization in contact zones must be preceded by a certain consensus between citizens and the local government sector, whose economic and administrative–legal basis should guarantee and stimulate economic development, with particular emphasis on the production of energy from renewable sources. It should probably not be done in the way of an experiment because, in the event of failure, it would generate significant costs but could be introduced as a proven and stable management principle. Such a solution has long been known in the market sector and is called prosumption. However, its implementation in the municipal government sector is slow and has encountered a number of obstacles, the most important of which is the low transparency of public administration. For this reason, there is a high asymmetry between the market sector and the public administration sector, which contributes to losses in economic growth at local, national, and international levels. It also negatively affects the development of energy production from renewable sources, thus delaying the energy transition. Without initiating prosumption in the local government sector, both contact zones, the human administrative legal contact zone and the more-than-human energy contact zone, will be dead and digitalization will not be possible, as it will not find a fulcrum there.

Prosumption is one of the most important business models of wikinomics [48]. Prosumers are a new type of economic entities who are not only consumers, but they are also actively involved in the design and creation of new products and services [49,50]. Their contribution to the whole business process is to provide unique ideas and non-standard solutions. Existing definitions of prosumption relate mainly to the market sector, while attempts to define this concept for the public administration sector are limited [51,52]. As part of the general concept of prosumption, ten detailed principles have been formulated, which are presented in Table 1 [34,53].

Table 1. The principles of prosumption.

Codes	Ten Principles of Prosumption
P.1	An official's efficiency of service
P.2	Expertise and competence of an official
P.3	The personal culture of an official
P.4	Technical conditions for providing of services
P.5	Availability of information, i.e., knowledge of where and how to deal with the matter
P.6	Consistent communication in the client–official relationship
P.7	How the service is performed as a factor that improves client–official relationships
P.8	A more favored attitude of the officials
P.9	Possibility of dealing with the matters at one branch of the office
P.10	The ability to deal with the matters online

In 2015, a direct survey was conducted in Poland. The aim was to rate the suitability of the ten principles of prosumption proposed by the authors (Table 1). The evaluation was made by the citizens of the Warmia and Mazury Province. When preparing the list of principles, the authors were guided by the knowledge of the most important elements of contact zones previously defined in the literature, in both theoretical and empirical references, economic and sociological determinants of efficient activities in communication between citizens and officials, and also issues related to the digitalization of public administration services. The present article examines the following research hypothesis:

society supports the idea of improving the local government sector by introducing digital prosumption and is ready to actively participate in this undertaking.

It can also be formulated in the form of the following equation:

$$\textit{Digitalization of contact zones} = \textit{ten principles of prosumption} + \textit{digital technology}. \quad (1)$$

From the point of view of the interactions taking place in the contact zones, answers will be important not only as to which rules of prosumption have been chosen by citizens but also as to which rules have not been chosen by them. The choices made are most often based on the knowledge possessed by the respondents, and the lack of response suggests that the respondents do not have adequate information. These shortcomings should prompt the municipal governments to fill the existing knowledge gaps of citizens, which will allow public administration to interact with people in order to jointly launch a virtuous cycle consisting of digitalization, decentralization, and electrification.

7.2. Multiple Correspondence Analysis

Multiple correspondence analysis is primarily used in research to examine issues with a number of categorical variables bigger than two [54–59]. It is frequently employed to analyze data from questionnaires [60]. This method is an extension of simple correspondence analysis and likewise it is used to study the co-occurrence of phenomena. This method uses the inner product of an indicator matrix called the Burt Table [61–63]. The table structure is based on frequency tables and consists of adding parts that represent the relationships of each variable to itself. The Burt Table is the symmetric matrix.

In recent years, multiple correspondence analysis has become increasingly popular in a variety of scientific disciplines, especially in medical and economic sciences. Regarding medical sciences, the usefulness of this method has been demonstrated in epidemiology [64], where it enabled in-depth research on epidemics of malaria [65,66]. Moreover, it has allowed researchers to achieve interesting results in the scope of cognitive aging [67]. It has also been used in studies on nonlinear relationships of the motor cortex and basal ganglia [68], and proved helpful in the identification of behavior patterns associated with overweight and obesity in adults [69]. In economic sciences, the multiple correspondence analysis is helpful in studying such issues as financial inclusion [70–73], multidimensional poverty [74–77], measuring the circular economy [78], and emergencies [79]. Based on results achieved with this method, the classification of residential electricity consumers has been developed [80]. Furthermore, this statistical method has considerable application potential in social sciences [81]. One interesting example is its use in exploring Nobel Prize data involving four categorical variables: the country of birth or residence of the laureate, the scientific discipline, the period in which the Nobel Prize was awarded, and the gender of the laureate [82]. Multiple correspondence analysis is also helpful for analyzing opinions of students about their school [83]. Moreover, it provides interesting knowledge on attitudes of Americans to life according to their marital status, people's opinion on abortion according to their religion and education, and association patterns developed by a mother and her infant during the first six months of life [84].

7.3. The Empirical Database

Empirical data collected during the direct survey were used to construct the 27 × 27 Burt Table. It was divided into three parts, which are presented in Tables 2–4. Part one of the Burt Table contains the number of respondents divided according to sex and age (Table 2). The second part contains the principles of prosumption that were not selected (symbol 0) or selected (symbol 1) by the respondents, taking the sex division of respondents into account (Table 3). The third part contains the principles of prosumption that were not selected (symbol 0) or selected (symbol 1) by the respondents, taking the age distribution into account (Table 4). The sampling was random.

Table 2. Number of respondents divided by age and sex.

Sex	Table of Observed Frequencies. Input Table (Row × Column): 27 × 27 Burt Table				
	AGE:0 (18–20)	**AGE:1 (21–35)**	**AGE:2 (36–45)**	**AGE:3 (46–60)**	**AGE:4 Up 60**
Women (W)	77	800	89	72	39
Men (M)	93	403	117	96	26

Table 3. The principles of prosumption that were not selected (symbol 0) and selected (symbol 1) by the respondents divided by sex.

Sex	Table of Observed Frequencies. Input Table (Row × Column): 27 × 27 Burt Table																			
	P1		**P2**		**P3**		**P4**		**P5**		**P6**		**P7**		**P8**		**P9**		**P10**	
	0	**1**	**0**	**1**	**0**	**1**	**0**	**1**	**0**	**1**	**0**	**1**	**0**	**1**	**0**	**1**	**0**	**1**	**0**	**1**
Woman (W)	456	621	428	649	767	310	957	120	755	322	640	437	466	611	799	278	780	297	703	374
Men (M)	326	409	242	493	461	274	597	138	432	303	481	254	255	480	463	272	474	261	419	316

Table 4. The principles of prosumption that were not selected (symbol 0) and selected (symbol 1) by respondents divided by age.

Age Division	Table of Observed Frequencies. Input Table (Row × Column): 27 × 27 Burt Table																			
	P1		**P2**		**P3**		**P4**		**P5**		**P6**		**P7**		**P8**		**P9**		**P10**	
	0	**1**	**0**	**1**	**0**	**1**	**0**	**1**	**0**	**1**	**0**	**1**	**0**	**1**	**0**	**1**	**0**	**1**	**0**	**1**
AGE:0 (18–20)	65	105	50	120	107	63	134	36	105	65	98	72	52	118	121	49	108	62	137	33
AGE:1 (21–35)	497	706	459	744	798	405	1052	151	797	406	711	492	483	720	822	381	817	386	726	477
AGE:2 (36–45)	92	114	83	123	152	54	170	36	152	54	163	43	76	130	161	45	156	50	118	88
AGE:3 (46–60)	94	74	55	113	123	45	149	19	90	78	104	64	82	86	111	57	129	39	102	66
AGE:4 Up 60	34	31	23	42	48	17	49	16	43	22	45	20	28	37	47	18	44	21	39	26

The following formula was used to determine the minimum sample size N_{min}:

$$N_{min} \geq \frac{(Z_{\alpha/2})^2}{4E^2}, \tag{2}$$

where $Z_{\alpha/2}$ is the number determined by the desired level of confidence and E denotes the margin of error. Usually, surveys assume that the confidence level is 95%, which results in $Z_{0.025} = 1.96$, and that the margin of error is $E = 0.03$. Based on the inequality (2), $N_{min} \geq 1067$ is obtained. In this study, the sample size equals $N = 1821$, which at the confidence level of 95% reduces the margin of error to $E \approx 0.023$.

7.4. Calculations

The three-part Burt Table (Tables 2–4) has become the basis for multiple correspondence analysis. Table 5 shows the quantitative characteristics of the Burt Table for fifteen dimensions: singular values, eigenvalues, percent of inertia, cumulated percent, and chi-square distances. The values for the first three dimensions have the largest contribution to inertia. Further dimensions explain the ever-smaller percentage of total inertia. The essence of multiple correspondence analysis is to move the relationships between variables from the original multidimensional space to the space of smaller number of dimensions as accurately as possible. Typically, the reduction of the dimension of a studied issue is carried out up to two or three dimensions, which allows the problem to be graphically analyzed.

Table 5. Quantitative characteristics of Burt Table.

Number of Dimensions	Eigenvalues and Inertia for All Dimensions. Input Table (Row × Column) = 27 × 27 Burt Table. Total Inertia = 1.2500				
	Singular Value	Eigenvalue	Percent of Inertia	Cumulated Percent	Chi-Square Distance (χ^2)
1	0.452265	0.204544	16.36351	16.3635	5426.368
2	0.337560	0.113946	9.11572	25.4792	3022.898
3	0.334371	0.111804	8.94430	34.4235	2966.056
4	0.309036	0.095503	7.64026	42.0638	2533.615
5	0.298270	0.088965	7.11722	49.1810	2360.169
6	0.288973	0.083505	6.68042	55.8614	2215.320
7	0.285695	0.081622	6.52974	62.3912	2165.352
8	0.282019	0.079535	6.36279	68.7540	2109.991
9	0.272199	0.074092	5.92736	74.6813	1965.596
10	0.257154	0.066128	5.29027	79.9716	1754.326
11	0.250720	0.062860	5.02883	85.0004	1667.629
12	0.237375	0.056347	4.50775	89.5082	1494.831
13	0.222021	0.049293	3.94346	93.4516	1307.707
14	0.205041	0.042042	3.36334	96.8150	1115.329
15	0.199532	0.039813	3.18504	100.0000	1056.202

In multiple correspondence analysis, the distribution of the Burt matrix by singular value decomposition method is essential. The eigenvalues obtained from the analysis of the Burt matrix are a better approximation of the inertia explained by the factors than the eigenvalues of the indicator matrix. Inertia is analogous to the concept of the moment of inertia used in physics, which denotes the mass of body product and the squared distance to the axis of rotation. In other words, inertia is the overall value of chi-square statistic divided by the total sum of the observations. Table 5 shows that the issue of prosumption in the public administration sector is fifteen-dimensional. In order to visualize the structure of links between categorical variables in a space with a smaller number of dimensions as closely as possible, the directions of the axes (dimensions) should be determined so that subsequent dimensions explain the smaller part of the total inertia (or the total value of chi-square statistic). The cumulative percentages of inertia explained by the dimension of the assumed space are used to determine the number of dimensions. In this case, the three-dimensional and two-dimensional space is adequate to represent the original structure of variables. Significantly, the chi-square metric is used in the study.

Table 6 shows the coordinates of twenty-seven columns, and it also includes the statistics of the quality of the solution. This information is needed to graphically illustrate the co-occurrence of such phenomena as sex and age of respondents with the non-selection (symbol 0) or the selection (symbol 1) of ten postulated principles of prosumption applicable in the local government sector.

In Table 6, separately for each of the 3 dimensions, there are 27 points representing all the categorical variables present in the studied issue. The distances between these points were calculated using the chi-square metric. The mass of the point indicates its validity in the conducted research study. It was found that the largest masses were points P4:0 (0.071468), P8:0 (0.058039), and P9:0 (0.057671). Therefore, the non-selection by some of the respondents of principles P4 (technical conditions for providing of services), P8 (a more favorable attitude of the officials), and P9 (possibility of dealing with the matters at one branch of the office) may indicate that they attach less importance to digital prosumption. This might prove that they are to some extent unaware of the immense prosumption power that lies in them. This may also partly confirm Ritzer's statement that prosumers do not realize that they can play a productive role in the economy and that companies have already made a profit out of it [85,86].

Table 6. Coordinates of columns and the statistics of the quality of the solution.

Column Coordinates and Contributions to Inertia. Input Table (Row × Column) = 27 × 27 Burt Table. Total Inertia = 1.2500													
Name of Row	Number of Rows	Mass	Quality	Relative Inertia	Coordinates			Dimension 1		Dimension 2		Dimension 3	
					Dimension 1	Dimension 2	Dimension 3	Inertia	Cos^2	Inertia	Cos^2	Inertia	Cos^2
W	1	0.049531	0.418352	0.027042	−0.201513	0.142077	−0.474037	0.009833	0.059502	0.008775	0.029578	0.099551	0.329271
M	2	0.033802	0.418352	0.039625	0.295278	−0.208186	0.694610	0.014409	0.059502	0.012857	0.029578	0.145872	0.329271
AGE:0	3	0.007818	0.031209	0.060412	0.293364	−0.441625	0.142650	0.003290	0.008910	0.013382	0.020192	0.001423	0.002107
AGE:1	4	0.055326	0.316010	0.022406	0.004152	0.118305	−0.382050	0.000005	0.000034	0.006796	0.027647	0.072229	0.288329
AGE:2	5	0.009474	0.250850	0.059088	−0.213760	−0.927081	1.024933	0.002116	0.005861	0.071460	0.110244	0.089015	0.134745
AGE:3	6	0.007726	0.177235	0.060486	−0.056762	0.671453	1.131503	0.000122	0.000329	0.030570	0.046072	0.088476	0.130834
AGE:4	7	0.002989	0.011323	0.064275	−0.019942	0.168154	0.525024	0.000006	0.000015	0.000742	0.001052	0.007370	0.010256
P1:0	8	0.035964	0.558106	0.037896	−0.437269	0.457399	0.578518	0.033618	0.145166	0.066032	0.158840	0.107658	0.254099
P1:1	9	0.047369	0.558106	0.028771	0.331985	−0.347268	−0.439225	0.025524	0.145166	0.050133	0.158840	0.081736	0.254099
P2:0	10	0.030813	0.338779	0.042016	−0.368145	−0.457242	−0.482534	0.020417	0.079515	0.056536	0.122660	0.064170	0.136604
P2:1	11	0.052520	0.338779	0.024650	0.215987	0.268259	0.283098	0.011978	0.079515	0.033169	0.122660	0.037648	0.136604
P3:0	12	0.056475	0.443814	0.021486	−0.444083	0.110580	0.040338	0.054450	0.414680	0.006061	0.025712	0.000822	0.003421
P3:1	13	0.026858	0.443814	0.045180	0.933791	−0.232522	−0.084820	0.114495	0.414680	0.012744	0.025712	0.001728	0.003421
P4:0	14	0.071468	0.478745	0.009492	−0.281691	0.010469	−0.004813	0.027725	0.477945	0.000069	0.000660	0.000015	0.000140
P4:1	15	0.011865	0.478745	0.057174	1.696698	−0.063059	0.028991	0.166995	0.477945	0.000414	0.000660	0.000089	0.000140
P5:0	16	0.054590	0.417843	0.022995	−0.435559	−0.133878	−0.111243	0.050631	0.360301	0.008587	0.034040	0.006042	0.023503
P5:1	17	0.028744	0.417843	0.043672	0.827214	0.254260	0.211273	0.096159	0.360301	0.016308	0.034040	0.011475	0.023503
P6:0	18	0.051554	0.599893	0.025423	−0.276565	−0.464062	0.279178	0.019279	0.124086	0.097436	0.349365	0.035940	0.126442
P6:1	19	0.031779	0.599893	0.041244	0.448667	0.752842	−0.452907	0.031275	0.124086	0.158068	0.349365	0.058304	0.126442
P7:0	20	0.033159	0.581321	0.040140	−0.475084	0.805986	−0.065752	0.036589	0.149159	0.189038	0.429304	0.001282	0.002857
P7:1	21	0.050175	0.581321	0.026527	0.313964	−0.532645	0.043453	0.024180	0.149159	0.124928	0.429304	0.000847	0.002857
P8:0	22	0.058039	0.280895	0.020235	−0.332833	−0.019703	0.106077	0.031433	0.254185	0.000198	0.000891	0.005841	0.025819
P8:1	23	0.025294	0.280895	0.046431	0.763702	0.045210	−0.243398	0.072125	0.254185	0.000454	0.000891	0.013403	0.025819
P9:0	24	0.057671	0.287232	0.020530	−0.345172	−0.093093	0.001282	0.033593	0.267753	0.004386	0.019476	0.000001	0.000004
P9:1	25	0.025662	0.287232	0.046137	0.775709	0.209208	−0.002882	0.075493	0.267753	0.009857	0.019476	0.000002	0.000004
P10:0	26	0.051600	0.230009	0.025386	−0.258478	−0.132888	−0.238704	0.016854	0.108640	0.007997	0.028715	0.026298	0.092654
P10:1	27	0.031733	0.230009	0.041280	0.420307	0.216087	0.388154	0.027407	0.108640	0.013004	0.028715	0.042762	0.092654

Table 6 also shows the quality of all points after the reduction of the dimension of the studied issue. This is to assess whether they are properly represented in the selected coordinate system. In other words, the distance between points should be approximated with satisfactory accuracy in the space with fewer number of dimensions. The quality of a given point is the square quotient of its distance from the center of the selected coordinate system and the square of the same distance in the coordinate system with a maximum number of dimensions (here fifteen). The definition of quality implies that this measure is a number that accepts values between 0 and 1. The closer the unity is, the better the point is represented. If the quality of a given point is less than 0.1, then it is not very well represented in the selected coordinate system. In Table 6, each principle of prosumption has a quality value, regardless of whether the respondents chose it or not. Table 6 shows that the points representing the principles of prosumption—P.6 (0.599893), P.7 (0.581321), and P.1 (0.558106)—had the best quality. Thus, the points associated with P.6 (consistent communication in the client–official relationships), P.7 (how the service is performed as a factor that improves client–official relationships), and P.1 (an official's efficiency of service) are best represented in three-dimensional and two-dimensional spaces. The lowest quality, below 0.1, was only seen in points AGE:4 (0.011323) and AGE:0 (0.031209), so these age groups are moderately represented in the selected coordinate system. The quality of a given point may also be considered as its share in the overall inertia at the chosen number of dimensions.

The next indispensable concepts in the analysis of Table 6 are the relative inertia and the relative inertia for each dimension. The relative inertia represents the proportion of a given point in the total inertia, which is measured irrespective of the number of selected dimensions. The following age groups had the biggest influence on the overall inertia: AGE:4 (0.064275), AGE:3 (0.060486), and AGE:0 (0.060412), although the first and third groups had slightly lower quality. This means that the opinions of these respondents on the principles of prosumption in the local government sector are important for the purpose of research study. This is consistent with common sense, as these groups include both people who have personal and professional experience and people who have just started their professional life (but who are important from the perspective of the future). The columns presenting the relative inertia for each dimension contain the relative contributions of the individual points to the inertia, which is generated by the given dimension. Therefore, these values are given separately for each dimension.

The last value given in Table 6 is cosine^2, which should be understood as quality or squared correlations with each dimension. These columns contain quality measures for each point in relation to each dimension. The sum of the values of these columns calculated for all dimensions is equal to the total quality of a given point. For example, consider the P.1 principle of prosumption (regardless of whether or not it is selected by respondents). By summing 0.145166 + 0.158840 + 0.254099 we obtain 0.5581, which is the value found in the column describing the quality of the P.1 point. The cosine^2 values can also be interpreted as the proportion of the dimension in the inertia point. This share is equal to the squared cosine value of the angle between the vector determined by a given point and the corresponding axis of the coordinate system.

Based on Table 6, the points representing all 27 categorical variables in three-dimensional and two-dimensional spaces are drawn. This table contains a lot of information that is not directly transferred to charts. Therefore, the graphical analysis shown below (Figures 4–7), which is based solely on the coordinates of the points representing the variables, should be supplemented by the other information included in Table 6. This obviously provides a very broad possibilities of interpretation of the obtained results. This paper is limited to the findings that are most strongly linked to the objectives of the research, and therefore to the importance of prosumption in the public administration and the possible role of this sector of the economy in accelerating the energy transition.

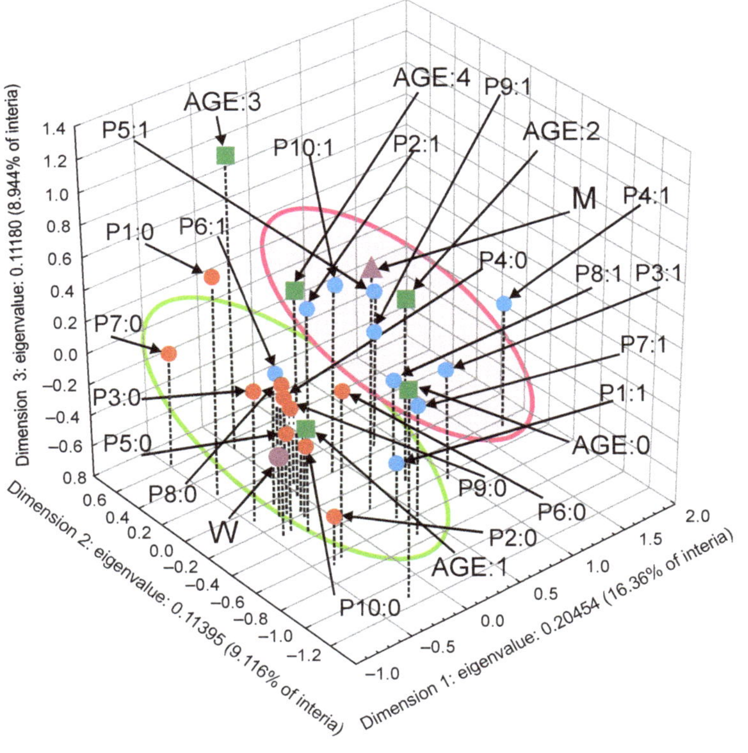

Figure 4. Three-dimensional plot of column coordinates—dimension 1 × 2 × 3. Input table (rows × columns): 27 × 27 Burt Table.

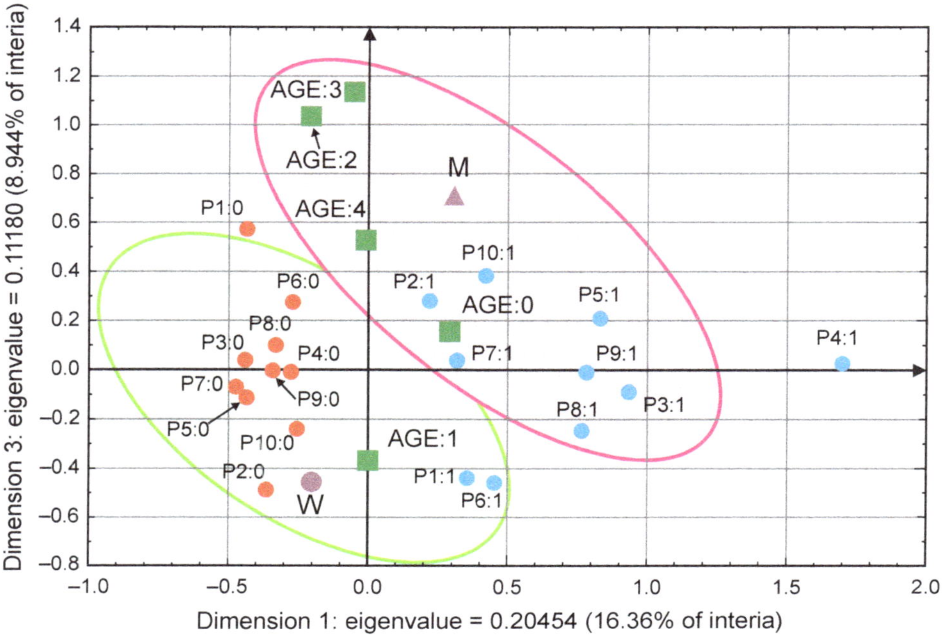

Figure 5. Two-dimensional plot of column coordinates—dimension 1 × 3. Input table (rows × columns): 27 × 27 Burt Table.

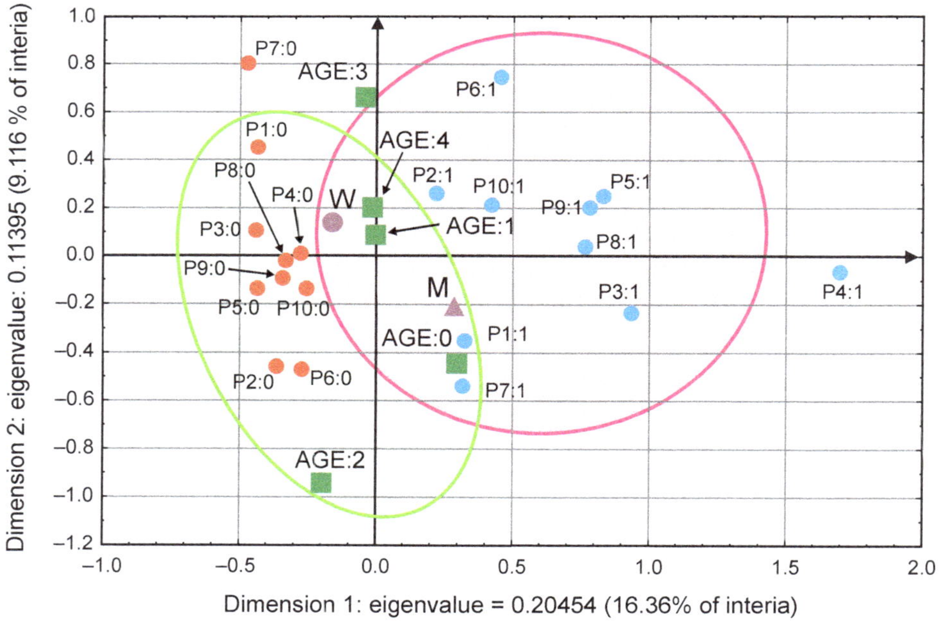

Figure 6. Two-dimensional plot of column coordinates—dimension 1 × 2. Input table (rows × columns): 27 × 27 Burt Table.

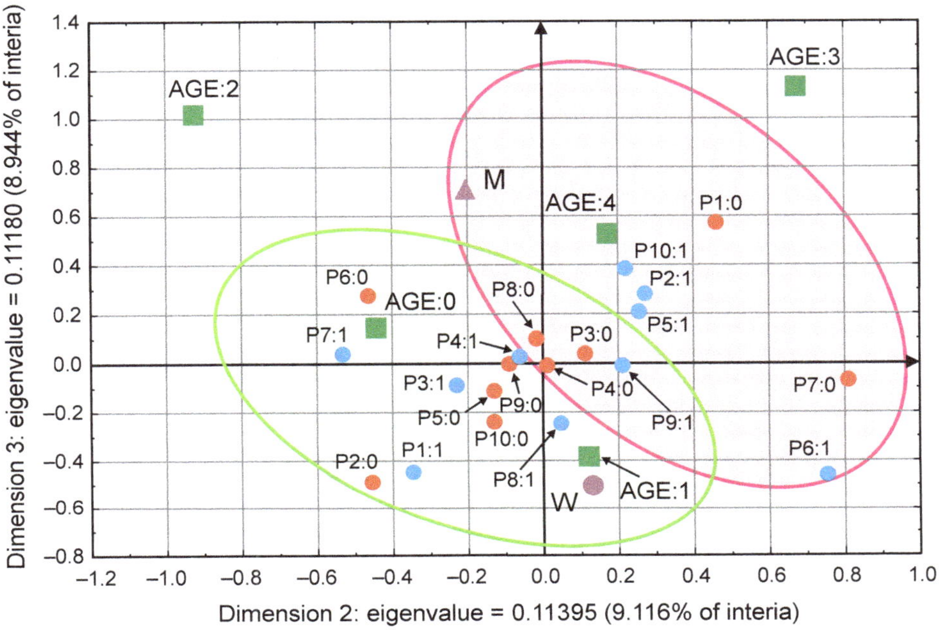

Figure 7. Two-dimensional plot of column coordinates—dimension 2 × 3. Input table (rows × columns): 27 × 27 Burt Table.

8. Results

8.1. Digital Prosumption as the Basic Principle of the Functioning of Contact Zones

The essence of multiple correspondence analysis is the reduction of the dimension of the studied issue, while maintaining the exact structure of the links between the categorical variables. As can be seen from Table 5, the issue of prosumption in the public administration is by its nature fifteen-dimensional. In order to present them graphically, it is necessary to reduce the number of dimensions from fifteen to three or two. Figures 4–7 present the studied issue in three-dimensional or two-dimensional sections, while ensuring the highest possible accuracy of preserving the original structure of the relationships between the variables.

Figure 4 shows the coexistence of phenomena in a three-dimensional space. If the respondents did not choose a particular principle of prosumption, the point representing that fact was marked with a red circle, and if selected the corresponding point was marked with a blue circle. The age groups of respondents are symbolized by the green squares. The sex of respondents was colored purple—circle for women and triangle for men. Two ellipses—purple and green-yellow—were used to visualize the respective structures in the data. Note that Figure 4 has an illustrative character. More detailed information is provided by simultaneous analysis of the three two-dimensional cross-sections shown in Figures 5–7.

In Figure 5 we have a more accurate confirmation of the observations made in the three-dimensional space (Figure 4). All the respondents, regardless of sex, who belong to groups AGE:0, AGE:2, AGE:3, and AGE:4 are aware of the fact that the introduction of prosumption in the local government sector is beneficial both to households and to businesses. The similar phenomenon—asymmetry in the evaluation of individual principles of prosumption by women and men—can be seen in the two-dimensional cross-section shown in Figure 6. The most interesting information, however, is shown in Figure 7. It appears that prosumption attitudes are widespread in the age groups AGE:0, AGE:1, and AGE:4

regardless of sex. Thus, the introduction of prosumption in the municipal governments will be welcomed and expected by the public.

It is possible to analyze Figures 4–7 in more detail. From the perspective of the digital prosumption, the most important principles of prosumption are: P4:1 (technical conditions for providing service), P5:1 (availability of information), P6:1 (consistent communication), P7:1 (how the service is performed), P9:1 (possibility of dealing with the matters at one branch of the office), and P10:1 (the ability to deal with the matters online). In Figure 4, all of these principles, apart from P4:1, are in the areas defined by ellipses. This rule is repeated in Figures 5 and 6. In Figure 7, all the principles, including P4:1, are inside the ellipses.

There is a very interesting situation connected with point P4:1. Figures 5 and 6 show that this is an isolated point. However, the calculations in Table 6 show that the P4:1 principle has a relatively high impact on the overall inertia (0.057174). It is therefore important for the point of view regarding the aim of this study. Moreover, the point that represents it has a relatively high quality (0.478745). At the same time, it is clear that the contribution of this point to the inertia generated by the first dimension is large (0.166995), significantly higher than the corresponding contribution to the inertia generated by the second (0.000414) and third dimension (0.000089). It also turns out that the quality of this point for the first dimension—represented by cosine^2—is 0.477945, and therefore it is much higher than for the other two dimensions (0.000660 and 0.000140). Figures 5 and 6 show that P4:1 has the highest correlation with the axis representing the first dimension, since the vector joining the origin point of the coordinate system to point P4:1 has the smallest slope in relation to that axis. In other words, point P4:1 is literally on the axis corresponding to the first dimension. Therefore, the rule P4:1 must represent some hidden property. Figure 5 and Table 6 show that similar properties are shown by points P9:1 and P7:1 in relation to the first dimension. In contrast, Figure 6 shows that the vector inclination angles P5:1, P9:1, and P.10:1 (the variables related to the digital prosumption) to the axis of the first dimension are also small. In turn, in Figure 7, point P4:1 is near the origin of the coordinate system, so it creates something similar to a prosumption standard. This information is sufficient to state that the hidden dimension correlated with the first axis (dimension) must be the digital prosumption. Thus, the respondents point to very important information related to prosumption in the local government sector—the only acceptable form of prosumption is digital prosumption. This fully confirms Ritzer's conclusion that digital prosumption is the most balanced form of prosumption [85,86]. It seems that its proper application in the municipal governments can be an effective tool to offset the negative effects of uncontrolled prosumption in the market sector. These threats arise mainly from economic development being based on the free-of-charge work of prosumers.

Table 5 shows that the first three dimensions have the greatest share in explaining inertia; hence, the research study was conducted in a three-dimensional space (Figure 4). Figures 4–7 illustrate the results obtained using multiple correspondence analysis. On the basis of Figure 4, it can be said that there is a great difference between men and women in their approach to prosumption in the public administration sector. Nonetheless, both approaches have complementary character. The highest number of non-selected answers is characteristic for women in the age group AGE:1. However, the aforementioned group see the importance of the principles P1:1 and P6:1. For men and women in groups AGE:0, AGE:2, and AGE:4, almost all the principles of prosumption are valid. The exception is P4:1, which in Figures 5 and 6 is an isolated point, but contains a hidden dimension—digital prosumption. A more detailed analysis of co-occurrence phenomena is possible using the two-dimensional cross-sections (Figures 5–7). The general conclusion is that the vast majority of respondents expect the introduction of postulated principles of prosumption to the public administration sector. At the same time, they are aware that the prosumption should be supported by ICT, and should therefore take the form of digital prosumption. This is evidenced by the central position of the P10:1 point in all the figures. This is probably due to the fact that respondents already have some experience with the digital prosumption taken from the market sector. They also seem to be aware of the importance of the P4:1

principle (technical conditions for providing of services). As shown in Figures 4–6, it is seemingly an isolated point, but in Figure 7 it is located near the origin of the coordinate system. This rule may therefore be considered a certain standard of digital prosumption to some extent emphasized by the respondents.

The conclusions from the research study on prosumption in the public administration clearly indicate that in this sector of the economy, respondents will only accept digital prosumption. The implementation of prosumption must therefore be supported by digital technologies. Thus, the research hypothesis concerning the implementation of digitalization in contact zones was verified positively. Society voiced their opinion, supporting the selected prosumption principles, although the survey revealed some gaps in their knowledge as certain prosumption principles were not selected. The latter are critical for initiating meetings and interactions between citizens and the public administration. Citizens' opinions are already known, which means that the development of solutions promoting climate protection and generation of energy from renewable sources in the more-than-human energy contact zone will require specific measures to be taken by the municipal governments in the near future.

8.2. Characteristic Features of Warmia and Mazury Contact Zones

The Warmia and Mazury Province is a region with relatively weak socio-economic potential in both Poland and the European Union [87,88]. The level of information infrastructure there is low, but changes in the synthetic ICT index determined for the province in comparison with the national average indicate the occurrence of a catching-up effect [89]. A negative phenomenon of shrinking of municipalities is observed in the region due to adverse demographic changes. A decrease in the population has occurred in more than 80% of municipalities in the province, which causes unfavorable economic effects in the form of a decrease in the budgetary revenues [90]. One of the most interesting initiatives undertaken in this region is the accession of many cities to the Polish National Cittàslow Network [91]. This serves the development of prosumer attitudes in society important from the point of view of the local government sector, because one of the postulates contained in the organization's program is precisely to improve the functioning of this sector.

In 2015, the construction of a fiber-optic backbone distribution network was completed in Warmia and Mazury, with the aim to enable broadband Internet access for residents, entrepreneurs, and government institutions from this region. This investment was accomplished as part of a public–private partnership, which signifies that the province's authorities must have considerable experience in this type of venture [92]. Therefore, the development of IT infrastructure in this region should be continued on similar principles in the future.

All the rules of prosumption not chosen by citizens remain a potential challenge for the local government sector. A closer look at this problem indicates the existence in the human administrative legal contact zone of all the features of a typical contact zone, such as autoethnographic texts, transculturation, colonialism, anti-conquest, radical heterogeneities, fundamental inequalities, coercion, violence, ambiguities, insurmountable conflicts, and the struggle for interpretive power.

A survey is a kind of autoethnographic text, a response to many earlier official documents, which are, in fact, ethnographic texts in which municipal governments present their views on citizens and perceptions of the problems of local communities. A large number of unselected rules of prosumption prove that people rarely meet with representatives of public administration, and even if they do, there are fundamental inconsistencies between the idioms of the metropolis and indigenous idioms. The only source of citizens' knowledge about prosumption and the possibilities of its application in the local government sector is the market sector, whereas municipal governments do not take sufficient measures to adapt to the requirements of the modern economy nor support the initiatives of local communities. The response of local public administration to such research as presented in this article leaves much to be desired; hence, the mirror dance which Pratt mentions cannot yet take place. Two facts can be cited in support of these conclusions. The first is a very worrying demographic phenomenon occurring in Warmia and Mazury, which has already been mentioned. In search of better living and working conditions, people have moved to other regions of Poland or even to other European or non-European countries. The second fact is the large potential for the development of micro- and small hydropower plants in the region [34]. Unfortunately, it is not utilized due to the lengthy administrative and legal procedures. According to the current regulations, hydroelectric power stations are classified as economic activities with a significant potential impact on the environment. In consequence, a potential investor must obtain not only an environmental approval, but also an approval for building conditions, a water permit, an approval for the acquisition of rights to use real estate owned by the State Treasury, and a building permit. Briefly, the construction of a small hydropower plant with a nameplate capacity in the tens of kilowatts requires the same number of permits and administrative approvals as the construction of a larger hydropower plant with a capacity of several megawatts [93]. It usually takes two to five years to settle all administrative and legal issues before the investor can proceed to the more-than-human energy contact zone [94].

Human administrative–legal contact zones are also places of transculturation, as citizens are forced to use what the metropolitan culture offers, which in this case is represented by the municipal governments. To explain this problem, reference must be made to the historical context. The lands of contemporary Warmia and Mazury used to belong to East Prussia, part of the German Empire and then the Third Reich, and they were incorporated into Poland as part of the Recovered Territories by the 1945 Potsdam Conference. In addition, until 1989, Poland had a centrally planned economy, and the whole country was in the sphere of influence of the Soviet Union. Only the economic changes initiated in the early 1990s resulted in the emergence of a capitalist economy. In the twentieth century, the province underwent German colonialism, which was later replaced by Soviet colonialism. In both cases, local governments were far from transparent, mostly based on a command-and-control system of management. The local public administration still bears the marks of these colonial influences, as they have penetrated deeply into the regional culture. Warmia and Mazury still has remnants of the centralized public administration of the Third Reich and the socialist economy imposed immediately after the Second World War. A common element of both these colonialisms was the enormous power of local authorities, which allowed for the effective suppression of any civic initiatives. This perfectly explains the current difficulties in adapting municipal governments to the requirements of the digital economy. Over the past few decades, citizens have tried to appropriate certain elements of dominant cultures, German and Soviet, and use them under a centrally planned economy, but after the advent of the digital economy, these have turned out to be contrary to the new market logic. Therefore, there has been no permanent cooperation between two metropolitan cultures and the Polish indigenous culture and, as a result, no new economic and legal reality was created. During the process of cultural change and mutual give and take, the two sides of the equation were not modified, as Ortiz assumed, but inequalities in many dimensions were created and increased, so in this case, we can only talk about failed transculturation. The remains of both colonialisms are visible in the survey in the form of the rules of prosumption which were not selected by citizens.

The survey is also a reflection of conquest and anti-conquest. The history of Warmia and Mazury over the last several hundred years is extremely complex, which to some extent still affects the current situation of the region. These areas were previously part of Royal Prussia and the Duchy of Prussia and, as such, belonged to the Crown of the Kingdom of Poland, the Duchy of Prussia remaining in fiefdom dependence [95,96]. The Treaties of Wehlau and Bromberg of 1657 led to the loss of the Duchy of Prussia, while the annexation of Royal Prussia by the Kingdom of Prussia took place in 1772 as a result of the First Partition of Poland. It can therefore be said that the region was lost as a result of conquest and remained so until the end of World War II. In the years 1945–1989, Warmia and Mazury, as well as Poland at large, had a command economy, which for the local society was an alien system of management imposed by the Soviet Union. During this period, the phenomenon of anti-conquest was perfectly visible in the region, as a representation strategy including both gestures of innocence and the simultaneous pursuit of hegemony was pursued. On the one hand, the region found itself again within the borders of Poland, but obedience to the colonizer ruled out any major changes in the social, economic, and legal spheres.

The current activities of society in the field of energy transition, particularly participation in the virtuous cycle of digitalization, decentralization, and electrification, may require the use of administrative–legal tools and discourses both within and outside formal institutional channels. In some cases, needing to protect the environment may force citizens to correct the mistakes of local governments through marches, demonstrations, rallies, pickets, strikes, roadblocks, or other forms of civil disobedience. In Poland, in recent years, two such cases in the form of grassroots movements deserve special attention, as they show how a local protest may transition into proactive energy policy formation at the national level. In the first case, citizens of the village of Żurawlów located in the Zamość region stopped the prospecting and extraction of shale gas. The second is the fight against air pollution in Kraków [41]. It is true that both examples come from outside Warmia and Mazury, but they are worth mentioning here for the sake of comparing contact zones in the studied province with the equivalents existing in other regions of the country.

8.3. The Rationale for the Need for Further Comparative Research

A prosumer sentiment survey showed that many of the prosumption principles listed in Table 1 were not chosen by respondents. Their ignorance and uncertainty in this regard indicate that something interesting may be happening in the contact zones in Warmia and Mazury. The questionnaire survey identified in this region some elements of contact zones known from the humanistic, social, and environmental literature cited earlier, such as ethnographic and autoethnographic texts, transculturation, and inconsistencies between the idioms of the metropolis and indigenous idioms. To end the empirical research at this point would mean that it would remain significantly incomplete. This has necessitated the introduction of two additional sections in the article (9 and 10), which will ensure the completeness of the research presented here. They are needed for the following reasons:

1. The issue of the impact of contact zones on energy transition is addressed here for the first time in science, which means that there is a lack of a frame of reference necessary to evaluate the results of the survey. To date, no other scientific papers have addressed this topic. The only paper is our previous study, which we mention in Section 6, but this insufficient, as its goals were somewhat different.
2. Explaining the reasons why many prosumption principles have not been chosen by citizens points to the need to find a reference system in the form of other local contact zones, preferably in Poland. Such zones should be developmental in nature and thus potentially have an impact on national energy policy. This will enrich the interpretation of the survey results. Two such contact zones were found in Żurawlów and Kraków. Without additional comparative analysis of all three contact zones, it would not be possible to draw up a list of their characteristics, which are critical for energy transition.
3. A full explanation of the importance of survey research is not possible without reference to such factors as the socio-economic history of the region, its current level of development, and specific features, as well as geographical and environmental conditions.

9. Two Selected Reference Systems: The Human Administrative Legal and the More-than-Human Energy Contact Zones Located Outside Warmia and Mazury

9.1. The Fight of Farmers from Żurawlów against Shale Gas Extraction

In the opinion of the Polish Geological Institute, the prospective resources (onshore and offshore) of shale gas in Poland range from 346 to 768 Bcm (billion cubic meters) [97] (p. 4). Some of the deposits are located in the vicinity of Żurawlów, located in the municipality of Grabowiec, Zamość District, a village of only 96 inhabitants, including children and the elderly. Between 3 June 2013 and 7 July 2014, residents and farmers from the vicinity protested against the activity of Chevron Corporation, an American multinational energy corporation, which intended to prospect for shale gas in the nearby areas [98]. By taking over the local company Lublin Energy Resources, Chevron succeeded in obtaining license number 30/2007/p, issued in a non-bidding procedure by the Ministry of Environment on 6 December 2007, for the exploration and prospection of deposits of coalbed methane and natural gas in the Grabowiec area. The licensed area of 1195 km^2 included the city of Krasnystaw and areas in 18 municipalities, including Grabowiec, Miączyn, Sitno, Skierbieszów, and Zamość [99] (pp. 20–24). The protest was caused by the numerous threats to the environment and life of citizens associated with the extraction of shale gas by hydraulic fracturing, which uses toxic chemical compounds and can release naturally occurring radioactive material. Residents feared the loss of their farms, which provided them with livelihoods, damage to houses and roads, as well as the irreversible degradation of the unique ecosystem and adverse changes in its biodiversity. Several dozen meters from the area leased by Chevron is the Wolica Valley, named for the Wolica River flowing through this area, constituting the right-hand tributary of the Wieprz River, which covers 938.28 ha and is a special area of habitat protection belonging to the Natura 2000 network [100]. Aware of the risks associated with fracking, citizens blocked access to the site where the exploratory drilling rig was to be built with tractors and agricultural machinery and then proceeded to continuously occupy it. As it soon turned out, Chevron did not have a sufficient legal basis for the actions that it intended to take in Żurawlów [99] (pp. 108–116). On one of the protesters' banners appeared an inscription in the form of an equation:

$$\text{Fracking} = \text{death of agriculture.} \tag{3}$$

Another banner indicated that the inhabitants of Żurawlów were keenly aware of international energy politics: *Chevron + Gazprom are a deadly threat to energy security*. In this way, they probably also pre-empted accusations of acting in favor of the Russian Gazprom corporation. The farmers' protest quickly gained publicity, not only in Poland but also abroad [101,102]. Billboards appeared in the center of Warsaw, which read: *Żurawlów in defence of land and water against the world's most toxic company. The fight continues. Now* [103]. In addition, support for the inhabitants of Żurawlów came from all over the world, and from the first days of the blockade they were supported by French committees and associations fighting against shale gas [104]. The protesters were also supported by Catholic priests from nearby parishes [105]. After 400 days, the occupation protest was successful, Chevron left the leased land, and the residents saved their soil, as well as drinking water resources including the Major Underground Water Reservoirs in Poland numbered 405, 406, and 407 [106] (pp. 30, 275–281). This action was the longest local protest against the exploration and exploitation of shale gas deposits in the world [107]. It was also a valuable lesson given by citizens to central and local authorities, who had to seriously revise their existing energy policy.

9.2. Citizen Initiatives against Air Pollution in Kraków

Air pollution is a serious problem in many cities around the world, and low emission is one of its major sources. Low emission refers to the emission of products of the combustion of solid, liquid, and gaseous fuels from sources of emission located at a height of no more than 40 m. Depending on the sources, three main types of emissions can be distinguished: from the generation of heat in household boilers and local boiler facilities, from transportation, and from industrial sources. Low emission is composed of suspended particulates PM10 and PM2.5, sulfur dioxide SO_2, nitrogen oxides NO_x, heavy metals, polycyclic aromatic hydrocarbons, and dioxins [108]. In large concentrations, they pose a threat to human health and life. In many large urban agglomerations in Poland, suspended particulates PM10 and PM2.5 are particularly hazardous because their daily allowed levels are exceeded. As regards PM10, its daily permissible threshold is 50 μg/m^3 and it may not be exceeded more than 35 times in a calendar year, while the permitted annual mean value is set at 40 μg/m^3. For PM2.5, the limit to be achieved by 1 January 2020 was 20 μg/m^3 (previously 25 μg/m^3). In 2019, the alert threshold for PM10 was lowered from 300 to 150 μg/m^3, while the information threshold went down from 200 to 100 μg/m^3, both being daily average values [109].

Among the Polish cities, Kraków stands out as the worst case, with the concentrations of suspended particulates PM10 in 1998–2008 approximating 70–95 μg/m^3 in a year [110] (p. 11). In 2017, the average yearly concentration of PM2.5 in that city was the highest in Poland, and equaled 28.4–40.1 μg/m^3 depending on the location of the measuring station [111] (p. 17). The situation has now improved slightly, but the thresholds are still exceeded and Kraków has been one of the world's most badly polluted cities for years. As a result, a civic organization called the Kraków Smog Alert was launched in December 2012. It was the first anti-smog civic movement in Poland [112]. The Kraków Smog Alert began to inform the city's residents, via social media, about air quality measurements commissioned by the local authorities and about threats to health due to air pollution. The subsequent measures included anti-smog campaigns and marches, whose main objective was to make citizens and local governments aware of the problem they are faced with and to suggest the best solutions [113]. Since the findings reported by the local authorities suggested that the main source of air pollution in the city was low emission, the movement recommended the imposition of some restrictions on heating buildings with low quality coal or wood [112]. These steps resonated increasingly within the community. While the first anti-smog march gathered around 300 people, the next one attracted around 2000 participants [114]. Over time, the Kraków Smog Alert has gained fame across Poland. It initiated the foundation of the Polish Smog Alert in 2015, which associates over 40 local movements from all over the country [112]. These initiatives have been noticed by foreign mass media [115–117].

The major successes of the Kraków Smog Alert include persuading local governments to adopt the country's first anti-smog resolutions that aimed at improving the air quality in Kraków and in the entire Małopolska Province [112,118]. Consequently, a total ban on burning coal and wood in boilers, stoves and fireplaces was imposed in Kraków on 1 September 2019, leaving gas and light fuel oil as the only allowed fuels now. The regulations are less stringent in the Małopolska Province. The burning of coal sludge and coal flotation concentrates has been banned since 1 July 2017, while wood and biomass have been allowed as long as their moisture content does not exceed 20%. Moreover, coal- and wood-fired boilers or a wood-burning fireplace may be used if their emission parameters comply with the EU eco-design regulations. However, local residents are obliged to replace coal- and wood-fired boilers by the end of 2022 if these devices do not meet any emission norms, and by the end of 2026 also those boilers that currently meet only basic emission requirements [119]. With time, other provinces in Poland have adopted similar anti-smog resolutions, which enforce the replacement of old boilers, stoves, and fireplaces with modern devices at different times, and sometimes impose a partial or complete ban on coal or wood burning. To this day, only two provinces have not passed such resolutions: Warmia and Mazury Province and Podlasie Province [120].

A more thorough inspection of the Kraków contact zones demonstrates that they are far more complex than it might appear at first glance. It turns out that there is another civic movement, called the Social Initiative Kraków Alert Stop the Ban, which was set up in 2013 and now associates around 300 persons who oppose the ban to burn coal and wood in stoves; hence, it is a countermove against the Kraków Smog Alert [118]. Those who protest against the ban argue that the Kraków Smog Alert takes advantage of doctored research on air pollution sources and draws erroneous conclusions regarding the environmental protection policy. They maintain that data from air monitoring stations of the Regional Inspectorate for Environmental Protection are only able to detect the presence and concentration of pollutants in the air, but fail to identify where these pollutants originate from. The new initiative admits that smog is a significant issue but emphasizes the role of pollutants produced by vehicles, and suggests completely different corrective measures.

Among the suggestions raised by this movement, there is an option to allow the use of modern, automated coal-fired retort boilers, which could be fed with eco-pea coal: high-quality coal fuel emitting far less pollutants than ordinary coal, hence the prefix 'eco'. Other types of fuels allowed could be renewable biofuel composed of wooden pellets. The proponents of this initiative also suggest reducing air pollution due to road traffic by implementing so-called green waves: smart traffic management systems in cities that facilitate the flow of vehicles. Essentially, such systems ensure the synchronization of traffic lights at all intersections so that vehicles travelling at a certain speed are ensured green lights at every intersection they pass through. Thus, the speed of vehicles is more stable, which reduces fuel consumption and thereby generates less pollution [121]. In other words, while the mentioned new group aims to improve the quality of air, it claims the goal cannot be achieved by imposing a total ban on the burning of solid fuels, but will require a series of more complex and fairer measures [122].

Further research on the air pollution in Kraków has shown that this problem is more complex than expected. A new actor appeared in the contact zone: a team of researchers from the Department of Environmental Protection and Design of the AGH University of Science and Technology in Kraków, who questioned the claim that low emissions were the main source of air contamination, and highlighted the need to re-examine sources of particulate matter and other harmful chemical compounds in the city's air [122]. The researchers maintained that the data collated by the Marshal's Office, which served to develop air protection programs for the entire province, were dubious. According to this information, the structure of air contamination with PM10 dust in 2005 was as follows: 68% originated from transportation and traffic, 24% came from solid fuel furnaces (mainly fed with coal), and 8% was generated by industries. However, in 2013, the data provided by the same office were completely different: the sources of smog consisted of vehicle exhaust fumes (21%), furnaces fed with coal and wood (53%), and industries (26%). These data instantly reveal distinct anomalies, as the number of vehicles registered in Kraków and in the region of Małopolska (figures in brackets) increased rapidly, from 383,000 (1,056,755) in 2005 to 503,287 (1,615,661) in 2013. Additionally, the higher percentage of air contaminants originating from industries in 2013 is puzzling. In fact, the number of industrial plants in Kraków decreased considerably between 2005 and 2013. One would therefore expect a reverse tendency; that is, the percentage of transportation emissions should be higher in 2013 than in 2005, while the percentage of industrial emissions should be lower. Should these observations be true, solid fuel furnaces might turn out to be a less serious threat to air quality than previously thought [123]. Despite the lack of adequate research, the above opinions were criticized by both the Kraków Smog Alert and the local authorities, and the researchers were accused of hindering the fight against smog [124].

Nevertheless, the hypotheses put forth by the Kraków-based scholars were confirmed by Stanisław W. Gawroński from the Independent Department of Natural Foundations of Horticulture at the Warsaw University of Life Sciences, who suggests that the major source of particulate matter in air is mechanical vehicles. These pollutants come from exhaust fumes and from tires, asphalt, brake discs, and brake pads as they are worn out. The available data show the following structure of air pollutants in Warsaw: over 70% of particulate matter originates from transport emissions, 20% from industries, and 10% from households. Gawroński points to an important, though often underestimated method for smog elimination, such as adequately designed urban green areas, which are an excellent trap for airborne pollutants. There is a list of selected plant species, comprising over 25 species of trees, over 25 species of shrubs, and 6 species of creeping and climbing plants, which are recommended for growing in towns because of their ability to arrest particulate matter. The plants which thrive well in polluted urban areas include birches, oaks, yews, ash trees, and purple willows [125]. The plant detoxification mechanism works very well even when the contamination is caused by heavy metals and polycyclic aromatic hydrocarbons. The composition of green belts along roads with heavy traffic should comprise plants with high phytoremediation capabilities [126].

The latest studies by the Kraków-based scientists point to other, previously unrecognized causes of air pollution in that city; namely, changeable meteorological conditions. The low wind velocity and low mixing-layer height could contribute to elevated concentrations of airborne dust particles and other air contaminants, especially in the winter months [127]. Moreover, it was found that mechanical street cleaning might lead to a short-lived, around 3 h, increase in the concentrations of PM10 dust in air because the garbage trucks used for this purpose cause air turbulence and lift pollutants from the streets, thereby re-introducing them to the atmosphere. As a result, the concentration of PM10 in air can double, and the duration of this effect depends on wind velocity and the atmospheric stability class [128]. Furthermore, a similar effect was discovered in connection with the traffic of vehicles, which also lifts dust from streets and increases the particulate matter concentration in air by about 50%. Thus, improper cleaning of streets is a considerable source of air contamination. This effect could be averted if streets were washed regularly [129]. It was also determined that the emissions of PM10 which originate from re-entrained road dust could comprise up to 25% in winter and 50% in summer of the total PM10 concentrations in the air near roads [130]. Thus, a conclusion can be drawn that the PM10 amounts determined in air can be counted several times, which means that the data obtained from air monitoring stations are burdened with a certain error. The above studies cast more light on the origins of air pollution and can provide at least a certain measure of explanation why—despite the numerous initiatives, bans on using coal and wood for heating homes, and programs of replacing old boilers—Kraków is still among the most polluted cities in the world. On 14 December 2021, at 8.40 a.m. local time, Kraków ranked as the most polluted city in terms of air quality and pollution, and its U.S. AQI (Air Quality Index) was 316, which corresponds to a level hazardous to health. In response, the city's authorities provided free public transport to everyone, which proves that traffic is taken into account as a serious source of pollution, just like coal and wood stoves [131]. When the U.S. AQI value is as high as mentioned above, anyone exposed to such air pollution can experience considerable deterioration in health [132].

In response to the discrepancies indicated above, a new civic legislative initiative was undertaken by the Open Wieliczka Association, where the principal desideratum was to say 'yes to wood burning and no to the gas lobby'. The authors demand amendments in the anti-smog resolution for Małopolska, such as lifting the ban on using dry wood and pellet for heating homes in Kraków, and postponing the deadline for replacing uncertified fireplaces and boilers in the whole Małopolska Province for 3 years. They emphasize that citizens should be given a choice of ecological methods for heating of their homes, which should not be constrained to the option offered by the gas lobby comprising German boilers and Russian gas. The main goal of this initiative is 'to stop forcing people to use expensive gas' [133]. On 1 March 2022, the association submitted to the Małopolska Regional Assembly a civic draft resolution on this matter signed by 2500 people [134]. The activists cite results of analyses which prove that the ban on using coal and wood in Kraków has not contributed to a significant reduction in the concentration of benzo(a)pyrene in air, which in 2016–2020 was persistently the highest in Poland, even though it decreased by 21%. In the same time period, the results obtained in the other fifteen Polish provincial capital cities were much better, and the cumulative numerical average of the air concentration of benzo(a)pyrene in these cities declined by 34%, even though they did not impose a total ban on using solid fuels [135]. In addition, the inconclusiveness of the results of earlier studies on relationships between an increase in air pollution and higher mortality or shorter average longevity was emphasized [136]. It was also hypothesized that activities carried out under the umbrella of smog alerts is very profitable for some individuals. Focusing only on the fight against coal and wood redirects financial resources, both public and private, to specific business environments while discarding some other valuable solutions able to improve air quality, such as chimney electrostatic precipitators, which can arrest from 50 to 90% of pollutants even if the fireplaces or furnaces using coal or wood do not have proper certificates [137]. It is worth mentioning that in the years 2012–2019 over PLN

320 million, approximately EUR 75,301,200 (according to the monthly exchange rate of July 2019), was allocated in Kraków to the program of replacing boilers [138]. Furthermore, the VAT tax rate on firewood was raised from 8 to 23% in 2020, which presumably was expected to improve the competitiveness of gas heating [139]. The hypotheses put forth by the activists from Wieliczka are confirmed by other sources, which implicate some links between the Kraków Smog Alert and organizations associating entrepreneurs interested in the replacement of old boilers and in the modernization of buildings to improve their energy efficiency [140].

The information provided above attests to the extraordinary complexity of Kraków's contact zones arising from both administrative and legal questions, as well as energy issues. There are many clashing opinions of citizens, associations and grassroots movements, and local governments, in addition to which there are many more-than-human elements in our environments, such as polluted air, sources of renewable and non-renewable energy, or heating equipment. In this environment, a smog alert can be perceived as a perfect example of anti-conquest because of innocence gestures when activists assure us of their selfless intention for clean air in the country. On the other hand, only one 'true' solution to the problem of smog is exposed while other measures, equally good and effective, are stifled, and huge business ventures are hidden under the guise of ecology, monopolizing energy markets and heating equipment as well as thermomodernization of buildings.

A comparison of the above case with the previously described protests against shale gas can substantiate the conclusion that—considering both contact zones—the only winners so far on the side of the citizens are the farmers from the village Żurawlów, who retained their farmland.

10. Discussion

A large number of prosumption principles not indicated by the citizens responding to our questionnaire indicates that the human administrative legal and the more-than-human energy contact zones in the region of Warmia and Mazury differ significantly from the reference contact zones in Żurawlów and Kraków. All this suggests that in recent years, there have not been such civic movements active in the analyzed region, focusing on issues connected with the energy transition and environmental protection, that could boast of sufficiently significant achievements to have an impact, if not on national policy, then at least on provincial policy. Farmers' protests have stood out in recent years among the civic initiatives in the Warmia and Mazury Province. They mostly consist of roadblocks and slowing down the flow of traffic on roads, although they may assume harsher forms, such as the blockade of the Province Office in Olsztyn [141]. The protesters emphasized the need to amend the agricultural policy so that it would favor domestic agricultural production and improve its profitability. Nowadays, most farmers' protests in Warmia and Mazury are coordinated by the Agrounia—the National Farmers' Association, an organization started in 2018. This association put forward five demands against the central authorities in Poland [142,143]: the government must compensate producers for the errors made when developing the agricultural policy (making intervention purchases of agricultural products and paying compensation for losses); 70% of food products in shops should be made in Poland; Polish farmers must not be discriminated against in the European Union; liquidation of profitable branches in agricultural production must be forbidden; and using the competitive advantage of retail chains must be stopped. Additionally, two local civic movements in the region are worth mention: Social Initiative Save Mazury and the Sadyba Association for the Protection of the Cultural Landscape of Mazury, which protest against the construction of the S16 expressway, which would cut through the Land of the Great Mazury Lakes, the most valuable part of the region in terms of natural, landscape, and tourist assets, as well as the Biebrza National Park in Podlasie Province. The S16 is planned as a four-lane road with parameters close to those of a motorway. It will belong to the Trans-European Transport Network (TEN-T) and will connect two international transport corridors: Via Carpatia and Via Baltica [144]. Activists from both movements carry out information campaigns on social media, submit petitions to offices, and inform the general public about the ecological threats associated with the construction of this road. Interestingly, the activists from both movements are being supported in their stance by an increasing number of local governments from the municipalities and districts which this road is planned to cross [145]. In brief, a daily flow of 23,000 motor vehicles, mainly trucks, will cause irreparable ecological damage in this area. Local inhabitants maintain that sacrificing the natural environment of Mazury is too high a price to pay for Russian and German transit. This trend in the development of civic initiatives encompasses the aforementioned activity of the 21 cities in the region which belong to the Cittàslow network, where the priority is given to sustainable development, protection of the natural environment, and generation of clean energy [146].

Studies carried out in selected municipalities in Warmia and Mazury suggest the presence of an autocratic management style in these administrative units. Strong power is held in the hands of one person because the citizens are unprepared to deal with public matters due to the lack of adequate knowledge. Self-government is understood there as the highest possible decision-making and financial independence of the mayor, and the mayor's smallest possible dependence on higher levels of authorities. The municipal community tends to be perceived as inert, helpless, and unprepared to solve common problems. In such a system, initiatives undertaken by local residents or activities pursued by associations, such as the Citizens' Committees created in 1989–1990, have been always treated as manifestations of political fractiousness. A distinguishing feature of the local communities in this region is that they were formed anew after the Second World War from new settlers, displaced persons, and few autochthonous people. A sense of social unity is hard to expect under such circumstances, and thus while the local inhabitants admit that the participation of citizens in making decisions about the community's matters is proper, in their situation it is essentially unachievable. The reason is the lack of adequate social and cultural infrastructure for the development of civil society. As a result, independent public opinions cannot be voiced. Many municipalities have no local newspapers, while some have had poor access to the Internet until recently. The local authorities are free from social control, which means that they are tempted to act beyond their competences established in legislation. This deepens the inertia of the local community while some sort of mild despotism or overprotective power evolves. The situation is certainly aggravated by the distrust and hostility of society towards authorities inherited from the time of socialism in Poland, 1945–1989, which is reciprocated by the authorities suppressing any grassroots initiatives [147]. The above considerations lead to the conclusion that the large number of the prosumption principles not indicated by the survey respondents is due to the deficit of social capital, which is founded on trust [118]. The relationships between citizens and local governments have been overwhelmed by the culture of distrust (cynicism), which should be understood as the general sense of suspicion in interpersonal relations, leading to constant monitoring of the actions of the other party in the hope of preventing abuse, lies, and conspiracies [148] (p. 326).

Currently, the residents of Warmia and Mazury are facing an extremely important test of their civic strength, which arises from anti-smog resolutions, only recently drafted and practically ready for implementation. As usual, the implementation was supposed to proceed according to the autocratic model, where the authorities know best what the people need. These resolutions, the same as in other provinces, would impose a ban on burning low-quality solid fuels, including biomass with a moisture content of over 20%, that would enter into force on the date of adopting the resolution in question. Different deadlines were also given for the replacement of old boilers powered with solid fuels, depending on their class and location (in or outside urban areas). The only acceptable fuels are to be gas and light heating oil. In the future, it will be prohibited to heat homes with coal or wood in buildings which are situated within a piped gas grid or heating network [149,150]. The content of these resolutions has caused astonishment and disbelief for many reasons.

Warmia and Mazury is the least polluted region in Poland [151], with the least developed gas grid [152], but the sixth-highest province in Poland in terms of total forest cover [153], and the region with the smallest socio-economic potential in Poland [87,88]. In the current economic and political conditions, the abandonment of the option of heating homes with wood, which is an inexpensive ecological fuel available in abundance in this region, in favor of expensive and imported fossil fuels, such as natural gas, is a highly ill-considered move. This decision may force the local inhabitants to burn expensive Russian gas imported from Germany. The anti-smog resolutions in Poland propose such actions that are an exact opposite to the energy policies conducted by West European countries. For example, in Germany, from January to November 2021, the Federal Funding for Efficient Buildings (BEG) received numerous applications for co-funding of renewable heating systems. The statistics distinguish the following renewable heating types [154]:

- Gas hybrid and gas renewable ready heaters—32,991 grant applications;
- Solar thermal heating systems—39,338 grant applications;
- Biomass heating systems—62,967 grant applications;
- Environmental heaters (heat pump heaters)—56,075 grant applications;
- Heat network heaters—8349 grant applications.

These data show that most applications were submitted to co-fund biomass heating systems, as they made up 31.5% of all applications, and the second most popular were heat pump heaters, with a contribution of 28%. Moreover, using gas is possible only in combination with another source of heat, and a gas installation must be adopted to be powered with a renewable fuel, which should encourage a step-wise transition towards electricity [155]. Last but not least, the report commissioned by Greenpeace and developed by the Wuppertal Institute for Climate, Environment and Energy assures that the heat supply to buildings in Germany will be fully secured by renewable energy by the year 2035, and this will allow a complete departure from oil or natural gas [156]; thus, the installation of new oil and gas heating systems should be banned from 2024, and the existing systems should be gradually phased out by 2035 [157]. Similar solutions with the aim to discontinue the heating of buildings with natural gas have been implemented in other countries of Western Europe. In Denmark, for example, a ban on installing fossil oil and gas boilers in new buildings came into force in 2013. Similar restrictions on natural gas used for heating have been imposed in the Netherlands, where new buildings have been prohibited from being connected to a gas grid since July 2018. In addition, a local planning approach has been introduced in this country, where gas heating is phased out gradually at district level, and the aim is to create gas-free districts. Austria has banned the installation of central heating boilers powered with liquid or solid fossil fuels in new buildings, and the current government program includes further plans to phase out fossil oil and gas boilers. In France, a regulation was to come into force in 2022 which would forbid heating systems based on fossil fuels only. In Belgium, in the Flemish Region, a ban was introduced in 2021 to connect large building projects to a gas grid. In Germany, it will be prohibited to install mono-fuel fossil boilers from 2026. Ireland, in turn, is planning to introduce legal solutions prohibiting the installation of fossil gas boilers in new buildings from 2025 [158] (pp. 18–21).

The import of gas to Poland is influenced by both economic and political factors. One of the most important economic aspects is inflation, which is forecast by the National Bank of Poland to increase in the coming quarters of 2022, and will exceed 12% on a year-to-year basis in the third quarter of this year. One of the major causes of inflation is the rise in prices of energy resources on global markets, including natural gas, the price of which in February 2022 was three times as high as a year earlier [159]. Political factors play a significant role as well, and the war waged by Russia against Ukraine comes to the fore. In these circumstances, the German press appeals to the German authorities to stop imports of gas and crude oil from Russia, because paying for these imports means financing the war [160,161]. Interestingly, similar demands were addressed to the Polish government by the Polish Smog Alert, demanding the cessation of imports of coal, gas, and oil from

Russia [162]. This is largely contrary to the anti-smog resolutions adopted recently in Poland by local governments, where using gas for heating buildings is mandatory. It is not known, however, where this gas should come from, since the Baltic Pipe, which was to deliver natural gas from Norwegian fields to Denmark and Poland, will not be launched until October 2022, and it is still not rather unclear if and to what extent it will ensure the country's energy security. Furthermore, the amounts of liquefied natural gas delivered to the Świnoujście LNG terminal by tankers from Qatar and other countries do not seem to be certain.

In the Warmia and Mazury Province, anti-smog resolutions are observed to be proceeding in an autocratic regime. For instance, an invitation to open consultations about draft resolutions was posted on the website of the Public Information Bulletin on 22 December 2021, and the deadline was set for 21 January 2022 [163]. This time period was most probably chosen deliberately because it includes the Christmas and New Year season, followed by Epiphany. During this festive time, most citizens do not deal with official matters and therefore the actual time for submitting comments and motions was shortened by almost half. The websites of municipal offices displayed this information much later; for example, the Miłki municipality announced this invitation in the Current Events section as late as 17 January 2022, thereby reducing the time for social consultations to just four days [164]. The authorities know better what citizens need. Despite these obstacles, the local governments were astonished by the active stance taken by citizens, who submitted around 500 comments to the proposed legislation [165]. Activists of the Warmia and Mazury Region branch of the National Movement, a nationwide political party in Poland, noticed the problem in time, publicized it and appealed to the councilors of the Warmia and Mazury Regional Assembly to reject the anti-smog draft resolutions [166]. In justification of their protest, they point to the fact that the ban on using wood for heating houses can ruin the local economy, where the timber industry generates a considerable revenue in the region. Moreover, the condition of the natural environment is good enough to deem restrictions and bans envisaged in the proposed regulations as unjustified. Another argument was that, in compliance with the EU and national law, products obtained from woody biomass are counted as renewable energy sources, while natural gas is a fossil fuel being gradually phased out in West European countries. In light of these facts, the resolutions which put a preference on gas and enforce the installation of expensive fossil fuel powered heating devices are short-sighted and may considerably worsen the living standards of local residents in the future. The path towards clean air should proceed through a rise in the wealth and awareness of citizens rather than through the use of force [167]. The local authorities responded as might have been expected. Rather than providing substantiated arguments, they accused the opponents of incompetence and misleading the region's inhabitants. It seems that the true motivation for adopting anti-smog resolutions was the fact that most other Polish provinces have already adopted such laws and we (Warmia and Mazury Province) must not lag behind. Additionally, the arguments concerning the sources of air pollution appear peculiar because the authorities claim that as much as 89–99% of the pollutants originate from the heating of houses. Hence, it would seem that there was hardly any traffic or industry in the region. However, some of the provisions proposed in these resolutions are acceptable; for example, the ban on burning trash in heating installations or the obligation to gradually phase out boilers that do not meet the essential combustion standards, i.e., the so-called 'smokers' [168].

The above discussion demonstrates that the contact zones in Warmia and Mazury are in the stage of gradual transformation from a purely autocratic model to a model appropriate for civil society. Thus far, the civic initiatives in this region have been encouraged by national organizations, such as farmers' unions or political parties. This type of activities deserve to be treated as extremely valuable because they give the region's inhabitants a model of how to shape and employ prosumer attitudes in the field of energy transition and in other areas. As regards anti-smog resolutions, they are most clearly an outcome of astroturf campaigns; that is, only seemingly arising from grassroots citizen movements, but actually shaped by lobbying actions undertaken by corporations and various interest groups [169]. This is a characteristic feature of a contact zone described previously as anti-conquest. In this situation, the residents of the Warmia and Mazury Province are yet to face a true test of their prosumer abilities in the spheres of economics and politics, and the outcome of this test is difficult to predict.

The actual energy transition takes place in the contact zone, where the thoughts and expectations of citizens clash with the solutions developed by the local government sector. An important factor in building a civil society is prosumption in public administration, which is a prerequisite for the revival of the human administrative legal contact zone. It is not until this zone develops that a step towards the more-than-human energy contact zone can be taken, in which the energy prosumers and the building owners will be able to participate in the development of renewable energy sources, reduction of pollutants and protection of the environment. There are several obstacles in Warmia and Mazury preventing the implementation of prosumption in the public sphere. Some elements of past German and Russian colonialism continue to linger in the awareness of the local residents, and the legacy of socialist economy still weighs heavily on contacts between citizens and municipal governments. In these circumstances, an autocratic system of public matters management can grow and consolidate easily. In recent years, however, some signs of energy prosumerism have appeared in the region although, notably, most measures taken in this area have been initiated by Poland's nationwide organizations. Local grassroots movements still seem weak, but they may turn into a permanent element of contact zones in this province with the increase in the awareness and wealth of the region's inhabitants. The role of civic initiatives following the stagnation period of 1995–2015 should gain more importance. The above research proves that in the social sphere, in this terra incognita illustrated in Figure 1, people who compose the middle class make an increasing number of autoethnographic gestures, which will cause transculturation of elements of metropolitan discourses. In this way, citizens, who in the current situation play the role of pawns subjugated by conquest, strive to free themselves from the excessive subordination to the authorities by creating self-affirmations addressed to the metropolis. This is a form of becoming engaged in the metropolis' constructions, where the aim is to neutralize autocratic elements in the social sphere, as these are perceived negatively by the majority of society. The propensity of the respondents to digital prosumption, revealed by this study, is one such autoethnographic gesture which is expected to weaken the adverse consequence of the conquest.

One of the principal goals of digitalization is to improve the flow of energy and information in both local and national economy. Thus far, the digital transformation in the local government sector has been poorly researched, and this is where the causes of failure and delays in the energy transition lie. These problems require in-depth research because citizens' daily contact with the municipal public administration offices determine the implementation of digital technologies [170]. The results of the study presented in this article shed new light on the conditions underlying digitalization on the level of municipalities and define the division of roles between citizens and local governments. Prosumption attitudes among society in Warmia and Mazury have been initiated by the digitalization of the market sector, but due to the conditions resulting from the autocratic management style they are yet in the early stage of development. In general, the municipal governments are alienated from their local communities, which forces people to look for better living conditions elsewhere because, for the time being, they are unattainable in their current place of residence. The best proof supporting this conclusion is the fact that our respondents can refer to some of the prosumption principles but fail to recognize some others. This implicates the need for the inclusion of society in the sphere of public administration so as to set up institutional practices based on participative citizenship [171]. This paper demonstrates that the starting point should be the ability of people to define a set of autoethnographic gestures indicating their willingness to use digital prosumption in public administration. In this way, society could voice its expectations, and the local governments must now act in the field of implementing digital technologies. Digitalization of the public administration is a factor accelerating the inclusion of people in this sector of economy.

Changes in the contemporary world economy tend to move the business models of wikinomics from the private sector to the public sector, and especially to the local government sector [51]. An important role in improving the functioning of public administration is played by the principles of prosumption and the platforms for participation, which can contribute both to the emergence of completely new public services and to the improvement of existing services.

The platforms for participation are another wikinomics model which is developing thanks to the rapid progress in ICT. These are the products and technological infrastructure shared by modern businesses to partner communities in order to create new economic values and initiate innovative projects. The platforms for participation are based on all four principles of wikinomics: openness, peering, sharing, and acting globally. This business model is slowly supplanting the conventional business model, because as a result of moving parts of the business to the Internet, they can operate on a broader stage. This enables faster and more effective collaboration with relevant partners to improve existing solutions. The company's productive capacity can thus be increased without incurring additional fixed costs. In the private sector, the platforms for participation are usually websites that have an e-commerce system for warehousing, purchasing, and distributing goods [48].

The results of the survey of public administration offices websites indicate that in their present form, they are not suitable for the platforms for participation [46,47]. This implies a delay in implementing the principles of prosumption in this sector. Thus, a serious barrier to growth and economic development is revealed. This also contributes to difficulties in foreign trade and lowers the country's position in the world economy, which can only be overcome by the appropriate economic policy of the state. Thus, the digital prosumption is, to a very small extent, present in the public administration sector in Poland.

In the public administration sector, it is necessary to implement the ten principles of prosumption which are discussed in this paper. This will require simultaneous transformation of the existing municipal office websites into wikinomics platforms for participation. These should be the platforms for grassroots action based on public disclosure and neighborhood knowledge [48]. In terms of the principles of prosumption and the platforms for participation, the Polish public administration sector is much less developed than the market sector. The reason for this is the low transparency of the entire public sector.

The harmonization of the public administration sector with the market sector encounters a number of barriers that can be divided into technological and mental barriers. The former concern the technical requirements of the platforms for participation. Therefore, they can easily be overcome with the right inputs, while the latter are not. Breaking mental barriers is only possible through the transformation of clients into prosumers, which requires simultaneous bottom-up actions of citizens and the implementation of the principle of openness in the public administration itself.

11. Conclusions

In today's world, the role of prosumer energy is steadily increasing, and this trend is expected to continue in the future [172,173]. This article shows that a success factor for the energy transition at the local level is the stimulation of the prosumption potential of societies by local municipal governments. At the same time, many changes in the contact zones are desirable, which are detailed below.

Two types of conclusions can be drawn from this study. The first type concerns the characteristic features of the contact zones in Warmia and Mazury, Poland, whereas the others are of general economic importance and they can span the entire country, international communities and the whole world. Local, international, or global, the contact zones have the same common features, and they can explain the inevitable conflicts of interest, clashes of opposing views, and the fight for economic and political power and domination in all aspects of reality. First, conclusions on the contact zones in Warmia and Mazury will be presented, arising both from the survey concerning the acceptance in society of the rules of prosumption proposed by the authors and from a comparative analysis aimed at determining the differences and similarities between the contact zone under study and the two reference contact zones. Subsequently, the paper presents general reflections on the study and the extrapolation of some discovered regularities onto higher, non-local levels of the economic reality. The paper will be concluded by identifying the directions of future research.

The empirical analysis performed with the use of multiple correspondence analysis shows that the contact zones in Warmia and Mazury exhibit a majority of typical features discussed earlier with examples from humanities and environmental science. However—obviously—these features are exemplified differently. The conclusions can be summarized as follows:

1. The research hypothesis was confirmed, which means that the society in the province under examination uses digital prosumption as the basic tool for communication with the municipal governments;
2. The rules of prosumption not chosen by the citizens mean that they lack the knowledge necessary to cooperate with the local governments. This also means that it is now the municipal governments' turn to act. However, developing common solutions concerning climate protection and energy generation from renewable sources should be preceded by education and building mutual trust.
3. The respondents' answers in the survey questionnaire are a kind of autoethnographic text, which is a list of citizens' experiences from previous contacts with the local government sector using official documents, i.e., ethnographic texts. The rules of prosumption that are not chosen mean that people's meetings with public officials are rare and that there are certain differences between the idioms of the metropolis and the indigenous idioms. This means that there is no mirror dance.
4. Failed transculturation is present in the contact zones in Warmia and Mazury. A metropolitan culture, represented now by the municipal governments, is based on the remnants of German and Soviet colonialism, which led to the huge competence of the local governments at the expense of civic initiatives. However, citizens did not adopt almost any elements of the dominant cultures—neither at the times of a centrally planned economy, nor during the subsequent economic transformation and the return to capitalism. Polish indigenous culture remained permanently alienated, which means that the Ortiz equation must be—for the time being—replaced with an inequality. This is shown by the rules of prosumption not chosen by citizens.
5. There are visible effects of the conquest and anti-conquest in the contact zones under study. Their remnants have survived until now as the autocratic style of municipal management. Warmia and Mazury were lost by Poland in the 17th and 18th centuries. The land was regained after 1945, but together with the command economy imposed by the Soviet Union.
6. There is still a representation strategy in the region, which simultaneously includes gestures of innocence and striving for hegemony. These territories were formally incorporated into Poland after 1945, but the need to cooperate with the colonizing forces ruled out any significant social and economic reforms. This strategy of representation is still followed by the local governments in a slightly modified form.
7. Not much can be inferred from the survey about the competing perspectives. Their existence is indicated by the rules of prosumption not chosen by citizens. This can also be a form of a silent protest against the relations of power in the province.
8. There is no visible platform for public consultations and negotiations, which is a direct result of the inherited colonialism and the autocratic style of the municipality management. This may be one of the factors responsible for the dwindling population in the region, as it forces economic migration. At the same time, it explains the low level of use of the existing hydropower potential.

The results of the comparative analysis of the contact zones in Warmia and Mazury with the two Polish contact zones—Żurawlów and Kraków—deepen and broaden these conclusions. Table 7 shows major similarities and differences between the contact zones under study, considering such features as the influence on the national energy policy, involved social strata, the method for initiating grassroots movements, dominating style of management, astroturf campaigns, radical heterogeneities, negotiations and civil dialogue, languages of contact zones, struggle for interpretive power, competing perspectives, neocolonial nature of social relations, conquest and anti-conquest, representation strategy, transculturation, autoethnographic and ethnographic texts, mirror dance, and oikophilia. This list shows that the contact zones in Warmia and Mazury are developing gradually and, with time, they can reach a level high enough to exert an impact on the domestic energy policy.

Table 7. Comparison of the most important features of the contact zones in Warmia and Mazury with those of the two reference contact zones in Żurawlów and Kraków.

Key Features of the Studied Contact Zones	Warmia and Mazury	Żurawlów	Kraków
Impact on national energy policy	None	Yes	Yes
Involved social strata	Middle class	Farmers	Middle class
Initiating grassroots movements	Top-down (by national organizations)	Bottom-up	Bottom-up
Prevalent management style	Autocratic	Democratic	Democratic
Astroturf campaigns	Likely	Currently lacking	Likely
Radical heterogeneities	Yes	Yes	Yes
Negotiations and civil dialogue	Initial phase	Advanced phase	Advanced phase
Languages of contact zones	Dominant groups	Mix: native and dominant groups	Mix: native and dominant groups
Struggle for interpretive power (marches, roadblocks, demonstrations)	Initial phase	Advanced phase	Advanced phase
Competing perspectives	Initial phase	Developed in all parties to the conflict	Developed in all parties to the conflict
Neocolonial nature of social relations	Yes	No	No
Conquest and anti-conquest	Yes	Yes	Yes
Representation strategy	Innocence and hegemony	Innocence and hegemony	Complex
Transculturation	Failed: Ortiz's inequality	Successful: Ortiz's equality	Successful: Ortiz's equality
Autoethnographic and ethnographic texts	Yes	Yes	Yes
Mirror dance	No	Yes	Yes
Oikophilia	Yes	Yes	Yes

The phenomena taking place in the contact zones examined in this study are of a general nature, and they are present on the local, international, and global levels. In Poland, the European Union and many other countries around the world, there are similar contact zones as in Warmia and Mazury. Most of them have the features mentioned in Table 7. Making use of their potential and filling them in with cooperation, despite the conflicts, opposite interests, and large contrasts, is a condition for sustainable development. There are not only problems but also solutions to be found in the contact zones.

This study examines the impact of the phenomena taking place in the contact zones on the shape of the energy policy oriented towards a reduction of global warming, which is regarded as the cause of catastrophic climate change. A virtuous cycle is sought, which includes digitalization, decentralization, and electrification, and which stimulates the use of energy from renewable sources. One of the trends in future research will be to focus on contact zones as a factor critical to the success or failure of the energy transition. The division of a contact zone into two sub-zones, the human administrative legal contact zone and the more-than-human energy contact zone, can also be useful, as it allows for categorizing phenomena into those shaped by people and those arising from human–nature relations. This involves local contact zones and investigating how to make use of their potential for green energy sector development. The energy transition is of a global nature as it was conceived of as a remedy for a global threat, but its success will be a consequence of a favorable outcome of the glocalization processes in various places around the world. Studies of the contact zones on various levels, i.e., those covering the development of the regional, but also the national and international energy policy, can be of particular importance. Such contact zones exist, and complex processes occur in them every day. One example is the contact zone covering the whole of the European Union, in which each member state has to submit its plan concerning energy and climate [174]. The authors hope that this work will encourage other scholars to continue similar research.

Author Contributions: Conceptualization, A.J. and D.R.; methodology, A.J. and D.R.; software, A.J. and D.R.; validation, A.J. and D.R.; formal analysis, A.J. and D.R.; investigation, A.J. and D.R.; resources, A.J. and D.R.; data curation, A.J. and D.R.; writing—original draft preparation, A.J. and D.R.; writing—review and editing, A.J. and D.R.; visualization, A.J. and D.R.; supervision, A.J. and D.R.; project administration, A.J. and D.R.; funding acquisition, A.J. and D.R. All authors have read and agreed to the published version of the manuscript.

Funding: This research received no external funding.

Data Availability Statement: Data are contained within the article.

Conflicts of Interest: The authors declare no conflict of interest.

References

1. Scruton, R. *How to Think Seriously about the Planet: The Case for an Environmental Conservatism*; Oxford University Press: New York, NY, USA, 2012; ISBN 978-0-19-989557-1.
2. Scruton, R. Environmentalism Starts with Loving Our Own. *The Conservative*, 1 October 2018. Available online: https://theconservative.online/article/environmentalism-starts-with-loving-our-own(accessed on 28 January 2022).
3. Directive (EU) 2018/844 of the European Parliament and of the Council of 30 May 2018 Amending Directive 2010/31/EU on the Energy Performance of Buildings and Directive 2012/27/EU on Energy Efficiency (Text with EEA Relevance). *Off. J. EU* **2018**, L156. Available online: https://eur-lex.europa.eu/eli/dir/2018/844/oj (accessed on 29 January 2022).
4. In Focus: Energy Efficiency in Buildings. European Commission, Brussels, 17 February 2020. Available online: https://ec.europa.eu/info/news/focus-energy-efficiency-buildings-2020-feb-17_en (accessed on 30 January 2022).
5. Buildings Performance Institute Europe: 97% of Buildings in the EU Need to be Upgraded. Available online: https://www.bpie.eu/publication/97-of-buildings-in-the-eu-need-to-be-upgraded/ (accessed on 31 January 2022).
6. Obwieszczenie Ministra Inwestycji i Rozwoju Z Dnia 8 Kwietnia 2019 r. w Sprawie Ogłoszenia Jednolitego Tekstu Rozporządzenia Ministra Infrastruktury w Sprawie Warunków Technicznych, Jakim Powinny Odpowiadać Budynki i Ich Usytuowanie. Dziennik Ustaw RP 2019, poz. 1065. Available online: https://isap.sejm.gov.pl/isap.nsf/DocDetails.xsp?id=WDU20190001065 (accessed on 31 January 2022).
7. Svobodová, I. Energy efficiency in buildings: Middle-class status as a trigger for the European fight against climate change. *Eur. View* **2019**, *18*, 163–170. [CrossRef]
8. Pratt, M.L. Arts of the Contact Zone. *Profession* **1991**, *1991*, 33–40. Available online: https://www.jstor.org/stable/25595469 (accessed on 15 January 2022).
9. Pratt, M.L. *Imperial Eyes: Travel Writing and Transculturation*; Routledge—Taylor & Francis Group: New York, NY, USA, 2003; ISBN 0-203-16309-5.
10. Guaman Poma, *Nueva Corónica y Buen Gobierno* (1615); GKS 2232 4°. København, Danmark, Det Kongelige Bibliotek. Available online: http://www5.kb.dk/permalink/2006/poma/info/es/frontpage.htm (accessed on 28 January 2022).
11. Guamán Poma de Ayala, F. *The First New Chronicle and Good Government: On the History of the World and the Incas up to 1615*; Hamilton, R., Ed.; Hamilton, R., Translator; University of Texas Press: Austin, TX, USA, 2009; ISBN 978-0-292-71959-0.
12. Guamán Poma de Ayala, F. *Nueva Corónica y Buen Gobierno; Pease García, F.—Transcripción, Prólogo, Notas y Cronología*; Fundación Biblioteca Ayacucho: Caracas, Venezuela, 1980; Volume 1, ISBN 84-660-0053-4. Volume 2, ISBN 84-660-0054-2; Available online: https://www.clacso.org.ar/biblioteca_ayacucho/ (accessed on 31 January 2022).

13. Adorno, R. Felipe Guaman Poma de Ayala's Nueva corónica y Buen Gobierno (New Chronicle and Good Government). In A Multi-Lingual Anthology of Texts and Links Compiled for the First Early Ibero/Anglo Americanist Summit in Tucson, AZ, USA, 16–19 May, 2002. Panel 10: New World Antiquities and Histories. Available online: https://archive.mith.umd.edu/summit/Proceedings/Adorno.htm (accessed on 31 January 2022).
14. Adorno, R. *Guaman Poma: Writing and Resistance in Colonial Peru*, 2nd ed.; With a New Introduction; University of Texas Press, Institute of Latin American Studies: Austin, TX, USA, 2000; ISBN 0-292-70503-4.
15. Ortiz, F. *Cuban Counterpoint: Tobacco and Sugar*; Alfred, A. Knopf, Inc.: New York, NY, USA, 1947; ISBN 978-0-307-82026-6.
16. Pratt, M.L. Apocalypse in the Andes: Contact Zones and the Struggle for Interpretive Power. Encuentros, March 1996, No. 15. Inter-American Development Bank, Cultural Center: Washington, D.C.; USA. Available online: https://publications.iadb.org/publications/english/document/Apocalypse-in-the-Andes-Contact-Zones-and-the-Struggle-for-Interpretive-Power.pdf (accessed on 22 January 2022).
17. Harris, J. Negotiating the contact zone. *J. Basic Writ.* **1995**, *14*, 27–42. Available online: https://wac.colostate.edu/docs/jbw/v14n1/harris.pdf (accessed on 2 February 2022). [CrossRef]
18. Miller, R.E. Fault lines in the contact zone. In *Professing in the Contact Zone: Bringing Theory and Practice Together*; Wolff, J.M., Ed.; National Council of Teachers of English: Urbana, IL, USA, 2002; pp. 121–146, ISBN 0-8141-3740-7.
19. West, T. Differencing negotiation. *Compos. Stud.* **1997**, *25*, 7–18. Available online: https://compstudiesjournal.com/archive/ (accessed on 5 February 2022).
20. Kingwell, M. *A Civil Tongue: Justice, Dialogue, and the Politics of Pluralism*; The Pennsylvania State University Press: University Park, PA, USA, 1995; ISBN 978-0-271-02773-9.
21. Key, D. Safe houses and sacrifices: Filling the rooms with precious riches. In *Professing in the Contact Zone: Bringing Theory and Practice Together*; Wolff, J.M., Ed.; National Council of Teachers of English: Urbana, IL, USA, 2002; pp. 102–118. ISBN 0-8141-3740-7.
22. Edelstein, M. Multiculturalisms past, present, and future. *Coll. Engl.* **2005**, *68*, 14–41. [CrossRef]
23. Lu, M.-Z. Professing multiculturalism: The politics of style in the contact zone. *Coll. Compos. Commun.* **1994**, *45*, 442–458. [CrossRef]
24. Bizzell, P. "Contact zones" and English studies. *Coll. Engl.* **1994**, *56*, 163–169. [CrossRef]
25. Clifford, J. *Routes: Travel and Translation in the Late Twentieth Century*; Harvard University Press: Cambridge, MA, USA, 1997; ISBN 0-674-77960-6.
26. Boast, R. Neocolonial collaboration: Museum as contact zone revisited. *Mus. Anthropol.* **2011**, *34*, 56–70. [CrossRef]
27. Radermacher, M. *Devotional Fitness: An Analysis of Contemporary Christian Dieting and Fitness Programs*; Springer International Publishing AG: Cham, Switzerland, 2017; ISBN 978-3-319-49823-2. [CrossRef]
28. Isaacs, J.R.; Otruba, A. Guest Introduction: More-than-human contact zones. *Environ. Plan. E Nat. Space* **2019**, *2*, 697–711. [CrossRef]
29. Skillington, T. Democracy, biodiversity and more than human justice imperatives: Institutional responses to crisis. *Soc. Sci.* **2020**, *9*, 166. [CrossRef]
30. Wilson, H.F. Contact zones: Multispecies scholarship through "Imperial Eyes". *Environ. Plan. E Nat. Space* **2019**, *2*, 712–731. [CrossRef]
31. Isaacs, J.R. The "bander's grip": Reading zones of human—Shorebird contact. *Environ. Plan. E Nat. Space* **2019**, *2*, 732–760. [CrossRef]
32. Fredriksen, A. Encounters in the ebb and flood: Knowing marine ecologies in the intertidal contact zone. *Environ. Plan. E Nat. Space* **2019**, *2*, 761–780. [CrossRef]
33. Sutherland, C.R. Encountering the burn: Prescribed burns as contact zones. *Environ. Plan. E Nat. Space* **2019**, *2*, 781–798. [CrossRef]
34. Jakimowicz, A.; Rzeczkowski, D. The impact of public administration digitalization on the decarbonization of the economy. *Energies* **2021**, *14*, 5739. [CrossRef]
35. Cont, W.; Romero, C.; Lleras, G.; Unda, R.; Celani, M.; Gartner, A.; Capelli, L.; Zipitría, L.; Besfamille, M.; Figueroa, N.; et al. *IDEAL 2021: Reducing Service Gaps: How Digitalization Can Improve the Use of Infrastructure*; Corporación Andina de Fomento—Development Bank of Latin America: Caracas, Venezuela, 2021; ISBN 978-980-422-223-8. Available online: http://scioteca.caf.com/handle/123456789/1763 (accessed on 3 January 2022).
36. Astarloa, B.; Kaakeh, A.; Lombardi, M.; Scalise, J. *The Future of Electricity: New Technologies Transforming the Grid Edge*; World Economic Forum: Cologny/Geneva, Switzerland, 2017; Available online: https://www3.weforum.org/docs/WEF_Future_of_Electricity_2017.pdf (accessed on 5 January 2022).
37. Collin, J.; Hiekkanen, K.; Korhonen, J.J.; Halén, M.; Itälä, T.; Helenius, M. (Eds.) *IT Leadership in Transition—The Impact of Digitalization on Finnish Organizations*; Aalto University publication series: Science + Technology, 7/2015; Aalto University, School of Science, Department of Computer Science, ACIO Research Program: Helsinki, Finland, 2015; ISBN 978-952-60-6243-3. Available online: http://urn.fi/URN:ISBN:978-952-60-6243-3 (accessed on 4 January 2022).
38. Kotilainen, K. Energy prosumers' role in the sustainable energy system. In *Affordable and Clean Energy. Encyclopedia of the UN Sustainable Development Goals*; Leal Filho, W., Azul, A.M., Brandli, L., Salvia, A.L., Wall, T., Eds.; Springer Nature Switzerland AG: Cham, Switzerland, 2021; pp. 1–14. ISBN 978-3-319-95863-7. [CrossRef]

39. Kucharska, A. Społeczny wymiar "Energiewende"—Rozwój energetyki prosumenckiej w Austrii. *Stud. Humanist. AGH* **2018**, *17*, 67–84. [CrossRef]
40. De Sousa Santos, B. The counter-hegemonic use of law in the struggle for a globalization from below. *An. De La Cátedra Fr. Suárez* **2005**, *39*, 421–474. Available online: https://revistaseug.ugr.es/index.php/acfs/article/view/1035/1225 (accessed on 21 February 2022). [CrossRef]
41. Grobelski, T. Becoming a Side: Legal Mobilization and Environmental Protection in Poland. A Dissertation Submitted in Partial Fulfillment of the Requirements for the Degree of Doctor of Philosophy, University of Washington, Seattle, WA, USA, 2016. Available online: https://digital.lib.washington.edu/researchworks/handle/1773/38149 (accessed on 26 April 2022).
42. Archimedes of Syracuse. On the equilibrium of planes or the centres of gravity of planes. In *The Works of Archimedes*; Ed. in Modern Notation, with Introductory Chapters by T.L. Heath; Cambridge University Press: Cambridge, UK, 1897; Book I; pp. 189–202.
43. Pappus of Alexandria. *Pappi Alexandrini Mathematicae Collectiones a Federico Commandino Urbinate. In Latinum Conuersae, & Commentarijs Illustratae. In Hac Nostra Editione ab Innumeris, Quibus Scatebant Mendis, & Praecipuè in Graeco Contextu Diligenter Vindicatae. Et Serenissimo Principi Leopoldo Gvlielmo Archidvci Avstriae, &c. Dicatae*; Ex Typographia, H.H., Ed.; de Duccijs: Bononiae, Italy, 1660; Available online: https://books.google.pl/books?id=KCvY8DKof7AC&hl=eng (accessed on 19 February 2022).
44. Jakimowicz, A. Wikinomiczne wzorce biznesowe w warunkach polskich. *Stud. I Pr. WNEiZ US* **2017**, *47*, 25–34. [CrossRef]
45. Jakimowicz, A.; Rzeczkowski, D. Application of the four colour theorem to identify spatial regional poles and turnpikes of economic growth. *Acta Phys. Pol. A* **2018**, *133*, 1362–1370. [CrossRef]
46. Jakimowicz, A.; Rzeczkowski, D. New measure of economic development based on the four-colour theorem. *Entropy* **2021**, *23*, 61. [CrossRef] [PubMed]
47. Jakimowicz, A.; Rzeczkowski, D. Municipality digital platforms and local development in the Warmia and Mazury region: Implications for smart specialization. In *Partnerships for Regional Innovation and Development: Implementing Smart Specialization in Europe*; Gancarczyk, M., Ujwary-Gil, A., González-López, M., Eds.; Routledge—Taylor & Francis Group: New York, NY, USA, 2021; pp. 246–270, ISBN 9780429330261. [CrossRef]
48. Tapscott, D.; Williams, A.D. *Wikinomics: How Mass Collaboration Changes Everything*; Portfolio/Penguin: New York, NY, USA, 2006; ISBN 978-1-59184-138-8.
49. Toffler, A. *The Third Wave*; William Morrow and Company: New York, NY, USA, 1980; ISBN 0-688-03597-3.
50. Toffler, A.; Toffler, H. *Revolutionary Wealth: How It Will Be Created and How It Will Change Our Lives*; Doubleday: New York, NY, USA, 2007; ISBN 978-0-385-52207-6.
51. Tapscott, D.; Williams, A.D. *Macrowikinomics: Rebooting Business and the World*; Portfolio/Penguin: New York, NY, USA, 2012; ISBN 978-1-59184-356-6.
52. Jakimowicz, A.; Rzeczkowski, D. Prosumption in the public administration sector. *Acta Phys. Pol. A* **2016**, *129*, 1011–1017. [CrossRef]
53. Jakimowicz, A.; Rzeczkowski, D. Perspektywy realizacji zasad wikinomii w administracji publicznej w Polsce. *Stud. I Pr. WNEiZ US* **2017**, *47*, 221–232. [CrossRef]
54. Abdi, H.; Valentin, D. Multiple correspondence analysis. In *Encyclopedia of Measurement and Statistics*; Salkind, N.J., Ed.; Sage Publications: Thousand Oaks, CA, USA, 2007; Volume 2, pp. 651–656, ISBN 9781412916110.
55. Greenacre, M. From simple to multiple correspondence analysis. In *Multiple Correspondence Analysis and Related Methods*; Greenacre, M., Blasius, J., Eds.; Chapman & Hall/CRC—Taylor & Francis Group: Boca Raton, FL, USA, 2006; pp. 41–76, ISBN 978-1-58488-628-0.
56. Nenadić, O.; Greenacre, M. *Computation of Multiple Correspondence Analysis, with Code in R. UPF Working Paper No. 887*; Universitat Pompeu Fabra: Barcelona, Spain, 2005. [CrossRef]
57. Hill, T.; Lewicki, P. *Statistics: Methods and Applications. A Comprehensive Reference for Science, Industry, and Data Mining*; StatSoft, Inc.: Tulsa, OK, USA, 2006; ISBN 1-884233-59-7.
58. Le Roux, B.; Rouanet, H. *Geometric Data Analysis: From Correspondence Analysis to Structured Data Analysis*; Kluwer Academic Publishers: Dordrecht, The Netherlands, 2004; ISBN 1-4020-2235-2.
59. Le Roux, B.; Rouanet, H. *Multiple Correspondence Analysis*; Sage Publications, Inc.: Los Angeles, CA, USA, 2010; ISBN 978-1-4129-6897-3.
60. Husson, F.; Josse, J. Multiple Correspondence Analysis. In *Visualization and Verbalization of Data. Computer Science and Data Analysis Series*; Blasius, J., Greenacre, M., Eds.; CRC Press—Taylor & Francis Group: Boca Raton, FL, USA, 2014; pp. 165–184, ISBN 978-1-4665-8981-0. [CrossRef]
61. Burt, C. The factorial analysis of qualitative data. *Br. J. Math. Stat. Psychol.* **1950**, *3*, 166–185. [CrossRef]
62. Lebart, L.; Saporta, G. Historical elements of correspondence analysis and multiple correspondence analysis. In *Visualization and Verbalization of Data. Computer Science and Data Analysis Series*; Blasius, J., Greenacre, M., Eds.; CRC Press—Taylor & Francis Group: Boca Raton, FL, USA, 2014; pp. 31–44, ISBN 978-1-4665-8981-0. [CrossRef]
63. Beh, E.J.; Lombardo, R. *Correspondence Analysis: Theory, Practice and New Strategies*; John Wiley & Sons: Chichester, UK, 2014; ISBN 978-1-119-95324-1.
64. Sourial, N.; Wolfson, C.; Zhu, B.; Quail, J.; Fletcher, J.; Karunananthan, S.; Bandeen-Roche, K.; Béland, F.; Bergman, H. Correspondence analysis is a useful tool to uncover the relationships among categorical variables. *J. Clin. Epidemiol.* **2010**, *63*, 638–646. [CrossRef]

65. Ayele, D.; Zewotir, T.; Mwambi, H. Multiple correspondence analysis as a tool for analysis of large health surveys in African settings. *Afr. Health Sci.* **2014**, *14*, 1036–1045. [CrossRef]
66. Lana, R.M.; Riback, T.I.S.; Lima, T.F.M.; da Silva-Nunes, M.; Cruz, O.G.; Oliveira, F.G.S.; Moresco, G.G.; Honório, N.A.; Codeço, C.T. Socioeconomic and demographic characterization of an endemic malaria region in Brazil by multiple correspondence analysis. *Malar. J.* **2017**, *16*, 397. [CrossRef]
67. Costa, P.S.; Santos, N.C.; Cunha, P.; Cotter, J.; Sousa, N. The use of multiple correspondence analysis to explore associations between categories of qualitative variables in healthy ageing. *J. Aging Res.* **2013**, *2013*, 302163. [CrossRef]
68. Rodriguez-Sabate, C.; Morales, I.; Sanchez, A.; Rodriguez, M. The multiple correspondence analysis method and brain functional connectivity: Its application to the study of the non-linear relationships of motor cortex and basal ganglia. *Front. Neurosci.* **2017**, *11*, 345. [CrossRef]
69. van Horn, A.; Weitz, C.A.; Olszowy, K.M.; Dancause, K.N.; Sun, C.; Pomer, A.; Silverman, H.; Lee, G.; Tarivonda, L.; Chan, C.W.; et al. Using multiple correspondence analysis to identify behaviour patterns associated with overweight and obesity in Vanuatu adults. *Public Health Nutr.* **2019**, *22*, 1533–1544. [CrossRef]
70. Balliester Reis, T. A Multi-Dimensional Index of Financial Inclusion: A Micro-Data Approach. In 18° Seminário de Diamantina, Universidade Federal de Minas Gerais—Centro de Desenvolvimento e Planejamento Regional da Faculdade de Ciências Econômicas (UFMG—Cedeplar): Belo Horizonte, Minas Gerais, Brazil, 19–23 August 2019. Available online: https://diamantina.cedeplar.ufmg.br/portal/download/diamantina-2019/D18_398.pdf (accessed on 28 September 2021).
71. Dungey, M.; Tchatoka, F.D.; Yanotti, M.B. Using multiple correspondence analysis for finance: A tool for assessing financial inclusion. *Int. Rev. Financ. Anal.* **2018**, *59*, 212–222. [CrossRef]
72. Toukam, C.K.; Sezine, S.T.; Tiomela, A. Socio-economic and geographical determinants of financial inclusion in Cameroon. *Int. J. Res. Sci. Manag.* **2020**, *7*, 17–33. [CrossRef]
73. Kambewa, P.; Seyama, A.; Mazalale, J.; Chaweza, R. A Comparative analysis of financial inclusion indices: Evidence from Malawi. *Asian J. Econ. Financ.* **2020**, *2*, 309–320. Available online: http://arfjournals.com/index.php?route=product/product/volumearticle&vissue_id=34&issue_id=73&product_id=393 (accessed on 24 September 2021).
74. Asselin, L.-M.; Anh, V.T. Multidimensional poverty and multiple correspondence analysis. In *Quantitative Approaches to Multidimensional Poverty Measurement*; Kakwani, N., Silber, J., Eds.; Palgrave Macmillan: Basingstoke, UK, 2008; pp. 80–103, ISBN 978-1-349-28165-7. [CrossRef]
75. Asselin, L.-M. *Analysis of Multidimensional Poverty: Theory and Case Studies*; Springer Science + Business Media: New York, NY, USA, 2009; ISBN 978-1-4419-0842-1. [CrossRef]
76. Ezzrari, A.; Verme, P. A Multiple Correspondence Analysis Approach to the Measurement of Multidimensional Poverty in Morocco, 2001–2007. In *Policy Research Working Paper, No. 6087*; The World Bank: Washington, DC, USA, 2012; Available online: http://hdl.handle.net/10986/9330 (accessed on 29 September 2021).
77. Ambapour, S. Using multiple correspondence analysis to measure multidimensional poverty in Congo. *J. Data Anal. Inf. Process.* **2020**, *8*, 241–266. [CrossRef]
78. Parchomenko, A.; Nelen, D.; Gillabel, J.; Rechberger, H. Measuring the circular economy—A multiple correspondence analysis of 63 metrics. *J. Clean. Prod.* **2019**, *210*, 200–216. [CrossRef]
79. Corral-De-Witt, D.; Carrera, E.V.; Muñoz-Romero, S.; Tepe, K.; Rojo-Álvarez, J.L. Multiple correspondence analysis of emergencies attended by integrated security services. *Appl. Sci.* **2019**, *9*, 1396. [CrossRef]
80. Bastos, R.R.; Hippert, H.S.; Souza, A.C. An Application of Multiple Correspondence Analysis to the Classification of Residential Electricity Consumers. In Proceedings of the XIII International Conference on Industrial Engineering and Operations Management. Energy that Moves Production: A Dialogue among Integration, Project and Sustainability, Iguassu Falls, PR, Brazil, 9–11 October 2007; pp. 1–9.
81. Hjellbrekke, J. *Multiple Correspondence Analysis for the Social Sciences*; Routledge—Taylor & Francis Group: New York, NY, USA, 2019; ISBN 978-1-138-69968-7.
82. Alhuzali, T.; Beh, E.J.; Stojanovski, E. Multiple correspondence analysis as a tool for examining Nobel Prize data from 1901 to 2018. *PLoS ONE* **2022**, *17*, e0265929. [CrossRef]
83. Mahdiyasa, A.W.; Pasaribu, U.S. Multiple correspondence analysis using Burt matrix: A study of Bandung Institute of Technology student characteristics. *IOP Conf. Ser. Mater. Sci. Eng.* **2019**, *598*, 012012. [CrossRef]
84. Khangar, N.V.; Kamalja, K.K. Multiple Correspondence Analysis and its applications. *Electron. J. App. Stat. Anal.* **2017**, *10*, 432–462. [CrossRef]
85. Ritzer, G. Prosumer capitalism. *Sociol. Q.* **2015**, *56*, 413–445. [CrossRef]
86. Ritzer, G. Prosumption: Contemporary capitalism and the "new" prosumer. In *The Oxford Handbook of Consumption*; Wherry, F.F., Woodward, I., Eds.; Oxford University Press: New York, NY, USA, 2019; pp. 75–93, ISBN 978-019-069-558-3.
87. Czyżewska-Misztal, D.; Golejewska, A. The least innovative regions in Poland in the process of smart specialization. *Optimum* **2016**, *5*, 123–137. [CrossRef]
88. Kogut-Jaworska, M.; Ociepa-Kicińska, E. Smart specialisation as a strategy for implementing the regional innovation development policy—Poland case study. *Sustainability* **2020**, *12*, 7986. [CrossRef]
89. Wierzbicka, W. Information infrastructure as a pillar of the knowledge-based economy—An analysis of regional differentiation in Poland. *Equilibrium* **2018**, *13*, 123–139. [CrossRef]

90. Wichowska, A. Shrinking municipalities and their budgetary revenues on the example of the Warmian-Masurian Voivodeship in Poland. *Oecon. Copernic.* **2019**, *10*, 419–432. [CrossRef]
91. Wierzbicka, W. Socio-economic potential of cities belonging to the Polish National Cittaslow Network. *Oeconomia Copernic.* **2020**, *11*, 203–224. [CrossRef]
92. Węgrzyn, J. Does experience exert impact on a public-private partnership performance? The case of Poland. *Equilibrium* **2018**, *13*, 509–522. [CrossRef]
93. Kuchmacz, J.; Mika, Ł. Description of development of prosumer energy sector in Poland. *Polityka Energetyczna Energy Policy J.* **2018**, *21*, 5–20. [CrossRef]
94. Kotłowski, R.; Szmidtke, L. Elektrownie Wodne—Dziś Trzeba Budować Od Podstaw. Pomorski Przegląd Gospodarczy 2013. Available online: https://ppg.ibngr.pl/pomorski-przeglad-gospodarczy/elektrownie-wodne-dzis-trzeba-budowac-od-podstaw (accessed on 2 March 2022).
95. Achremczyk, S. *Historia Warmii i Mazur*; Ośrodek Badań Naukowych im. Wojciecha Kętrzyńskiego w Olsztynie: Olsztyn, Poland, 2010; Volume 1: Pradzieje—1772, ISBN 978-83-60839-33-1. Volume 2: 1772–2010, ISBN 978-83-60839-46-1.
96. Małłek, J. *Dwie Części Prus. Studia z dziejów Prus Książęcych i Prus Królewskich w XVI i XVII Wieku*; Wydawnictwo Pojezierze: Olsztyn, Poland, 1987; ISBN 83-7002-302-9.
97. Kiersnowski, H.; Dyrka, I.; Janas, M.; Sidorczuk, M.; Rudnicki, A.; Jureczka, J. Oil & Gas in Poland: New Opportunities; 4 September 2015. Project: "Support to Informational, Analytical and Implementation Shale Oil and Gas Activities Aimed at Ensuring Poland's Energy Security and Protection of the Environment, including Public Participation in the Licensing Process". Available online: https://infolupki.pgi.gov.pl/en/library/folder-oilgas-poland-new-opportunities (accessed on 3 March 2022).
98. Occupy Chevron (Poland). Protest Mieszkańców Żurawlowa Przeciwko Inwestycjom Koncernu Chevron. Available online: https://occupychevron.tumblr.com (accessed on 3 March 2022).
99. Szołucha, A. *Gaz Łupkowy w Polsce. Historia, Magia, Protest*; Wydawnictwo Naukowe PWN SA: Warsaw, Poland, 2021; ISBN 978-83-01-22065-5. Available online: https://ruj.uj.edu.pl/xmlui/bitstream/handle/item/287497/szolucha_gaz_lupkowy_2021.pdf (accessed on 5 March 2022). [CrossRef]
100. Dolina Wolicy. Natura 2000 a Turystyka—Portal Informacyjno-Edukacyjny. Available online: https://ine.eko.org.pl/index_areas.php?rek=669 (accessed on 3 March 2022).
101. Neslen, A. Poland's Shale Gas Revolution Evaporates in Face of Environmental Protests. *The Guardian*, 12 January 2015. Available online: https://www.theguardian.com/environment/2015/jan/12/polands-shale-gas-revolution-evaporates-in-face-of-environmental-protests?CMP=share_btn_fb#img-3(accessed on 4 March 2022).
102. Reed, S. Chevron to Abandon Shale Natural Gas Venture in Poland. *The New York Times*, 30 January 2015. Available online: https://www.nytimes.com/2015/01/31/business/international/chevron-to-abandon-shale-venture-in-poland-a-setback-to-fracking-europe.html(accessed on 4 March 2022).
103. Protest w Żurawlowie wspiera Warszawa. Available online: https://zamosc.naszemiasto.pl/protest-w-zurawlowie-wspiera-warszawa/ar/c1-1904427 (accessed on 4 March 2022).
104. Brutalne działania korporacji Chevron w Polsce. Solidarnościowe Wsparcie Dla Mieszkańców Żurawlowa od Francuskich Komitetów i Stowarzyszeń Walczących z Gazem Łupkowym, 10 June 2013. Available online: http://eko.org.pl/index_news.php?dzial=2&kat=19&art=985 (accessed on 5 March 2022).
105. Żurawlów: Drugi Tydzień Protestu i Żadnych Widoków Na Ugodę, 14 June 2014. Available online: https://www.lublin112.pl/zurawlow-drugi-tydzien-protestu-i-zadnych-widokow-na-ugode-zdjecia/ (accessed on 5 March 2022).
106. Mikołajków, J.; Sadurski, A. (Eds.) *Informator PSH (Państwowej Służby Hydrogeologicznej): Główne Zbiorniki Wód Podziemnych w Polsce*; Państwowy Instytut Geologiczny—Państwowy Instytut Badawczy: Warsaw, Poland, 2017; ISBN 978-83-7863-720-2. Available online: https://www.pgi.gov.pl/psh/dane-hydrogeologiczne-psh/947-bazy-danych-hydrogeologiczne/8890-gzwp.html (accessed on 5 March 2022).
107. Leman, H. Przynajmniej nie staliśmy bezczynnie. *Tygodnik Przegląd*, 11 August 2014. Available online: https://www.tygodnikprzeglad.pl/przynajmniej-nie-stalismy-bezczynnie/(accessed on 6 March 2022).
108. Kaczmarczyk, M. (Ed.) *Niska Emisja: Od Przyczyn Występowania do Sposobów Eliminacji*; Redakcja Globenergia—Geosystem Burek, Kotyza, S.C.: Kraków, Poland, 2015; ISBN 978-83-64339-02-8.
109. Obwieszczenie Ministra Klimatu i Środowiska z Dnia 12 Kwietnia 2021 r. w Sprawie Ogłoszenia Jednolitego Tekstu Rozporządzenia Ministra Środowiska w Sprawie Poziomów Niektórych Substancji w Powietrzu. Dziennik Ustaw RP 2021, poz. 845. Available online: http://isap.sejm.gov.pl/isap.nsf/DocDetails.xsp?id=WDU20210000845 (accessed on 6 March 2022).
110. Graboś, A.; Żymankowska-Kumon, S.; Sadlok, J.; Sadlok, R. Przeciwdziałanie Niskiej Emisji Na Terenach Zwartej Zabudowy Mieszkalnej. Stowarzyszenie Na Rzecz Efektywności Energetycznej i Rozwoju Odnawialnych Źródeł Energii "Helios": Bochnia, Poland, 2014. Available online: http://home.agh.edu.pl/~szk/files/docs/niska_emisja.pdf (accessed on 6 March 2022).
111. Toczko, B. Pomiary i oceny zanieczyszczenia powietrza w Polsce. Pył zawieszony. In *Zanieczyszczenia Powietrza w Polsce: Stan, Przyczyny, Skutki. Raport*; Pawłowski, L., Ed.; Wydawnictwo Polskiej Akademii Nauk—Komitet Inżynierii Środowiska: Lublin, Poland, 2019; pp. 8–23, ISBN 978-83-63714-50-5. Available online: https://kis.pan.pl/images/stories/pliki/pdf/Monografie/Monografie-vol-144.pdf (accessed on 6 March 2022).

112. Medoń, S. Problem smogu w świetle socjologii translacji. Studium przypadku zanieczyszczeń powietrza w Krakowskim Obszarze Metropolitalnym. *Przegląd Socjol.* **2021**, *70*, 41–64. Available online: http://journals.ltn.lodz.pl/index.php/Przeglad-Socjologiczny/article/view/1198/1109 (accessed on 8 March 2022). [CrossRef]
113. Grzechnik, J. *Rola nowych Mediów w Aktywizacji Społeczności Lokalnej Na Przykładzie Ruchów Miejskich w Krakowie. Studium Przypadku: Krakowski Alarm Smogowy*; Wydawnictwo ToC: Gdańsk, Poland, 2018; ISBN 978-83-952278-6-8. Available online: https://mediastery.ug.edu.pl/publikacje/ (accessed on 8 March 2022).
114. Lutomska, E. Krakowski Alarm Smogowy. Historia Ludzi, Którzy Się Wreszcie Zorganizowali, 30 September 2017. Available online: https://wszystkoconajwazniejsze.pl/ewa-lutomska-krakowski-alarm-smogowy/ (accessed on 9 March 2022).
115. Gardiner, B. Air of Revolution: How Activists and Social Media Scrutinise City Pollution. *The Guardian*, 31 January 2014. Available online: https://www.theguardian.com/cities/2014/jan/31/air-activists-social-media-pollution-city(accessed on 10 March 2022).
116. Gardiner, B. 'The Air is Stinking, it's Dirty': The Fight against Pollution in Kraków. *The Guardian*, 13 April 2015. Available online: https://www.theguardian.com/cities/2015/apr/13/air-dirty-fight-pollution-krakow-poland-ban-wood-coal(accessed on 10 March 2022).
117. Barteczko, A. Activists Warn Smog in Poland's Top Winter Resort Denser than in Delhi. *Reuters—World News*, 8 January 2016. Available online: https://www.reuters.com/article/poland-climatechange-pollution-idINKBN0UM1MZ20160108(accessed on 10 March 2022).
118. Łabędź, K. Nowe ruchy społeczne w przestrzeni miejskiej (na przykładzie Krakowa). *Ann. Univ. Mariae Curie Skłodowska Lub. Pol. Sect. K Politol.* **2017**, *24*, 187–204. Available online: https://journals.umcs.pl/k/article/view/3641/5038 (accessed on 11 March 2022). [CrossRef]
119. Antysmogowe ABC. Available online: https://www.malopolska.pl/dla-mieszkanca/srodowisko/antysmogoweabc (accessed on 11 March 2022).
120. Przegląd uchwał antysmogowych, 4 February 2022. Available online: https://czysteogrzewanie.pl/podstawy/uchwaly-antysmogowe/ (accessed on 12 March 2022).
121. Kelly, B. A 'Green Wave' Reprieve. *Traffic Eng. Control.* **2012**, *53*, 1–15. Available online: https://www.semanticscholar.org/paper/A-%E2%80%98Green-Wave%E2%80%99-Reprieve-Kelly/bc795655b1e9b8e78e1e41d81a8bb0103f93cb0d (accessed on 14 March 2022).
122. Bojadżijewa, I. Ekologia polityczna powietrza: O uwidzialnianiu miejskiego ryzyka. *Kult. i Społeczeństwo* **2016**, *60*, 35–53. Available online: http://czasopisma.isppan.waw.pl/index.php/kis/article/download/1242/1015/ (accessed on 14 March 2022). [CrossRef]
123. Rąpalski, P. Smog w Krakowie. Naukowcy z AGH Podważają Wcześniejsze Wyliczenia. *Gazeta Krakowska*, 11 December 2014. Available online: https://gazetakrakowska.pl/smog-w-krakowie-naukowcy-z-agh-podwazaja-wczesniejsze-wyliczenia/ar/3680048(accessed on 14 March 2022).
124. Maj, A. Teoria Naukowca z AGH Utrudni Walkę Ze Smogiem. *Dziennik Polski*, 13 December 2014. Available online: https://dziennikpolski24.pl/teoria-naukowca-z-agh-utrudni-walke-ze-smogiem/ar/3682398(accessed on 14 March 2022).
125. Serbeńska, A. Zieleń Receptą Na Zanieczyszczenia i Odtrutką Dla Cywilizacyjnych Toksyn cz. I, 21 August 2012. Available online: https://edroga.pl/ochrona-srodowiska/zielen-recepta-na-zanieczyszczenia-i-odtrutka-dla-cywilizacyjnych-toksyn-cz-i-21086390 (accessed on 15 March 2022).
126. Serbeńska, A. Zieleń Receptą Na Zanieczyszczenia i Odtrutką Dla Cywilizacyjnych Toksyn cz. II, 22 August 2012. Available online: https://edroga.pl/ochrona-srodowiska/zielen-recepta-na-zanieczyszczenia-i-odtrutka-dla-cywilizacyjnych-toksyn-cz-ii-22086397 (accessed on 15 March 2022).
127. Oleniacz, R.; Bogacki, M.; Szulecka, A.; Rzeszutek, M.; Mazur, M. Assessing the impact of wind speed and mixing-layer height on air quality in Krakow (Poland) in the years 2014–2015. *Czas. Inżynierii Lądowej Sr. I Archit. JCEEA* **2016**, *33*, 315–342. [CrossRef]
128. Bogacki, M.; Oleniacz, R.; Rzeszutek, M.; Szulecka, A.; Mazur, M. The impact of street cleaning on particulate matter air concentrations: A case study of a street canyon in Krakow (Poland). *E3S Web Conf.* **2018**, *45*, 00009. [CrossRef]
129. Jaskulska, A. Kraków: Nowe Pomysły Na Walkę z Zanieczyszczeniem Powietrza. Pomysł Czyszczenia Ulic Pojawił Się Po ogłoszeniu Wyników Badań Akademii Górniczo-Hutniczej. *Inżynieria*, 9 November 2016. Available online: https://inzynieria.com/energetyka/smog/wiadomosci/46540,krakow-nowe-pomysly-na-walke-z-zanieczyszczeniem-powietrza(accessed on 15 March 2022).
130. Bogacki, M.; Mazur, M.; Oleniacz, R.; Rzeszutek, M.; Szulecka, A. Re-entrained road dust PM10 emission from selected streets of Krakow and its impact on air quality. *E3S Web Conf.* **2018**, *28*, 01003. [CrossRef]
131. Sowa, P. Kraków Najbardziej Zanieczyszczonym Miastem Świata. Fatalna Jakość Powietrza. Interia Wydarzenia—Wiadomości Lokalne, 14 December 2021. Available online: https://wydarzenia.interia.pl/malopolskie/news-krakow-najbardziej-zanieczyszczonym-miastem-swiata-fatalna-j,nId,5706742 (accessed on 16 March 2022).
132. Air Quality Index (AQI) Basics. Available online: https://www.airnow.gov/aqi/aqi-basics/ (accessed on 16 March 2022).
133. "Wybór, nie zakaz"—Apelują Autorzy Obywatelskiej Inicjatywy, Domagając Się Zmian w Uchwale Antysmogowej dla Małopolski, 21 February 2022. Available online: https://nowytarg24.tv/wybor-nie-zakaz-apeluja-autorzy-obywatelskiej-inicjatywy-uchwalodawczej-domagajac-sie-zmian-w-uchwale-antysmogowej-dla-malopolski/ (accessed on 16 March 2022).
134. Obywatelski Projekt TAK Dla Drewna Złożony. Podpisało się 2500 Osób, 2 March 2022. Available online: https://panbartek.pl/obywatelski-projekt-tak-dla-drewna-zlozony-podpisalo-sie-2500-osob/ (accessed on 16 March 2022).

135. Analiza Wpływu Zakazu Palenia w Krakowie Na Zmiany Stężeń Benzo(a)pirenu Na Tle Innych Miast Wojewódzkich. Stowarzyszenie Otwarta Wieliczka, January 2022. Available online: http://badtek.webd.pro/wp-content/uploads/2022/01/analiza-wplywu-zakazu-palenia-w-Krakowie-na-zmiany-stezen-zanieczyszczen-powietrza-na-tle-innych-miast-wojewodzkich.pdf (accessed on 17 March 2022).
136. Uproszczona Analiza Wpływu Zanieczyszczeń Powietrza Na Zdrowie w Miastach "Smogowych" i "Nie Smogowych" Na Podstawie Oficjalnych Danych Statystycznych w Polsce. Stowarzyszenie Otwarta Wieliczka, January 2022. Available online: https://panbartek.pl/wp-content/uploads/2022/01/Uproszczona-analiza-wplywu-zanieczyszczen-powietrza-na-zdrowie-w-miastach-1.pdf (accessed on 17 March 2022).
137. Walka ze Smogiem: Walka o Nasze Zdrowie, Czy o Nasze Pieniądze? Pierwszy w Polsce Obywatelski Raport o Smogu, Który nie Został Sfinansowany Przez Żadne Instytucje ani Środowiska Biznesowe. Stowarzyszenie Otwarta Wieliczka, January 2022. Available online: https://panbartek.pl/wp-content/uploads/2022/01/Obywatelski-raport-o-smogu-styczen-2022.pdf (accessed on 17 March 2022).
138. Rekordowe Zainteresowanie Wymianą Pieców, 2 July 2019. Available online: https://www.krakow.pl/213212,1962,231438,powietrze,rekordowe_zainteresowanie_wymiana_piecow_.html (accessed on 18 March 2022).
139. Stawka VAT Na Drewno Opałowe Wzrosła Kilkanaście Procent! Magazyn Biomasa, 2 July 2020. Available online: https://magazynbiomasa.pl/stawka-vat-na-drewno-opalowe-wzrosla-kilkanascie-procent/ (accessed on 18 March 2022).
140. Brzoza, W. Ujawniamy. Kto i Co Stoi Za Alarmem Smogowym? Ekologia, Czy Biznes? Gazeta Bałtycka—Polski Serwis Informacyjny, 17 January 2017. Available online: http://gazetabaltycka.pl/promowane/ujawniamy-kto-i-co-stoi-za-alarmem-smogowym-ekologia-czy-biznes (accessed on 18 March 2022).
141. Stankiewicz, C. Blokada Urzędu Wojewódzkiego w Olsztynie Zakończona. *Rolnicy Zablokują Mrągowo*, 12 February 2015. Available online: https://olsztyn.naszemiasto.pl/blokada-urzedu-wojewodzkiego-w-olsztynie-zakonczona-rolnicy/ar/c8-3278853(accessed on 18 March 2022).
142. Szałaj, K. "Piątka AgroUnii". Rolnicy Przedstawili Postulaty. Sad Nowoczesny, 30 November 2020. Available online: https://www.sadnowoczesny.pl/artykuly/prawo-dla-sadownikow/agrounia-przedstawiala-swoje-postulaty/ (accessed on 18 March 2022).
143. Oleszczyński, I. Jak AgroUnia Chce Poprawić Sytuację Polskiego Rolnictwa? *Tygodnik Poradnik Rolniczy*, 7 December 2020. Available online: https://www.tygodnik-rolniczy.pl/articles/aktualnosci_/jak-agrounia-chce-poprawic-sytuacje-polskiego-rolnictwa/(accessed on 18 March 2022).
144. Worobiec, K.A. S16—Kontrowersyjna Inwestycja Drogowa Na Mazurach i Nie Tylko. Miesięcznik Dzikie Życie 2019, 7–8/301–302. Available online: https://dzikiezycie.pl/archiwum/2019/lipiec-i-sierpien-2019/s16-kontrowersyjna-inwestycja-drogowa-na-mazurach-i-nie-tylko (accessed on 18 March 2022).
145. Ziomek, W. Mazury Nie Chcą Nowej Drogi. Protesty Przeciwko Budowie S16. Money.pl, 6 May 2019. Available online: https://www.money.pl/gospodarka/mazury-nie-chca-nowej-drogi-protesty-przeciwko-budowie-s16-6378058578208385a.html (accessed on 18 March 2022).
146. Miasta Dobrego Życia 2030. Program Rozwoju Miast Sieci Cittaslow Województwa Warmińsko-Mazurskiego. Projekt; Cittaslow, 8 November 2021. Available online: https://cittaslowpolska.pl/images/stowarzyszenie/miasta_2030/CITTASLOW_2030_Program_2021_PROJEKT_1111.pdf (accessed on 19 March 2022).
147. Malewska–Szałygin, A. *Wiedza Potoczna o Sprawach Publicznych. Rozmowy o Władzy Lokalnej w Wybranej Gminie Mazurskiej w Latach 1994–1995*; Instytut Etnologii i Antropologii Kulturowej Uniwersytetu Warszawskiego & Wydawnictwo DiG: Warsaw, Poland, 2002; ISBN 83-7181-252-3.
148. Sztompka, P. *Socjologia. Analiza Społeczeństwa*; Wydawnictwo Znak: Kraków, Poland, 2002; ISBN 83-240-0218-9.
149. Treter, W. Warmińsko-Mazurskie: Przymus Ogrzewania Gazem w Krainie Taniego Drewna? 17 January 2022. Available online: https://czysteogrzewanie.pl/2021/12/warminsko-mazurskie-przymus-ogrzewania-gazem-w-krainie-taniego-drewna/ (accessed on 19 March 2022).
150. "Uchwały Antysmogowe" Dla Obszarów Miast Województwa Warmińsko-Mazurskiego/Oraz Dla Obszarów Poza Miastami. Available online: http://wfosigw.olsztyn.pl/wp-content/uploads/2022/01/Streszczenie-uchwal.pdf (accessed on 19 March 2022).
151. Ocena Jakości Powietrza w Strefach w Polsce za rok 2020. Zbiorczy Raport Krajowy z Rocznej Oceny Jakości Powietrza w Strefach Wykonanej Przez GIOŚ Według Zasad Określonych w Art. 89 ustawy—Prawo ochrony środowiska. Główny Inspektorat Ochrony Środowiska, Narodowy Fundusz Ochrony Środowiska i Gospodarki Wodnej: Warsaw, Poland, 2021. Available online: https://powietrze.gios.gov.pl/pjp/documents/download/109787 (accessed on 19 March 2022).
152. Warmińsko-Mazurskie 2030. Strategia Rozwoju Społeczno-Gospodarczego, Olsztyn, Poland, 18 February 2020. Available online: https://strategia2030.warmia.mazury.pl/wp-content/uploads/2020/02/Strategia-W-M-2030.pdf (accessed on 19 March 2022).
153. Leśnictwo w Województwie Warmińsko-Mazurskim w 2017 r. Urząd Statystyczny w Olsztynie, 31 August 2018. Available online: https://olsztyn.stat.gov.pl/download/gfx/olsztyn/pl/defaultaktualnosci/1290/2/4/1/lesnictwo_2017.pdf (accessed on 19 March 2022).
154. Brumme, D. BAFA—Förderung für Solarthermie: Nachfrage legt im November zu, 13 December 2021. Available online: https://blog.paradigma.de/bafa-foerderung-fuer-solarthermie-nachfrage-legt-im-november-zu/ (accessed on 20 March 2022).

155. Britto, A.; Dehler-Holland, J. Optimal investment in energy efficiency as a problem of growth-rate maximisation: Evidence and policy implications. In *ECEEE 2021 Summer Study on Energy Efficiency: A New Reality*; Lindström, T.L., Blume, Y., Hampus, N., Eds.; European Council for an Energy Efficient Economy: Stockholm, Sweden, 2021; pp. 855–864, ISBN 978-91-983878-9-6.
156. Thomas, S.; Schüwer, D.; Vondung, F.; Wagner, O. *Heizen ohne Öl und Gas bis 2035—Ein Sofortprogramm für erneuerbare Wärme und effiziente Gebäude. Im Auftrag von Greenpeace e.V.*; Wuppertal Institut: Wuppertal, Germany, 2022; Available online: https://www.greenpeace.de/publikationen/Heizen%20ohne%20%C3%961%20und%20Gas%20bis%202035.pdf (accessed on 20 March 2022).
157. Greenpeace—Studie: Gebäudewärme ab 2035 vollständig ohne Öl und Gas möglich. Ausstiegsgesetz für Öl- und Gasheizungen kann Umstieg auf Erneuerbare beschleunigen, 2 March 2022. Available online: https://presseportal.greenpeace.de/211054-greenpeace-studie-gebaudewarme-ab-2035-vollstandig-ohne-ol-und-gas-moglich (accessed on 20 March 2022).
158. Braungardt, S.; Keimeyer, F.; Bürger, V.; Tezak, B.; Klinski, S. *Phase-Out Regulations for Fossil Fuel Boilers at EU and National Level*; Öko-Institut e.V. Institut für angewandte Ökologie: Freiburg im Breisgau, Germany, 2021; Available online: https://www.oeko.de/fileadmin/oekodoc/Phase-out_fossil_heating.pdf (accessed on 21 March 2022).
159. Raport o Inflacji—Marzec 2022 r. Narodowy Bank Polski—Rada Polityki Pieniężnej: Warsaw, Poland, 2022. Available online: https://www.nbp.pl/home.aspx?f=/polityka_pieniezna/dokumenty/raport_o_inflacji.html (accessed on 20 March 2022).
160. Brinkmann, B. Krieg in Europa: Deutschland kann auf Gas und Öl aus Russland verzichten. *Süddeutsche Zeitung*, 8 March 2022. Available online: https://www.sueddeutsche.de/meinung/russland-deutschland-ukraine-gas-oel-importstopp-1.5543329?reduced=true(accessed on 21 March 2022).
161. Dudek, B. Niemiecka Prasa: Zachód Finansuje Wojnę Putina, 9 March 2022. Available online: https://www.dw.com/pl/niemiecka-prasa-zach%C3%B3d-finansuje-wojn%C4%99-putina/a-61058620 (accessed on 21 March 2022).
162. PAS Podpisał Apel do Rządu o Zaprzestanie Importu Surowców Energetycznych z Rosji, 28 February 2022. Available online: https://polskialarmsmogowy.pl/2022/02/pas-podpisal-apel-do-rzadu-o-zaprzestanie-importu-surowcow-energetycznych-z-rosji/ (accessed on 21 March 2022).
163. Zarząd Województwa Warmińsko-Mazurskiego Ogłasza Konsultacje Projektów Uchwał w Sprawie Wprowadzenia Na Terenie Województwa Warmińsko-Mazurskiego Ograniczeń Lub Zakazów w Zakresie Eksploatacji Instalacji, w Których Następuje Spalanie Paliw, o Których Mowa w Art. 96 Ustawy z Dnia 27 Kwietnia 2001 r. Prawo Ochrony Środowiska, 22 December 2021. Available online: https://bip.warmia.mazury.pl/kategoria/59/konsultacje-spoleczne.html (accessed on 22 March 2022).
164. Kusznerczuk, P. Spotkanie Konsultacyjne w Sprawie Projektów tzw. *Uchwał Antysmogowych*, 17 January 2022. Available online: http://gminamilki.pl/2022/01/17/spotkanie-konsultacyjne-w-sprawie-projektow-tzw-uchwal-antysmogowych/(accessed on 22 March 2022).
165. Treter, W. Konsultacje i Zmiany Uchwał Antysmogowych—Luty 2022, 20 February 2022. Available online: https://czysteogrzewanie.pl/2022/02/konsultacje-i-zmiany-uchwal-antysmogowych-luty-2022/ (accessed on 24 March 2022).
166. Grosz, K. Ruch Narodowy z Olsztyna Stanowczo Sprzeciwia Się Uchwale Antysmogowej. Przewodnicząca Sejmiku: To Puste Słowa. Gazeta Olsztyńska, 13 January 2022. Available online: https://gazetaolsztynska.pl/797746,Ruch-Narodowy-z-Olsztyna-stanowczo-sprzeciwia-sie-uchwale-antysmogowej-Przewodniczaca-Sejmiku-to-puste-slowa.html (accessed on 24 March 2022).
167. Wysocki, K.M. Przymus? Nie Tędy Droga. *Gazeta Olsztyńska*, 15 February 2022. Available online: https://gazetaolsztynska.pl/806770,Przymus-Nie-tedy-droga.html(accessed on 24 March 2022).
168. Szczepanik, R. Zielone Płuca Polski Zagrożone Smogiem. Co z Uchwałami Antysmogowymi? *Gazeta Olsztyńska*, 10 February 2022. Available online: https://gazetaolsztynska.pl/805467,Zielone-Pluca-Polski-zagrozone-smogiem.html(accessed on 24 March 2022).
169. Lyon, T.P.; Maxwell, J.W. Astroturf: Interest group lobbying and corporate strategy. *J. Econ. Manag. Strategy* **2004**, *13*, 561–597. [CrossRef]
170. Bousdekis, A.; Kardaras, D. Digital transformation of local government: A case study from Greece. In *Proceedings of the 2020 IEEE 22nd Conference on Business Informatics (CBI 2020), Antwerp, Belgium, 22–24 June 2020*; IEEE Computer Society: Los Alamitos, CA, USA, 2020; Volume 2, pp. 131–140, ISBN 978-1-7281-9926-9. [CrossRef]
171. dos Santos, M.P.; de Melo, S.C.; Santiago, M.C.; Nazareth, P. Inclusion in public administration: Developing the concept of inclusion within a school of accounts and administration. *Int. J. Educ. Admin. Pol. Stud.* **2017**, *9*, 35–46. Available online: https://files.eric.ed.gov/fulltext/EJ1136777.pdf (accessed on 24 March 2022). [CrossRef]
172. Jakimowicz, A. The energy transition as a super wicked problem: The energy sector in the era of prosumer capitalism. *Energies* **2022**, *15*, 9109. [CrossRef]
173. Jakimowicz, A. The future of the energy sector and the global economy: Prosumer capitalism and what comes next. *Energies* **2022**, *15*, 9120. [CrossRef]
174. Maris, G.; Flouros, F. The Green Deal, National Energy and Climate Plans in Europe: Member States' compliance and strategies. *Adm. Sci.* **2021**, *11*, 75. [CrossRef]

energies

MDPI

Perspective

The Material Entropy and the Fourth Law of Thermodynamics in the Evaluation of Energy Technologies of the Future

Aleksander Jakimowicz

Department of Economic Policy, Institute of Economics, Polish Academy of Sciences, Palace of Culture and Science, 00-901 Warsaw, Poland; ajakimowicz@inepan.waw.pl

Abstract: The primary purpose of this article is to use the laws of thermodynamics, mainly the second and fourth laws, to evaluate three energy technologies of the future: fusion, solar, and fission. Among the criteria used to evaluate them, the most important are the amount of matter needed to sustain the technology itself and the environmental impact. Much emphasis is placed here on the fourth law of thermodynamics, which introduces the concept of material entropy. Zemansky–Georgescu-Roegen's Law of Inevitable Dissipation of Useful Concentrated Matter states that, in the economic process, some matter is inevitably degraded and becomes unavailable matter. This has tremendous implications for humanity as a whole since the Earth is thermodynamically a closed system, meaning that it cannot exchange matter with space but is open to the flow of solar energy. This results in the need to conserve matter and natural resources. This law can be used as an important criterion for the selection of energy technology. Moreover, the flow–fund model, which was proposed by Georgescu-Roegen, was used to assess the viability of energy technologies. The final conclusion is that there is no Promethean technology of the third kind yet, but the closest to meeting this condition is solar technology. Technology based on nuclear fission has been rejected due to its adverse ecological effects, while fusion technology has proven to be less useful due to the matter criterion, the negative environmental impact, since radioactive waste only becomes safe for humans after 500 years, and the risks associated with nuclear proliferation. Solar technology can become Prometheus III only after all of humanity is involved with this project, which requires profound social changes, widespread demilitarization, and the development of organic agriculture. This implies the necessity of the emergence of a global solar society based on an economic system called solar communism.

Keywords: entropy of matter; fourth law of thermodynamics; flow–fund model; energetic dogma; viability of energy technology; solar energy; nuclear fission energy; controlled fusion energy; Promethean recipe; entropic debt

1. Introduction

Nicolas Georgescu-Roegen is a well-known but underappreciated pioneer of applications of the laws of thermodynamics to economics. Over time, his concepts have become an important trend in research known as ecological economics or bioeconomics. He is known for his many controversial views, which to this day have not been satisfactorily explained, although they are critical to the survival of humanity. The main purpose of this article is to demonstrate that Georgescu-Roegen's key thermodynamic views were correct and can be applied to the evaluation of three energy technologies of the future: fusion, solar, and fission. The choice between these technologies is particularly important now, at a time of global warming, which is causing catastrophic climate change. The consequences of this choice could help the survival of the human species, enable economic growth and development, or lead to a true climate catastrophe known from various dystopias [1].

Having clarified the uncertainties surrounding the fourth law of thermodynamics, I was able to proceed with the main goal of this work, which is to use this law to evaluate three of humanity's energy technologies: fusion, solar, and fission. The choice of these technologies as the most important for humanity's survival was made by Georgescu-Roegen himself. He also formulated the concept of a Promethean energy technology that meets the conditions for viability, meaning that it is capable of sustaining all the processes of the economic system. Accordingly, selected energy technologies must be evaluated within the flow–fund matrix. In addition to the material entropy criterion, I also use an ecological criterion based on the other laws of thermodynamics. Moreover, I take into account the entropic debt, the average LCOE, the fulfillment of the Promethean condition regarding the viability of the technology, and the type of energy conversion. There have only been two Promethean technologies in human history so far, the mastery of fire and the steam engine, while a third such technology is still a matter of the future. The in-depth analysis in this article indicates that Prometheus III will soon be solar technology.

2. Zemansky–Georgescu-Roegen's Law of Inevitable Dissipation of Useful Concentrated Matter

In Nicholas Georgescu-Roegen's view, the fourth law of thermodynamics applies to closed systems that can exchange energy with the external environment but are closed due to the flow of matter. In such systems, material entropy gradually increases until it reaches the maximum level associated with a homogeneous mixture of matter and, thus, its greatest disorder. This state is characterized by the absence of qualitative differences between materials so that all matter eventually becomes unavailable. In this state, we address the material death of the system. This complements Kelvin's hypothesis of the heat death of the universe resulting from the operation of the second law of thermodynamics, which is valid for isolated systems [2].

The source of the fourth law of thermodynamics is phenomena such as friction, viscosity, inelasticity, electric resistance, and magnetic hysteresis [3] (p. 193). This results in the dissipation of matter in the environment, best exemplified by the abraded tires of vehicles moving on the road. This dissipated dust cannot be recycled due to the finitude of human existence. This type of matter is irrevocably lost to humans. Only secondary raw materials such as waste paper, scrap metal, or broken glass can be recycled. Georgescu-Roegen refers to this matter as garbojunk. It is still accessible to man, but it no longer has a usable form. No single general mathematical formula can be given to describe the entropy of matter, as is the case with energy degradation, because matter in bulk is heterogeneous, and the dissipation factors vary from one type of matter to another.

The fourth law of thermodynamics can be formulated in several equivalent ways:

1. A closed system (i.e., a system that cannot exchange matter with the environment) cannot perform work indefinitely at a constant rate [4] (p. 304);
2. In a closed system, matter continuously and irrevocably degrades from an available to unavailable state [5] (p. 121, footnote 24);
3. Unavailable matter cannot be recycled [4] (p. 304);
4. Complete recycling is impossible [6] (p. 60);
5. A closed system that can perform mechanical work steadily constitutes the perpetual motion of the third kind [5] (p. 121, footnote 24).

The equivalent of the fourth law of thermodynamics is even found in the Bible. During the Sermon on the Mount, the Lord Jesus Christ delivers the following words:

> *Do not store up for yourselves treasures on earth, where moths and vermin destroy, and where thieves break in and steal. But store up for yourselves treasures in heaven, where moths and vermin do not destroy, and where thieves do not break in and steal. For where your treasure is, there your heart will be also.* (Matthew, 6: 19–21, NIV)

After in-depth research, it was found that the forerunner of the fourth law of thermodynamics was the well-known physicist Mark W. Zemansky, who, decades before Georgescu-Roegen, had formulated the concept of a perpetual motion machine of the third kind in an identical way [3] (p. 193). He was engaged in the study of dissipative effects occurring in machines as a result of friction and similar phenomena. This led him to discover the phenomenon of external mechanical irreversibility, which involves the dissipation of work into internal energy. If there were no friction in nature, machines could operate indefinitely without violating the laws of thermodynamics, the first and second. In that case, they would run, but they would not perform work. The lack of dissipation of work in machines precisely means the existence of a perpetual motion machine of the third kind. This means that the thermodynamic effects of friction cannot be identified with the first two laws of thermodynamics, and a qualitatively new law is needed.

An unintended consequence of Zemansky and Georgescu-Roegen's work is the confusion over the existence of different kinds of perpetual motion machines in physics. By convention, each such machine is associated with a specific law of thermodynamics: zeroth, first, and second. There is also the concept of a perpetual motion machine of the third kind associated with the third law of thermodynamics [7] (pp. 283–284). Therefore, what Zemansky and Georgescu-Roegen discovered should be classified as a perpetual motion machine of the fourth kind. Their discovery proves not only the existence of the fourth law of thermodynamics based on dissipative effects but also its independence from the other laws of thermodynamics.

Paul A. Samuelson, who received the Sveriges Riksbank Prize in Economic Sciences in Memory of Alfred Nobel in 1970, referred to the law discovered by Georgescu-Roegen as Georgescu-Roegen's Law of Inevitable Dissipation of Useful Concentrated Matter [8] (p. XVII). In light of the above, one should refer to Zemansky–Georgescu-Roegen's Law.

3. Energetic Dogma versus the Actual Relationship between the Economic Process and the Environment

There is a notion in the main science current on the relationship between energy and matter, which can be called the energetic dogma. It states that the dissipation of matter, predicted by the fourth law of thermodynamics, can be completely eliminated, provided a sufficient amount of energy is available. Therefore, matter is not important; only energy counts. In other words, according to the energetic dogma, a perpetual motion machine of the third kind is possible [6] (pp. 53–56). Such views have been expressed by many researchers. Here, one will be shown, presented by a well-known economist, Kenneth E. Boulding:

> *In the case of material systems, we can distinguish between entropic processes, which take concentrated materials and diffuse them through the oceans or over the earth's surface or into the atmosphere, and anti-entropic processes, which take diffuse materials and concentrate them. Material entropy can be taken as a measure of the uniformity of the distribution of elements and, more uncertainly, compounds and other structures on the earth's surface. There is, fortunately, no law of increasing material entropy, as there is in the corresponding case of energy, as it is quite possible to concentrate diffused materials if energy inputs are allowed. Thus the processes for the fixation of nitrogen from the air, processes for the extraction of magnesium or other elements from the sea, and processes for the desalinization of sea water are anti-entropic in the material sense, though the reduction of material entropy has to be paid for by inputs of energy and also inputs of information, or at least a stock of information in the system. In regard to matter, therefore, a closed system is conceivable, that is, a system in which there is neither an increase nor a decrease in material entropy. In such a system, all outputs from consumption would constantly be recycled to become inputs for production, as, for instance, nitrogen in the nitrogen cycle of the natural ecosystem.* [9] (Boulding, 2011, p. 7)

It is astonishing that the energetic dogma is applied to the Earth, which—thermodynamically—is a closed system, i.e., it is open to the flux of solar energy but closed to the flux of matter. If the Earth was an open system, which could exchange not only matter but also energy with the environment, the energetic dogma would be justified.

In order to justify the truth of the fourth law of thermodynamics, Georgescu-Roegen prepared the flow–fund model, which he used both for studies of the standard economic process as part of the energetic dogma but also to examine the process under conditions that take into consideration its actual relations with the environment [6] (pp. 56–58, 63–64). The fund factors (agents) include the capital, people, and Ricardian land. They are used during the process, but their size remains constant. On the other hand, the flow factors undergo transformation.

3.1. The Flow–Fund Model under the Conditions of the Energetic Dogma

Table 1 shows the flow–fund matrix, which analytically describes the economic process in line with the energetic dogma [6] (pp. 56–58). Complete matter recycling is possible in this situation as long as people have a sufficient amount of energy. In other words, energy, not matter, is important in the struggle for humanity's economic survival. The positive coordinates of the matrix are outflows of any kind, whereas negative coordinates represent the inflows. Subsequently, the economic activity is divided into consolidated processes and aggregated commodities in the following manner:

- P_1: produces controlled energy, CE, from energy in situ, ES;
- P_2: produces capital goods, K;
- P_3: produces consumer goods, C;
- P_4: completely recycles the material wastes, W, of all processes into recycled matter, RM;
- P_5: maintains the population, H.

Table 1. The economic process in relation to the environment according to the energetic dogma.

Product/Process	P_1 (ES → CE)	P_2 (Produces K)	P_3 (Produces C)	P_4 (W → RM)	P_5 (Maintains H)
			Flow Coordinates		
Controlled energy (CE)	x_{11}	$-x_{12}$	$-x_{13}$	$-x_{14}$	$-x_{15}$
Capital equipment (K)	$-x_{21}$	x_{22}	$-x_{23}$	$-x_{24}$	$-x_{25}$
Consumer goods (C)	*	*	x_{33}	*	$-x_{35}$
Recycled matter (RM)	*	$-x_{42}$	$-x_{43}$	x_{44}	*
Energy in situ (ES)	$-e_1$	*	*	*	*
Wastes (W)	w_1	w_2	w_3	$-w_4$	w_5
Dissipated energy (DE)	d_1	d_2	d_3	d_4	d_5
			Fund Coordinates		
Capital	K_1	K_2	K_3	K_4	K_5
People	H_1	H_2	H_3	H_4	H_5
Ricardian land	L_1	L_2	L_3	L_4	L_4

During each of these processes, part of the energy transforms into dissipated heat, and thereby it becomes unavailable. This dissipated energy, DE, is channeled back into the environment. The energy flows between the economy and the environment consist of the input flow, e_1, and the output flow:

$$d = \sum_{i=1}^{5} d_i \quad (1)$$

According to energetic dogma, matter cannot be transferred to an economic process from the environment, nor can it leave the process. The whole matter in the economy is totally recycled. Neither a growing nor a declining economy can be a test for the energetic dogma but a stationary process whose characteristic features include reproducibility. In other words, material growth cannot be based on an environmental flow of energy alone, whereas a declining economy may not need a flow of environmental matter. One should not forget that each energy flow requires a material transmitter. This means that an economic process cannot be conducted without a material scaffold, which is represented by its agents, i.e., the fund coordinates. One should mention here the capital equipment, K, people, H, and Ricardian land, L. At the same time, in order to meet the reproducibility condition, the output flow of capital, x_{22}, is intended to maintain the capital funds, K_i; therefore, its wear and tear must be compensated for with the maintenance flows x_{2i}. This means that the consumed capital, K, is fully restored during the economic process. In the same manner, the flows x_{i5} maintain the population, H, on a constant level. After expressing all flows in the physical units, the conservation laws on the macro level have the following form:

$$d_1 = e_1 - x_{11}\,, \tag{2}$$

$$d_i = x_{1i}\,,\ (i = 2,\ 3,\ 4,\ 5)\,, \tag{3}$$

$$w_1 = x_{21}\,, \tag{4}$$

$$w_2 = x_{42} - x_{22}\,, \tag{5}$$

$$w_3 = x_{23} + x_{43} - x_{33}\,, \tag{6}$$

$$w_4 = x_{44} - x_{24}\,, \tag{7}$$

$$w_5 = x_{25} + x_{35}\,, \tag{8}$$

According to Georgescu-Roegen, this model assumes that every recipe is feasible, i.e., it creates its own product when it is supported by specific funds and when specific inputs are directed to it. The feasibility of every recipe does not have to entail the viability of the technology, which depends on all the processes taken together [6] (p. 58). The necessary and sufficient conditions for the viability of the technology of a reproducible economic system shown in Table 1 are as follows:

$$x_{i5} \geq x_{i5}^{0}\,,\ (i = 1,\ 2,\ 3), \tag{9}$$

$$\sum_{i=2}^{5} x_{1i} = x_{11}\,, \tag{10}$$

$$x_{35} = x_{33}\,, \tag{11}$$

$$\sum_{i=1,2,3,5} w_i = w_4\,, \tag{12}$$

$$\sum_{i=1,\ 3,\ 4,\ 5} x_{2i} = x_{22}\,, \tag{13}$$

$$\sum_{i=2,3} x_{4i} = x_{44}\,. \tag{14}$$

where x_{i5}^{0} denotes the minimums determined on the basis of the normal standard of living, and it is evident that the variable subscript differed significantly from the fixed one.

Georgescu-Roegen points out that the theoretical foundation for complete recycling is provided by the quasi-static process, which uses the van't Hoff box (or van't Hoff equilibrium box) [10] (pp. 139–141). It is a vessel of unchangeable volume, which contains various substances taking part in the reaction. These substances are in the equilibrium state [11]. The vessel walls are permeable to some substances and impermeable to others. The application of this vessel is limited to reactions in a homogeneous system, e.g., to gases or dilute solutions [12,13] (pp. 118–119).

Usually, a reaction has the following form:

$$A + B = C + D. \tag{15}$$

The box contains four substances $A,\ B,\ C,\ D$ in a state of equilibrium. Subsequently, 1 mol of substance A and 1 mol of substance B are introduced into the box isothermally and reversibly through walls permeable only to one of them at the equilibrium concentration or pressure. Later, assuming that a chemical reaction occurred, 1 mol of substance C and one mol of substance D are removed through walls permeable only to them while the equilibrium concentration or pressure is maintained. It is found that the transformation of A and B into C and D under conditions of the equilibrium concentration and pressure goes on, and it occurs without mechanical work, as the box volume is constant. At the end, the box and its contents are in exactly the same state as at the beginning. Under these conditions, no actual external work was performed during the chemical transformation. If substances A and B are under an arbitrary pressure or concentration at the beginning, then work will be performed to bring them isothermally and reversibly to the concentration corresponding to the equilibrium. After this stage is reached, they can be introduced into the equilibrium box, where they will come into reaction and turn into substances C and D, which in turn can be removed at the equilibrium concentration value, with the in and out operation not requiring any work [14] (p. 103).

The aim of the van't Hoff box is to obtain useful work, which arises from the following diagram:

$$\text{reaction} \rightarrow \text{energy} \rightarrow \text{heat} \rightarrow \text{useful work}, \tag{16}$$

without violating the second law of thermodynamics, which is possible even at a constant temperature, using the heat generated during the reaction [15]. After dividing the van't Hoff equilibrium box into two parts, it can generate heat and useful work at the same time, without violating the first law of thermodynamics [16]. However, Kozo Mayumi points out that an ideal box uses a process which actually needs an infinite time for an infinitesimal movement. This shows that complete recycling, i.e., separation of the mixed materials completely, would also have to take place in an infinite time. Therefore, in reality, there has to be a stage at which the recycling process would have to stop. This proves that the energetic dogma cannot be implemented in the real world [17], [18] (pp. 56–61).

The economic process shown in Table 1 presents only one elementary aspect of reality, i.e., that one cannot address energy without a material lever. Therefore, despite the flow complex existing in the Western intellect, it is not assumed that the energetic dogma is so unreal to suggest that the actual processes do not require any material structures accompanying energy at the macro level.

3.2. *Flow of Matter and Energy in the Real Economic Process*

The Law of Inevitable Dissipation of Useful Concentrated Matter leads one to slightly different conclusion regarding the finitude of natural resources than those reached by Robert M. Solow, another winner of the Sveriges Riksbank Prize in Economic Sciences in Memory of Alfred Nobel, this time in 1987. He claims that the productivity of natural resources will grow exponentially over time due to the substitutability relation with human-made capital [19]. However, the fourth law proves absolute in this regard. A total of 2 billion tons of iron were produced during the period of 1870–1950 in the United States. If all of it had been used at the end of this period, its inventory would have amounted to 13.5 tons per person. However, the total amount of iron in circulation at the time amounted to 60% of the original production, which corresponded to approximately 8 tons per person. The remaining iron was lost irretrievably as a result of oxidation by the air, corrosion by liquids, general wear, and production loss [20] (p. 192). It should also be stressed that approximately 10% of the iron used as the charge in a steel furnace is lost irrecoverably, and the iron which is not converted to steel is lost irrecoverably within about 100 years due to corrosion and other loss factors [21] (p. 156). Therefore, the fourth law of thermodynamics is largely responsible for an increase in the productivity of iron and can be similar with other natural resources.

The flow–fund model can be used to show real relations between the economic process and the natural environment [6] (pp. 63–64). Table 2, which is expanded from Table 1, shows an additional process, P_0, which transforms matter in situ, MS, into controlled matter, CM. It can be seen that even a stationary economy cannot function without a constant inflow of matter from the environment. There are new flows here, s_i, denoting dissipated matter, and DM, which is produced by every process and released to the environment. Moreover, the recycling process, P_4, due to the fourth law, no longer transforms all material waste, as some matter is irreversibly dissipated in the environment and becomes unavailable. Only trash and waste, earlier referred to as garbojunk, GJ, can be recycled. Moreover, there is also a flow of refuse, R, which can largely consist of available matter and available energy, but whose form is not potentially useful.

Table 2. The actual relationship between the economic process and the environment.

Product/Process	P_0 (MS → CM)	P_1 (ES → CE)	P_2 (Produces K)	P_3 (Produces C)	P_4 (W → RM)	P_5 (Maintains H)
			Flow Coordinates			
Controlled matter (CM)	x_{00}	*	$-x_{02}$	$-x_{03}$	$-x_{04}$	*
Controlled energy (CE)	$-x_{10}$	x_{11}	$-x_{12}$	$-x_{13}$	$-x_{14}$	$-x_{15}$
Capital equipment (K)	$-x_{20}$	$-x_{21}$	x_{22}	$-x_{23}$	$-x_{24}$	$-x_{25}$
Consumer goods (C)	*	*	*	x_{33}	*	$-x_{35}$
Recycled matter (RM)	*	*	$-x_{42}$	$-x_{43}$	x_{44}	*
Energy in situ (ES)	*	$-e_1$	*	*	*	*
Matter in situ (MS)	$-M_0$	*	*	*	*	*
Garbojunk (GJ)	w_0	w_1	w_2	w_3	$-w_4$	w_5
Dissipated energy (DE)	d_0	d_1	d_2	d_3	d_4	d_5
Dissipated matter (DM)	s_0	s_1	s_2	s_3	s_4	s_5
Refuse (R)	r_0	r_1	r_2	r_3	r_4	r_5
			Fund Coordinates			
Capital	K_0	K_1	K_2	K_3	K_4	K_5
People	H_0	H_1	H_2	H_3	H_4	H_5
Ricardian land	L_0	L_1	L_2	L_3	L_4	L_4

The viability of the steady state is represented by the following relations:

$$\sum_{i=2,3,4} x_{0i} = x_{00}\,, \quad (17)$$

$$\sum_{i=0,2,3,4,5} x_{1i} = x_{11}\,, \quad (18)$$

$$\sum_{i=0,1,3,4,5} x_{2i} = x_{22}\,, \quad (19)$$

$$x_{35} = x_{33}\,, \quad (20)$$

$$\sum_{i=2,3} x_{4i} = x_{44}\,, \quad (21)$$

$$\sum_{i=0,1,2,3,5} w_i = w_4\,, \quad (22)$$

Since R can include both energy and matter, one cannot note the relations for the conservation of these items separately, like in the previous case.

4. Evaluation of Energy Technologies under the Flow-Fund Model

Georgescu-Roegen defines technology as a set of feasible recipes, with one recipe existing for producing each commodity. Heating a house with photovoltaic panels is an example of a feasible recipe, whereas producing electricity with thermonuclear fusion is not.

Moreover, it has been suggested that technologies can be viable or not viable, depending on whether Equations (17)–(22) are satisfied [6] (p. 70).

With two viable technologies, $T^1(e_1^1, M_0^1)$ and $T^2(e_1^2, M_0^2)$, which provide the same net real income, one will have to be chosen which will prove more beneficial from the environmental perspective. The choice between them is reasonable only when inequalities $e_1^1 < e_1^2$ and $M_0^1 > M_0^2$ exist. If both technologies use a certain terrestrial energy, then an analysis of energy and matter will not provide an answer, but if they use solar radiation, then matter prevails, which shows that T^2 must be chosen. When T^2 is solar technology, and T^1 is terrestrial technology, the answer is clear—T^2 is the preferable one. In the opposite case, the terrestrial source will be used as long as the whole energy in situ is exhausted [6] (p. 70).

This analysis shows the considerable importance of solar energy for the global economy. However, Georgescu-Roegen suggests that there are only feasible recipes in this area, whereas viable technologies are still beyond humanity's reach. The implementation of a viable technology requires developing a recipe for the production of converters of solar radiation, which would allow for producing a sufficient number of new converters only with the use of solar energy converted by other devices of this type. Currently, this is not possible, which leads one to the conclusion that the existing solar technologies are parasites of fossil fuel-based technologies, and they will disappear once these fuels are exhausted. However, Georgescu-Roegen admits in his analysis of the matter and technology that solar energy may be the best option for humanity [6] (pp. 70–71).

When reviewing humanity's entropic problem, Georgescu-Roegen points out that harnessing fire was the first viable technology. Referring to the Greek myth of Prometheus, who stole fire from gods and gave it to people, he calls controlling fire the first Promethean recipe. This resulted in a classification of energy technologies, which—counting from the beginnings of civilization until now—includes only two items, whereas the third is—for now—hypothetical, and it has been demonstrated earlier that it applies to the technology of acquiring solar energy. They will be discussed here individually [6] (pp. 71–74).

1. The technology of fire control is a Promethean recipe of the first kind. Nowadays, it is one of the most ordinary phenomena, but making use of it changed the whole earlier world. Fire is associated with qualitative energy conversion, as it allows one to convert the chemical energy of flammable materials into caloric power. Moreover, fire is a source of a chain reaction, as a small flame allows for the burning of not only one forest but even all the forests. Harnessing fire allowed people to heat shelters, cook food, smelt and shape metals, and bake bricks and ceramics. This technology is associated with the Wood Age, as wood was the basic fuel at the time and was used until its shortage started to be felt. This happened in the second half of the seventeenth century and made it necessary to seek other sources of energy.
2. A Promethean recipe of the second kind is a coal-fired heat engine, which enables a new qualitative energy conversion, allowing for the transformation of caloric power into motor power. The first heat engine, called an aeolipile, was developed by a Greek-Egyptian mathematician and engineer, Hero of Alexandria, in the first century AD [22]. It was a kind of bladeless radial steam turbine in the form of a ball fixed on an axis with two nozzles in opposite directions [23] (p. 72). It was powered by steam fed from a boiler with water heated by fire, situated under the turbine [24] (pp. 228–232). However, Heron's engine is not regarded as the forerunner of the steam engine, as it was a machine exhibiting technical ingenuity rather than technological progress [25] (pp. 55–56). With respect to the flow–fund model, it would be a feasible technology rather than a viable technology. The steam engine was invented by Thomas Savery and Thomas Newcomen much later, in the pre-industrial revolution times (1760–1840), and the device was improved by James Watt in 1776, which considerably accelerated global economic growth [26,27]. As with fire control, it is a chain reaction in this case, too. With a specific amount of coal and a heat engine, one can excavate more coal and other minerals, which will enable one to produce more heat engines, and

engaging these new devices to work will yield the number of heat engines needed at the moment.

3. A Promethean recipe of the third kind is for a solar collector, which could meet the condition of a chain reaction, i.e., provide enough energy to produce additional collectors of the same type or even more perfect. Currently, there is no Promethean recipe which would allow the global economy, powered with solar energy, to be energy self-sufficient. Nevertheless, methods of solar energy conversion into electricity are still improving, which may suggest that a technological breakthrough is not far away. One should also note that the implementation of solar communism does not only depend on the technological progress alone, i.e., solarization of the global energy infrastructure, but also on demilitarization and the development of organic farming [28,29]. Therefore, the improvement of solar technologies should be simultaneous with profound economic, social, and cultural transformations for which humanity—it seems—is not yet prepared.

Promethean recipes of the second kind are still dominating energy technologies. Steam turbines in thermal power stations, which generate rotary motion, are coupled with electric generators to make use of the motion to produce electricity. Such turbines can be powered by fossil fuels, nuclear fuels, geothermal energy and, in some cases, concentrated solar power (which is the optimum solution due to the heat budget of the Earth) [30]. Even nuclear submarines are powered by steam turbines. It is estimated that 90% of the electricity in the USA is produced by steam turbines, whose maximum efficiency cannot exceed 38%, which is a consequence of the upper limit imposed by the Carnot cycle. Systems of electricity production which do not utilize steam expanding in a turbine include hydropower plants and photovoltaic cells. Therefore, this limitation does not apply to them [31] (pp. 189–190). Moreover, the development of our civilization is still based on fossil fuels. In 2015, at least 85% of the world's total commercial primary energy supply—except the share of traditional biofuel, whose consumption cannot be measured exactly—came from fossil fuels [32] (p. 195). This shows that, despite the continuous perfection of technology, and the appearance of many impressive inventions, such as computers, smartphones, jet aircraft, spacecraft, and modern information and communication technologies, we are still in the 19th century with respect to energy generation technologies, that is, in the embrace of Prometheus II.

5. Possible Candidates for a Promethean Recipe of the Third Kind

Nowadays, there are many feasible energy technologies which are likely to become at least partial solutions to humanity's energy problems in the future, but they cannot be regarded as viable technologies at present. Usually, breeder reactors and controlled thermonuclear energy are mentioned along with solar energy [6] (pp. 70–74).

1. A breeder reactor is a nuclear reactor which converts fertile materials into fissionable fuels and generates heat, which is used for electricity production and, at the same time, produces more fissile material than it uses [33,34] (pp. 53–62). It mainly uses two fertile isotopes, such as uranium-238, which is transformed into fissile plutonium, and thorium-232, which is transformed into fissile uranium. One of the measures of the efficiency of such a reactor is the breeding ratio, meaning the ratio of fissile material produced per cycle to fissile material destroyed per cycle. For breeder reactors, this number is always greater than one [35] (pp. 8–10, 234–236). Theoretical models of breeders indicate that high breeding ratios, reaching as much as 1.84, can be achieved under commercial conditions [36]. The Soviet BR-1 (Bystry Reactor-1) test reactor, put to use in 1955, with a compact plutonium core and uranium blanket, fueled with metallic plutonium and operating without a coolant, reached, depending on the source, a breeding ratio of 1.8 to 2.5 [37,38]. It turns out that the uranium present in seawater provides an amount of fuel for breeder reactors, which will satisfy humanity's demand for energy for the next 5 billion years [39]. However, due to the

threat to the environment and technical issues associated with the operation of such reactors, the hopes associated with them did not materialize [40].

2. A thermonuclear reaction is a very promising source of energy, but the need to control this naturally unstable source is the main issue. One cannot totally exclude the possibility of thermonuclear energy being used only as a weapon, as is the case with gunpowder and dynamite. Nuclear fusion involves the joining of two or more atomic nuclei, producing one or more atomic nuclei and releasing subatomic particles, such as neutrons or protons. Many various atomic nuclei can be used as nuclear fuel, but the synthesis of deuterium and tritium nuclei (hydrogen isotopes) is the most feasible. The largest body of research focuses on this case as it requires a lower—compared with other reactions of this type—plasma temperature to overcome the Coulomb barrier problem. The following reaction then occurs [41] (p. 150):

$$^{2}_{1}\mathrm{D} + ^{3}_{1}\mathrm{T} \rightarrow ^{4}_{2}\mathrm{He}\ (3.54\ \mathrm{MeV}) + ^{1}_{0}n\ (14.05\ \mathrm{MeV}), \tag{23}$$

where the fusion of deuterium $\left(^{2}_{1}\mathrm{D}\right)$ with tritium $\left(^{3}_{1}\mathrm{T}\right)$ produces helium-4 $\left(^{4}_{2}\mathrm{He}\right)$, while releasing a neutron $\left(^{1}_{0}n\right)$ and emitting 17.59 MeV of kinetic energy, with a loss of a certain amount of mass. Thermonuclear reactions run at very high temperatures, on the order of $10 - 300 \times 10^{6}$ K. At this temperature, matter exists in the form of plasma, containing ionized atoms and free electrons, which is the fourth state of matter. The potential energy that can be obtained from an earth-bound fusion system is huge. Deuterium accounts for 0.015% of all hydrogen isotopes. There is 1.4×10^{18} m^{3} water on Earth, containing 1.4×10^{43} atoms of deuterium. Considering the D-D thermonuclear reaction

$$^{2}_{1}\mathrm{D} + ^{2}_{1}\mathrm{D} \rightarrow ^{4}_{2}\mathrm{He} + 23.84\ \mathrm{MeV}, \tag{24}$$

for which there is 11.9 MeV/D, this number of deuterium atoms can provide energy in the amount of 1.8×10^{31} J. At the world's current rate of use, this amount of energy will satisfy all of humanity's energy needs for the next 50 billion years. This huge potential provides encouragement to obtain useful amounts of energy from thermonuclear reactions [41] (pp. 149, 151–152).

3. Solar technologies unquestionably stand the greatest chance of becoming the Promethean recipe of the third kind. They are constantly improving, which gives one hope for a rapid breakthrough. The effectiveness of solar cells is constantly increasing, which, in fact, does not depend on the technology applied, although the fastest growth has been observed for relatively new technologies, where the following results were achieved: quantum dot cells (various types)—18.1%, organic cells—18.2%, perovskite cells—25.7%, perovskite/Si tandem (monolithic)—32.5%, two-junction (non-concentrator)—32.9%, three-junction (non-concentrator)—39.5%, four-junction or more (non-concentrator)—39.2%, four-junction or more (concentrator)—47.6% [42]. Technological progress in this area is extremely fast. The world's most efficient four-junction solar cell, with an efficiency of 47.6% at a concentration of 665 suns, was developed in 2022 [43]. New solutions are appearing at a nearly exponential rate. Three examples can be mentioned. A solar cell can be integrated with a triboelectric nanogenerator, which converts mechanical energy into electricity. Thus, electricity is produced even when it rains [44,45]. A photovoltaic panel can be fitted out with a thermoelectric generator, which—by making use of the temperature difference between the cell and the surroundings—generates additional electricity, both at night and during the day [46]. Furthermore, a nighttime photovoltaic cell (a thermoradiative cell) generates electricity at night by using infrared radiation (heat) emitted from the Earth's surface (a heat source) toward deep space (a heat sink) [47]. Solar energy has a huge potential for humanity, as one hour of solar flux falling on the Earth's surface provides an amount of energy equal to its annual consumption by the global economy [48] (p. 14). It is noteworthy that the free energy received from the

sun within four days is comparable with the highest estimates of terrestrial energy resources [49] (pp. 303–304).

Georgescu-Roegen points out that the technologies that make use of low- or high-intensity energy need large amounts of matter, which in the former case is needed to concentrate the weak flow (solar energy), while in the latter it is needed to contain potentially dangerous and unstable sources (thermonuclear energy) [6] (p. 70). This indicates the need to take into account material entropy, i.e., the fourth law, in designing future energy systems, regardless of whether they are based on solar or thermonuclear technologies. In the late 1970s and early 1980s, the role of matter, both in solar and thermonuclear technologies, was rather important (and in the latter case—even huge). Nowadays, however, the development paths for these two recipes have parted significantly. The material entropy associated with obtaining solar energy decreases gradually as photovoltaic panels become increasingly thin and efficient, whereas it increases rapidly with each attempt at harnessing thermonuclear energy.

6. Promethean Recipes of the Third Kind and the Criterion of Matter

With the basic knowledge of candidates for Promethean recipes of the third kind, one can formulate forecasts on the potential winner in this energy race. It seems that breeder reactors have the smallest chance due to still unresolved technology issues and a considerable environmental hazard [40]. Therefore, two options remain: solar and thermonuclear energy. Using the criterion developed by Georgescu-Roegen, one should choose a technology of solar energy acquisition, although the choice could be difficult nowadays because—as has been shown above—thermonuclear energy in situ can be practically inexhaustible. However, the fourth law and the matter criterion can be helpful in this case.

Currently, there are two major areas of research on thermonuclear energy: inertial confinement fusion and magnetic confinement fusion [34] (pp. 75–83) and [50]. In an inertial confinement fusion reactor, a small pellet of thermonuclear fuel, usually with a diameter of several millimeters, is compressed to a huge density and temperature by focusing a very strong laser beam on it. Thermonuclear fuel consists of several milligrams of deuterium $^{2}_{1}H$ and tritium $^{3}_{1}H$. Thermonuclear energy is produced for a very short time, only for several nanoseconds, until the reaction ends. The time needed for burning the pellet must be shorter than the disassembly time, which is a condition for an efficient thermonuclear burn. The experiments conducted at the National Ignition Facility (NIF), which owns a laser with the highest energy in the world, involve directing up to 192 powerful laser beams into a centimeter-sized cylinder, called a hohlraum, which contains a pellet of the thermonuclear fuel. Laser beams enter the hohlraum through laser entrance holes and hit its interior, generating X-rays. This results in an implosion, which compresses and heats partially frozen hydrogen isotopes to extreme pressures and temperatures so that fusion reactions can take place. NIF generates temperatures in excess of 180 million degrees Fahrenheit and pressures in excess of 100 billion Earth atmospheres [51]. Ignition is reached when a self-sustaining fusion reaction produces a greater amount of energy than that provided by the laser. The implosion speed exceeds 400 km/s, which allows for a fusion reaction before the fuel is disassembled. A nuclear fusion in a small capsule of the reactants is so fast that they cannot escape due to their own inertia. Hence, the name of the method—the fuel is entrapped in the reactor by its own inertia. On 5 December 2022, a breakthrough in the history of research took place, as an experiment with the NIF laser far surpassed the ignition threshold. As a result, 3.15 megajoules (MJ) of fusion energy output was produced from 2.05 MJ of laser energy delivered to the target [52,53]. On 13 December 2022, this achievement was confirmed by the United States Department of Energy [54].

Magnetic confinement fusion is an approach which uses the electrical conductivity of plasma to contain its expansion with magnetic fields. This purpose is usually attained with tokamaks, in which a combination of the toroidal and poloidal fields creates a magnetic field with helical paths around the torus. In this manner, plasma is entrapped in the tokamak [55].

The tokamak, called ITER (International Thermonuclear Experimental Reactor), which is being built in France, and whose main goal is to create burning, or self-sustaining, fusion plasma, is the greatest international project in the field. Moreover, it will generate 500 MW of fusion power for 400 to 600 s while using only 50 MW of power supplied to the tokamak by the systems that heat the plasma. In other words, a fusion energy gain factor (Q), i.e., a ratio of the fusion power generated in the reactor to the power required to maintain plasma in the steady state, will be $Q \geq 10$. The condition $Q = 1$ is called a breakeven, or in some sources, scientific breakeven. The current record of energy production with the use of nuclear fusion is held by the NIF reactor, mentioned earlier, which reached $Q = 1.54$ in December 2022. The other goals of the ITER reactor include demonstrating and testing the technologies which will be needed to operate fusion power plants in future, testing tritium breeding, and demonstrating the safety of a fusion plant [56].

Let it be assumed that among all the available possibilities, one must choose an energy technology which is to become the Promethean recipe of the third kind. As the review above shows, one can choose between solar technology, $T^2(e_1^2, M_0^2)$, and thermonuclear technology $T^1(e_1^1, M_0^1)$ as the NIF reactor or the ITER reactor (terrestrial in Georgescu-Roegen's nomenclature). It can be easily demonstrated that both thermonuclear technologies require, for now, huge material scaffolds and—being experimental installations—they do not yet provide energy for consumer purposes, i.e., $e_1^1 = 0$. Solar technologies do not require such extreme material scaffolds and, what is more, they already provide considerable amounts of energy, i.e., $e_1^2 \gg 0$. Therefore, there is $e_1^2 \gg e_1^1 = 0$ and $M_0^1 \gg M_0^2$. If $T^1(e_1^1, M_0^1)$ is, for evident reasons, treated as a pseudo-solar technology, as the thermonuclear reaction is the source of energy in both cases, then the conclusion is evident and $T^2(e_1^2, M_0^2)$ is the preferred technology. Even if one takes into consideration that nuclear fusion uses unlimited energy in situ, the environmental criterion still makes one prefer solar technologies, seen as safer [6] (p. 70). The waste produced during thermonuclear fusion may also pose a threat to humanity and the environment.

To conclude the issue, it is worth analyzing the problem of the economic profitability of both technology types, if the thermonuclear technology in energy generation can be regarded as real. Let it be assumed that it can. Although there are still no thermonuclear power plants, advanced research programs exploring the issue do exist [57]. However, it should be pointed out that the characteristic features of present fusion technologies include a huge material scaffold and high cost, counted in billions of USD or EUR.

For example, the National Ignition Facility occupies a ten-story building the length of three football fields. In fact, the building consists of three interconnected ones: the Optics Assembly Building, the Laser and Target Area Building, and the Diagnostics Building. It was completed on 31 March 2009, and the total cost amounted to USD 3.5 billion [58,59]. The success of exceeding the ignition threshold is only the first step toward the commercial use of inertial confinement fusion, and there are at least decades of further studies and experiments ahead [60,61].

The construction of the ITER complex started in 2013, and the tokamak assembly has been ongoing since 2020 [62,63]. Plans for 2025 include the completion of the assembly, the beginning of the start-up phase, and the production of the first plasma, although the deuterium–tritium operation will not start until 2035 [64]. The whole complex occupies an area of 180 hectares, with its main part being a human-made level platform with an area of 42 hectares, completed in 2009, 1 km long and 400 m wide, which is equivalent to 60 soccer fields [65]. The ITER Tokamak and its plant systems are located in 39 buildings and technical areas. The central object is the Tokamak Building, constructed from reinforced concrete, which is to have seven stories, stand 73 m tall, and it will be situated 13 m below the platform level and 60 m above it [66]. The ITER machine's weight will be 23,000 tons, i.e., three times as much as the Eiffel Tower. Each of the 18 toroidal field coils has a weight of 310 tons. A total weight of 400,000 tons will be placed in the lower basement of the Tokamak Complex, including the buildings, the main machine, and the equipment [65]. Initially, the project budget amounted to EUR 5 billion, but it soon appeared that the cost of construction and operation could be as high as EUR 18–22 billion [67]. According to other sources, the total cost could reach USD 45–65 billion [68,69]. Nowadays, the ITER reactor is regarded as one of the most complicated engineering projects in the history of humanity and one of the most ambitious enterprises undertaken by multiple cooperating countries, along with the International Space Station and the Large Hadron Collider [70]. This very complicated megaproject involves the cooperation of 35 countries, which represent over half of the world's population [71,72]. To make the picture complete, one should note that the whole device will be experimental, and the net production of electricity will not be its task.

To evaluate the efficiency of future commercial thermonuclear power plants, the engineering gain factor or engineering breakeven is used, defined as the ratio of a plant's electrical power output to the electrical power input of all plants' internal systems [73]. The condition $Q = 100$ must be met to make the thermonuclear technology viable. However, starting a thermonuclear power plant requires another several dozen years of research and experiments. The planned successor of ITER—the DEMO tokamak—will not be limited to sustaining the burning plasma, like its predecessor, but it will be the first real nuclear fusion power station, which will supply electricity to the grid. The linear dimensions of the DEMO reactor will be about 15% greater than ITER, and its vacuum chamber will comprise 2200 m^3 compared with 800 m^3 in the case of ITER [74]. Therefore, Georgescu-Roegen was not very mistaken when he wrote in 1981 that a feasible thermonuclear reactor could be the size of Manhattan [6] (p. 70).

The importance of matter and material entropy increases rapidly under the conditions presented here. Therefore, these elements must be taken into account as criteria for the choice of the right technology. The costs of both energy sources must also be taken into account. When it comes to solar technologies, the average LCOE (levelized cost of electricity) for solar PV crystalline decreased from 359 USD/MWh in 2009 to 37 USD/MWh in 2020. At present, there are no commercial thermonuclear plants, which makes comparisons difficult, but the existing nuclear power plants can be taken into consideration. In this case, an increase in the mean LCOE from 123 USD/MWh in 2009 to 163 USD/MWh in 2020 was observed [75] (p. 8). Moreover, the efficiency of solar technologies is increasing steadily, and it reaches 50% for concentrated solar radiation [43]. What is more, this increase in efficiency is accompanied by a gradual decrease in the material scaffold.

Significant progress (in fact, a true breakthrough) has occurred in the field of scalable techniques for producing ultrathin, lightweight solar cells that are tens of micrometers thick. These innovative solar cells can be easily attached to any surface and used as a power source. The power of these cells per kilogram is 18 times greater, and the weight per m^2 is 100 times smaller compared with silicon PV modules on glass substrates. These achievements are even better if a comparison is made against CdTe thin film PVs on glass substrates because, in this case, the power per kilogram is 28 times greater and the weight per m^2—130 times smaller [76].

Considering the fact that multiple technical issues associated with the startup and operation of thermonuclear technologies are still unresolved, the fourth law indicates that the development of solar power plants is the most prudent choice, which is highly likely to bring us to the production of clean energy by the year 2050. When it comes to thermonuclear technologies, this is a thing of the future, and it is not even certain whether they will be viable. There is no need to spend a significant amount of money for the construction of thermonuclear reactors on Earth when there is already a reactor in operation at a safe distance from us.

7. Is the Present Solar Technology Viable?

In order to prove that the present solar technology is not viable, Georgescu-Roegen proposes that the schematic flow matrix presented in Table 3 should be considered, which represents the structure of solar energy technology [6] (pp. 197–198). There are four processes here:

- The process P_1 converts solar energy, SE, into controlled energy, CSE, with the aid of some collectors, CL, and some other capital equipment, K;
- Process P_2 produces collectors with the aid of the energy controlled by P_1, and also some capital equipment;
- Process P_3 uses CSE to produce capital equipment for all purposes;
- Process P_4 supports all other activities of production and consumption with the necessary CSE and capital equipment as well.

Table 3. The reduced flow matrix of a technology.

Elements	P_1 (SE → CSE)	P_2 (Produces CL)	P_3 (CSE → K)	P_4 (CSE+K) → ALL
Controlled solar energy (CSE)	x_{11}	$-x_{12}$	$-x_{13}$	$-x_{14}$
Solar collectors (CL)	$-x_{21}$	x_{22}	*	*
Capital equipment (K)	$-x_{31}$	$-x_{32}$	x_{33}	$-x_{34}$

The Promethean conditions for viability, (9)–(14), are as follows:

$$x_{11} = x_{12} + x_{13} + x_{14}\,, \tag{25}$$

$$x_{22} = x_{21}\,, \tag{26}$$

$$x_{33} = x_{31} + x_{32} + x_{34}\,. \tag{27}$$

A technology based on the conversion of energy in the process P_1 is not viable if

$$x_{11} < x_{12} + x_{13} + x_{14}\,, \tag{28}$$

or even

$$x_{11} < x_{12}\,. \tag{29}$$

The last inequality underscores the current situation, reminiscent of Georgescu-Roegen's time: no pilot facilities exclusively manufacture solar collectors using solar energy conversion (SE → CSE). As a result, while practical solar recipes are feasible, there is a dearth of viable solar technology [6] (p. 198).

Georgescu-Roegen points out that a viable technology, like a viable species, must maintain itself after it emerges from the previous technology. The threshold between the old and the new technology can be overcome only at an additional cost from the old prices. The viability of solar technology only requires that its material scaffold should be self-supporting, which is a condition independent of the current profitability of the production of solar collectors [77]. Therefore, developing a pilot of full-fledged solar technology should be the nearest goal in perfecting the solar technology to prove its viability independent of prices [78] (pp. 1052–1053).

The next stage of Georgescu-Roegen's reasoning involves demonstrating that contemporary recipes for using solar energy are parasites of fossil fuel-based technologies [6] (pp. 199–200). Solar collectors were used mainly for heating at his time. In order to determine the technical relations of solar collectors with the associated processes, which existed in the late 1970s and early 1980s, he refers to Table 4, where all the processes make use of only non-solar energy mainly obtained from fossil fuels. A new process appeared, P_0, which represents the production of fossil energy, FE. In a simplified approach, the process P_4 receives only heat from the production complex.

Table 4 shows that

$$x_{00} = x_{01} + x_{02} + x_{03}\,. \tag{30}$$

Table 4. The past and present technical relations of solar collectors with the associated processes.

Elements	P_1 (SE → CSE)	P_0 (Produces FE)	P_2 (Produces CL)	P_3 (CSE → K)	P_4 (CSE+K) → Heat
Controlled solar energy (CSE) [1]	x_{11} $[x_{11}]$	* $[-x_{10}]$	* $[-x_{12}]$	* $[-x_{13}]$	$-x_{11}$ $[-x_{14}]$
Solar collectors (CL)	$-x_{21}$	*	x_{22}	*	*
Capital equipment (K)	$-x_{31}$	$-x_{30}$	$-x_{32}$	x_{33}	*
Fossil energy (FE)	$-x_{01}$	$+x_{00}$	$-x_{02}$	$-x_{03}$	*

[1] Brackets indicate the present state of controlled solar energy.

Moreover, x_{02} must be of the same order of magnitude as x_{12} in Table 3 ($x_{02} \approx x_{12}$). According to Equation (30) and Inequality (29),

$$x_{00} \gg x_{11}\,, \tag{31}$$

which confirms the parasitic nature of the recipes for harnessing solar radiation. They could not exist without technologies based on other kinds of energy.

8. Discussion

Undoubtedly, solar technologies at the times of Georgescu-Roegen parasitized other energy sources, mainly fossil energy, but now Table 4 can be modified slightly by introducing additional flow factors provided in brackets to it. This means that controlled solar energy (CSE) also powers all economic processes, including P_0, which produces energy from fossil fuels. Evidently, the Promethean condition of viability

$$[x_{11}] = [x_{10}] + [x_{12}] + [x_{13}] + [x_{14}]\,, \tag{32}$$

cannot be met even nowadays, and, evidently, solar collectors cannot be produced without the use of other energy sources in the current state of technology. However, in some areas, such as construction and transport, there are pilot installations in operation which are powered fully or in large part with solar energy, so instead of Inequality (29), there is the following relationship:

$$[x_{11}] \cong [x_{12}]\,. \tag{33}$$

Moreover, the situation nowadays is a little better than in the late 1970s and early 1980s, because certainly

$$[x_{11}] \gg x_{11}\,. \tag{34}$$

Georgescu-Roegen sees the benefits of solar energy, as he regards the Sun as the only steady and completely healthy source of energy, but he doubts whether it will enable people to drive cars at the speed of one hundred kilometers per hours, to live in skyscrapers, or to fly in jets [78] (p. 1055). However, what used to be impossible slowly begins to come true. Electric cars driving at the speed mentioned above are no longer something extraordinary, and they can be powered by electricity produced by photovoltaic panels. Several prototypes of solar cars have been constructed in recent years. They are intended for use on public roads, and they use self-contained solar cells for full or partial powering with solar radiation. One of them, Lightyear 0, entered volume production in December 2022 [79]. Production of another car of this type, the Aptera solar-powered EV, is set to start in 2023 [80,81]. With vehicles of this kind, users can skip the charging grid and drive in a more sustainable manner. The moment is probably approaching when skyscrapers become energy self-sufficient owing to see-through solar cells [82]. There are skyscrapers and office buildings mostly powered by solar energy, and the next ambitious project in the field is planned. Apple's Spaceship Headquarters in Cupertino, California, whose construction cost amounted to USD 5 billion, is equipped with a 17-megawatt onsite rooftop solar installation, which satisfies 75% of its energy demand [83,84]. A skyscraper has been designed in Australia called the Sol Invictus Tower, which is to have 60 floors and acquire more than 50 percent of the building's baseload power from solar panels [85]. Moreover, self-sufficient solar homes have been a real thing for some time [86]. What is more, a prototype airplane called Solar Impulse 2, powered by solar energy, had its first solar flight around the world in 2015–2016 [87]. Therefore, a technological barrier in the form of an operating pilot solar installation has been overcome in some areas, which at least cancels Inequality (29).

In modern times, the digitalization of power systems has facilitated the decentralization of power generation and the integration of solar radiation and other renewable sources into smart grids [88,89]. For the current technologies to be regarded as Promethean recipes of the third kind, they must certainly be perfected, which is why humanity's considerable effort is directed toward pursuing this goal. However, this is not a sufficient condition. The Promethean condition for viability will be met when the solarization of the global power infrastructure takes place. Therefore, not only changes arising from technological progress are required, but—most importantly—socioeconomic changes will make introducing solar communism possible [90]. Only after solar technologies trigger a chain reaction, as was the case with fire and the steam engine, can one talk about the real Promethean recipe of the third kind. For this to happen, all of humanity must participate in obtaining solar energy due to its dispersed nature. This requires common demilitarization as well as developing and implementing strict plans for biosphere conservation and the propagation of organic farming.

Tables of technological processes, such as Tables 3 and 4, cannot be prepared for thermonuclear energy because there has been no active (or even pilot) fusion power plant, and scientific effort has been focused on attempts at controlling energy generated from the fusion of atomic nuclei. It is only known that according to the matter criterion, solar technologies should be chosen, and fusion energy should not be used by humanity. However, one can go a step further in the research and verify the ecological criterion thoroughly. Even now, it is possible to evaluate the ecological effects of fusion energy technologies, in which the second law of thermodynamics (the entropy law) can be used. The point is that relations between fusion energy and the mean entropy budget of the Earth should be established.

The entropy budget describes the entropy flow between the Earth and space [91,92]. Thermodynamically, the Earth is a closed system, i.e., it does not exchange matter with space, except meteorites and spacecraft, but it is open to energy flow. The flux of solar radiation with a high temperature and low entropy reaches its surface, and waste heat is emitted to space, which plays the role of a heat sink. It is mainly infrared radiation, which has low temperature and high entropy. For the climate system equilibrium to be maintained, the long-term entropy balance of the Earth must be equal to zero. In this case, the internal production of entropy in the system is offset by the entropic outflow to the environment (deep space) [93] (pp. 161–172).

When one considers the long-term mean, the incoming flux of solar energy is equal to the outgoing flux of longwave radiation to space. However, the corresponding entropy fluxes are completely different. The entropy in the longwave radiation which leaves the Earth is about 22 times greater than the entropy in the incoming solar radiation. The total amount of entropy exported by the climate system to space is -925 mW $m^{-2}K^{-1}$, whereas the amount of entropy imported by the incoming solar radiation is 41.3 mW $m^{-2}K^{-1}$. The irreversible processes in the climate system are responsible for the internal production of entropy, which is equal to the net export of entropy to space effected by the system [91]. The difference between the incoming entropy in solar radiation and the entropy associated with the outgoing radiative heat flow is a natural resource in the form of potential for entropy production, which can be used properly or wasted. After creating a solar society, humanity will have a huge potential for increasing its negentropy resources without increasing the global rate of entropy production [94].

Greenhouse gases, which stop longwave (mainly infrared) radiation emitted from the surface, have been present in the atmosphere of the Earth from time immemorial. They contribute to the formation of the natural greenhouse effect, which helps to maintain the optimum temperature on the surface for biosphere development. The situation has become complicated as a result of human economic activities, which is the cause of what is known as the anthropogenic greenhouse effect. It is caused by increasing amounts of greenhouse gases in the Earth's atmosphere, which is caused, in turn, by burning fossil fuels. These extra amounts of gases capture greater amounts of heat and contribute to an increase in the average temperature above the level which is an effect of the natural greenhouse effect. This results in global warming, which contributes to catastrophic climate change, upsets the biosphere balance, and makes maintaining biodiversity difficult.

Let it be assumed that technological progress enabled people to use thermonuclear energy in economic processes. The only difference between fossil energy and thermonuclear energy is that the latter does not cause greenhouse gas emissions. However, it is worth noting that—like energy from fossil fuels—thermonuclear energy causes the emission of additional waste heat to the atmosphere. If the global economy is not totally based on a direct solar flux, but it uses even relatively clean thermonuclear technologies, then the heat budget of the Earth will change irreversibly, as heat emission will exceed the natural flux from the surface. If the atmosphere contains anthropogenic greenhouse gases, thermonuclear energy will still increase the existing global warming, raising the average temperature on the Earth even further. Therefore, perfecting thermonuclear technologies should be followed by progress in developing and implementing methods of capturing anthropogenic greenhouse gases, mainly carbon dioxide and methane, from the atmosphere. Irrespective of this, excess waste heat from fusion power plants above the natural flux from the surface could become a serious problem in the future.

Since the industrial revolution, with the burning of fossil fuels, our thermal debt toward the environment in a broad sense, i.e., space, which is called entropic debt, has been increasing. Restoring the balance of the natural mean entropy budget requires for energy to no longer be acquired from burning fossil fuels, nuclear fission, and similar processes, as they produce surplus waste heat above the natural flux from the surface. This applies to future thermonuclear power plants—if ones are constructed—as their operation will increase our thermal debt toward space. It is not known exactly how the growing entropic debt can affect the global climate system, biosphere organization and what happens in the universe. Only an economy based on direct solar flux will not upset the heat budget of the Earth [30].

Moreover, thermonuclear power supply entails two other types of hazards. Firstly, one should not underestimate radioactive waste, which will be produced by future fusion plants. Radioactive materials are produced as a result of the inevitable exposure of some equipment elements to a flux of neutrons. The radiotoxicity of fusion waste decreases rapidly compared with the waste from a PWR (pressurized water reactor) fission plant, but it reaches the radiotoxicity of ash in a coal-fired power plant only after 500 years [95]. Secondly, thermonuclear plants could be used for military purposes, as there is a considerable risk that they may contribute to nuclear proliferation [96]. Hence, the construction of extremely costly thermonuclear power plants, which require vast amounts of material and pose a significant safety risk on Earth, should be questioned. Instead, we ought to consider the Sun as a source of clean and healthy energy, which has been operational for an extended period and is situated at a safe distance from Earth.

9. Conclusions

Table 5 compares three selected future energy technologies: fusion (T^1), solar (T^2), and fission (T^3). Criteria such as energy in situ (e_1), controlled energy $(x_{11}^1, x_{11}^2, x_{11}^3)$, material requirements (M_0), entropic debt, average LCOE, degree of environmental risk, the possibility of being considered Prometheus III, and type of energy conversion were used. It appears that the best energy technology is solar technology, as it is associated with an inexhaustible, natural, and healthy source of renewable energy, no waste heat production over and above the natural flux from the surface, the lowest material entropy $\left(M_0^2 \ll M_0^3 \ll M_0^1\right)$, non-increasing entropic debt, the smallest and steadily declining average LCOE (37 USD/MWh), the greatest environmental safety, the achievement of viability in the near future, as signaled by the existence of pilot installations in transportation and construction, and the easiest conversion of solar energy into electricity through the use of the photovoltaic effect. The inequality $x_{11}^3 > x_{11}^2$ is due to the fact that, in 2021, the share of fission energy in global electricity production was 9.9%, while the share of solar energy was slightly smaller, equal to 3.7% [97]. These relationships are expected to change dramatically in the coming years. There are already concrete plans for Europe to import solar electricity from the Middle East and North Africa [98]. There is also a serious contradiction between solar technology and fission and fusion technologies when it comes to universal demilitarization and ensuring lasting world peace. As is well known, the latter two technologies are linked to the military–industrial complex and can be used for nuclear proliferation. Some of the most important challenges facing humanity in the coming years include stopping the expansion of the entropic debt and then gradually reducing it by developing solar technology and removing anthropogenic greenhouse gases from the atmosphere.

Table 5. Comparison of energy technologies of the future.

Type of Technology	Energy/Matter Criterion	Entropic Debt	Average LCOE 2009–2020 (USD/MWh)	Ecological Criterion	Prometheus III	Type of Energy Conversion
Fusion $T^1(e_1^1, M_0^1)$	$e_1^1 \gg 0$ $x_{11}^1 = 0$ $M_0^1 \gg 0$	Yes (↑)	0→0	Medium environmental risk; after 500 years, the waste has the radioactivity of coal ash	Not viable	Steam turbine
Solar $T^2(e_1^2, M_0^2)$	$e_1^2 \to \infty$ $x_{11}^2 \gg 0$ $M_0^2 \ll M_0^1$	No	359→37 (↓)	Environmentally safe provided waste recycling is developed	Approaching viability	Photovoltaic effect, sometimes steam turbine
Fission $T^3(e_1^3, M_0^3)$	$e_1^2 \gg e_1^1 \gg e_1^3$ $x_{11}^3 > x_{11}^2 \gg 0$ $M_0^2 \ll M_0^3 \ll M_0^1$	Yes (↑)	123→163 (↑)	Very high environmental risk in the extremely long term	Not viable	Steam turbine

Entropic debt is closely related to the energy technologies chosen by humanity. It arises not only from the second, but also from the fourth principle of thermodynamics, as it relates to both energy degradation and matter dissipation. Let us first consider the effects of the second law of thermodynamics. Any economy that uses energy sources other than direct solar flux alters the Earth's heat budget, as it emits thermal radiation in excess of the natural flux from the surface. This is true for fossil fuel-based technologies, as well as for fission and fusion technologies. It is also known that there is a stock of anthropogenic greenhouse gases in the Earth's atmosphere, which has been steadily increasing since the beginning of the industrial revolution. Under such conditions, fission and fusion technologies, even if they do not emit an additional portion of greenhouse gases, still contribute to the intensification of global warming. Greenhouse gases capture some of the long-wave radiation (mainly infrared) emitted from the Earth's surface as waste heat and radiate it back to the surface, resulting in an increase in its global mean temperature. In turn, entropy in the environment increases as a debt for any internal process taking place in the system [30]. In other words, as a result of the second law of thermodynamics, fission and fusion technologies may contribute to the intensification of global warming and, at the same time, increase the entropic debt to the cosmos as an external environment due to the emission of additional waste heat, which may contribute to accelerating the heat death of the universe. This worsens the situation not only on Earth but also in space. Only global solar power can pay off its entropic debt in the form of non-incremental waste heat without leading to catastrophic climate change [99] (p. 21).

On top of all this is the operation of the fourth law of thermodynamics, which talks about the entropy of matter. The entropy debt of matter is the sum of the waste produced by the results of irreversible internal processes and the dissipation of natural resources, or what Georgescu-Roegen calls unavailable matter. It can, over time, lead to the material death of ecosystems on Earth. Sometimes, the sum of the entropy debt of energy and the entropy debt of matter is referred to as the cost of complexity of an economic system, the incurring of which is necessary to maintain its internal complexity in the form of a certain social order, energy technologies, infrastructure, and communication networks. This cost is related to the disruption of the natural circulation of energy and entropy in the biosphere and the violation of the ecological integrity of natural resources, without which humanity has no chance of survival. It includes all that is associated with anthropogenic environmental degradation [100].

The multifaceted battle regarding humanity's choice of the dominant energy technology of the future is being fought today in various contact zones: global, international, national, and even local. It involves energy law and administrative issues, as well as available technical solutions. Therefore, it is taking place simultaneously in the two most important contact zones: human administrative legal contact zones and more-than-human energy contact zones [101].

Jeremy Rifkin believes that the current accumulation of entropic debt far exceeds the capacity of the biosphere to absorb it. The effects of the fourth law are increasingly evident as a result of the depletion of fossil fuels and rare earth minerals, so he proposes that thermodynamic efficiencies and entropic consequences should be taken into account when measuring productivity. In his view, the entropy bill does not take into account the full cost of matter dissipation because if it did, fossil capitalism would have to collapse rapidly [102] (pp. 207–208). Rifkin assumes that the entropic debt can be overcome by humanity's empathic dimension, which could result in a global, biosphere-wide consciousness and distributed capitalism based on renewable energy [103].

In the future, the problem of humanity's entropic debt can be expected to emerge as one of the fundamental issues of international politics. Already in global discussions on climate change, such concepts as greenhouse gas stocks and flows are beginning to be used. The issue of flows is often emphasized by industrialized countries, while the issue of stocks is important for non-industrialized or newly industrialized countries. The stockpile of greenhouse gases began to accumulate in the atmosphere with the advent of the industrial revolution, and the countries where it began incurred the greatest entropic debt and thus contributed the most to the persistence of current levels of carbon dioxide in the air and the resulting global damage. Even a significant reduction in current greenhouse gas emissions into the atmosphere cannot reduce the stock that is already there. Accordingly, James B. Quilligan predicts that the future international system will have to combine the stocks and flows that exist in current accounts between surplus and deficit countries with the stocks and flows of the global commons, which will take into account both renewable and depletable resources [104]. With the goodwill of all stakeholders, this could restore fairness to the world's balance-of-payments adjustment system.

Funding: This research received no external funding.

Data Availability Statement: Not applicable.

Conflicts of Interest: The author declares no conflict of interest.

References

1. Kaplan, E.A. *Climate Trauma: Foreseeing the Future in Dystopian Film and Fiction*; Rutgers University Press: New Brunswick, NJ, USA, 2016; ISBN 978-0-8135-6400-5.
2. Thomson, W. On a Universal Tendency in Nature to the Dissipation of Mechanical Energy. *Proc. R. Soc. Edinb.* **1857**, *3*, 139–142. [CrossRef]
3. Zemansky, M.W. *Heat and Thermodynamics: An Intermediate Textbook*, 5th ed.; McGraw-Hill Kogakusha, Ltd.: New Delhi, India, 1957.

4. Georgescu-Roegen, N. Afterword. In *Entropy: Into the Greenhouse World, Revised ed.*; Rifkin, J., Howard, T., Eds.; Bantam Books: New York, NY, USA, 1989; pp. 299–307; ISBN 0-553-34717-9.
5. Georgescu-Roegen, N. Energy and matter in mankind's technological circuit. In *Energy Policy: The Global Challenge*; Nemetz, P.N., Ed.; Institute for Research on Public Policy: Montreal, QC, Canada, 1979; pp. 107–127; ISBN 0-920380-30-1.
6. Georgescu-Roegen, N. Energy, matter, and economic valuation: Where do we stand?/Reply. In *Energy, Economics, and the Environment: Conflicting Views of an Essential Interrelationship*; Daly, H.E., Umaña, A.F., Eds.; Westview Press, Inc.: Boulder, CO, USA, 1981; pp. 43–79, 193–200; ISBN 0-86531-282-6. [CrossRef]
7. Prasad, R. *Classical and Quantum Thermal Physics*; Cambridge University Press: Cambridge, UK, 2016; ISBN 978-1-107-17288-3.
8. Samuelson, P.A. Foreword. In *Bioeconomics and Sustainability: Essays in Honor of Nicholas Georgescu-Roegen*; Mayumi, K., Gowdy, J.M., Eds.; Edward Elgar Publishing Limited: Cheltenham, UK, 1999; pp. XIII–XVII; ISBN 978-1-85898-667-8.
9. Boulding, K.E. The economics of the coming spaceship Earth. In *Environmental Quality in a Growing Economy: Essays from the Sixth RFF Forum*; Jarrett, H., Ed.; RFF Press—Resources for the Future: New York, NY, USA, 2011; pp. 3–14; ISBN 978-1-61726-027-8.
10. Georgescu-Roegen, N. *La Décroissance. Entropie—Écologie—Économie*; Nouvelle édition; Présentation et traduction de Jacques Grinevald et Ivo Rens; Éditions Sang de la terre: Paris, France, 1995. [CrossRef]
11. van't Hoff, J. The function of osmotic pressure in the analogy between solutions and gases. *Lond. Edinb. Dublin Philos. Mag. J. Sci.* **1888**, *26*, 81–105. Available online: https://archive.org/details/s5philosophicalm26lond (accessed on 5 January 2023). [CrossRef]
12. Iniguez, J.C. Work Degrading and the Reversible to Irreversible Transition in Chemical Reactions. *Int. J. Chem.* **2014**, *6*, 1–13. [CrossRef]
13. Laidler, K.J. *The World of Physical Chemistry*; Oxford University Press Inc.: New York, NY, USA, 1995; ISBN 0-19-855597-0.
14. Lewis, W.C.; Mc, C. *A System of Physical Chemistry in Three Volumes*, 3rd ed.; Longmans, Green and Co.: London, UK, 1922; Volume 2; Thermodynamics.
15. Thims, L. *Hmolpedia: A–Z Encyclopedia of Human Thermodynamics, Human Chemistry, and Human Physics*; Lulu Press, Inc.: Morrisville, NC, USA, 2016; Volume 8, pp. 5193–5196; ISBN 978-0-9973844-7-5.
16. Bazhin, N.M.; Parmon, V.N. Conversion of Chemical Reaction Energy into Useful Work in the Van't Hoff Equilibrium Box. *J. Chem. Educ.* **2007**, *84*, 1053–1055. [CrossRef]
17. Mayumi, K. Georgescu-Roegen's 'Fourth Law of Thermodynamics' and the flow-fund model. In *Entropy and Bioeconomics: Proceedings of the First International Conference of the European Association for Bioeconomic Studies—E.A.B.S. Rome, Italy, 28–30 November 1991*; Dragan, J.C., Seifert, E.K., Demetrescu, M.C., Eds.; Nagard Publisher: Milano, Italy, 1993; pp. 399–413; ISBN 88-85010-11-3.
18. Mayumi, K. *The Origins of Ecological Economics: The Bioeconomics of Georgescu-Roegen*; Routledge—Taylor & Francis Group: New York, NY, USA, 2001; ISBN 0-415-23523-5.
19. Solow, R.M. Is the end of the world at hand? *Challenge* **1973**, *16*, 39–50. Available online: https://www.jstor.org/stable/40719094 (accessed on 5 February 2023). [CrossRef]
20. Brown, H. *The Challenge of Man's Future: An Inquiry Concerning the Condition of Man during the Years That Lie Ahead*; The Viking Press: New York, NY, USA, 1956.
21. Brown, H.; Bonner, J.; Weir, J. *The Next Hundred Years. Man's Natural and Technological Resources: A Discussion Prepared for Leaders of American Industry*, 2nd ed.; The Viking Press: New York, NY, USA, 1964.
22. Papadopoulos, E. Heron of Alexandria (c. 10–85 AD). In *Distinguished Figures in Mechanism and Machine Science. Their Contributions and Legacies. Part 1*; Ceccarelli, M., Ed.; Springer: Dordrecht, The Netherlands, 2007; Volume 1; History of Mechanism and Machine Science; pp. 217–245; ISBN 978-1-4020-6365-7. [CrossRef]
23. *The Pneumatics of Hero of Alexandria. From the Original Greek Translated for and Edited by Bennet Woodcroft*; Taylor Walton and Maberly: London, UK, 1851; Available online: https://books.google.pl/books?id=O8PVAAAAMAAJ&printsec=frontcover&redir_esc=y#v=onepage&q&f=false (accessed on 7 January 2023).
24. *Herons von Alexandria Druckwerke und Automatentheater. Griechisch und Deutsch Herausgegeben von Wilhelm Schmidt*; Druck und Verlag von B. G. Teubner: Leipzig, Germany, 1899; Available online: https://archive.org/details/heronsvonalexandhero/page/228/mode/2up (accessed on 7 January 2023).
25. Drachmann, A.G. The classical civilizations. In *Technology in Western Civilization*; Kranzberg, M., Pursell, C.W., Jr., Eds.; Oxford University Press: New York, NY, USA, 1967; Volume 1; The Emergence of Modern Industrial Society, Earliest Times to 1900; pp. 47–66.
26. Siegel, B. *The Steam Engine*; Walker and Company: New York, NY, USA, 1986; ISBN 0-8027-6655-2.
27. Kras, S.L. *The Steam Engine*; Chelsea House Publishers: Philadelphia, PA, USA, 2004; ISBN 0-7910-7453-6.
28. Jakimowicz, A. The Future of the Energy Sector and the Global Economy: Prosumer Capitalism and What Comes Next. *Energies* **2022**, *15*, 9120. [CrossRef]
29. Schwartzman, P.; Schwartzman, D. *The Earth Is Not for Sale: A Path Out of Fossil Capitalism to the Other World That Is Still Possible*; World Scientific Publishing Co. Pte. Ltd.: Singapore, 2019; ISBN 978-981-3234-24-6.
30. Schwartzman, D. Solar communism. *Sci. Soc.* **1996**, *60*, 307–331. Available online: https://www.jstor.org/stable/40403574 (accessed on 10 November 2021).
31. Wiser, W.H. *Energy Resources: Occurrence, Production, Conversion, Use*; Springer Science + Business Media: New York, NY, USA, 2000; ISBN 978-1-4612-7050-8. [CrossRef]
32. Smil, V. Examining energy transitions: A dozen insights based on performance. *Energy Res. Soc. Sci.* **2016**, *22*, 194–197. [CrossRef]

33. Judd, A.M. *Fast Breeder Reactors: An Engineering Introduction*; Pergamon Press: New York, NY, USA, 1981; ISBN 0-08-023220-5.
34. Breeze, P. *Nuclear Power*; Academic Press: London, UK, 2017; ISBN 978-0-08-101043-3.
35. Waltar, A.E.; Reynolds, A.B. *Fast Breeder Reactors*; Pergamon Press: New York, NY, USA, 1981; ISBN 0-08-025983-9.
36. Hiraoka, T.; Sako, K.; Takano, H.; Ishii, T.; Sato, M. A High-Breeding Fast Reactor with Fission Product Gas Purge/Tube-in-Shell Metallic Fuel Assemblies. *Nucl. Technol.* **1991**, *93*, 305–329. [CrossRef]
37. Russian Reactions. *Nuclear Engineering International*. 22 May 2016. Available online: https://www.neimagazine.com/features/featurerussian-reactions-4899799/ (accessed on 27 December 2022).
38. Korobeinikov, V. Innovative Concepts Based on Fast Reactor Technology. In Proceedings of the 1st Consultancy Meeting for Review of Innovative Reactor Concepts for Prevention of Severe Accidents and Mitigation of their Consequences (RISC), IAEA Headquarters, Vienna, Austria, 31 March–2 April 2014. Available online: https://web.archive.org/web/20160304072421/https://www.iaea.org/NuclearPower/Downloadable/Meetings/2014/2014-03-31-04-02-CM-INPRO/RUSSIA_PPT_MS_Presentations_1st_CM_on_RISC_.pdf (accessed on 28 December 2022).
39. Cohen, B.L. Breeder reactors: A renewable energy source. *Am. J. Phys.* **1983**, *51*, 75–76. [CrossRef]
40. Cochran, T.B.; Feiveson, H.A.; Mian, Z.; Ramana, M.V.; Schneider, M.; von Hippel, F.N. It's Time to Give Up on Breeder Reactors. *Bull. Sci.* **2010**, *66*, 50–56. [CrossRef]
41. Shultis, J.K.; Faw, R.E. *Fundamentals of Nuclear Science and Engineering*; Marcel Dekker, Inc.: New York, NY, USA, 2002; ISBN 0-8247-0834-2.
42. *Best Research-Cell Efficiency Chart*; The National Renewable Energy Laboratory: Golden, CO, USA. Available online: https://www.nrel.gov/pv/cell-efficiency.html (accessed on 22 December 2022).
43. *Fraunhofer ISE Develops the World's Most Efficient Solar Cell with 47.6 Percent Efficiency*; Fraunhofer Institute for Solar Energy Systems ISE: Freiburg im Breisgau, Germany, 2022. Available online: https://www.ise.fraunhofer.de/en/press-media/press-releases/2022/fraunhofer-ise-develops-the-worlds-most-efficient-solar-cell-with-47-comma-6-percent-efficiency.html (accessed on 27 December 2022).
44. Liu, Y.; Sun, N.; Liu, J.; Wen, Z.; Sun, X.; Lee, S.-T.; Sun, B. Integrating a Silicon Solar Cell with a Triboelectric Nanogenerator *via* a Mutual Electrode for Harvesting Energy from Sunlight and Raindrops. *ACS Nano* **2018**, *12*, 2893–2899. [CrossRef] [PubMed]
45. Wang, Y.; Yang, Y.; Wang, Z.L. Triboelectric nanogenerators as flexible power sources. *NPJ Flex. Electron.* **2017**, *1*, 10. [CrossRef]
46. Assawaworrarit, S.; Omair, Z.; Fan, S. Nighttime electric power generation at a density of 50 mW/m^2 via radiative cooling of a photovoltaic cell. *Appl. Phys. Lett.* **2022**, *120*, 143901. [CrossRef]
47. Deppe, T.; Munday, J.N. Nighttime Photovoltaic Cells: Electrical Power Generation by Optically Coupling with Deep Space. *ACS Photon.* **2020**, *7*, 1–9. [CrossRef]
48. Schwartzman, D. Ecosocialism or Ecocatastrophe? *Capital. Nat. Social.* **2009**, *20*, 6–33. [CrossRef]
49. Georgescu-Roegen, N. *The Entropy Law and the Economic Process*, 2nd ed.; Harvard University Press: Cambridge, MA, USA, 1974; ISBN 0-674-25781-2.
50. Meade, D. 50 years of fusion research. *Nucl. Fusion* **2010**, *50*, 014004:1–014004:14. [CrossRef]
51. *The National Ignition Facility (NIF) Is the Size of a Sports Stadium—Three Football Fields Could Fit Inside*; Society for Science at User Research Facilities: Lafayette, CO, USA. Available online: https://ssurf.org/did-you-know-the-national-ignition-facility-nif-is-the-size-of-a-sports-stadium-three-football-fields-could-fit-inside/ (accessed on 28 December 2022).
52. *Achieving Fusion Ignition*; Lawrence Livermore National Laboratory: Livermore, CA, USA. Available online: https://lasers.llnl.gov/science/pursuit-of-ignition (accessed on 29 December 2022).
53. Clery, D. Explosion Marks Laser Fusion Breakthrough. National Ignition Facility Achieves Net Energy "Gain", but Commercial Plants Remain a Distant Dream. *Science* **2022**, *378*, 1154–1155. [CrossRef]
54. *National Ignition Facility Achieves Fusion Ignition*; Lawrence Livermore National Laboratory: Livermore, CA, USA, 2022. Available online: https://www.llnl.gov/news/national-ignition-facility-achieves-fusion-ignition (accessed on 29 December 2022).
55. Wesson, J. *The Science of JET. The Achievements of the Scientists and Engineers Who Worked on the Joint European Torus 1973–1999*; JET Joint Undertaking: Abingdon, Oxon, UK, 2006. Available online: https://www.euro-fusion.org/fileadmin/user_upload/Archive/wp-content/uploads/2012/01/the-science-of-jet-2000.pdf (accessed on 29 December 2022).
56. *What Will ITER Do?* ITER Organization. Available online: https://www.iter.org/sci/goals (accessed on 30 December 2022).
57. El-Guebaly, L.A. History and evolution of fusion power plant studies: Past, present, and future prospects. In *Nuclear Reactors, Nuclear Fusion and Fusion Engineering*; Aasen, A., Olsson, P., Eds.; Nova Science Publishers, Inc.: New York, NY, USA, 2009; pp. 217–271; ISBN 978-1-60876-722-9.
58. Kohut, M. Lessons from the National Ignition Facility. *ASK Mag.* **2011**, 54–59. Available online: https://appel.nasa.gov/2011/02/02/lessons-from-the-national-ignition-facility/ (accessed on 29 December 2022).
59. Cartlidge, E. National Ignition Facility Demonstrates Net Fusion Energy Gain in World First. *Physics World*. 14 December 2022. Available online: https://physicsworld.com/a/national-ignition-facility-demonstrates-net-fusion-energy-gain-in-world-first/ (accessed on 30 December 2022).
60. Vogt, A.; Hayes, M.; Nilsen, E.; Hammond, E. US Officials Announce Nuclear Fusion Breakthrough. Available online: https://edition.cnn.com/us/live-news/nuclear-fusion-reaction-us-announcement-12-13-22/index.html (accessed on 27 December 2022).

61. Woodward, A.; Hiller, J. Nuclear-Fusion Breakthrough Accelerates Quest to Unlock Limitless Energy Source. *Wall Str. J.* **2022**. Available online: https://www.wsj.com/articles/nuclear-fusion-energy-breakthrough-reported-by-scientists-at-u-s-lab-11670944595?mod=hp_lead_pos10 (accessed on 30 December 2022).
62. ITER Tokamak Complex Construction Begins. *Nuclear Engineering International.* 24 June 2013. Available online: https://www.neimagazine.com/news/newsiter-tokamak-complex-construction-begins (accessed on 18 December 2022).
63. Cohen, A. ITER, the World's Largest Nuclear Fusion Project: A Big Step Forward. *Forbes*. 7 August 2020. Available online: https://www.forbes.com/sites/arielcohen/2020/08/07/iter-the-worlds-largest-nuclear-fusion-project-a-big-step-forward/?sh=1093f09d5e2c (accessed on 18 December 2022).
64. *What is ITER?* ITER Organization. Available online: https://www.iter.org/proj/inafewlines (accessed on 18 December 2022).
65. *Facts & Figures*; ITER Organization. Available online: https://www.iter.org/factsfigures (accessed on 22 December 2022).
66. *Building ITER*; ITER Organization. Available online: https://www.iter.org/construction/construction (accessed on 22 December 2022).
67. De Clercq, G. Nuclear Fusion Reactor ITER's Construction Accelerates as Cost Estimate Swells. *Reuters*. 7 October 2016. Available online: https://www.reuters.com/article/us-france-nuclear-iter-idUSKCN1271BC (accessed on 27 December 2022).
68. Claessens, M. Breakthrough at the ITER Fusion Reactor Paves Way for Energy Source That May Alter the Course of Civilization. *Newsweek*. 29 May 2020. Available online: https://www.newsweek.com/breakthrough-iter-fusion-reactor-paves-way-energy-source-that-may-alter-course-civilization-1507324 (accessed on 31 December 2022).
69. Kramer, D. ITER Disputes DOE's Cost Estimate of Fusion Project. *Physics Today*. 16 April 2018. Available online: https://physicstoday.scitation.org/do/10.1063/PT.6.2.20180416a/full/ (accessed on 31 December 2022). [CrossRef]
70. Parisi, J.; Ball, J. *The Future of Fusion Energy*; World Scientific Publishing Europe Ltd.: Singapore, 2019; ISBN 978-178-634-542-4.
71. Stacey, W.M. *The Quest for a Fusion Energy Reactor: An Insider's Account of the INTOR Workshop*; Oxford University Press: New York, NY, USA, 2010; ISBN 978-0-19-973384-2.
72. Clery, D. *A Piece of the Sun: The Quest for Fusion Energy*; Overlook Duckworth, Peter Mayer Publishers, Inc.: New York, NY, USA, 2013; ISBN 978-1-4683-1041-2.
73. *What is the Difference between Plasma Energy Breakeven and Engineering Breakeven?* ITER Organization. Available online: https://www.iter.org/faq#What_is_the_difference_between_plasma_energy_breakeven_and_engineering_breakeven (accessed on 30 December 2022).
74. Nathan, S. Beyond ITER—Next Steps in Fusion Power. *The Engineer*. 3 January 2019. Available online: https://www.theengineer.co.uk/content/in-depth/beyond-iter-next-steps-in-fusion-power (accessed on 30 December 2022).
75. Lazard's Levelized Cost of Energy Analysis—Version 14.0, Lazard. October 2020. Available online: https://www.lazard.com/media/451419/lazards-levelized-cost-of-energy-version-140.pdf (accessed on 19 December 2022).
76. Saravanapavanantham, M.; Mwaura, J.; Bulović, V. Printed Organic Photovoltaic Modules on Transferable Ultrathin Substrates as Additive Power Sources. *Small Methods* **2022**, *7*, 2200940:1–2200940:8. [CrossRef]
77. Couix, Q. Georgescu-Roegen's Flow-Fund Theory of Production in Retrospect. *Ecol. Econ.* **2020**, *176*, 106749. [CrossRef]
78. Georgescu-Roegen, N. Energy Analysis and Economic Valuation. *South. Econ. J.* **1979**, *45*, 1023–1058. [CrossRef]
79. Mihalascu, D. Lightyear 0 Solar EV Enters Production at a Rate of One Per Week. *InsideEVs*. 12 December 2022. Available online: https://insideevs.com/news/626100/lightyear-0-solar-ev-enters-production-at-rate-one-per-week/ (accessed on 18 January 2023).
80. Voelcker, J. Exclusive: 3-Wheeled Aptera Reboots as World's Most Efficient Electric Car: 1600 km per Charge Promised by the Team That Created a Lightweight Two-Seat EV a Decade Ago. *IEEE Spectr.* **2019**. Available online: https://spectrum.ieee.org/exclusive-3wheeled-aptera-reboots-as-worlds-most-efficient-electric-car (accessed on 18 January 2023).
81. Aptera Joins Forces with C.P.C. Group to Accelerate Production . 3 November 2022. Available online: https://aptera.us/cpc-group-agreement/ (accessed on 21 January 2023).
82. Service, R. Skyscrapers could soon generate their own power, thanks to see-through solar cells: Solar windows absorb ultraviolet and infrared light while letting visible light pass through. *Science* **2018**. [CrossRef]
83. Apple Now Globally Powered by 100 Percent Renewable Energy. 9 April 2018. Available online: https://www.apple.com/newsroom/2018/04/apple-now-globally-powered-by-100-percent-renewable-energy/ (accessed on 22 January 2023).
84. Apple's 'Spaceship' Headquarters Nears Completion. *The Express Tribune*. 11 November 2016. Available online: https://tribune.com.pk/story/1227783/apples-spaceship-headquarters-nears-completion (accessed on 22 January 2023).
85. Peterson, K. This Solar-Powered Skyscraper Is a Major Advance in Sustainable Design. 30 August 2016. Available online: https://www.architecturaldigest.com/story/solar-powered-skyscraper-sustainable-design (accessed on 22 January 2023).
86. Richardson, L. What Is a Solar Home? 4 July 2022. Available online: https://news.energysage.com/solar-homes-houses/ (accessed on 25 January 2023).
87. Around the World in a Solar Airplane. Available online: https://aroundtheworld.solarimpulse.com/adventure#route (accessed on 27 January 2023).
88. Jakimowicz, A.; Rzeczkowski, D. The Impact of Public Administration Digitalization on the Decarbonization of the Economy. *Energies* **2021**, *14*, 5739. [CrossRef]
89. Jakimowicz, A. The Energy Transition as a Super Wicked Problem: The Energy Sector in the Era of Prosumer Capitalism. *Energies* **2022**, *15*, 9109. [CrossRef]

90. Schwartzman, D. Solar communism. In *The Routledge Handbook on Ecosocialism*; Brownhill, L., Engel-Di Mauro, S., Giacomini, T., Isla, A., Löwy, M., Turner, T.E., Eds.; Routledge—Taylor & Francis Group: New York, NY, USA, 2022; pp. 280–289; ISBN 978-042-934-142-7. [CrossRef]
91. Peixoto, J.P.; Oort, A.H.; De Almeida, M.; Tomé, A. Entropy budget of the atmosphere. *J. Geophys. Res. Atmos.* **1991**, *96*, 10981–10988. [CrossRef]
92. Gan, Z.; Yan, Y.; Qi, Y. Entropy budget of the earth, atmosphere and ocean system. *Prog. Nat. Sci.* **2004**, *14*, 1088–1094. [CrossRef]
93. Schwartzman, D. *Life, Temperature, and the Earth: The Self-Organizing Biosphere*; Columbia University Press: New York, NY, USA, 1999; ISBN 0-231-10212-7.
94. Kåberger, T.; Månsson, B. Entropy and economic processes—physics perspectives. *Ecol. Econ.* **2001**, *36*, 165–179. [CrossRef]
95. Hamacher, T.; Bradshaw, A.M. Fusion as a Future Power Source: Recent Achievements and Prospects. Available online: https://fire.pppl.gov/energy_eu_wec01.pdf (accessed on 31 January 2023).
96. Goldston, R.J.; Glaser, A.; Ross, A.F. Proliferation Risks of Fusion Energy: Clandestine Production, Covert Production, and Breakout. Princeton Plasma Physics Laboratory, August, 2009. Prepared for the U.S. Department of Energy under Contract DE-AC02-09CH11466. Available online: https://www.osti.gov/servlets/purl/962921-w4qgvc (accessed on 31 January 2023).
97. Distribution of Electricity Generation Worldwide in 2021, by Energy Source. Statista Research Department. 16 January 2023. Available online: https://www.statista.com/statistics/269811/world-electricity-production-by-energy-source/ (accessed on 6 February 2023).
98. Trieb, F. *Trans-Mediterranean Interconnection for Concentrating Solar Power. Final Report*; German Aerospace Center (DLR), Institute of Technical Thermodynamics, Section Systems Analysis and Technology Assessment: Stuttgart, Germany, 2006. Available online: http://www.dlr.de/tt/Portaldata/41/Resources/dokumente/institut/system/projects/TRANS-CSP_Full_Report_Final.pdf (accessed on 7 February 2023).
99. Schwartzman, D. *The Global Solar Commons, the Future That is Still Possible: A Guide for 21st Century Activists*; The Solar Utopia.org Press: Galesburg, IL, USA, 2021. Available online: https://www.theearthisnotforsale.org/solarcommons2021.pdf (accessed on 30 January 2023).
100. von Schilling, C.; Straussfogel, D. Entropy Debt: A Link to Sustainability? In Proceedings of the 53rd Annual Meeting of the International Society for the Systems Sciences, ISSS 2009, Brisbane, Australia, 12–17 July 2009. Available online: https://journals.isss.org/index.php/proceedings53rd/article/view/1302 (accessed on 8 February 2023).
101. Jakimowicz, A.; Rzeczkowski, D. Contact Zones in the Energy Transition: A Transdisciplinary Complex Problem. *Energies* **2023**, *16*, 3560. [CrossRef]
102. Rifkin, J. *The Third Industrial Revolution: How Lateral Power is Transforming Energy, the Economy, and the World*; St. Martin's Press, LLC: New York, NY, USA, 2011; ISBN 978-0-230-11521-7.
103. Rifkin, J. *The Empathic Civilization: The Race to Global Consciousness in a World in Crisis*; Polity Press: Cambridge, UK, 2010; ISBN 978-0-7456-4145-4.
104. Quilligan, J.B. The commons of mind, life and matter: Toward a non-polar framework for global negotiations. *Kosm. J. Glob. Transform.* **2010**, 41–47. Available online: https://www.kosmosjournal.org/wp-content/article-pdfs/the-commons-of-mind-life-and-matter-toward-a-non-polar-framework-for-global-negotiations.pdf (accessed on 8 February 2023).

energies

MDPI

Article

Improving the Feed-in Tariff Policy for Renewable Energy Promotion in Ukraine's Households

Tetiana Kurbatova [1], Iryna Sotnyk [2], Olha Prokopenko [3,4,5,*], Iryna Bashynska [6] and Uliana Pysmenna [7]

1 International Economic Relations Department, Sumy State University, 40007 Sumy, Ukraine; t.kurbatova@biem.sumdu.edu.ua
2 Department of Economics, Entrepreneurship and Business Administration, Sumy State University, 40007 Sumy, Ukraine; i.sotnyk@biem.sumdu.edu.ua
3 Estonian Entrepreneurship University of Applied Sciences, 11415 Tallinn, Estonia
4 Department of Applied Sciences, Academy of Applied Sciences Mazovia, 08-110 Siedlce, Poland
5 Department of Business Economics and Administration, Sumy State Makarenko Pedagogical University, 40000 Sumy, Ukraine
6 Department of Enterprise Management, AGH University of Krakow, 30-059 Krakow, Poland; bashynska@agh.edu.pl
7 Department of Economics and Entrepreneurship, National Technical University of Ukraine "Igor Sikorsky Kyiv Polytechnic Institute", 03056 Kyiv, Ukraine; uliamyxa@gmail.com
* Correspondence: olha.prokopenko@eek.ee

Abstract: This paper deals with improving the feed-in tariff policy for green energy advancement in Ukraine's households based on assessing the economic efficiency of investment project implementation under the current state support mechanisms. This study was conducted for solar and wind power plants with different installed capacities. The Levelized Cost of Electricity and the payback period for such power plants were calculated considering the ongoing feed-in tariffs and discount rates determined by various equity and debt capital ratios. The results showed that the state support provides attractive payback periods for solar and wind power plants with an installed capacity of ≥30 kW. In comparison, 5 kW solar power plants and wind power plants with a capacity of up to 10 kW are not paid off during the power plants' lifecycle. It confirmed that the ongoing energy policy in Ukraine's residential sector is still designed to obtain profits by the owners of generating facilities by selling the excess electricity. In the meantime, its main goal—providing households energy independence—has been levelled. To resolve the issues caused by such a state support model, a methodical approach to improve the feed-in tariff calculation is proposed. In addition, recommendations for reconsidering other policy measures to ensure effective renewable energy development in the residential sector have been made.

Keywords: renewable energy; the feed-in tariff; investment; solar power plant; wind power plant; energy policy; households; Ukraine

1. Introduction

Countries worldwide consider renewable energy (RE) development as a crucial tool in resolving the problem of growing energy demand and mitigating climate change consequences. In addition, deploying RE facilities plays a significant role in ensuring the energy security of the world's countries, which is especially relevant under recent geopolitical events. Nowadays, achieving ambitious goals for RE transition directly depends on state support. The understanding of this is reflected in the establishment by the countries of indicative targets for increasing the green energy portion in the final energy consumption and the implementation of state support mechanisms to encourage energy production by renewable energy resources (RES) [1–3].

Along with promoting RE development in the industrial sector, countries' governments also seek to foster the use of RES in the residential sector. The most common motivational tool used in the residential sector for RE adoption is the feed-in tariff (FIT). The widespread use of the FIT and its adaptation to the goals and needs of national policies have led to the appearance of a number of its variations, in particular, the fixed FIT, front-end-loaded FIT, adjusted FIT, premium FIT, etc. [4,5]. The effectiveness of the FIT as an incentive tool is confirmed by the rapid deployment of RE facilities in countries where it is used. Thus, the introduction of the FIT into the household sector has had a remarkable impact on deploying solar energy in Germany [6], the UK [7], China [8], and other countries [9,10]. At the same time, despite the fact that the main goal of increasing green electricity has been achieved, RE promotion with the help of generous FITs has led to some challenges, the main of which include an increase in state expenditures for compensating payments for FITs and technical limitations of the electricity grids [6,7]. It has contributed to the revision of the FIT policy, which, in most countries, caused uncertainty regarding the future trends in RE development or a decrease in the pace of RE facilities installation in the household sector [7,11].

It should be noted that Ukraine was no exception, and, since 2014, it has implemented an energy policy for RE promotion in households as part of the energy strategy, where the primary economic tool is the fixed FIT [12,13]. However, despite certain successes in developing small-scale-RE distributed energy generation, state policy in this sector cannot be called effective. The result of its implementation was the rapid development of only solar photovoltaic (PV) systems, which became possible due to the high rates of the FIT. The pursuit of profits from selling the excess electricity at the high FIT has led to some abuses by the household owners. Cases of increasing the capacity of the generating facilities beyond the permitted value, setting the inverter equipment to the maximum output (which has negative consequences for the electrical equipment of neighboring consumers), and installing solar PV systems without or with low-level electricity consumption have been recorded repeatedly. Thus, the electricity consumption in about 40% of households with installed solar PV systems is less than 10 kWh/month, while the average electricity consumption in Ukrainian households, as of 2020, was 168 kWh (or 954 kWh with electricity heating) [14].

Because of the considerable growth of solar power plants and the aforementioned manipulations, the payments of the transmission system operator to cover the difference between the FIT and the market electricity price are increasing annually, which ultimately falls on other consumers as an additional financial burden. Therefore, in 2020, such expenses amounted to UAH 2.26 billion; moreover, in 2021, it increased to UAH 4.07 billion [15]. In addition, the over incentives for solar energy have led to disparities in RE facilities deployment in households. Thus, although the FIT also applies to the electricity produced by wind power plants, due to the higher rates of solar power (which for a long time ensured a payback period of 3.3–4.5 years [16,17]), households preferred to invest in solar power plants.

Considering the above, it can be concluded that the main focus of the RE policy in the residential sector has not been placed on ensuring household energy independence, due to their electricity production, but on obtaining significant profits by the owners of RE power plants owners by selling the unconsumed excess electricity. This energy policy model has resulted in several challenges and abuses in the household sector. Thus, it is evident that the current approaches for fostering RE in the residential sector should be revised for further effective green energy transition. Given the above, this paper's primary focus will be on evaluating the cost of the electricity produced by RE power plants and the payback period based on the current FIT rates and discount rates determined by the different ratios of equity and debt capital involved in investment project implementation. Based on the obtained results, recommendations regarding improving the methodology for calculating the FIT and the state policy for fostering RE facilities deployment in Ukraine's households will be given. Figure 1 presents the main steps of the research algorithm.

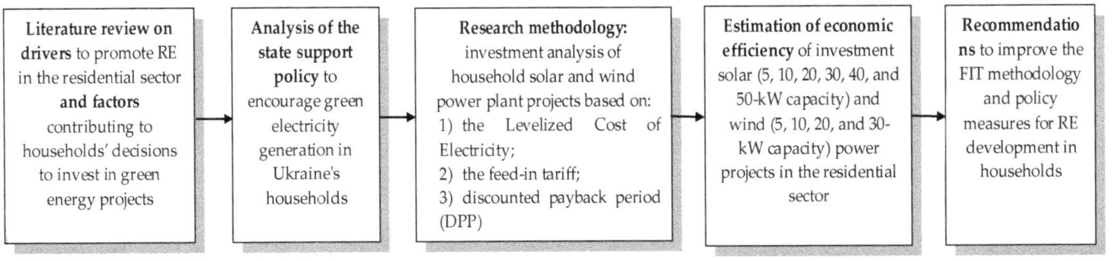

Figure 1. The research algorithm.

The article structure is as follows: Section 2 presents a literature review regarding the researched subject. In Section 3, we analyze policy for fostering RE in the residential sector of Ukraine and its impact on RE facilities deployment. Section 4 describes the research methods. In Section 5, we discuss the obtained research results and propose approaches for improving the methodical approaches for FIT rates calculation and give policy recommendations for effective RE development in the household sector. Section 6 contains conclusions and points to potential areas of further study.

2. Literature Review

Many research papers are dedicated to RE development, including investment issues [18–22], the identification of drivers and barriers for the effective deployment of RE facilities [23–25], the evaluation of economic, environmental, social, and other benefits from RE project implementation, etc. [26–28]. Two main research blocks can be distinguished when adopting RE in the residential sector. The first concerns the analysis of the support schemes for stimulating RE development, comparing their effectiveness, and assessing their impact on installing RE facilities in households. The second focuses on the role of socio-demographic, housing, environmental, and other factors influencing the decisions of the household owners to install RE power plants.

Many studies focus on the fact that the economic mechanisms aimed at encouraging green electricity production play a crucial role in RE development. Nowadays, tools such as FIT, net metering, and net billing are most widely used to promote green energy development in the household sector, while the RE portfolio standards, green auctions, and tender systems are primarily used in the industrial sector [29,30]. The state support schemes play a unique role at the initial stage of RE technology development, due to their inability to compete with conventional methods in market conditions. Considering the trends towards a gradual decrease in electricity production cost based on RES, the value of economic incentives is somewhat reduced; however, the state support for RE promotion in households still plays a considerable role in most countries.

Thus, the research results obtained by Jacksohn and et al. [31] prove that the decisions of German households to invest in RE projects are mainly influenced by economic factors. The authors found a clear relationship between the investment decision and the anticipated profit from the investment activity. At the same time, socio-demographic factors, housing characteristics, and environmental issues do not significantly impact household owners' decision making.

Lan et al. [9] investigated the impact of state policy measures, particularly the FIT, on stimulating solar energy deployment in Australian households. The study results show that the readiness of household owners to invest in solar PV systems increases with the growing FIT rates. The authors concluded that the FIT rates should be as high as possible in the early stages of RE technology development. In contrast, at later stages, they can be appropriately reduced. For more effective stimulation and a reduction in the financial pressure on the government budget, combining the FIT with the lending policy with low interest rates is advisable.

The study results of Guta [32] demonstrate that, in Ethiopia, high household revenue directly affects RE technology adoption in the residential sector, as financial ability allows household owners to pay for their initial cost. Among other factors, household membership in energy cooperatives, access to credit, and education policy increase the probability of investing in RE. These results are confirmed by a study conducted in Kenya [33], where the authors added to the above factors, other socio-economic and demographic characteristics, that positively influence decision making regarding RE projects implementation, in particular, increasing household size, family status, and age of the household owners.

In turn, a study by Etongo and Naidu [34] has proven that age, education, and family size, which were significant in the abovementioned studies, are not crucial for solar energy technology adoption in Seychelles' households. The primary motivating factors for solar PV systems installation in the Seychelles are cost saving, energy security, and environmentally friendly perceptions. Meanwhile, low electricity tariffs, the high investment cost of solar PV systems, credit conditions, and a long payback period for investment projects are the main obstacles to investing in solar energy. According to a study by Štreimikienė and Baležentis [35], despite the willingness of Lithuanian household owners with higher revenues and educational status to pay higher prices for green electricity, there are other factors besides these in making investment decisions. The main factors of the willingness of Lithuanian households to invest in green energy projects are the awareness of RES and environmental issues. Thus, there is a particular regional difference in the critical factors influencing households' investment in RE facilities. Therefore, when developing public policy, it is necessary to consider each country's peculiarities.

Braito and et al. [36] studied personal, social group, and psychological patterns of investment in RE within the framework of different energy policy models. The authors investigated contrasting state support policies for solar energy promotion in households, particularly the high financial initiatives in Italy's province versus the lower ones in Austria. The authors concluded that the financial incentives directly impact solar PV system development in the households of both countries, as both individual and collective investors seek to receive economic benefits. However, the authors emphasized the importance of balanced economic support and its combination with non-monetary initiatives, as overly generous financial incentives in Italy may supplant households' preferred non-financial motivation, which can lead to a decrease in solar PV systems installation after the expiration of the state incentives.

Solar energy in Ukraine's households may face similar risks, since the stimulation of its development for a long period was based on high financial support and virtually no non-monetary initiatives. As a result, there is a strong link between the economic support level and solar power plants deployment in households. At the same time, some economic entities, in particular, higher education institutions, do not have the opportunity to take advantage of such economic incentives. As a result, the level of RES involvement in electricity production by them is practically absent. However, even without state economic support, implementing such projects can ensure significant economic, ecological, and social gains [28]. Thus, generous economic incentives may carry potential risks with a sharp change in regulatory approaches in the future, which are inevitable amid the increasing green electricity share and financial burden to support such measures. In addition, they can reduce the motivation of other economic entities whose conviction to invest in RE is based on non-monetary initiatives. These studies confirm this statement [37,38], where the authors note that, in addition to the direct price effect, monetary support indirectly influences in the form of the crowding out specific groups or reducing the internal motivation to provide public goods constantly.

This study contributes to the scientific literature by conducting an in-depth cost evaluation of the electricity produced by small solar and wind power plants in the residential sector of Ukraine. The research examines the payback periods under various scenarios and considers different discount rates alongside the prevailing FITs. By offering a comprehensive assessment of these crucial factors, this paper enhances our understanding of the economic viability of RE investments in Ukraine's households.

Furthermore, this study sheds light on the efficiency of the economic support for RE development, particularly in the residential sector. It quantifies the benefits and highlights the potential risks associated with overpromotion, where economic incentives may be compromised. This nuanced analysis provides policymakers and stakeholders with valuable insights for refining RE policies and FIT methodologies.

Moreover, this research identifies the pivotal conditions that signal the need for timely adjustments in RE management, thereby helping to ensure the sustainability of green energy initiatives. By offering recommendations grounded in empirical data, this study aims to guide informed decision making, fostering responsible and sustainable RE development in Ukrainian households while minimizing potential pitfalls.

3. State Support Policy and Its Impact on Renewable Energy Facility Deployment in Ukraine's Households

The state policy for promoting RE development was introduced in Ukraine in 2009; however, the economic mechanisms to encourage green electricity production were only applied to the industrial sector. In 2014, the legislation was amended, and it was extended to include the household sector.

Thus, under the Law of Ukraine "On Alternative Energy Sources" [13], the primary motivational tool is the FIT—a tariff under which the state purchases the excess electricity produced by the RE power plants of households that is not consumed for their own needs.

Starting from 2014, households have been allowed to implement solar and wind power plants with a total capacity of ≤30 kW. Since the beginning of 2019, considering the substantial demand for installing solar PV systems, their allowed capacity has been reconsidered up to 50 kW, providing that they are placed on the roofs and/or building facades and other facilities without using agricultural land. However, the specified legal norm was valid only until the end of 2019, after which it was cancelled. Starting in 2019, the FIT began to apply to hybrid wind–solar power plants with a capacity of up to 50 kW [13].

To protect the owners of the RE facilities from possible inflation, the FITs are recalculated to the EUR exchange rate and are revised quarterly, pursuant to the official rate of the National Bank of Ukraine (NBU). It should be noted that the household revenue from selling the excess electricity at the FIT is taxable. The tax amount is 19.5%, of which 18% is personal revenue tax and 1.5% is military levy.

The FIT for promoting RE deployment in households has been established until 1 January 2030. The state guarantees to buy all of the electricity unconsumed for household needs.

Further to the FIT, households can use some tax and customs benefits during RE project implementation [39,40], namely, the waiver of payment of value-added tax (VAT) and customs fees when purchasing equipment for installing the RE power plants.

In addition, households can use special credit lines, particularly "Green Energy" of Oschadbank and "Eco Energy" of Ukrgasbank. Thus, under the above programs, households can receive a loan to purchase and install RE power plants. According to the credit terms, the minimum first instalment that households must pay is 15% of the cost of the goods and services. The credit term is from 1 to 6 years. The maximum loan amount is UAH 1 million. The loan rate depends on the loan term and the amount of the first instalment [41,42].

Implementing the above support mechanisms increased the household sector's investment attractiveness; however, only solar PV systems have acquired dynamic development (Table 1).

Table 1. The dynamics of RE plants development in the household sector of Ukraine in 2015–2021 [43,44].

RE Power Plants in the Household Sector	2015	2016	2017	2018	2019	2020	2021
			Number of RE power plants, units				
Solar power plants	244	1109	3010	7450	21,968	29,931	44,888
Wind power plants	0	0	3	3	4	4	4
			The installed capacity of RE power plants, MW				
Solar power plants	2	17	51	157	553	779	1205
Wind power plants	0	0	0.029	0.029	0.027	0.041	0.057
			Amount of electricity produced, million kWh				
Solar power plants	0.4	4	23	92	303	756	1151
Wind power plants	0	0	2.5	2.5	0.2	0.6	1.9

Thus, the data in Table 1 show that, as of the end of 2021, 44,888 small solar PV systems, with a cumulative installed capacity of 1205 MW, were installed. It should be noted that such growth was ensured by the high FITs for the solar PV systems, the payments for which are included in the electricity tariffs. Thus, considering that Ukraine's budget does not provide for special sources for financing the expenses under the FIT, the increase in green energy share places an additional financial burden on the electricity consumers.

At the same time, wind energy in households is practically not developing. By the end of 2021, only four wind power plants, with a cumulative installed capacity of 57 kW, were commissioned. The main reason for this was the higher FIT rate for solar power plants, which made the solar energy segment extremely attractive to investors. One more barrier to wind energy deployment in households was the long-term absence of the FIT for hybrid wind–solar power plants. Thus, until 2019, due to the difference in the FITs, installing hybrid wind–solar plants required the registration of two electricity production metering points, which was not profitable. Hence, households preferred solar power plants instead of hybrid wind–solar plants. To resolve this issue, the FIT for combined wind–solar power plants was introduced in 2019. However, after four months, its rate was reduced by 25%, making their installation economically unattractive and the plans for such energy facilities development declarative [45].

Given the above, it can be argued that, despite certain achievements, the state policy for RE promotion in Ukrainian households cannot be called effective. The main problem was the focus of the FIT on the purposeful sale of the excess electricity, which caused the rapid PV systems installations but, at the same time, led to problems with the financing of the payments under the FIT. In turn, due to the imperfect energy policy, wind energy in Ukraine's households is practically not developing.

Since green energy transition remains one of the state's priorities for the coming years, and its pace depends largely on the efficiency of the energy policy, the revision of approaches to forming support measures is becoming particularly relevant. In the household sector, where the primary motivational tool is the FIT, methodical approaches to calculating its rates for RE facilities considering their installed capacities, the RE technologies used, and the structure of the investment expenditures should be improved.

4. Methods

To balance the interests of households and the state, the FIT rates for RE power plants should be based on assessing the economic efficiency of such projects implementation. To determine the required level of state support, we will estimate the electricity production cost, the current rates of the FITs, and the RE projects' payback periods.

The cost of electricity produced by household RE power plants will be evaluated based on the Levelized Cost of Electricity (LCOE) method. The LCOE demonstrates the fixed electricity cost during a power plant's lifecycle, which equates the total discounted expenditures for its construction and operation to the total discounted revenue from selling electricity [46]. The calculation of LCOE will be based on the following indicators: investment, operating and decommissioning expenditures, the amount of electricity produced, and the discount rate. Considering these indicators, the formula will be as follows:

$$LCOE = \frac{\sum_{t=0}^{n} \left((I_t + Q_t + D_t) \cdot (1+r)^{-t} \right)}{\sum_{t=0}^{n} \left(E_t \cdot (1+r)^{-t} \right)}, \quad (1)$$

where $LCOE$ is the fixed cost of electricity production throughout the lifecycle of a power plant, UAH/kWh; E_t is the amount of produced electricity by the power plant in the t-th year, kWh; I_t is the investment expenditures in the t-th year, UAH; Q_t is the operating and maintenance expenditures in the t-th year, UAH; D_t is the decommissioning expenditures in the t-th year, UAH; n is the lifecycle of the power plant, years; r is the discount rate; and t is the year of realization of the investment project.

To calculate the discount rate, we will use the following formula of weight average cost of capital ($WACC$) [47]:

$$WACC = K_s \cdot W_s + K_d \cdot W_d, \quad (2)$$

where K_s is the equity cost for the investment project realization, unit share; W_s is the equity share by the balance, unit share; K_d is the debt cost for the investment project implementation, unit share; and W_d is the debt share by the balance, unit share.

Next, to assess the economic feasibility of the current FITs, we will calculate their rates and determine the projects' payback periods based on them.

The minimum FIT rate is calculated following the resolution of the NCSREPU No. 1817, dated 30 August 2019 [48]. Thus, according to [48], the fixed minimum FIT rate (FIT_{min}, UAH per 1 kWh without VAT) is determined by the following formula:

$$FIT_{min} = \frac{T_{2009} \cdot k_{FT}}{E_{2009}}, \quad (3)$$

where T_{2009} is the electricity tariff for households as of January 2009; k_{FT} is the FIT coefficients, which are defined by [12]; and E_{2009} is the exchange rate of UAH to EUR as of 1 January 2009.

Quarterly, the minimal FIT rates are reviewed by the NCSREPU by recalculating them relative to the EUR exchange rate as of 1 January 2009.

The discounted revenue for the project will be calculated based on the FIT validity period, the average amount of electricity consumption by the household, and the market price of electricity under the methodology presented in [49]. Next, these will be used for calculating the projects' payback periods according to the following formula [50]:

$$DPP = m + \frac{I_\Sigma - S_m}{Inc_{m+1}} \cdot (1+r)^{m+1}, \quad (4)$$

where DPP is the discounted payback period of the investment project, years; I_Σ is the total discounted investment expenditures for the project, given at the time of the start of the investment, UAH; S_m is the total discounted revenue determined as a cumulative sum until the inequality is satisfied: $S_m < I_\Sigma < S_m + 1$, where m is the number of complete years when the discounted revenues, determined as the cumulative sum, are less than the discounted investment expenditures; $(m + 1)$ is the year when the discounted revenue, determined as a cumulative sum, will be higher than that of the discounted investment expenditures, UAH; and Inc_{m+1} is the project's revenue in the $(m + 1)$-th year, UAH.

Based on assessing the economic efficiency of RE project implementation in households, we will offer an improved methodology to calculate the FIT rates and the terms of state support to ensure the balance of interests for the households and the state. A conceptual map of creating an effective FIT for green energy facilities in households is presented in Figure 2.

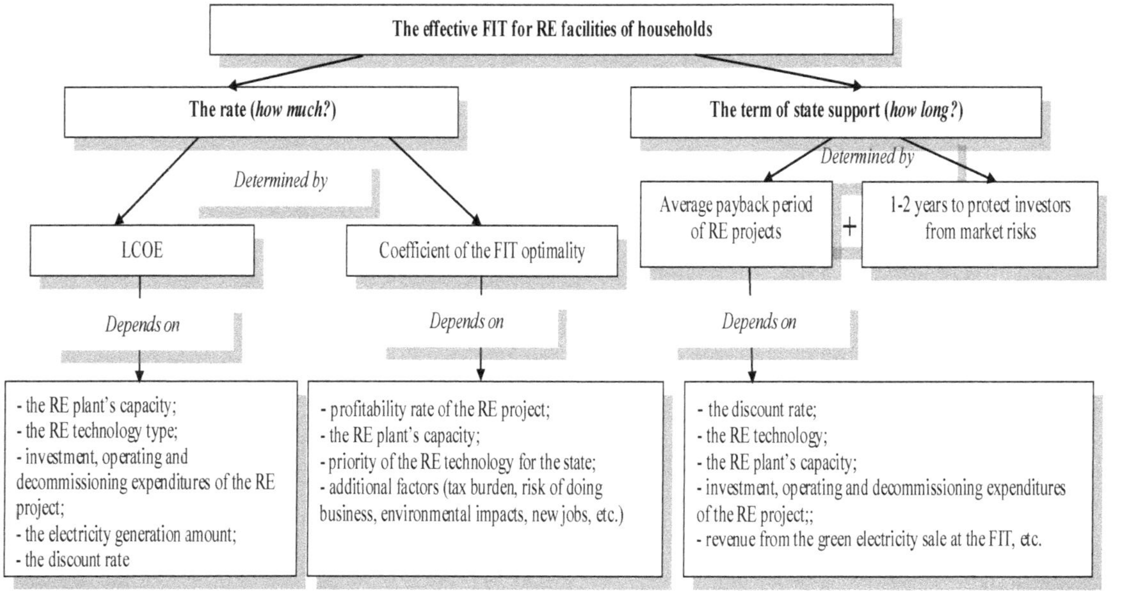

Figure 2. A conceptual map of forming an effective FIT for households.

5. Result and Discussion

5.1. Evaluation of the Cost of Electricity Produced by RE Power Plants in Ukraine's Households and Investment Projects' Payback Periods Based on the Current FIT Rates

This study will calculate the LCOE for household solar power plants with installed capacities of 5, 10, 20, 30, 40, and 50 kW and wind power plants with installed capacities of 5, 10, 20, and 30 kW. The specified capacity range corresponds to the households' most common power plants and is also determined by the current (0–30 kW) and former (0–50 kW) legal limits. The difference in the selected capacities of the RE facilities is due to the productivity factor of wind power plants, which is 2–3 times higher than that of productivity factor of solar power plants [51]. It is worth noting that the main focus of state support should be providing households with their own electricity and not obtaining profits from the sale of their surpluses at the FIT. Given this, extending state support mechanisms in the household sector to wind power plants with a capacity >30 kW is impractical. In the meantime, as mentioned above, starting from 2019, the FIT was applied to solar PV systems with a capacity of up to 50 kW, providing that they are located on building roofs or facades, without using agricultural land. However, at the end of 2019, the FIT for solar power plants whose installed capacity is between 30 and 50 kW was canceled. In this study, we will estimate the cost effectiveness of such RE projects in order to conclude the feasibility of repealing this legislative norm.

It is noteworthy that the datasets of the solar and wind power plant projects differ somewhat, as they depend on various factors (the solar insolation and wind speed in the region of the RE power plant installation, the technical characteristics of the solar panels and wind generators, etc.); therefore, for calculations in this research, their averaged data were taken, shown in Table 2 [52–59].

Table 2. The dataset of investment projects of household solar and wind power plants [52–59].

	Installed Capacity, kW					
	5	10	20	30	40	50
			Solar power plants			
Electricity production, kWh/year	4947	10,520	21,836	33,153	46,206	57,791
Investment expenditures, UAH	146,229	260,802	517,347	763,448	990,478	1,138,253
Operating expenditures, UAH/year	1462	2608	5173	7634	9905	11,383
Decommissioning expenditures, UAH	7311	13,040	25,867	38,172	49,523	56,913
			Wind power plants			
Electricity production, kWh/year	15,000	30,000	60,000	90,000	-	-
Investment expenditures, UAH	380,948	654,239	1,117,280	1,571,370	-	-
Operating expenditures, UAH/year	3809	6542	11,173	15,714	-	-
Decommissioning expenditures, UAH	19,047	32,712	55,864	78,569	-	-

It is worth mentioning that the data in Table 2 concerning the electricity production amount will be adjusted for the production reduction factor, which for solar power plants is 0.8% annually [60] for wind power plants is 0.2% annually [61]. The operating and decommissioning expenditures in this study were determined as 1% and 5% of the investment expenditures, respectively [62], and the lifecycle of both types of power plants was determined to be 25 years [63].

Under this study, the cost of electricity production by the RE power plants will be calculated considering various options for the structure of the investment expenditures, as follows: (1) on the condition that the investment projects are implemented entirely with the investor's funds; and (2) on the condition of attracting own and credit resources for five years within the framework of the Ukrgasbank "Eco Energy" program in the following ratios of 75:25, 50:50, and 25:75, respectively, which will affect the size of the discount rate.

To calculate the discount rate, the equity cost will be determined as the alternative investments in deposits for individuals in the national currency. For this aim, we analyzed the deposit rates in UAH in the most reliable banks in Ukraine based on their stability rating [64]. As of the beginning of June 2023, the following banks offered such annual rates for deposits in UAH for individuals: Credit Agricole—12%, Raiffeisen—12.5%, and Kredobank—14% [65]. Thus, Kredobank offered the highest rate, and it will be used to determine the equity cost.

The cost of the debt capital will be calculated according to the "Eco Energy" credit line. Therefore, subject to attracting credit resources for 5 years with the first instalment payments of 25%, 50%, and 75%, the annual interest rates on the loan are 20.99%, 19.59%, and 18.69%, respectively [42].

Thus, the discount rate calculated according to Formula (2), provided that the investment projects are implemented entirely with the investor's funds, is 14%, provided that own and credit funds are attracted in the following ratios: 25:75—19.2%, 50:50—16.8%, and 75:25—15.2%.

The LCOE values for the electricity produced by the RE power plants in households, considering different discount rates, are shown in Table 3.

Next, we will compare the obtained LCOE values and the current FIT rates at which the households sell the excess electricity. We will calculate the FIT rates using Formula (3). As mentioned above, the minimum FIT rate is based on the FIT coefficients for the RE power plants, which are shown in Table 4.

Table 3. LCOE for RE power plants of households (calculated by authors).

The Installed Capacity of RE Power Plants, kW		LCOE, UAH/kWh 5	10	20	30	40	50
				Solar power plants			
Discount rate, %	14	4.28	3.59	3.43	3.33	3.10	2.85
	15.2	4.52	3.79	3.62	3.52	3.28	3.01
	16.8	4.84	4.06	3.88	3.77	3.51	3.23
	19.2	5.32	4.46	4.27	4.15	3.86	3.55
				Wind power plants			
Discount rate, %	14	3.55	3.04	2.60	2.44	-	-
	15.2	3.75	3.22	2.75	2.58	-	-
	16.8	4.03	3.46	2.96	2.77	-	-
	19.2	4.44	3.81	3.26	3.05	-	-

Table 4. The FIT coefficients for RE power plants of households commissioned from 2019 to 2029 [12].

RE Power Plants of Households for which the FIT is Applied	The FIT Coefficients for Electricity Produced by RE Power Plants of Households, Commissioned		
	From 01.01.2019 to 31.12.2019	From 01.01.2020 to 31.12.2024	From 01.01.2025 to 31.12.2029
Solar power plants with installed capacity ≤30 kW	3.36	3.02	2.69
Solar power plants with installed ≤50 kW, provided they are placed on the roofs and/or buildings' facades	3.36	-	-
Wind power plants with installed capacity ≤30 kW	2.16	1.94	1.73

As seen in Table 4, the FIT coefficients depend on the generating facility's commissioning date. They tend to decrease due to the reduction in the cost of equipment, materials, and components for such power plants, which directly affects the electricity production cost.

In this study, the calculation will be carried out for the RE power plants commissioned in 2023; accordingly, the coefficient for the power plants commissioned from 1 January 2020 to 31 December 2024 will be used. As evidenced by the data in Table 4, from 2020 the FIT does not apply to solar power plants with installed capacity between 30 and 50 kW. It is worth noting that, in 2019, the FIT coefficients were the same for the solar PV systems with with an installed capacity of ≤30 kW and those whose capacity is between 30 and 50 kW. Given this, to conclude regarding the feasibility of cancelling the FIT for this category of generating facilities, a coefficient of 3.02 will be used for the 40- and 50-kW solar PV systems. To carry out the FITs recalculation relative to the EUR exchange according to algorithm given in [48], the exchange rate of UAH 3983 for EUR 100 will be used [66]. Table 5 includes the obtained results.

Table 5. The FIT rates for electricity produced by RE power plants of households commissioned in 2023 (calculated by the authors).

RE Power Plants of Households	The FIT, UAH/kWh
Solar power plants with installed capacity ≤30 kW	6.48
Solar power plants with installed capacity ≤ 50 kW, provided they are placed on the roofs and/or buildings' facades	6.48
Wind power plants with installed capacity ≤30 kW	4.16

Next, we will calculate investment projects' payback periods according to Formula (4), based on the following factual data and assumptions:

- A household consumes 168 kWh of electricity per month (corresponding to the average monthly electricity consumption by Ukrainian households) [14]) and sells the excess electricity at the FIT. Since the support scheme for RE promotion based on the FIT is valid until 31 December 2029, after its expiration, the electricity will be realized at the market price. When calculating cash flows, savings from the electricity expenditures received by the household as a result of their own production and consumption for their own needs will be considered. If the production is not enough to cover the household's own needs, then the revenue from the power plant is calculated as the amount of electricity cost savings based on the produced amount;
- The electricity tariff for households as of 1 June 2023 was UAH 2.64/kWh [67];
- As mentioned above, the revenue from selling electricity at the FIT is taxable at a cumulative rate of 19.5% [39].

The results of the calculations of the payback periods of RE investment projects are given in Table 6.

Table 6. Payback periods of RE investment projects, years (calculated by the authors).

		Payback Periods of RE Investment Projects, Years					
The Installed Capacity, kWh		5	10	20	30	40	50
				Solar power plants			
Discount rate, %	14	>25	13.73	8.53	6.98	6.09	5.30
	15.2	>25	17.94	9.73	7.80	6.32	5.45
	16.8	>25	>25	12.21	9.29	6.66	5.68
	19.2	>25	>25	>25 years	13.34	7.59	6.02
				Wind power plants			
Discount rate, %	14	>25	>25	13.00	10.19	-	-
	15.2	>25	>25	15.55	11.43	-	-
	16.8	>25	>25	>25	14.06	-	-
	19.2	>25	>25	>25	>25	-	-

The results of the calculation of the RE investment projects' payback periods in Ukraine's households proved the existence of an economic effect depending on the project's size. A reduction in investment and operating expenditures as the capacity of an RE power plants increases determines a more attractive payback period for projects with the maximum allowed capacity. Although it is an incentive for households that have the opportunity to implement large-scale projects, a certain category of households, in particular, those that do not have enough financial resources or the ability to install such generating facilities following the requirements of the law, cannot use this incentive. At the same time, for households with an electricity consumption of 2016 kWh/year, 5 kW solar power plants and wind power plants with a capacity of up to 10 kW are not paid off during their lifecycle. The payback period is unattractive for 10 kW solar and 20 kW wind power plants. When attracting a debt capital at 50% or more, the projects are not paid off. Given the above, it is reasonable to gradate the FIT rates in accordance with the RE power plants' installed capacity. This will allow households with different revenue levels to have equal opportunities to invest in RE power plants.

The primary purpose of deciding to invest in micro RE projects is to ensure household energy autonomy, since the excess electricity for 5–10 kW RE power plants, i.e., that not consumed for the households' needs, is insignificant. Therefore, supporting such projects should become a priority task of household energy policy. In the meantime, the payback periods of 20-, 30-, 40-, and 50-kW solar power plants and wind power plants with a capacity of 30 kW remain pretty attractive. Implementing solar power plant projects with 40–50 kW capacity is the most profitable. In our opinion, the decision to cancel the FIT for such generating facilities was not logical, because they can significantly contribute to achieving the indicative goals for increasing the green electricity share in the country's energy consumption. However, the FIT rates for such generating capacities must be significantly adjusted in order to avoid excess profits by their owners and an increase in the financial burden on the final consumers, at the expense of which payments are made under the FIT. At the same time, for identical reasons, the FIT rates for 20- and 30-kW solar power plants also need some adjustment.

In addition, the obtained results have demonstrated the existing imbalance in the payback periods of solar and wind power plants. The difference between their payback periods is quite significant, which will likely continue to contribute to the predominance of solar power plants in households. Therefore, the adjustment of state support should be based on the priority of particular RE technology development, considering the evaluation of the economic, ecological, and social gains from its implementation. The development of hybrid solar–wind power plants is optimal, which will allow the attraction of both energy resources to electricity production. Therefore, the economic justification of the FIT for such plants is highly relevant.

As for acceptable payback periods for RE investment projects in the residential sector, world experience shows that they are usually, at most, 10 years. For example, for solar power plants in the US, the indicators fluctuate within 10–12 years [68] and 16–22 years in the United Kingdom [69]. In the meantime, the average payback period of small wind power plants in the US is 15 years [70], and, in Poland, when they are located in areas with the best wind speed, it is 13 years [71]. In our opinion, Ukraine should be guided by world experience, ensuring, by economic incentives, a payback period of RE plants of different installed capacities of 10–13 years. This approach to investment policy is logical in terms of the gradual reorientation of state policy towards the energy autonomy of households and the reduction in financial pressure on the final electricity consumers amid the dynamically increasing green electricity share in the country's energy mix.

Thus, it can be concluded that effective RE development in households requires improving the methodology for the FIT calculation. We believe that the optimal FIT rates should ensure a payback period of investment projects of 10–13 years. With a lifecycle of RE power plants of 25 years, such a payback period will allow households to receive profits from the sale of the excess electricity for a reasonably long period after reaching the break-even point. On the one hand, it will allow to maintaine the investment attractiveness of the sector, and on the other hand, it will reduce the financial pressure on the final electricity consumers.

5.2. Methodical Approaches to Determine the Optimal FIT Rates and Policy Recommendations for Effective RE Development in Ukraine's Households

To improve the methodology of the FIT calculation, we suggest using modernized methodical approaches to determine the optimal FIT rates outlined in [72] and adjusted to the specifics of the residential sector. Thus, it is advisable to calculate the optimal rate of the FIT for the *i*-th RE technology in the *t*-th year (FIT_{opt_it}) for the household sector by the following formula:

$$FIT_{opt\_it} = LCOE_{it} \times k_{opt\_it}, \tag{5}$$

where $LCOE_{it}$ is the fixed electricity production cost throughout the power plant lifecycle, which uses the *i*-th RE technology commissioned in the *t*-th year, UAH/kWh (calculated according to Formula (1)), and k_{opt_it} is the coefficient of optimality of the FIT for the household sector for the *i*-th RE technology in the *t*-th year, which is determined according to the following formula:

$$k_{opt\_it} = \prod_{j=1}^{n} (1 + k_{itj}), \tag{6}$$

where k_{it} is a coefficient that considers the impact of the *j*-th factor, which determines the economic, social, and environmental effects of the *i*-th RE technology development in the *t*-th year for the households, and *n* is the number of factors considered.

It is reasonable to include the profitability of the RE power plant, its capacity, and the priority of the *i*-th RE technology development for the state as part of the main influencing factors. The range of recommended values for k_{itj}, from the main influencing factors, is presented in Table 7. The higher the k_{itj}, the greater the influence of the particular factor on the FIT rate.

Table 7. Ranges of recommended k_{itj} values for RE power plants depending on the main influencing factors (developed by the authors).

Influence Factor *j*	Recommended Range of k_{itj} Values by Factor *j*
Profitability of the RE power plant	0.05–0.25
The capacity of the RE power plant (higher values for smaller capacities)	0–0.5
Priority of the *i*-th RE technology deployment for the state	0–0.3

In addition to the abovementioned main factors, other factors may be considered, due to which the state can adjust the final FIT rates for various RE technologies, encouraging or restraining the development of some of them. Such factors may include economic and financial (tax burden and the risk of doing business for the *i*-th RE technology), environmental (environmental effect of the *i*-th RE technology), social (creation of new jobs), etc.

The advantage of the proposed approach is flexibility in adjusting the FITs for different RE technologies depending on the needs and goals of the state policy. The FITs, calculated according to the proposed methodological approach, must be reviewed periodically, at least once every three years, due to the trends in reducing the cost of RE technologies. Additionally, as in the case of the current methodology, the FITs must be converted into EUR to insure the RE plants' owners against possible inflation.

Along with determining the reasonable FIT rates in the household sector, the issue of the validity period of such a state support scheme is significant. It should be sufficient to ensure the return on the investment, but minimal to avoid overspending the state budget or/and that of the final electricity consumers. We believe such a validity period should be established according to the formula "average payback period of projects + 1–2 years" according to the particular RE technology and the group of installed capacity. Adherence to the average payback periods will allow us to return the main part of the funds to the RE power plants' owners, and an additional 1–2 years will protect the investors from market risks and guarantee confidence in the profitability of the started projects. It is worth noting that the updated methodology for the FIT calculation should only be used for new RE energy power projects in order to maintain the confidence of the existing RE power plants owners.

It is worth noting that, for a more effective implementation of the FIT policy for RE development in the residential sector, improving other policy measures, particularly financial and credit mechanisms, is advisable. As evidenced by the calculation results given in Section 5.1, the credit resources cost, which is reflected in the discount rate, significantly impacts the projects' payback periods. Despite the availability of credit programs for financing RE projects in Ukraine's households, the interest rates on loans are not affordable, especially in the absence of the ability of the households' owners to pay a significant part of the down payment. Given this, the government should pay attention to improving the existing and developing new affordable loan programs, in particular, the formation of syndicated lending lines that will allow the accumulation of finance of banks of Ukraine and the European Bank for Reconstruction and Development, whose credit lines are already are open in Ukraine for RE projects in the industrial sector, but currently do not extend to the residential sector [73].

Another approach that can ensure the accumulation of the necessary funds to implement RE projects is the formation of energy cooperatives. It should be noted that Ukraine has already created legislation for energy cooperative formation, and, since 2019, the FIT has been introduced for the RE power plants implemented under such energy associations [74]. However, implementing RE projects within the framework of such cooperation has not gained significant popularity. The main barrier is the population's lack of awareness of this issue. Therefore, the government should focus on the popularization of the advantages of creating energy cooperatives and informing about the economic, ecological, and social gains from implementing RE projects.

The cost of technologies directly affects the payback period of RE investment projects and, therefore, the FITs rate. One of the approaches to make them cheaper is state support for Ukrainian enterprises, which can produce equipment for the RE industry. It should be noted that two powerful factories in Ukraine have provided a full cycle of solar PV systems production—from growing the silicon to assembling the solar panels. However, the introduction in 2019 of benefits for the import of foreign-made solar panels negatively impacted their competitiveness. As a result of the introduction of such benefits, the prices of the imported panels became 20% lower than the domestic ones. As a result, enterprises were forced to reorient their activities mainly to the assembly of panels in Ukraine from imported equipment [75]. A stimulus for the development of the industry could be the spread of the legislative norm regarding the allowance to the FIT for the use of Ukrainian-made equipment for RE project implementation, which currently applies to industrial RE power plants but does not apply to RE facilities in the residential sector [12].

It should be noted that promoting RE development based on the FIT calculated according to the proposed methodology should be used in the transition period—from the current FIT model to a new support scheme for households to ensure further RE development without state support. Such mechanisms can be net metering or net billing, which are focused on covering exclusively a household's own electricity consumption and do not require additional financial expenses from the state or from other consumers.

Net metering is a mechanism that allows the owners of RE power plants to store the excess electricity in the country's energy grid and use it when the need arises. Net billing is a variation of net metering that works on a similar principle. The difference is that the excess electricity supplied to the grid, instead of kWh, is counted in monetary equivalents, according to the electricity price at the time of the supply. Thus, under net billing, the consumers receive a cash deposit for the excess electricity supplied into the grid, which can then be used to pay for the electricity consumed by them in subsequent periods, provided that the need for electricity exceeds the amount of its production.

A significant barrier to the effective implementation of the mentioned mechanisms can be the continuation of subsidizing electricity prices for Ukrainian households. This is because low electricity tariffs do not encourage household owners to invest in RE projects and negatively affect the profitability of their implementation under net metering/net billing support schemes. It is worth noting that, although, since June 2022, the electricity tariff for households in Ukraine has been increased to UAH 2.64/kWh (EUR 0.065/kWh), it remains extremely low compared to the tariffs in other European countries (for example, as of the beginning of 2023, in Greece it was EUR 0.18/kWh, in Italy—EUR 0.17/kWh, and in Switzerland—EUR 0.15/kWh [76]). Adopting decisions regarding the metering/net billing introduction should be preceded by creating favorable organizational and economic conditions under which these mechanisms will allow households to transit to prosumerism without state incentives.

6. Conclusions

RE development is one of the key priorities of government energy policy, therefore, the issue of its promotion, including that in the household sector, is given considerable attention at the national level. To foster electricity production from RES in the household sector, the fixed FIT was introduced, the high rates of which promoted the rapid deployment of RE facilities. However, in addition to the significant contribution of households to increasing the green energy share, several shortcomings of this approach to the promotion can be stated today. The foremost of these was the pursuit of additional profits from selling the excess electricity at the FIT and, as a result, a rapid increase in the financial pressure on the final consumers, by which payments under this tariff are compensated.

Thus, the current model of the FIT for households mainly encourages them to maximize the supply of electricity from the RES to the grid for profit, and the primary goal of the state policy—ensuring household energy autonomy—has become of secondary importance. Understanding this led to the decision of the state authorities to revise the FIT rates downwards. However, their economic validity today requires additional study, which formed the basis of this study.

The obtained results have proven that the current FIT rates must be adjusted, as they cannot ensure effective RE development in the residential sector. While the FIT does not ensure payback for 5- or 10-kW solar power plants and wind power plants with a capacity of up to 10 kW, it remains pretty attractive for 20-, 30-, 40-, and 50-kW solar power plants and 30-kW wind power plants. Given that 5–10 kW power plants best satisfy the energy needs of households, the current FITs continue to stimulate the achievement of their owners' profits from selling the excess electricity to the grid.

To resolve this issue, this paper proposes a methodical approach to improve the FIT calculation based on assessing the LCOE and influencing factors that determine the economic, social, and environmental effects of the specific RE technology. The main influencing factors include the profitability of the RE power plant, its capacity, and the priority of the particular RE technology development for the state. If necessary, the main factors can be supplemented with additional ones (level of tax burden, environmental effect, etc.) depending on the state policy goals.

Other essential directions toward the improvement of state incentives for RE development in households in the short term should be measures aimed at improving the access to attracting and reducing investment expenditures for RE project implementation, particularly for developing affordable loan programs, stimulating the creation of energy cooperatives, and the domestic production of equipment for the RE sector. The economic justification of the FIT rates for hybrid wind–solar power plants also needs special attention in order to maximize the involvement of both types of RES in electricity production. In the medium-term perspective, the key changes in the energy policy should be related to the creation of favorable conditions for the transition to the net metering/net billing support scheme and the development of conceptual foundations for its effective functioning in Ukraine.

Author Contributions: Conceptualization, T.K. and I.S.; methodology, T.K. and I.S.; validation, T.K., I.S., O.P., I.B. and U.P.; formal analysis, T.K. and I.S.; investigation, T.K. and I.S.; resources, O.P., I.B. and U.P.; data curation, T.K., I.S. and O.P.; original draft preparation, T.K., I.S. and O.P.; review and editing, T.K. and I.S.; visualization, T.K., I.S., O.P., I.B. and U.P.; funding acquisition, T.K., I.S. and U.P. All authors have read and agreed to the published version of the manuscript.

Funding: This publication was prepared under the research project "Formation of economic mechanisms to increase energy efficiency and provide sustainable development of renewable energy in Ukraine's households" (No. 0122U001233), funded by the National Research Foundation of Ukraine.

Conflicts of Interest: The authors declare no conflict of interest. The funders had no role in the study's design, in the collection, analyses, data interpretation, the writing of the manuscript, or the decision to publish the results.

References

1. Gencer, B.; Ackere, A. Achieving long-term renewable energy goals: Do intermediate targets matter? *Util. Policy* **2021**, *71*, 101243. [CrossRef]
2. Mišík, M. When the Accession Legacy Fades Away: Central and Eastern European Countries and the EU Renewables Targets. *Int. Spect.* **2021**, *56*, 56–70. [CrossRef]
3. Strunz, S.; Lehmann, P.; Gawel, E. Analysing the ambitions of renewable energy policy in the EU and its Member States. *Energy Policy* **2021**, *156*, 12447. [CrossRef]
4. Couture, T.; Gagnon, Y. An analysis of feed-in tariff remuneration models: Implications for renewable energy investment. *Energy Policy* **2010**, *38*, 955–965. [CrossRef]
5. Ramli, M.; Twaha, S. Analysis of renewable energy feed-in tariffs in selected regions of the globe: Lessons for Saudi Arabia. *Renew. Sustain. Energy Rev.* **2015**, *45*, 649–661. [CrossRef]

6. Galvin, R. Why German households will not cover their roofs in photovoltaic panels: And whether policy interventions, rebound effects and heat pumps might change their minds. *Renew. Energy Focus* **2022**, *42*, 236–252. [CrossRef]
7. Castaneda, M.; Zapata, S.; Cherni, J.; Aristizabal, A.; Dyner, I. The long-term effects of cautious feed-in tariff reductions on photovoltaic generation in the UK residential sector. *Renew. Energy* **2020**, *155*, 1432–1443. [CrossRef]
8. Zhang, Y.; Song, I.; Hamori, S. Impact of subsidy policies on diffusion of photovoltaic power generation. *Energy Policy* **2011**, *39*, 1958–1964. [CrossRef]
9. Lan, H.; Cheng, B.; Gou, Z.; Yu, R. An evaluation of feed-in tariffs for promoting household solar energy adoption in Southeast Queensland, Australia. *Sustain. Cities Soc.* **2020**, *53*, 101942. [CrossRef]
10. Li, H.; Zhang, Y.; Li, Y.; Huang, J.; Costin, G.; Zhang, P. Exploring payback-year based feed-in tariff mechanisms in Australia. *Energy Policy* **2021**, *50*, 112133. [CrossRef]
11. Dio, V.; Favuzza, S.; La Cascia, D.; Massaro, F.; Zizzo, G. Critical assessment of support for the evolution of photovoltaics and feed-in tariff(s) in Italy. *Sustain. Energy Technol. Assess.* **2015**, *9*, 95–104. [CrossRef]
12. Verkhovna Rada of Ukraine. Law of Ukraine "On the Electricity Market". 2017. Available online: https://zakon.rada.gov.ua/laws/show/2019-19#Text (accessed on 8 June 2023).
13. Verkhovna Rada of Ukraine. Law of Ukraine "On Alternative Energy Sources". 2003. Available online: https://zakon.rada.gov.ua/laws/show/555-15#Text (accessed on 8 June 2023).
14. NCSREPU. All-Ukrainian Information and Statistical Information of European Institutions in the Electricity Field. 2021. Available online: http://surl.li/jwghe (accessed on 18 June 2023).
15. Epravda. How to Provide 25% of Energy from RES without Support from the Budget. 2022. Available online: http://surl.li/jwghq (accessed on 18 June 2023).
16. Sun-Energy. 20 kW Medium Solar Power Plant 2023. Available online: http://surl.li/bfnsn (accessed on 15 June 2023).
17. Eco-Tech. Solar Power Plant of 30 kW for the Home under the Feed-In Tariff. 2023. Available online: http://surl.li/-jwgho (accessed on 15 June 2023).
18. Sala, D.; Bashynska, I.; Pavlova, O.; Pavlov, K.; Chorna, N.; Romanyuk, R. Investment and Innovation Activity of Renewable Energy Sources in the Electric Power Industry in the South-Eastern Region of Ukraine. *Energies* **2023**, *16*, 2363. [CrossRef]
19. Prokopenko, O.; Kurbatova, T.; Zerkal, A.; Khalilova, M.; Prause, G.; Binda, J.; Berdiyorov, T.; Klapkiv, Y.; Sanetra-Półgrabi, S.; Komarnitskyi, I. Impact of investments and R&D costs in renewable energy technologies on companies' profitability indicators: Assessment and forecast. *Energies* **2023**, *16*, 1021. [CrossRef]
20. Kwilinski, A.; Lyulyov, O.; Pimonenko, T. Greenfield Investment as a Catalyst of Green Economic Growth. *Energies* **2023**, *16*, 2372. [CrossRef]
21. Prokopenko, O.; Toktosunova, C.; Sharsheeva, N.; Zablotska, R.; Mazurenko, V.; Halaz, L. Prospects for the Reorientation of Investment Flows for Sustainable Development under the Influence of the COVID-19 Pandemic. *Probl. Ekorozwoju* **2021**, *16*, 7–17. [CrossRef]
22. Sotnyk, I.; Kurbatova, T.; Romaniyk, Y.; Prokopenko, O.; Gonchar, V.; Sayenko, Y.; Prause, G.; Sapiński, A. Determining the optimal directions of investment in regional renewable energy development. *Energies* **2022**, *15*, 3646. [CrossRef]
23. Chygryn, O.; Shevchenko, K. Energy industry development: Key trends and the core determinants. *Socioecon. Chall.* **2023**, *7*, 115–128. [CrossRef]
24. Kwilinski, A.; Lyulyov, O.; Pimonenko, T. Inclusive Economic Growth: Relationship between Energy and Governance Efficiency. *Energies* **2023**, *16*, 2511. [CrossRef]
25. Prokopenko, O.; Cebula, J.; Chayen, S.; Pimonenko, T. Wind energy in Israel, Poland and Ukraine: Features and opportunities. *Int. J. Ecol. Dev.* **2017**, *32*, 98–107.
26. Ksonzhyk, I.; Lavrushchenko, Y.; Oleksiuk, M.; Saienko, V.; Buryk, Z. Influence of renewable green energy on the economic development of the EU states. *Environ. Ecol. Res.* **2021**, *9*, 271–281. [CrossRef]
27. Tymoshenko, M.; Saienko, V.; Serbov, M.; Shashyna, M.; Slavkova, O. The impact of industry 4.0 on modelling energy scenarios of the developing economies. *Financ. Credit. Act.-Probl. Theory Pract.* **2023**, *1*, 336–350. [CrossRef]
28. Kurbatova, T.; Lysenko, D.; Trypolska, G.; Prokopenko, O.; Järvis, M.; Skibina, T. Solar energy for green university: Estimation of economic, environmental and image benefits. *Int. J. Glob. Environ. Issues* **2022**, *21*, 198–216. [CrossRef]
29. Raikar, S.; Adamson, S. Public policy mechanisms to support renewable energy. *Renew. Energy Financ.* **2020**, 9–19. [CrossRef]
30. Barnea, G.; Hagemann, C.; Wurster, S. Policy instruments matter: Support schemes for renewable energy capacity in worldwide comparison. *Energy Policy* **2022**, *168*, 113093. [CrossRef]
31. Jacksohn, A.; Grösche, P.; Rehdanz, K.; Schröder, C. Drivers of renewable technology adoption in the household sector. *Energy Econ.* **2019**, *81*, 216–226. [CrossRef]
32. Guta, D. Determinants of household use of energy-efficient and renewable energy technologies in rural Ethiopia. *Technol. Soc.* **2020**, *61*, 101249. [CrossRef]
33. Ang'u, C.; Muthama, N.; Mutuku, M.; M'IKiugu, M. Determinants of the sustained use of household clean fuels and technologies: Lessons from Vihiga county, Kenya. *Energy Rep.* **2023**, *9*, 1990–2001. [CrossRef]
34. Etongo, D.; Naidu, H. Determinants of household adoption of solar energy technology in Seychelles in a context of 100% access to electricity. *Discov. Sustain.* **2022**, *3*, 38. [CrossRef]

35. Štreimikienė, D.; Baležentis, A. Assessment of willingness to pay for renewables in Lithuanian households. *Clean Technol. Environ. Policy* **2015**, *17*, 515–531. [CrossRef]
36. Braito, M.; Flint, C.; Muhar, A.; Penker, M.; Vogel, S. Individual and collective socio-psychological patterns of photovoltaic investment under diverging policy regimes of Austria and Italy. *Energy Policy* **2017**, *109*, 141–153. [CrossRef]
37. Nyborg, K.; Rege, M. Does Public Policy Crowd Out Private Contributions to Public Goods. *Public Choice* **2003**, *115*, 397–418. [CrossRef]
38. Thøgersen, J. Monetary incentives and recycling: Behavioural and psychological reactions to a performance-dependent garbage fee. *J. Consum. Policy* **2003**, *26*, 197–228. [CrossRef]
39. Verkhovna Rada of Ukraine. Tax Code of Ukraine. 2010. Available online: https://zakon.rada.gov.ua/-laws/show/2755-17#Text (accessed on 22 June 2023).
40. Verkhovna Rada of Ukraine. Customs Code of Ukraine 2012. Available online: https://zakon.rada.gov.ua/-laws/show/4495-17#Text (accessed on 22 June 2023).
41. Oschadbank. Credit Program "Green energy". 2023. Available online: https://www.oschadbank.ua/credit/green-energy (accessed on 25 June 2023).
42. Ukrgasbank. Credit Program "Eco Energy". 2023. Available online: http://www.ukrgasbank.com/private/-credits/eco_energy/ (accessed on 25 June 2023).
43. SAEE. *Information Letter No. 16-01/17/31-22 about the Main Indicators of Renewable Energy Development in 2015–2021*; SAEE: Kyiv, Ukraine, 2022; 3p.
44. SAEE. *Information Letter No. 33-02/17/31 about the Main Indicators of Renewable Energy Development 2018–2022*; SAEE: Kyiv, Ukraine, 2023; 3p.
45. Ukrainian Energy. Ukrainian Small Wind Farms—Autonomous Saviors or Suitcases without a Handle? 2021. Available online: http://surl.li/adgrs (accessed on 25 June 2023).
46. Breeze, P. *Chapter 9—The Cost of Electricity*; Elsevier: Amsterdam, The Netherlands, 2021; pp. 117–136. [CrossRef]
47. Finswin. What Is WACC and How Is It Calculated? 2021. Available online: https://finswin.com/projects/ekonomika/-wacc.html (accessed on 3 July 2023).
48. NCSREPU. Resolution No. 1817 Dated 30.08.2019 On Approval of the Procedure for Establishing, Revising and Terminating the Feed-in Tariff for Electricity for Economic Entities, Electricity Consumers, Including Energy Cooperatives, and Households Whose Generating Plants Produce Electricity from Alternative Energy Sources. 2019. Available online: https://zakon.rada.gov.ua/laws/show/v1817874-19#Text (accessed on 3 July 2023).
49. Sotnyk, I.; Kurbatova, T.; Blumberga, A.; Kubatko, O.; Prokopenko, O. Solar business prosumers in Ukraine: Should we wait for them to appear? *Energy Policy* **2023**, *178*, 113585. [CrossRef]
50. Project-Management. Discounted Payback Period: Definition, Formula, Example & Calculator. 2021. Available online: http://surl.li/afigj (accessed on 3 July 2023).
51. IEA. Average Annual Capacity Factors by Technology. 2018. Available online: http://surl.li/jwgjm (accessed on 6 July 2023).
52. Sun-Energy. Solar Power Plant Electricity Production Calculator. 2023. Available online: https://sunenergy.com.ua/-sun-calculator (accessed on 6 July 2023).
53. Generacia. Calculation of Solar Generation for Power Plants. 2023. Available online: https://generacia.energy/-kalkuljator/ (accessed on 6 July 2023).
54. Sun-Energy. Sets of Solar Power Plants. 2023. Available online: https://sun-energy.com.ua/solar-power/solar-power-plants/ (accessed on 6 July 2023).
55. Solar-Tech. Solar Power Plant 50 kW for Own Consumption. 2023. Available online: http://surl.li/iknsi (accessed on 9 July 2023).
56. Energosave. Solar Power Plants. 2023. Available online: http://surl.li/iknsk (accessed on 9 July 2023).
57. Raiffeisen. Wind in Favor. 2023. Available online: http://surl.li/aybtf (accessed on 9 July 2023).
58. Avante. Wind Generator W10 20000 Bt. 2023. Available online: http://surl.li/iknue (accessed on 9 July 2023).
59. Kworum. Classic Wind Farms. 2023. Available online: https://kworum.com.ua/c/vitriaki-klasychni (accessed on 9 July 2023).
60. Jordan, D.; Kurtz, S. Photovoltaic Degradation Rates—An Analytical Review. NREL/JA-5200-51664. 2022. Available online: http://www.osti.gov/bridge (accessed on 14 July 2023).
61. Staffell, I.; Green, R. How does wind farm performance decline with age? *Renew. Energy* **2014**, *66*, 775–786. [CrossRef]
62. IEA. Projected Costs of Generating Electricity. 2010. Available online: http://surl.li/jwori (accessed on 14 July 2023).
63. Fischer, J. Comparing Wind and Solar Energy Impacts on the Environment: A LCA Approach Using openLCA Platform. 2021. Available online: http://surl.li/idyrf (accessed on 14 July 2023).
64. Minfin. Stability Rating of Banks as of the 1st Quarter of 2023. 2023. Available online: https://minfin.com.ua/ua/banks/rating/ (accessed on 17 July 2023).
65. Feensee. Deposits 2023: Calculator, Reliable Banks, Detailed Conditions. 2023. Available online: http://surl.li/uorg (accessed on 14 July 2023).
66. Minfin. The Official Average Exchange Rate of UAH to the EUR for May 2023. 2023. Available online: https://index.minfin.com.ua/exchange/nbu/curr/eur/ (accessed on 14 July 2023).
67. CMU. Resolution No. 544 Dated 30 May 2023, On Amendments to the Resolution of the Cabinet of Ministers of Ukraine Dated 5 June 2019. 2023. Available online: https://zakon.rada.gov.ua/laws/show/544-2023-%D0%BF#Text (accessed on 21 July 2023).

68. Solar Reviews. How Long Does It Take for Solar Panels to Pay for Themselves? Average Solar Panel Payback Period for Homes in the US in 2022. 2022. Available online: http://surl.li/bfjec (accessed on 21 July 2023).
69. Greenmatch. Are Solar Panels Worth It in the UK in 2023? 2023. Available online: http://surl.li/bfntu (accessed on 21 July 2023).
70. Attainablehome. Small Wind Turbines: What Is the Average Payback Period? 2022. Available online: http://surl.li/jwgji (accessed on 22 July 2023).
71. Zalewska, J.; Damaziak, K.; Malachowski, J. An Energy Efficiency Estimation Procedure for Small Wind Turbines at Chosen Locations in Poland. *Energies* **2021**, *14*, 3706. [CrossRef]
72. Sotnyk, I.; Kurbatova, T.; Kubatko, O.; Baranchenko, Y.; Li, R. The price for sustainable development of renewable energy sector: The case of Ukraine. *E3S Web Conf.* **2021**, *280*, 02006. [CrossRef]
73. Kropelnytska, S.; Mayorova, T. The financing of renewable energy development projects in Ukraine. *Polityka Energetyczna—Energy Policy J.* **2021**, *24*, 77–88. [CrossRef]
74. Verkhovna Rada of Ukraine. Law of Ukraine "On Amendments to Some Laws of Ukraine Regarding Ensuring Competitive Conditions for Producing Electricity from Alternative Energy Sources". 2019. Available online: https://zakon.rada.gov.ua/laws/show/2712-19#Text (accessed on 5 September 2023).
75. Energy Transition. Genealogy of Ukrainian Solar Panels: Where the Equipment Comes From. 2020. Available online: http://surl.li/kyokx (accessed on 5 September 2023).
76. Slovoidilo. How Much Will EU Residents Pay for Electricity in 2023: Comparison of Tariffs with Ukrainian Ones. 2023. Available online: http://surl.li/jwgjf (accessed on 22 July 2023).

energies

MDPI

Article

Energy Efficiency Forecast as an Inverse Stochastic Problem: A Cross-Entropy Econometrics Approach †

Second Bwanakare [1,2]

1 Faculty of Social and Economic Sciences, Institute of Economics and Finance, Cardinal Stephan Wyszynski University in Warsaw, 01-815 Warsaw, Poland; s.bwanakare@uksw.edu.pl

2 Institute of Statistics in Rzeszów, 35-959 Rzeszów, Poland

† This paper is an extended version of our paper published in 2015 "Greenhouse Emission Forecast as an Inverse Stochastic Stochastic Problem: A Cross-Entropy Econometrics Approach", Proceedings of the 7th Symposium FENS, Lublin, 14–17 May 2014, Acta physica Polonica A, Vol. 127, No. 3-A, 2015.

Abstract: This paper forecasts the energy efficiency coefficients at the Polish province level (NUT-2), based on imperfect and contradictory knowledge. On the one hand, we have information on the aggregated national energy efficiency coefficients in the industrial, transport, household, and service sectors. On the other hand, we also have information on the energy intensity at the Polish province level. Since the two samples are of different natures and known with uncertainty, we are obviously dealing with an inverse stochastic problem whose solution requires particular statistical devices. The applied technique of non-extensive cross-entropy econometrics generalizes the Shannon-Kullback-Leibler approach based on the Gaussian assumptions. Its justification is explained throughout this paper from both methodological and empirical points of view. The model forecasts lead to the high-value energy efficiency estimates from quasi-unstructured sets of information. This constitutes the main contribution of this research. These outputs should provide energy policy units with valuable new devices for the optimization of the energy management processes on a disaggregated local level where, by contrast, different agents and households act decisively. On a global level, the proposed technique can be applied in different EU countries and elsewhere, in the context of experimental official statistics.

Keywords: overall energy efficiency score; energetic intensity; non-extensive cross-entropy econometrics; stochastic inverse problems; regional innovation

1. Introduction

The oil shocks and market supply instabilities and uncertainties that have been occurring from 1973 until today have led to an increase in energy costs with negative consequences on the growth of different economies in different parts of the world. In addition, the 1997 Kyoto Protocol operationalized the United Nations Framework Convention on Climate Change by committing industrialized countries and economies in transition to limiting and reducing greenhouse gas (GHG) emissions in accordance with agreed-upon individual goals. It is therefore the occurrence of these two joint goals of the end of the last century that forced public decision-makers to initiate energy- and carbon-saving policies. Consequently, Europe defined the goals for 2030 [1], which are a 40% reduction in greenhouse gas emissions with respect to the 1990 values and an increment of energy saving of at least 32.5%.

Following IRENA [2], energy efficiency and renewable energy sources may provide more than 80% of the required emissions savings [3]. These saving-oriented objectives require a management system enabling the evaluation of progress with respect to these defined goals on different time and space scales. In fact, many authors are aware of the complexity of this problem. This is highlighted in recent publications either from a descriptive or prospective point of view (e.g., [4–7]) or from a methodological point of view (e.g., [8,9]). One of the issues is the lack of statistical data on a disaggregated scale, which contrasts with the pressing needs for information on a parallel level for such a strategic sector of the economy.

Therefore, this paper tries to address one of these issues. We will present and apply a recent statistical approach to recover statistical information in conditions when traditional mathematical or statistical techniques are unable to pass muster. This approach is related to the principle of maximum entropy, which is known to deal with ill-posed inverse problems such as the one to be solved in this article. The same approach has been applied in the recent past. The most noteworthy work is one that assesses the interregional distribution of greenhouse emissions in Poland by industry [10]. In this document, questions will arise when trying to recover the estimates from a dashboard of energy efficiency scores at a disaggregated national level where the statistical data does not exist. The case of Polish provinces will be methodologically illustrated for many similar countries, particularly those within the European Union. Many developed countries publish annually (e.g., [11]) sectoral energy efficiency scores aggregated at the national level. However, such aggregated information makes it very difficult to assess and plan the energy policies for which scope of operation happens to be at the sub-regional levels. To overcome this problem, statistical institutes instead calculate energy intensity scores, which are easier to estimate. The first question is whether the production of statistics on energy efficiency coefficients at a disaggregated level (in our case at NUT 2 level) deserves to be calculated alongside the energy intensity coefficients. The counter argument is that the energy intensity coefficients available should be sufficient to provide information on its efficiency. Nevertheless, the correlation between energy intensity and energy efficiency is generally far from perfect. For instance, a small service-based economy in a mild climate region will be characterized by a lower intensity than a large industry-based economy in a colder climate region, even though the latter uses energy more intensively and efficiently. In addition, other elements will also play a role in defining the efficiency levels and trends. Among these are regional economic structure (share of large energy-consuming industries), geographical characteristics (e.g., longer distances leading to higher demand for transport), and climatic and weather conditions (changing demand for heating or cooling) (e.g., [11]).

Based on what has just been presented above, let us now concretize the problem targeted by this paper as follows: we need to estimate the coefficients of energy efficiency for the sixteen province levels of Poland. The energy efficiency in question concerns the following four sectors: industry, transport, households, and services. The Odyssee-Mure project (e.g., [11]) publishes national statistics on the sectoral averages of these coefficients. Next, some institutes such as the Polish Institute of Statistics [12] publish the energy intensity coefficients at a disaggregated level, e.g., at the province level. The question posed in this article is, therefore, how to reconcile these two sources of information to forecast the sector ratios of energy efficiency at a province level. Mathematically, the problem as presented in a rectangular table (see Table 1) is ill-posed. On the one hand, we only have information on the averages of the energy intensity coefficients for each of the 16 provinces, while at the level of the four sectors, we have the averages of energy efficiency without, in both cases, any details at the individual interprovincial sectoral level. On the other hand, it follows that the total of the rows cannot correspond to the total of the columns. Finally, if we additionally take into consideration the fact that the systems generating these two sources of information are not known with precision, the problem to be solved turns out to be an ill-posed stochastic inverse problem for which the solution goes beyond traditional mathematical techniques.

Table 1. Sector and province energy efficiency ratios (2021).

Polish Province	Industry	Transport	Households	Services	Average Energy Intensity Coefficients (GWh\mln zl Value Added)
	$Y_{.1}P_{11}$	$Y_{.2}P_{12}$	$Y_{.l}P_{1l}$	$Y_{.L}P_{1L}$	0.08
Mazowieckie	$Y_{.1}P_{21}$	..	$Y_{.l}P_{2l}$	$Y_{.l}P_{2L}$	0.05
Malopolskie	..	..	..	..	0.06
Silesian	..	..	..	..	0.09
Lublin	..	..	..	..	0.07
Podkarpackie	..	..	..	..	0.06
Podlaskie	..	..	..	..	0.06
Swietokrzyskie	..	..	..	..	0.09
Lubuskie	$Y_{.1}P_{k1}$	..	$Y_{.l}P_{kl}$	$Y_{.L}P_{kL}$	0.07
Wielkopolska	..	..	..	..	0.05
Zachodnia pomors	..	..	..	..	0.07
Dolnoslaskie	..	..	..	..	0.07
Opole	..	..	..	..	0.11
Kujawska-pomorska	..	..	..	..	0.07
Pomorska	..	..	..	..	0.06
Warminsko-mazurskie	$Y_{.1}P_{K1}$	..	$Y_{.l}P_{Kl}$	$Y_{.L}P_{KL}$	0.06
Average energy efficiency ratio	0.372	0.612	0.621	0.689	

Sources: Own based on Odyssey-Mure and Polish Institute of Statistics (GUS).

The remaining part of this paper is organized as follows. Section 2 introduces the concepts and definitions related to optimal energy use and saving through recent literature on the subject. Section 3 present mathematical and statistical insights of the model in the context of the problem to be solved. In this context, the concept of inverse problem is defined, and this will be followed by a detailed presentation of non-extensive cross-entropy econometrics as the main technique applied by the model. Section 4 presents the model outputs and comments. At the end of this section, a sub-section is devoted to the main limitations of the proposed approach and future areas for further research. Section 5 draws conclusions and highlights potential outcomes related to the application of the presented method.

2. Energy Efficiency and Its Measurement

This section provides basic concepts to enable good understanding and interpretation of the computed outputs. There exists a vast literature on the energy efficiency definition and measurement (e.g., [13,14]). The reason may reside in the complexity of the energy efficiency technical evaluation [15]. The issues are the choice of the right aggregation level, the appropriate variables to construct a reference energy consumption trend, the energy units to be applied, and interaction between various effects. Uncertainty margins for the results lack in most presentations as well. Among a vast literature, we underscore some of these works that disserve a higher level of attention in the context of this paper. In particular, the authors [16] present two approaches to measuring energy efficiency. The bottom-up approach, to which the "Odex" index is linked, has been developed under the EU Odyssee-Mure programme. This approach will be developed in this article. The second approach is a top-down approach, which brings together the "Decomposition" methodologies as used by e.g., the Netherlands, Canada, and New Zealand. Likewise, the authors [17] propose the calculation of total factor energy efficiency (TFEE) using the concept of global production instead of domestic product, which excludes intermediate consumption. Another paper worthy to cite comes from the authors [18]. In that work, a literature review was carried out, and the authors found that the currently in use definitions of total-factor energy efficiency and total-factor carbon emissions efficiency are confusing and misleading.

Regarding institutional research, the Environmental and Energy Study Institute [19] uses a definition of energy efficiency, emphasizing using less energy to perform the same task, i.e., eliminating energy waste. The European Parliamentary Research Service (EPRS) definition [20] highlights the fact that energy efficiency should refer, in general terms, to the amount of output that can be produced with a given energy input. Most commonly, energy efficiency is measured as the amount of energy output for a given energy input. However, other kinds of output can also be used. The EU Energy Efficiency Directive uses a very broad definition: "energy efficiency" means the ratio of output of performance, service, goods, or energy to input of energy. Following the International Energy Agency [21], the energy efficiency ratio is a ratio between energy consumption (measured in energy units) and activity data (measured in physical units).

Finally, the UNDP, as an organ of the United Nations, has published a summary work that takes up different approaches to calculating energy efficiency. In fact, this institution, in the third chapter [22], begins by identifying the methodological challenges associated with defining and measuring energy efficiency. It then proposes a framework for understanding energy efficiency trends, integrating the current UNDP approach to energy efficiency developed by various international agencies and national institutions, and establishing a methodology to identify a starting point in relation to which future improvements in energy efficiency can be measured globally and at national levels.

It is worthwhile to pay attention to the related three concepts to be used in the next part of the paper. For an economy-wide measure, GDP is often compared to energy use, to give *the energy intensity* (measured, for example, in kilowatt-hours per euro). Next, *energy savings* are the reduction of energy use, without reference to output produced. Finally, as far as energy efficiency assessment is concerned, this can be done at different levels and according to different techniques. These levels may start from economy-wide and sectoral energy intensity to individual units of activity.

We now present some details regarding the ODEX energy efficiency index from which the data to be used in our model are extracted a priori. This index is published by the Odyssee-Mure project [23] to measure the energy efficiency progress of the main sectors (industry, transport, households, services) and for the whole economy (all final consumers). The ODEX composite indicator is calculated as a weighted average of sectoral indices.

For each sector, the index is calculated as a weighted average of sub-sectoral indices of energy efficiency progress. The sub-sectors stand for industrial branches, service sector branches, end-uses for households, or transport modes.

This project calculates the scores by including the following energy efficiency components:

- the energy efficiency level,
- the energy efficiency trends,
- the energy efficiency policies, and
- the overall energy efficiency.

These three first criteria are scored between 0 and 1 on the basis of a variety of indicators (extracted from the Odyssee Database) and of energy policies (extracted from the Mure Database). The overall energy efficiency score is obtained as an average of the three scores obtained for "energy efficiency level", "energy efficiency progress", and "energy efficiency policies" (i.e., one-third weighting). This work will use data representing the overall energy efficiency. Following [23], the energy efficiency scoring technique is based on the OECD Composite Indicator methodology. This method allows the countries or regions to be compared in a relevant range where minimum and maximum values indicators define the best and worst scores and countries or regions are ranked between these two extrema.

The indicators are calculated and normalized so that they range between 0 and 1, following this formula:

$$\text{Normalized score} = [\text{Indicator - Min indicator}\backslash((\text{Max indicator} - \text{Min indicator}) * \text{direction})] + 0.5 * (1 - \text{direction})$$

where

Indicator: The indicator value of the country/region.

Min indicator: The minimum indicator value across all countries/regions.

Max indicator: The maximum indicator value across all countries/regions.

Direction: The favored direction in the level of indicator; −1 if the decline is favored, 1 if the incline is favored.

In spite of a "one-third weighting", the most influential score is the one of the energy efficiency level to which the two remaining scores should be related. Its scoring, according to the Odyssee-Mure practice, is done as follows:

- Scoring is done separately for the four considered sectors (households, transport, industry, and services) and for all sectors together.
- The score by sector is based on scores computed for statistically selected indicators of end uses in buildings or modes in transport. For the industry sector, an aggregate score is obtained from various industrial branch scores that account for the energy efficiency characteristics of each of them.
- The score by sector is calculated as a weighted score of each indicator. The weights correspond to the average shares over the last 3 years of each end use or transport mode in the sector consumption.

Finally, for comparative reasons, sector score values are normalized into the interval 0–1 according to the next formula:

$$\text{Normalized score} = (\text{Indicator} - \text{Min indicator})/(\text{Max indicator} - \text{Min indicator})$$

where

Indicator: The indicator value of the sector.

Min indicator: The minimum indicator value across all sectors.

Max indicator: The maximum indicator value across all sectors.

To close this section, it is worthwhile to notice that the use of the above presented energy efficiency measurements will not apply in all cases. To illustrate this issue, one can mention difficulties of the energy efficiency estimation in the presence of the recent energy system known as the integrated energy system (IES). In the context of the energy crisis and environmental degradation, an integrated energy system (IES) based on the complementarity of multiple energy sources and the cascading use of energy is considered an effective way to mitigate these problems. Due to different forms of energy and different characteristics of IESs, the interrelationships between different forms of energy are complicated, which increases the difficulty of assessing the energy efficiency of IESs. A limited number of techniques exist. We can send the interested authors to the authors [15]. These authors have proposed a technique mixing energy use efficiency (EUE) and exergy efficiency (EXE) based on the first/second laws of thermodynamics [4].

3. Mathematical Problem Setting

(a) Inverse problem and the maximum entropy principal

In many real-world situations, theorists and empiricists observe at a given time two or more quantifiable multivariate stochastic systems and want to infer on an unknown cross-correlation between their random elements. To illustrate that, we implement a cross-entropy formalism to forecast an interprovincial sectoral energy efficiency score matrix based on imperfect and contradictory information from province or sector aggregates.

The basic model for (For example, by limiting itself to ordinary cases of signal or imaging, the basic equation (Equation (1)) can be extended to, for example, the impulse response of a measurement system.) dealing with poorly posed inverse problems is to solve the integral equation of the first type. We formulate this—in the context of a model that will be developed later—as follows:

$$G(\zeta) = \int_D f(\beta)h(\beta,\zeta)d\beta + b(\zeta) \tag{1}$$

- G is the amounts observed in rows or columns;
- f is the unknown regional cross-sectoral energy efficiency coefficient matrix;
- D defines the model Hilbert support space;
- h is the transformation kernel associating measures G and f; and
- b explains the random components.

The literature on various methodologies devoted to the recovery of ill-posed inverse problems [24] is expansive when dealing with empirical problems (e.g., [25–27]). In addition to the well-known Tikhonov regularization theory [28], the Gibbs–Shannon–Jaynes principle of maximal (minimum) entropy [29,30] and recent extensions [31,32], until recently, have remained the most commonly used techniques to solve this class of problems. The general rule that applies to both approaches is to link the linear or nonlinear problem of the least squares with the regularization rule (a priori or additional information) to get to a well-posed problem. Moreover, the Gibbs–Shannon–Jaynes principle of maximum (minimum) entropy formalism tries to search for global regularity—related to the second law of thermodynamics—while producing the smoothest reconstructions consistent with the data available in the Bayesian spirit. In this research, as is often the case in many empirical applications, the discrete form of equation 1 was implemented.

Focusing on social science research, a number of other techniques have been tried for this class of inverse problems. Examples include the pseudo-inverse Moore–Penrose problem approach or the bi-proportional RAS approach and its extension [33]. Although the latter technique requires an initial transaction matrix, it offers a worse solution when the model studied is stochastic. The authors [32] showed poorer performance of the Markov chain model compared to the generalized Gibbs–Shannon entropy for this class of inverse problems. Subsequently, the Bayesian approach showed its relative superiority, particularly that associated with the principle of maximal entropy. A neural class model can also be proposed. However, it is not based on a compact theory, its application takes time, and the results are not always guaranteed. Nevertheless, recent promising research studies have proposed new algorithms to solve multicriterial, dynamic inverse problems (e.g., [26,34,35]). The entropy model has been successfully applied to update and balance social accounting matrices [33]. However, on theoretical grounds, this assumes that entropy is a positive linear function of the number of possible states and therefore sets aside the possibility of interdependencies between states and their influence. In a recent paper, the authors [36] demonstrated the convergence of two standard regularization techniques to two special power law (PL)-related q Tsallis values. For q = 2, a Tikhonov regularization is obtained, and for q = 1, the classical formulation of the Boltzmann–Gibbs–Shannon entropy is obtained. The central point is that in addition to the well-known law of scale, PL exhibits a series of interesting characteristics related to its aggregation properties, in that it is preserved under minimum and maximum, addition, multiplication, and polynomial transformation ([37,38]).

Since we are dealing with a poorly conditioned inverse problem, we must meet all three conditions of regularity (existence, uniqueness, and stability of the solution) at the same time.

If, in general, the conditions of existence and uniqueness remain available due to regular a priori constraints, the stability of the optimal solution, by random or systematic errors, is much more difficult to reach. In short, the problem is, among an infinite number of distributions that meet all the restrictions imposed, to find the one that best replicates the data generation system (DGS). As for the formalism of maximum entropy, thanks to Jaynes' contribution [5], the reasonable candidate should be the one that reduces uncertainty about the system the most. Similarly, according to Kullback-Leibler and information divergence metrics [29,39], the best candidate should be those posteriors who meet all the binding conditions and deviate the least from the priors.

(b) Non-extensive cross-entropy energy model and confidence interval area

Kullback-Leibler information divergence metrics are best known when they apply to the Gaussian attractor phenomena for parameter estimation, while dealing with an insufficient sample. In the case of PL phenomena—of which the Gaussian law is a particular case, the divergence metric applied is the q-Generalized Kullback-Leibler relative entropy ([40,41]). Formally, this metric constitutes a junction of both the power law and the Kullback-Leibler relative entropy formalisms. The first is known to analytically solve non-linear—and/or fractal—systems [42], and the second is known to find discriminative information from two or more hypotheses in the context of insufficient information. Accordingly, let us present below the main conditions following which emerges competing superiority of the q-generalized Kullback-Leibler relative entropy with respect to traditional techniques. Following (e.g., [32,43,44]), insufficient information means that we are trying to solve an ill-posed problem that is likely to arise in the following cases:

- sample statistics are linear or collinear for various reasons;
- non-stationary or non-cointegrating variables result from poor model specification;
- data from the sampling plan are insufficient and/or incomplete due to technical or financial constraints—official statistics on small areas could illustrate this situation;
- The Gaussian properties of random disturbances are questioned, among others, due to systematic errors resulting from the research process;
- the model is not linear, and the last option is an approximate linearization; and
- observations of aggregated data (in time or space) may hide a very complex system represented, for example, by a PL distribution, and there may be multifractal properties of the system.

Thus, it results from the above that the q-Generalized Kullback-Leibler relative entropy will be conceptually free from the traditional hypotheses, mainly those related to the least squares method. Among these, we can list the sphericity or collinearity attributes of the model.

Regarding the model specification, we estimate, using the proposed cross-entropy technique, the random parameters of the model from the cross-sectional data [45] of two distinct periods (years 2020 and 2021). In the context of that technique, authors [32] proposed the solutions to panel models based on various statistical hypotheses. The outputs of those models revealed higher precision in comparison with traditional techniques. Nevertheless, since we deal with a stochastic inverse problem, treating the problem as related to a panel structure could lead to important theoretical and computational problems related to the hypothesis of the non-extensivity of entropy.

The model related to that metric has been extensively presented in different publications (e.g., [44]) in the context of macroeconomic analysis. Since this work deals with energy management, it is worthwhile to reformulate the model in the energy management context to enable interpretation of the outputs.

We implement (the generalized Bregman Kullback-Leibler may be an alternative version of this model) the usual (the generalized Bregman Kullback-Leibler may be the alternative version of this model) discrete form of the q-Generalized Kullback-Leibler relative entropy (Equation (5)). With the cross-entropy new constraining data, the model updates the initial information (priors in Table 1) and provides new outputs (posteriors).

It is necessary to redefine the parameterization of the generalized linear model (Equation (2)), which plays the role of constraints. The inside table elements to be forecasted can be meaningfully presented by columns as the discrete Bayesian joint probabilities explaining each region's average cross-sector weight or probabilities corresponding to individual energy efficiency ratio. The ratio total per column will sum up to unit. We recall that energy efficiency coefficients to be forecasted are in the form of normalized indicators, thus belonging within the space zero and one. In this case, the parameter processing space coincides with the probability space. Under these conditions, the accuracy of the estimated parameters is greater, since a priori there is no loss of information from these data [46]. In any case, let us briefly present the general procedure of reparameterization in the case of a general linear inverse model:

$$Y = X' \cdot \beta + \varepsilon, \tag{2}$$

where the values of unknown parameters β are not necessarily bounded between 0 and 1, indicating the need for reparameterization. This ε term is an unobservable random term for perturbations, plausibly with finite variance exhibiting observation errors from empirical measurement or random shocks that may be driven by PL. The variable Y consists of data observed—with errors—of an unknown data-generating system of energy efficiency coefficients by sector, and X may represent known average regional energy intensity coefficients—with uncertainty—through the relational parameter matrix β and the unobservable disturbance ε to be estimated through the observable error components e. Unlike classical econometric models, no binding assumptions are required—for example, regarding the distribution of random errors. In particular, as we deal with an ill-behaved inverse problem, the number of parameters to be estimated should be greater than the observed data points, and the quality of the informative data collected should be low. The process of the true system recovery requires the entropy objective function to include all of the interacting constraining consistency moments. Thus, referring to the properties of the relative entropy principle, each new piece of constraining information will reduce the entropy level of the system in accordance with the degree of data consistency with the system. For this multi-dimensional space inverse problem, among an unlimited number of model solution candidates, the best solution will result from identifying the one that—in terms of probability—best simulates the data-generating system. By taking each of β_{kl} $(k = 1 \ldots K, l = 1 \ldots L)$ as a discrete and random variable with a compact support (e.g., [32]) and $2 < M < \infty$ possible outcomes, it can be estimated using B_{kl}, i.e.,

$$B_{kl} = \sum_{m=1}^{M} p_{klm} v_{klm}, \tag{3}$$

where p_{klm} is the probability of the outcome v_{klm}, and the probabilities must be non-negative and added to one. Similarly, by treating each element $e_{.l}$ (which affects the total uncertainty of the sector efficiency) as an e finite, discrete random variable with a compact support and $2 < J < \infty$ possible results centred at zero, we can express $e_{.l}$ as

$$e_{.l} = \sum_{j=1..J} r_{.lj} z_{.lj}, \tag{4}$$

As mentioned, it can be assumed that each previous entire row of errors has been evaluated, $\omega_{k.}$ and a similar support space should be constructed as follows:

$$\omega_{k.} = \sum_{s=1}^{S} \mu_{k \cdot s} v_{k \cdot s}$$

where $r_{.lj}$ and $\mu_{k \cdot s}$ are the outcome probabilities in the support spaces, respectively. z_{klj} $v_{!s}$ $j = 1 \ldots J$ $s = 1 \ldots S$. Therefore, $k = 1 \ldots K$ and $l = 1 \ldots L$ represent, respectively, the indexes of the number of rows and columns whose coefficient sums were estimated with errors. Moreover, e the term error ω is empirically set around the empirical standard error of the stated variables and a priori represents the Bayesian hypothesis. The choice of error limits, of course, depends on their own properties. In this study, their sets were determined by Chebyshev's inequality [47], with the boundaries of the support space ranging from -3 to $+3$. Notice that in spite of this Gaussian property of the priors, posterior probabilities in the support space may represent a class of a non-Gaussian distribution, in particular a PL.

The element v_{klm} constitutes a priori information provided by the researcher, while p_{klm} is an unknown probability generating the true parameter β_{kl}, the value of which must be determined by solving a non-extensive cross-entropy econometrics problem. In matrix notation, let us rewrite $\beta = V \cdot P$ with $p_{klm} \geq 0$ and $\sum_{k=1}^{K} \sum_{j>2...J} p_{klm} = 1$. Also, let $e = r \cdot z$ with $r_{\cdot lj} \geq 0$ and $\sum_{j>2...J} r_{\cdot lj} = 1$ for K and L, the number of rows and columns, and J, the number of data points over the support space for the error terms inside the regional cross-sector matrix. The same conditions of normality can be easily formulated for any vector of column sums. Next, the cross-entropy econometric estimator of the Tsallis entropy can be presented as follows:

$$
\begin{aligned}
& Min H_q(p//p^0, r//r^0, \mu//\mu^0) \\
& \qquad \equiv \alpha \sum p_{klm} \frac{[p_{klm}/p^o{}_{klm}]^{q-1}-1}{q-1} \\
& \qquad + \beta \sum r_{\cdot lj} \frac{[r_{\cdot lj}/r^o{}_{\cdot lj}]^{q-1}-1}{q-1} + \ldots + \delta \sum \mu_{k \cdot s} \frac{[\mu_{k \cdot s}/\mu^o{}_{k \cdot s}]^{q-1}-1}{q-1}
\end{aligned} \tag{5}
$$

Subject to

$$
Y_{\cdot l} = ccj.l \sum_k \left([Y_{\cdot l} P_{kl}]' + e_{\cdot l} \right) = \left[ccj.l \sum_k^K \left[\left(\sum_{m \succ 2}^{M} Y_{\cdot l}' v_{klm} (p_{klm}{}^q) \right]' + \sum_{j=1..J} r^q{}_{\cdot lj} z_{\cdot lj} \right) \right] \tag{6}
$$

$$
H_{k\cdot} = C_{k\cdot} (X_{k\cdot} + \omega_{k\cdot}) = C_{k\cdot} \left(X_{k\cdot} + \sum_{s=1}^{S} \mu^q{}_{k \cdot s} v_{k \cdot s} \right) \tag{7}
$$

$$
\sum_{k..K} H_{k\cdot} = \sum_{l..L} Y_{\cdot l} \tag{8}
$$

$$
\sum_{k=1}^{K} \sum_{j>2...J} p_{klj} = 1 \tag{9}
$$

$$
\sum_{j>2...J}^{J} r_{\cdot lj} = 1 \tag{10}
$$

$$
\sum_{s>2...S}^{S} \mu_{k \cdot s} = 1
$$

where

$Y_{\cdot l}$: indicates each sum per column l (values observed by energy sector l, including unknown errors);

$H_{k\cdot}$: each row total (observed values per province k) corrected for $C_{k\cdot}$;

$X_{k\cdot}$: total energy intensity indicators by region affected by unknown errors; and

p_{kl}: probabilistic structure of energy efficiency ratios by sector and region.

$C_{k\cdot}$ is a positive scaling factor related to province climatic factor (average annual temperature by province) to match the totals of energy intensity and energy efficiency indicator levels,

$CC_{\cdot j}$ is the arbitrary additional scaling factor representing other random variables, but $C_{k\cdot}$ to balance row and column sums "$\cdot$" refers to a variable bound to a row or column total, depending on the context.

Non-extensive statistics use a number of binding forms in which expectations can be set. The above model uses Curado-Tsallis (C-T) constraints [40,48], the general form of which is as follows:

$$
\langle y_q \rangle = \sum_i p_i{}^q y_i
$$

The parameter q, as already mentioned, represents the Tsallis parameter. Theoretically, the values of this parameter vary between 0 and 3 [49]. Once again, its value equal to 1 corresponds to the Gaussian attractor. In the real world, the values of this parameter should evolve between 1 and 5/3, thus including spontaneous or man-made stable fractal structures such as in most financial or economic markets ([50,51]) indices. As Table 1 suggests, the $C_{k\cdot}$ remains a scaling factor related to province climatic factor contributing to match the totals of energy intensity and energy efficiency indicator levels that are known with uncertainty. Unlike the factor $C_{k\cdot}$, whose role has just been expressed above, the factor $CC_{\cdot j}$ is the arbitrary additional scaling factor representing other random variables but $C_{k\cdot}$ to complementarily contribute to balancing row and column sums. This is because the $C_{k\cdot}$, which explains that climatic differences within provinces could not play the balancing role alone since other factors may exist in explaining the difference between energy intensity and energy efficiency. Among these, as already said in the introductory section, are the regional economic structure (e.g., share of large energy-consuming industries) and the geographical characteristics of provinces (e.g., longer distances leading to higher demand for transport). Still, the sum per column (i.e.,per sector) of the posterior probabilities is constrained to unity, given $CC_{\cdot j}$. In the above model (Equations (5)–(10)), their values will play the role of the new Bayesian data discriminating in favor of the new inferential evidence. We must recall that when no new data are included in the cross-entropy model, the results correspond to those from formulating a maximum entropy principle, without additional conditions, except those of normality.

The above $H_q(p//p^0, r//r^0, \mu//\mu^0)$ is nonlinear and measures entropy in the model. The relative entropies of the three independent terms (respectively the three posteriors p, r, μ and corresponding priors p^0 r^0 μ^0) are then added together using the weights. These are the real positives that amount to unity within the mentioned constraints. The first term known as the one of "parameter precision" takes into account the discrepancies between the estimated parameters α, β, δ and the prior parameters (usually defined in the support space). The second and third terms "ex-post predictions" include the empirical error term as the difference between the predicted and observed data values (see the last row and column in Table 1) in the model. The first component of the criterion function can relate to the structure of the table parameters, the second component to the errors in the row totals, and the last component to the errors in the column totals.

It should be noted that the estimates of the model and their variances should be influenced not only by the length of the support space but also by the spatial scale effect, i.e., the number of affected point values [32]. The greater the number of these points, the better the prior information, i.e., nonlinear starting points, about the system.

Next, the random errors ω_i (see Equation (7)) explain the errors in data collection and processing and are not necessarily related to the Gaussian distribution being itself a particular case of a PL. Traditionally, regarding Bayesian formulations and relative entropy, it should be noted that both models will lead to similar results if and only if the real expected errors associated with the data generation system are zero in the symmetric support space around zero (see, e.g., [44]). Similarly, the results of the Gibbs-Shannon cross-entropy and the Tsallis' non-extensive cross-entropy will match when errors included in the model are not correlated, and the system distribution will then evolve towards the Gaussian attractor (e.g., [52–54]).

With respect to the confidence interval of the parameters, Equation (11) shows the non-additivity of the Tsallis entropy for two—probable—independent systems, one related to the probability distribution of the parameters and the other to the probability distribution of error perturbations:

$$S(\hat{P}r) = [S(\hat{p} + \hat{r})] = \{[S(\hat{p}) + S(\hat{r})] + (1-q) \cdot S(\hat{p}) \cdot S(\hat{r})\} \tag{11}$$

where

$$S(\hat{p}) = -\left[1 - \textstyle\sum_k \sum_m (p_{klm})^q\right] / \left[K \cdot \left(M^{1-q} - 1\right)\right]$$

$$S(\hat{r}) = -\left[\left(1 - \sum_l \sum_j r^q{}_{\cdot lj}\right)\right] / \left[L \cdot \left(J^{1-q} - 1\right)\right]$$

$S(\hat{P}r)$ is then the sum of the normalized entropy associated with the parameters of the $S(\hat{p})$ model and the term of perturbation $S(\hat{r})$. Similarly, the last value is $S(\hat{r})$ obtained for all observations l, with J the number of data points above the support of the estimated probabilities r related to the time length of errors.

The values of these normalized entropy indices range from zero to one. The values close to one indicate a weak information variable, while lower values indicate a parameter that is estimated to be more informative $\hat{\beta}_{kl}$ in the model.

4. Outputs and Comment

We will illustrate the empirical basis of the theoretical model developed in the above sections by applying the Polish energy efficiency coefficients to the case. Table 1—whose row and column symbols correspond to Equation (6)—is presented to illustrate the extent to which a researcher can have limited information about quantity and quality before solving an inverse and ill-posed inverse problem such as the one in this article.

We have two aggregated energy indicators. each from the Polish energy demand sectors per province. The problem in hands stands as [29] a discrete problem in a multidimensional space, leading to $(k-1) \times (l-1)$ degree of freedom that would illustrate the case of a standard inverse problem. In this problem, as illustrated in Table 1 ([55,56]), we have $3 \times 15 = 45$ degrees of freedom related to the number of energy efficiency indicators per sector and province. Moreover, as alluded to before, the row total and the column total are different, probably as a result of the two separate data sources with different nature and scale.

As the above pages made us aware of, this kind of problem corresponds the best to the philosophy naturally contained in the principle of maximum entropy that we have implemented to forecast the energy efficiency ratios displayed in Table 2.

Table 2. Post-entropic 2021 energy efficiency ratio forecasts and efficiency progress in 2020/2021(%).

	Industry	Transport	Households	Services	Average Efficiency Ratio per Province	Efficiency Ratio Change_2021/2020
Lodz	0.4	0.667	0.677	0.754	0.625	−0.12
Slaskie	0.429	0.729	0.741	0.83	0.682	0.81
Mazowieckie	0.302	0.471	0.477	0.521	0.443	−3.16
Wielkopolska	0.333	0.526	0.533	0.583	0.494	7.49
Dolnoslaskie	0.369	0.602	0.61	0.676	0.564	−1.29
Opole	0.482	0.847	0.861	0.974	0.791	2.43
Malopolskie	0.336	0.536	0.543	0.596	0.503	−2.44
Swietokrzyskie	0.445	0.764	0.777	0.873	0.715	5.32
Zachodnia pomors	0.373	0.61	0.618	0.685	0.572	0.04
Kujawska-pomorska	0.37	0.604	0.613	0.678	0.566	−0.88
Pomorska	0.336	0.536	0.543	0.597	0.503	−2.44
Lublin	0.369	0.602	0.61	0.675	0.564	−1.29
Podkarpackie	0.336	0.536	0.543	0.597	0.503	−2.33
Lubuskie	0.386	0.636	0.645	0.716	0.596	4.07
Podlaskie	0.336	0.536	0.543	0.597	0.503	−2.39
Warminsko-mazurskie	0.339	0.541	0.548	0.602	0.507	−1.53
Sector average ratio	0.371	0.609	0.618	0.685		
Progress_2021/2020(in %) ratio	0.067	0.369	0.385	0.548		

Let us now comment on the model outputs presenting the final solution about the cross "energy efficiency ratios" (column 5 of Table 2) by sector and province. As presented in the theoretical model (Equations (5)–(10)), in addition to the classical normality conditions and moment consistency, the earlier explained random factors $CC_{.j}$ and $C_{k.}$ allowed the system to balance. The first factor is related to various variables differentiating energy intensity from energy efficiency except the climatical factor represented by $C_{k.}$. As already said, these calculated ratios in the above table are not deterministic, since the priors are not known with certainty. Therefore, we treat them as random variables, from which the posteriors result from the optimization process. It is worthwhile to recall that the presented energy efficiency scores stand for the overall energy efficiency scores, i.e., a combination of the three already presented components: the energy efficiency level, the energy efficiency progress (i.e., energy efficiency trends), and the energy efficiency policies. The prior matrix of energy efficiency ratios, which is not presented in this paper, has been initiated on the basis of knowledge of the province energy intensity index and the sector energy efficiency ratio averages. Next, using proportions of the energy intensity index (the last column of Table 1) of different provinces, we computed the initial energy efficiency ratios by sector to sum up to the total of the energy efficiency ratios—known with error—available from the last row of the same Table 1. In probabilistic terms, we have assumed the energy efficiency ratio to be a uniform distribution across the provinces, and thus the last column of the energy intensity is a marginal probability. The assumption behind this procedure is that the province with a lower or a higher average energy intensity index will have a lower or a higher average energy efficiency ratio, respectively, irrespective of the considered sector. Doing so, we enabled the nonlinear mathematical system to start the search of the model global optimal solution from the best starting points leading to a quick convergence. The post-entropy posterior ratios in Table 2 are normalised, and the higher the value, the lower the energy efficiency level for a given sector and/or province. The obtained forecasts are empirically close to real world expectations of energy efficiency ranking within Polish provinces. We notice that in the industry sector, the provinces Mazowieckie (including Warsaw) and Wielkopolski (including Poznan) have the lowest energy efficiency ratios, respectively, around 0.320 and 0.333. The highest ratio is shown in the provinces Swietokrzyskie and Opole, with ratios around to 0.445 and 0.482, respectively. Globally, we notice that Mazowieckie displays the highest efficiency in all sectors, while Opole displays the lowest efficiency. We notice too that the industry sector globally remains the most efficient among all sectors.

Table 3 presents the level at which the model has discriminated the prior in favor of the post cross-entropy solution given the new model consistency moments and normality conditions. Consequently, we notice in that table the highest post cross-entropy discriminating values through all sectors in the case of Opole in comparison with the remaining sectors. It is worthy to recall that, as described in the precedent paragraphs, the prior values presented the same structure for different sectors. The cross-entropy formalism has discounted the non-valuable information to just retain the information fitting to corresponding sectors and provinces given the initial information on the energy intensity indicator and the climatic factor.

Following the formulation of Equation (11), the confidence interval of the model is a normalized value ranging between zero to one. In the present case, the comparable to the corrected classical coefficient of determination R^2 is equal to 0.011, much closer to zero (lowest entropy) than one (highest entropy). Then, the proposed model system has discriminated optimally from the priors, given all the constraints encountering the energy efficiency system and defined by the equations. Nevertheless, it is important to mention that the cross-entropy information metric does not conceptually conform to the property of triangular inequality inherent in Euclidean distance.

Table 3. Model cross matrix increment (in %) between priors and posteriors (2021).

	Industry	Transport	Households	Services
Lodz	−4.901	−4.569	−4.566	−4.558
Slaskie	−9.481	−8.137	−8.098	−7.812
Mazowieckie	14.89	11.26	11.13	10.104
Wielkopolska	14.89	11.26	11.13	10.104
Dolnoslaskie	0.405	−0.418	−0.455	−0.76
Opole	−17.153	−14.151	−14.051	−13.297
Malopolskie	6.77	4.634	4.552	3.9
Swietokrzyskie	−9.481	−8.137	−8.098	−7.812
Zachodnia pomors	0.405	−0.418	−0.455	−0.76
Kujawska-pomorska	0.405	−0.418	−0.455	−0.76
Pomorska	6.77	4.634	4.552	3.9
Lublin	0.405	−0.418	−0.455	−0.76
Podkarpackie	6.77	4.634	4.552	3.9
Lubuskie	0.405	−0.418	−0.455	−0.76
Podlaskie	6.77	4.634	4.552	3.9
Warminsko-mazurskie	6.77	4.634	4.552	3.9
AEE ratio	−0.289	−0.908	−0.937	−1.178

Source: Own calculations.

Limitations of the Study and Prospective Research Area

Let us start with the strong side of the model and then discuss its limitations. Next, we will point out the specificity of the approach of the relative non-extensive entropy econometrics. We talk about it in the context of solving a random inverse problem consisting of the estimation of energy efficiency coefficients by sector and province in Poland. As explained in the article, this approach inherits its relative strength and originality from the combination of the following three attributes:

- The power law distribution generalizes most of the known statistical laws and has proven to be designed for analytically resolving non-stationary functions. Its application leads to the analytical closed form outputs of the model.
- The Kullback-Leibler relative entropy, being a combination between the properties of entropy and Bayesian formalism, stands for a strong information metric, particularly in the case of inverse problem modelling. Thanks to this formalism, several related hypotheses using the least squares method become obsolete.
- While these two scientific sub-disciplines are based on solid hypotheses, joining traditional econometrics to them leads to the model proposed in this paper.

As a matter of fact, this approach will combine the advantages listed above: a minimum of hypotheses, the generalization of the normal law, and high precision of estimated parameters.

Now let us talk about the limitations. The major limitation of the approach is owing to computational complications that may arise while solving nonlinear systems. We used GAMS (General Algebraic Modelling System) software 43.4.1) to perform computations. Based on our own experience, the solution to this problem was to use one of the recent programs (connected to GAMS) designed to calculate the global optimum of nonlinear systems as well as to find the most informative initial points of the solution (the priors). In this case, we used "Knitro". The point concerning the choice of the appropriate starting solution was explained when we presented the model.

In recent years, the non-extensive entropy approach has been extended to two parameters (in addition to q), and this opens up greater ease of modelling as well as faster convergence towards the global optimum from any point space and with a minimum of additional information about the system. This line of research could make this approach easier to apply to different models hitherto considered insoluble or soluble due to non-realistic or unverifiable hypotheses.

The next point of improvement of the model should be to analyze by simulation techniques the individual impact of the main factors impacting the energy efficiency coefficients. Once again, it is about climatic conditions, the regional economic structure, and the geographical characteristics of provinces. The model presented in this paper is limited to just estimating the sectoral coefficients by province. Therefore, certain factors have been grouped into the same variable "cc.j".

Next, it would be worthwhile for the future to compare the relative entropy approach presented here with other powerful forecasting techniques for the same specific case. We could cite here, among others, the grey systems ([8,57]) based on solid theoretical foundations, the approach of neural networks successfully used in many fields of sciences, or the Fuzzy Time Series technique [58] plus its different versions, known for its relatively good precision in terms of predictions. The comparison of all these techniques requires a good understanding of the theoretical background of each of them and then the conceptually targeted area of their applications.

Finally, thanks to this approach, this problem could also be solved in the case of further disaggregation, for example, to the NUTS 3 level. Once again, this is of capital importance because it is at the more disaggregated levels that economic actors act.

5. Conclusions

The objective of this article was to carry out a forecast of energy efficiency ratios in conditions where statistical data are not only missing but also from contradictory sources. In empirical research, this case is rather a rule than an exception. In most situations, this will lead to a nonlinear ill-posed inverse problem for which traditional statistics fail to find an analytical solution unless commodity additional hypotheses are made, as is often the case in empirical econometrics.

In the present paper, we have analysed this problem at the level of the 16 Polish regions for four sectors of energy consumption. We started with sparse information concerning the annual sectoral energy efficiency ratio averages aggregated at the national level. We disposed as well of the annual averages of energy intensity aggregated at the provincial level. The problem, therefore, consisted of bringing together these two aggregates of different natures to produce reliable information on interprovincial sectoral energy efficiency coefficients.

The forecasts obtained seem to be in line with the expectations of those who know the realities behind the Polish energy sector.

This work more or less described in detail the generalized cross-entropy technique and illustrated its applicability in the field of the energy experimental statistics.

Depending on the countries studied, the energy efficiency factors presented in this paper could differently impact the deviation between the energy intensity and energy efficiency indicators. Next, a comparison of information on the energy efficiency across sectors and regions will be of significant benefit to target grey areas and redefine more appropriate energy policies fitting to realistic social parameters inside a disaggregated locality. In particular, the outputs of the model could allow public energy institutions to monitor the relative progress of each locality in their endeavour to optimize energy use given their own specifical conditions. Taking things on a province scale, the provinces like Opole, Swietokrzyskie or Lubuskie with the lowest efficiency in all sectors should try to allocate higher investment budgets to increasing parallelly to energy intensity and efficiency through new technology and province industry geographical reorganisation. This may take time but will pay off in the future. Finally, as it results from the above comments, an extensive forecast on the European or larger scale of these indicators through experimental official statistics could allow for better management of energy saving as targeted by the European or worldwide climate change mitigation institutions or projects.

Funding: This research is supported by the Cardinal Stephan Wyszynski University of Warsaw.

Data Availability Statement: The data is available and explained in this article; details of calculations may be available on request.

Conflicts of Interest: The author declares no conflict of interest.

References

1. Tsemekidi Tzeiranaki, S.; Bertoldi, P.; Paci, D.; Castellazzi, L.; Serrenho, T.; Economidou, M.; Zangheri, P. *Energy Consumption and Energy Efficiency trends in the EU-28, 2000–2018*; Office of the European Union: Luxembourg, 2020.
2. IRENA. *Global Energy Transformation: A Roadmap to 2050 (2019 Edition)*; International Renewable Energy Agency: Abu Dhabi, United Arab Emirates, 2019; ISBN 978-92-9260-121-8.
3. IRENA. *Reaching Zero with Renewables: Eliminating CO2 Emissions from Industry and Transport in Line with the 1.5 Degree, Climate Goal*; International Renewable Energy Agency: Abu Dhabi, United Arab Emirates, 2020; ISBN 978-92-9260-269-7.
4. Jakimowicz, A. The Material Entropy and the Fourth Law of Thermodynamics in the Evaluation of Energy Technologies of the Future. *Energies* **2023**, *16*, 3861. [CrossRef]
5. Jakimowicz, A. The future of the energy sector and the global economy: Prosumer capitalism and what comes next. *Energies* **2022**, *15*, 9120. [CrossRef]
6. Foster, J. Energy, aesthetics and knowledge in complex economic systems. *J. Econ. Behav. Organ.* **2011**, *80*, 88–100. [CrossRef]
7. Batrancea, L.; Pop, M.C.; Rathnaswamy, M.M.; Batrancea, I.; Rus, M.I. An empirical investigation on the transition process toward a green economy. *Sustainability* **2021**, *13*, 13151. [CrossRef]
8. Wang, Q.; Li, S.; Li, R. Forecasting energy demand in China and India: Using single-linear, hybrid-linear, and non-linear time series forecast techniques. *Energy* **2018**, *161*, 821–831. [CrossRef]
9. Wang, Q.; Li, S.; Li, R.; Jiang, F. Underestimated impact of the COVID-19 on carbon emission reduction in developing countries–a novel assessment based on scenario analysis. *Environ. Res.* **2022**, *204*, 111990. [CrossRef]
10. Bwanakare, S. Greenhouse Emission Forecast as an Inverse Stochastic Stochastic Problem: A Cross-Entropy Econometrics Approach. *Acta Phisica Pol. A* **2015**, *127*, A-13–A-20. [CrossRef]
11. Odyssee-Mure. Indicator Tools. Available online: https://www.odyssee-mure.eu/ (accessed on 24 April 2023).
12. Polish Institute of Statistics. Regional Statistics. 2020–2021. Available online: https://bdl.stat.gov.pl/bdl/pomoc/stanzasilenia?active=0# (accessed on 16 November 2022).
13. Proskuryakova, L.; Kovalev, A. Measuring energy efficiency: Is energy intensity a good evidence base? *Appl. Energy* **2015**, *138*, 450–459. [CrossRef]
14. Office of Energy Efficiency & Renewable. *Energy Efficiency vs. Energy Intensity Analysis*; Office of Energy Efficiency & Renewable: Waszyngton, DC, USA, 2023.
15. Su, H.; Huang, Q.; Wang, Z. An energy efficiency index formation and analysis of integrated energy system based on exergy efficiency. *Front. Energy Res.* **2021**, *9*, 723647. [CrossRef]
16. Reuter, M.; Narula, K.; Patel, M.K.; Eichhammer, W. Linking energy efficiency indicators with policy evaluation—A combined top-down and bottom-up analysis of space heating consumption in residential buildings. *Energy Build.* **2021**, *224*, 110987. [CrossRef]
17. Li, S.; Diao, H.; Wang, L.; Li, C. Energy Efficiency Measurement: A VO TFEE Approach and Its Application. *Sustainability* **2021**, *13*, 1605. [CrossRef]
18. Li, S.; Wang, W.; Diao, H.; Wang, L. Measuring the efficiency of energy and carbon emissions: A review of definitions, models, and input-output variables. *Energies* **2022**, *15*, 962. [CrossRef]
19. EESI. Environmental and Energy Study Institute. Available online: https://www.eesi.org/ (accessed on 28 July 2023).
20. European Union. EPRS. EU. Available online: https://www.europarl.europa.eu/at-your-service/en/stay-informed/factsheets-on-the-european-union (accessed on 8 November 2023).
21. IEA. *Energy Efficiency Indicators: Fundamentals on Statistics*; IEA: Paris, France, 2014.
22. UNDP. Chapter 3: Energy Efficiency. In *Global Tracking Framework*; United Nations: San Francisco, CA, USA, 2017.
23. Odyssee-Mure. *Methodology European Energy Efficiency Scoreboard*; Odyssee-Mure: Luxemburg, 2020.
24. Kirsch, A. *An Introduction to the Mathematical, Theory of Inverse Problems*; Springer International Publishing: New York, NY, USA, 1996.
25. Zhang, Y.; Zhang, Y.; Shi, W.; Shang, R.; Cheng, R.; Wang, X. A new approach, based on the inverse problem and variation method, for solving building energy and environment problems: Preliminary study and illustrative examples. *Build. Environ.* **2015**, *91*, 204–218. [CrossRef]
26. Zio, E. Approaching the inverse problem of parameter estimation in groundwater models by means of artificial neural networks. *Prog. Nucl. Energy* **1997**, *31*, 303–315. [CrossRef]

27. Zhang, Y.; O'Neill, Z.; Dong, B.; Augenbroe, G. Comparisons of inverse modeling approaches for predicting building energy performance. *Build. Environ.* **2015**, *86*, 177–190. [CrossRef]
28. Tikhonov, A.N.; Arsenin, V.I. *Solutions to Poorly Conditioned Problems*; John Wiley & Sons: New York, NY, USA, 1977.
29. Jaynes, E.T. *Probability Theory: The Logic Of Science*; Washington University: Washington, DC, USA, 1944.
30. Shannon, C.E. A mathematical theory of communication. *Bell Syst. Tech. J.* **1948**, *27*, 379–423. [CrossRef]
31. Bwanakare, S. Econometrics of non-extended entropy: New statistical characteristics of constant elasticity of substitution-related models. *Entropy* **2014**, *16*, 2713–2728. [CrossRef]
32. Golan, A.; Juge, G.; Miller, D. *Maximum Entropy Econometrics: Robust Estimation with Limited Data*; Wiley: Chichester, UK, 1996.
33. Robinson, S.; Cattaneo, A.; El-Said, M. Updating and estimating a social accounting matrix using cross entropy methods. *Econ. Syst. Res.* **2001**, *13*, 47–64. [CrossRef]
34 Gee, S.B.; Tan, K.C.; Alippi, C. Solving multiobjective optimization problems in unknown dynamic environments: An inverse modeling approach. *IEEE Trans. Cybern.* **2017**, *47*, 4223–4234. [CrossRef]
35. Perera, R.; Fang, S.E.; Ruiz, A. Application of particle swarm optimization and genetic algorithms to multiobjective damage identification inverse problems with modelling errors. *Meccanica* **2010**, *45*, 723–734. [CrossRef]
36. Velho, F.; de Campos, H.; Elcio, H.; Shiguemori; Carvalho, F.M.; Ramos, J.C.A. Theory of Unified Regularization: The Principle of Non-Extensive Maximum Entropiy. Differential equation. *Appl. Math.* **2006**, *25*, 307–330.
37. Jessen, H.A.; Mikosch, T. Regularly varying functions. *Publ. De L'institut Math.* **2006**, *80*, 171–192. [CrossRef]
38. Gabaix, X. *The Laws of Power in Economics and Finance*; BER: Washington, DC, USA, 2008.
39. Kullback, S. *Information Theory and Statistics*; John Wiley and Sons: New York, NY, USA, 1959.
40. Plastino, A.R.; Venkatesan, R.C. Deformed Statistics K-L Divergence Minimization within a Scaled Bregman Framework. *Phys. Lett. A* **2011**, *375*, 4237–4243.
41. Bağci, G.B.; Arda, A.; Sever, R. On the problem of constraints in nonextensive formalism: A quantum mechanical treatment. *Int. J. Mod. Phys. B* **2006**, *20*, 2085–2092. [CrossRef]
42. Nielsen, F.; Nock, R. A closed-form expression for the Sharma–Mittal entropy of exponential families. *J. Phys. A Math. Theor.* **2012**, *45*, 32003. [CrossRef]
43. Jakimowicz, A. The role of entropy in the development of economics. *Entropy* **2020**, *22*, 452. [CrossRef]
44. Bwanakare, S. *Non-Extensive Entropy Econometrics for Low Frequency Series: National Accounts-Based Inverse Problems*; De Gruyter Open Poland: Warsaw, Poland, 2018.
45. Wooldridge, J.M. *Econometric Analysis of Cross Section and Panel Data*, 2nd ed.; MIT Press: Cambridge, MA, USA, 2010; ISBN 100262232588.
46. Shen, E.Z.; Perloff, J.M. Maximum entropy and Bayesian approaches to the ratio problem. *J. Econom.* **2001**, *104*, 289–313. [CrossRef]
47. Pukelsheim, F. *The Three Sigma Rule*; American Statistical Association: Alexandria, VA, USA, 1994; Volume 48, pp. 88–91.
48. Abe, S.; Bagci, G.B. Constraints and Relative Entropies in Nonextensive Statistical Mechanics. *arXiv* **2004**, arXiv:cond-mat/0404253.
49. Gell-Mann, M.; Tsallis, C. *Nonextensive Entropy, Interdisciplinary Applications*; Oxford University Press: New York, NY, USA, 2004.
50. Rak, R.; Drożdż, S.; Kwapień, J. Nonextensive statistical features of the Polish stock market fluctuations. *Phys. A Stat. Mech. Its Appl.* **2007**, *374*, 315–324. [CrossRef]
51. Drăgulescu, A.; Yakovenko, V.M. Exponential and power-law probability distributions of wealth and income in the United Kingdom and the United States. *Phys. A Stat. Mech. Its Appl.* **2001**, *299*, 213–221. [CrossRef]
52. Tsallis, C. *Introduction to Non-extensive Statistical Mechanics: Approaching a Complex World*; Springer: Berlin, Germany, 2009.
53. Bwanakare, S. Solving a generalized constant elasticity of substitution function of production: A non ergodic maximum enropy principle. *Acta Phys. Pol.* **2013**, *123*, 502. [CrossRef]
54. Bwanakare, S. Non-extensive econometric model of entropy (NO): A case of labour demand in Podkarpackie Voivodeship. *Acta Phys. Pol.* **2010**, *117*, 647. [CrossRef]
55. EU. Eurostat. 2021. Available online: https://ec.europa.eu/eurostat/web/main/data/statistical-themes (accessed on 19 December 2022).
56. GUS. GUS. 2021. Available online: http://old.stat.gov.pl/gus/publikacje_a_z_PLK_HTML.htm (accessed on 30 June 2022).
57. Deng, J. Introduction to Grey System Theory. *J. Grey Syst.* **1989**, *1*, 1–24.
58. Singh, P.; Dhiman, G. A hybrid fuzzy time series forecasting model based on granular computing and bio-inspired optimization approaches. *J. Comput. Sci.* **2018**, *27*, 370–385. [CrossRef]

energies

MDPI

Article

Renewables, Energy Mix and Inflation in the European Union Countries

Łukasz Markowski and Kamil Kotliński *

Institute of Economics and Finance, University of Warmia and Mazury in Olsztyn, Oczapowskiego 4, 10-719 Olsztyn, Poland; lukasz.markowski@uwm.edu.pl
* Correspondence: k.kotlinski@uwm.edu.pl

Abstract: One of the effects of the 2022 energy crisis was a high increase in inflation. According to a review of the literature on the subject, a factor that may increase the economy's resistance to external price shocks is the relatively high share of renewable sources in energy production. The aim of this paper is to assess the relationship between the share of renewables in energy production and the level of HICP and core inflation in a cross-sectional approach in the crisis year of 2022 within the EU countries. This study used regression methods, cluster analysis and dispersion measures. As a result of the analysis, it was found that in 2022 (using a cross-sectional approach), there was a statistically significant relationship between the above-mentioned variables. A higher share of renewables by 1 p.p. meant lower HICP by 0.13 p.p. and lower core inflation by 0.1 p.p. "Wind" and "solar" had the greatest importance in the multiple regression equation. It can also be stated that in 2022, countries with a similar energy mix structure were characterized as having a similar rate of inflation. It is therefore justified to increase the share of renewables in energy production, because it brings not only positive environmental but also economic effects.

Keywords: renewables; inflation; energy mix

1. Introduction

The energy crisis of 2022 was a shock of unprecedented breadth and complexity. Pressures on markets predated Russia's invasion of Ukraine, which turned a rapid economic recovery from the pandemic—which strained all manner of global supply chains, including energy—into a full-blown energy turmoil. Russia has been by far the world's largest exporter of fossil fuels, so its curtailments of natural gas supply to Europe and European sanctions on imports of oil and coal from Russia have severed one of the main arteries of global energy trade [1]. The result of this extremely complex situation was a spike in inflation in almost every country in the world [2,3]. Neri et al. [4] estimated the contribution of energy shock to headline inflation at around 60% in the fourth quarter of 2022, while its contribution to core inflation ranged from 20 to 50%, depending on the model.

The impact of energy costs on the price index (CPI) is direct and indirect. Of course, the price of electricity, heating and transport costs contribute to the CPI, but energy is a cost for the production of many other products included in the CPI. This applies to both agriculture goods and the industrial sector. Rising production costs push the prices of consumer products up. Economists name this phenomenon cost-push inflation. When the economy is hit by aggregate supply or energy price (cost-push) shocks, the central bank faces a trade-off, as countering the impact on inflation would amplify the negative effects on economic activity [4,5] Households do not directly consume crude oil. The first stage of pass-through is reflected in oil-related products and other energy goods and services that are part of the consumer basket, whose price developments are summarized by HICP energy. Indirect inflationary pressures may emerge as higher energy prices increase input costs and the production of non-energy goods and services, which, if passed on to consumers, will be reflected in core inflation (see, e.g., [6–8]).

The energy shocks are widely described in the literature (see, e.g., [9–15]). Over most of the last two decades, the main source of energy price shocks has been the price of crude oil (gas prices have been much more stable than oil prices), but the 2022 crisis differs because of the extraordinary increase in the price of natural gas, which has had a differential and very strong impact on European countries [16,17]. Natural gas, for the purpose of heating and electricity production, provides a significant contribution to the European energy mix.

The problem is that the EU countries are not abundant in fossil fuel energy sources and therefore depend largely on imports. These types of energy sources have seen a steady increase in price as fossil fuels become depleted and the cost of extracting them increases. This also has a negative impact on macroeconomic variables within the economy. The system of supplying energy or fossil fuels to Europe may also be destabilized by all kinds of conflicts or may be used as a tool of political influence [18,19]. This is effected, for example, by Russia, which seeks leverage, exposing consumers to higher energy bills and supply shortages [1].

Aware of such threats, the European Commission took some steps several years ago to reduce this dependence through an ambitious plan to make Europe the first climate-neutral continent in the world [20]. One of the elements of these changes is the increase in the share of renewables in energy production which is in line with the sustainable development goals implemented in 2016 [21]. The current EU target for 2030 assumes the use of energy from renewable sources at a level of 32% [22].

The existing literature describing renewable energy sources emphasizes environmental aspects, such as zero emission of air pollutants, emissions of greenhouse gases, global warming, environmental degradation and public health (see, e.g., [23–27]). The economic aspect is usually analyzed from a microeconomic aspect (see, e.g., [28,29]). What is important from the point of view of this study is the relationship between renewables and macroeconomic variables, namely high inflation in 2022. Renewable energy sources produce energy domestically, and do not use imported fossil fuels, which makes the economy independent of their fluctuations on world markets and makes it more resistant to external shocks.

Generally, it should be noted that the impact of renewables on inflation is not clear, and research results often lead to different conclusions (compare, e.g., [19,30–32]). First, it should be pointed out that implementing investments in renewables is expensive and the initial costs of their use are high. This leads to cost inflation, especially since the technologies used to produce renewable energy sources have not yet reached the "maturity level" that would allow for cost reduction [19,33,34]. So, in countries that are abundant in fossil fuels, using renewable energy sources can be expensive. However, in countries such as EU members that do not have sufficient oil, gas and coal resources, it may be more beneficial to use renewables and limit the import of fossil fuels. Moreover, the EU's sustainable development policy means that the costs of renewables are decreasing in the long term. The investment cost decreases with the development of technology because advanced technological infrastructure is developed. Examples include a better way to produce photovoltaic panels, new concepts for installing PV systems on rivers or reservoirs, the development of the energy storage industry or the growth of off-grid installations. The efficiency of the systems results in the commercialization of renewable energy sources and their economic justification. This is due to energy savings, increased efficiency and less damage to the environment. In the long term, unlike the short term, renewables result in lower costs and then inflation [19,35]. Moreover, renewable energy is expected to become more preferred [18,19]. According to a Lazard analysis [36], certain renewable energy generation technologies are already cost-competitive with conventional generation technologies. Key factors regarding the continued cost decline of renewable energy generation technologies are the ability of technological development, industry scale and increased competition. The Lazard study shows the significant historical cost declines for utility-scale renewable energy generation technologies driven by, among other factors, decreasing capital costs that should also translate into lower inflation. Timmons et al. [18] points out that after some time, renewable energy becomes cheaper than fossil fuels, and therefore market forces will effect the transition to renewable energy sources. In countries where this has already happened, renewable energy sources dominate. Therefore, it follows that those countries with a higher share of wind, solar, hydro and other renewable energy sources in the energy mix should also be more resistant to shocks, i.e., the price shock effect should be lower there than in countries with a smaller share of renewables.

It is therefore worth making an attempt to empirically verify whether, within the EU, in countries where the share of renewables is relatively high (i.e., where we can already discuss a developed technological infrastructure in this area allowing for cost reduction), in the face of the shock of 2022, we can observe relatively lower inflation. In other words, it can be treated as the result of many years of accumulating investments in renewables in these countries in the form of resistance to external shocks. The aim of this paper is to assess the relationship between the share of renewables in energy production and the level of HICP and core inflation using a cross-sectional approach in the crisis year of 2022 in the European Union countries. It is important to verify whether this relationship is statistically significant, which would mean that it is not a coincidence. Achieving the aim is related to answering the following research questions:

- Will the higher share of renewables in energy production be accompanied by lower inflation in 2022?
- Is there an inverse relationship between the share of renewables in energy production and the level of inflation?
- Do countries with a similar energy mix structure have similar levels of inflation?
- Which energy mix variants promote resistance to energy shocks?

The issue of renewables is very important and topical. However, the literature on the relationship between these energy sources and macroeconomic variables is relatively less extensive. They mainly focus on time series analysis. This study contributes to the ongoing discussion through a static analysis, i.e., an assessment of whether the relatively high share of renewables is a factor that increases the resilience of economies to external energy shocks. This conducted study is also part of an extremely important debate on the effects of the energy crisis of 2022.

This paper has the following structure. After the above introduction, Section 2 presents the materials, research methods and data sources. Section 3 contains the research results. Section 4 is devoted to a discussion presenting other research on the relationship between inflation and renewables and discussion of the specificity of 2022 in the context of the presented issue. Section 5 contains conclusions and recommendations.

2. Materials and Methods

Data for research on the inflation level of European Union countries were downloaded from the Eurostat statistical database [37]. Due to the availability of information for 2022, data on the energy mix structure were obtained from the "Our World in Data" website. This platform contains electronic publications on differences and changes in living conditions, global issues and problems (poverty, diseases, hunger, climate change, wars, existential threats and social inequalities). It also provides statistical data regarding, among others, energy aspects, including energy sources [38].

According to "Our World in Data", five categories of energy sources are specified in the energy mix: nuclear, coal, gas, oil and renewables. This structure was adopted in this study. Renewable energy, the importance of which is the main consideration in the context of inflation levels in EU countries, is a collective term used to describe many different energy sources. In this study, in accordance with the structure presented in "Our World in Data", the individual renewable components include biofuel, hydro, solar, wind (both onshore and offshore) and other renewables.

The year under study is 2022, which had a particular impact with regards to the energy crisis. The spatial scope of the research is the countries of the European Union. Malta and Cyprus were excluded from the analysis due to lack of data on energy sources. The analysis is therefore cross-sectional.

In order to assess the importance of the level of the share of renewable energy sources for the level of HICP and core inflation, a simple regression method was used. It should be emphasized here that the aim of this study is not to build a model that would explain inflation in 2022 to the greatest extent possible. Striving to obtain the maximum R^2 would be slightly different from the issues discussed in this paper. Because the intention is to verify in 2022, in spatial terms, we can discuss the relationship between two variables: renewables and inflation. In this study, it was crucial to check whether the relationship was positive and statistically significant, which means that it was not a result of chance. In other words, the aim is to answer the question whether, in addition to the climate protection function, higher renewables can also limit the inflationary effects of an energy shock.

In order to determine the hierarchy of the impact of individual types of renewable energy sources, multiple regression was used, carried out on variables normalized via the standard standardization rule.

The next stage of this study was to assess the similarity of EU countries in terms of their energy mix. It was crucial to answer the question of whether countries with a similar energy mix structure were characterized as having a similar level of inflation and to identify which energy mix variant could be most conducive to resistance to energy shocks. For this purpose, EU countries were divided into subgroups using classic cluster analysis. In the first stage of this study, variables were normalized (standardization rule). Its aim was to bring the variables to comparability by depriving the titre of the measurement results and unifying their orders of magnitude [39,40]. In the next step, the Euclidean distances between individual units were estimated. Cluster analysis started by selecting the smallest value in the matrix that represented the most similar objects in a given metric (Figure 1).

$$D = \begin{bmatrix} d_{11} & d_{12} & \dots & d_{1n} \\ d_{21} & d_{22} & \dots & d_{2n} \\ \vdots & \vdots & \vdots & \vdots \\ d_{n1} & d_{n2} & \dots & d_{nn} \end{bmatrix}$$

Figure 1. Distance matrix where $d_{i,j}$ $(i, j = 1, \dots, n)$ is the Euclidean distance between country i and j. Source: own study.

The Ward method was used to determine new distances for the reduced set of units [41]. The measure of the variation is the Error of Sum Square (ESS):

$$\text{ESS} = \sum\nolimits_{i=1}^{k} x_i^2 - \frac{1}{k}\left(\sum\nolimits_{i=1}^{k} x_i\right)^2$$

where:

x_i—value of the feature constituting the segmentation criterion that characterizes the i-th unit.

k—number of objects in the group.

The precise division of countries into groups was based on the method of maximum difference in agglomeration distance [42]. Inflation data from individual clusters were compared with data for the entire EU. The degree of differentiation was measured using classical measures of dispersion (standard deviation and coefficient of variation).

3. Results

The starting point in the adopted research procedure was to assess whether there was any relationship at all between inflation and the share of renewables in energy production in 2022. For this purpose, these variables are presented in the form of a scatterplot (Figure 2).

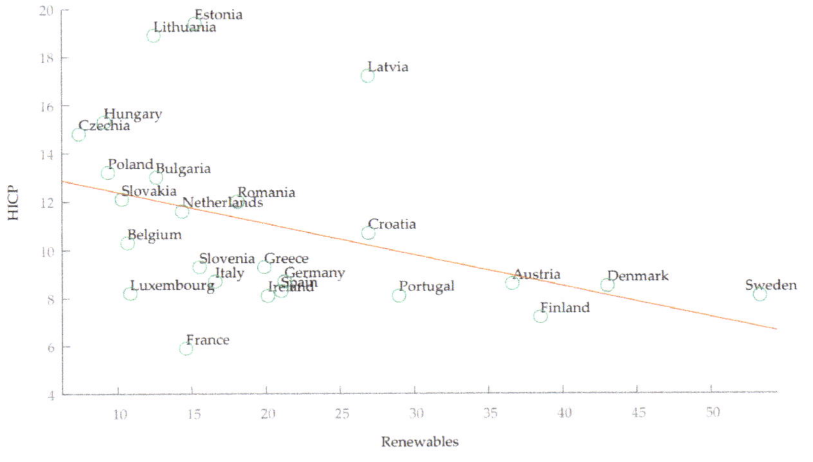

Figure 2. Scatterplot—HICP inflation (%) and the share (%) of renewables in primary energy in EU countries in 2022. The orange line is a trend. Source: Own study based on Eurostat and Our World in Data [37,38].

Based on the above chart, a certain negative relationship can be observed. Countries with a high share of renewables were characterized as having relatively lower inflation (e.g., Sweden, Denmark, Finland) and vice versa (e.g., Estonia, Lithuania, Hungary). The situation is similar in the case of core inflation (Figure 3).

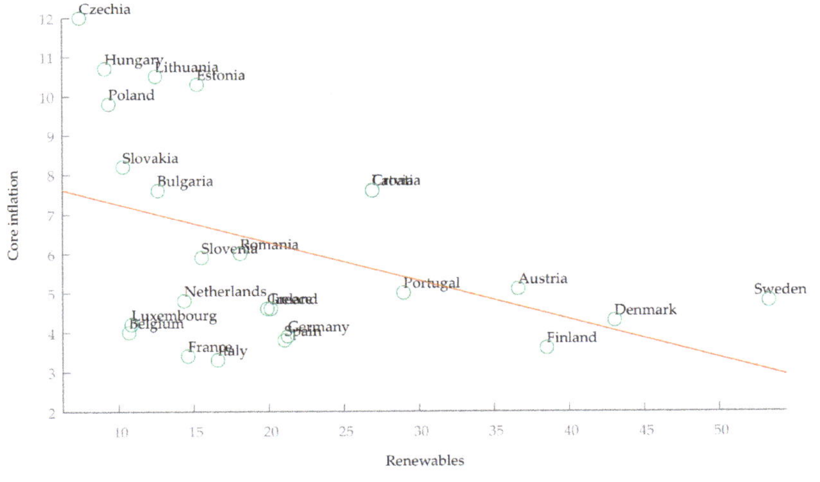

Figure 3. Scatterplot—Core inflation (%) and the share (%) of renewables in primary energy in EU countries in 2022. The orange line is a trend. Source: Own study based on Eurostat and Our World in Data [37,38].

The next element of this study was to verify whether the observed relationships were statistically significant. Table 1 presents the results of a simple regression that examined the relationship between the share of renewables in energy production (the share of individual types of this energy) and the inflation rate (HICP and core inflation) in EU countries in 2022. Due to the observed dispersion of variables in Figures 1 and 2, linear and exponential regression were estimated. For this kind of model, linearization was performed and the so-called linear model with respect to the parameters was obtained:

$$\hat{V}_t = \beta_0 + \beta_1 X_1$$

where:

$\hat{V}_t = \ln Y_t$

$\beta_0 = \ln \alpha_0$

$\beta_1 = \ln \alpha_1$

This model is verified as strictly linear models, and the values of all coefficients calculated for the auxiliary model apply to the original model [43].

Table 1. Simple regression—renewable energy sources and inflation.

Variables	Linear Model $\hat{Y}=\alpha+\beta X$		Exponential Model $\hat{Y}=\alpha_0 \times \alpha_1^x \rightarrow \hat{V}_t=\beta_0+\beta_1 X$	
	HICP	**Core Inflation**	**HICP**	**Core Inflation**
total renewable energy sources	α = 13.68 $\beta = -0.13$ R^2 = 0.17 SER = 3.43 SE (α) = 1.40 SE (β) = 0.06 *p*-value (β) = 0.04	α = 8.22 $\beta = -0.10$ R^2 = 0.18 SER = 2.46 SE (α) = 1.01 SE (β) = 0.04 *p*-value (β) = 0.03	α_0 = 13.31 α_1 = 0.9885 R^2 = 0.19 SER ($\hat{V}_t$ model) = 0.29 SE (β_0) = 0.12 SE (β_1) = 0.005 *p*-value (β_1) = 0.03	α_0 = 7.56 α_1 = 0.9867 R^2 = 0.15 SER ($\hat{V}_t$ model) = 0.38 SE (β_0) = 0.16 SE (β_1) = 0.007 *p*-value (β_1) = 0.054
biofuel	α = 12.95 $\beta = -1.28$ R^2 = 0.10 SER = 3.57 SE (α) = 1.39 SE (β) = 0.79 *p*-value (β) = 0.12	α = 7.44 $\beta = -0.81$ R^2 = 0.08 SER = 2.61 SE (α) = 1.02 SE (β) = 0.58 *p*-value (β) = 0.18	α_0 = 12.69 α_1 = 0.8814 R^2 = 0.14 SER ($\hat{V}_t$ model) = 0.3 SE (β_0) = 0.12 SE (β_1) = 0.07 *p*-value (β_1) = 0.07	α_0 = 6.96 α_1 = 0.8789 R^2 = 0.09 SER ($\hat{V}_t$ model) = 0.39 SE (β_0) = 0.15 SE (β_1) = 0.09 *p*-value (β_1) = 0.16
hydro	α = 11.56 $\beta = -0.09$ R^2 = 0.03 SER = 3.7 SE (α) = 0.95 SE (β) = 0.1 *p*-value (β) = 0.38	α = 6.55 $\beta = -0.05$ R^2 = 0.02 SER = 2.69 SE (α) = 0.69 SE (β) = 0.07 *p*-value (β) = 0.46	α_0 = 10.99 α_1 = 0.9927 R^2 = 0.03 SER ($\hat{V}_t$ model) = 0.31 SE (β_0) = 0.08 SE (β_1) = 0.008 *p*-value (β_1) = 0.37	α_0 = 5.9 α_1 = 0.9956 R^2 = 0.01 SER ($\hat{V}_t$ model) = 0.41 SE (β_0) = 0.11 SE (β_1) = 0.01 *p*-value (β_1) = 0.68
solar	α = 11.92 $\beta = -0.38$ R^2 = 0.03 SER = 3.70 SE (α) = 1.27 SE (β) = 0.44 *p*-value (β) = 0.4	α = 7.19 $\beta = -0.41$ R^2 = 0.07 SER = 2.62 SE (α) = 0.90 SE (β) = 0.31 *p*-value (β) = 0.19	α_0 = 11.20 α_1 = 0.9726 R^2 = 0.02 SER ($\hat{V}_t$ model) = 0.32 SE (β_0) = 0.11 SE (β_1) = 0.04 *p*-value (β_1) = 0.47	α_0 = 6.77 α_1 = 0.9313 R^2 = 0.09 SER ($\hat{V}_t$ model) = 0.39 SE (β_0) = 0.14 SE (β_1) = 0.05 *p*-value (β_1) = 0.14
wind	α = 12.78 $\beta = -0.27$ R^2 = 0.20 SER = 3.36 SE (α) = 0.99 SE (β) = 0.11 *p*-value (β) = 0.02	α = 7.52 $\beta = -0.20$ R^2 = 0.21 SER = 2.41 SE (α) = 0.71 SE (β) = 0.08 *p*-value (β) = 0.02	α_0 = 12.24 α_1 = 0.9762 R^2 = 0.22 SER ($\hat{V}_t$ model) = 0.28 SE (β_0) = 0.08 SE (β_1) = 0.01 *p*-value (β_1) = 0.02	α_0 = 6.96 α_1 = 0.9703 R^2 = 0.20 SER ($\hat{V}_t$ model) = 0.37 SE (β_0) = 0.11 SE (β_1) = 0.01 *p*-value (β_1) = 0.02

Table 1. *Cont.*

Variables	Linear Model $\hat{Y}=\alpha+\beta X$		Exponential Model $\hat{Y}=\alpha_0\times\alpha_1^x \rightarrow \hat{V}_t=\beta_0+\beta_1 X$	
	HICP	Core Inflation	HICP	Core Inflation
other renewables	α = 11.14 β = −0.03 R^2 = 0.001 SER = 3.76 SE (α) = 1.18 SE (β) = 0.22 *p*-value (β) = 0.9	α = 6.52 β = −0.07 R^2 = 0.01 SER = 2.71 SE (α) = 0.85 SE (β) = 0.16 *p*-value (β) = 0.65	α_0 = 10.77 α_1 = 0.9936 R^2 = 0.005 SER ($\hat{V}_t$ model) = 0.32 SE (β_0) = 0.1 SE (β_1) = 0.02 *p*-value (β_1) = 0.74	α_0 = 6.06 α_1 = 0.9865 R^2 = 0.01 SER ($\hat{V}_t$ model) = 0.41 SE (β_0) = 0.13 SE (β_1) = 0.02 *p*-value (β_1) = 0.58

Source: Own study based on Eurostat and Our World in Data [37,38].

The research results allow for an affirmative answer to the first research question. In the case of "total renewable energy sources", the results are statistically significant. In 2022 (in cross-sectional terms), a higher share of renewable energy meant a lower level of HICP inflation by 0.13 p.p. and core inflation by 0.1 p.p.ong the individual components of renewable energy sources, the results are statistically significant only in the case of "wind". It is worth emphasizing, however, that in all cases, the direction of the relationship was consistent with expectations. The exponential model leads to similar conclusions. The parameter α_1 less than 1 indicates a negative relationship. If the share of renewables increases by 1 p.p. (in cross-sectional terms), the value of HICP inflation will decrease by approximately 1.15% and the base inflation by 1.33% $((\alpha_1 - 1) \times 100\%)$. The important role of "wind" has been confirmed.

What may be of particular interest is the interpretation of the effect sizes of individual variables and the determination of the hierarchy of their influence. However, the above values of unstandardized coefficients are not a good indicator of the strength of the effect. Even in the case of multiple regression, they show the absolute importance of the variable in the function. To know the real strength of the effects of individual variables and to indicate which of them contributes the most to the function, standardized coefficients are necessary [44]. For this purpose, the variables were normalized (using the standardization rule) and the parameters of multiple regression were estimated. The renewable energy sources are marked as follows: x_1—biofuel, x_2—hydro, x_3—solar, x_4—wind, x_5—other. The equation for which HICP is the dependent variable is as follows:

$$\hat{y} = -0.43x_1 - 0.25x_2 - 0.35x_3 - 0.46x_4 + 0.31x_5$$

The equation for core inflation is as follows:

$$\hat{y} = -0.40x_1 - 0.25x_2 - 0.46x_3 - 0.42x_4 + 0.18x_5$$

Based on the above equations, it can be concluded that wind (x_4) had the greatest importance for the HICP level. The variables x_2, x_3 and x_5 turned out to be statistically insignificant (at the significance level of 0.05). The R^2 coefficient was 0.45.

Solar (x_3) had the greatest importance for the level of core inflation. The variables x_1 (although the *p*-value was 0.06), x_2 and x_5 turned out to be statistically insignificant. The R^2 coefficient was also 0.45. It is worth noting that the last variable (x_5) indicated the direction of the relationship in the two equations, which is inconsistent with the theory.

The next stage of this study was to verify whether countries with a similar energy mix structure were characterized by a similar rate of price growth in 2022. For this purpose, the 25 EU countries surveyed were divided into clusters using hierarchical segmentation (classic cluster analysis). The results are shown in Figure 4.

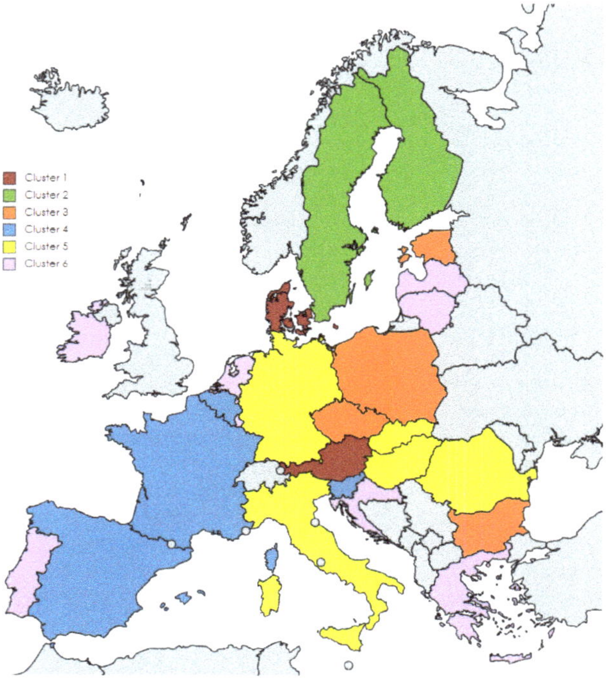

Figure 4. Similarity of countries due to the structure of the energy mix—separate clusters for 2022. Source: own work using MapChart [45]. Gray dots are micro-states (Andora, Monako, Lichtenstain, San Marino, Vatican) that are beyond the scope of study. Countries marked in gray on the map were not included in the study.

Table 2 contains the names of the countries in each cluster and the corresponding HICP and core inflation rates.

Table 2. Separate clusters due to the structure of the energy mix and the level of HICP and core inflation in 2022.

Cluster Number	Country	HICP	Core Inflation
1	Austria	8.6	5.1
	Denmark	8.5	4.3
2	Finland	7.2	3.6
	Sweden	8.1	4.8
3	Bulgaria	13	7.6
	Czechia	14.8	12
	Estonia	19.4	10.3
	Poland	13.2	9.8
4	Belgium	10.3	4
	Spain	8.3	3.8
	Slovenia	9.3	5.9
	France	5.9	3.4
5	Germany	8.7	3.9
	Romania	12	6
	Italy	8.7	3.3
	Hungary	15.3	10.7
	Slovakia	12.1	8.2

Table 2. *Cont.*

Cluster Number	Country	HICP	Core Inflation
6	Croatia	10.7	7.6
	Ireland	8.1	4.6
	The Netherlands	11.6	4.8
	Greece	9.3	4.6
	Lithuania	18.9	10.5
	Latvia	17.2	7.6
	Portugal	8.1	5
	Luxembourg	8.2	4.2

Source: Own study based on Eurostat and Our World in Data [37,38].

Based on such a preliminary analysis, it is possible to substantiate the assumption that countries with a similar energy mix structure were characterized by similar price growth dynamics in 2022. In order to deepen the analysis, the results regarding individual clusters should be compared to the data of the entire European Union (Table 3).

Table 3. Mean, standard deviation and coefficient of variation in HICP and core inflation in each cluster compared to the EU.

Cluster Number		1	2	3	4	5	6	
Countries		Austria, Denmark	Finland, Sweden	Bulgaria, Czechia, Estonia, Poland	Belgium, Spain, Slovenia, France	Germany, Romania, Italy, Hungary, Slovakia	Croatia, Ireland, The Netherlands, Greece, Lithuania, Latvia, Portugal, Luxembourg	European Union (25)
HICP	mean	8.6	7.7	15.1	8.5	11.4	11.5	11
	standard deviation	0.0	0.5	2.6	1.6	2.5	4.0	3.6
	coefficient of variation	0.6	5.9	17.1	19.3	21.8	34.6	32.7
Core inflation	mean	4.7	4.2	9.9	4.3	6.4	6.1	6.2
	standard deviation	0.4	0.6	1.6	1.0	2.7	2.1	2.6
	coefficient of variation	8.5	14.3	15.8	22.5	42.8	34.2	41.9

Source: Own study based on Eurostat and Our World in Data [37,38].

The first cluster (Austria, Denmark) consists of only two countries. It is characterized by a relatively high share of renewables in energy production (average 39.8%). Oil has a slightly lower share (38.4%). These countries were very similar in terms of inflation levels and were characterized by the lowest differentiation among all the identified clusters.

The next cluster includes two Scandinavian countries (Finland and Sweden) with a very large share of renewables (average 45.9%) and nuclear (average 19.86%). In this separate group, the average HICP and core inflation was the lowest. In this case, we can also discuss the similarity of the price growth rate compared to other clusters.

The third cluster consists of four Central and Eastern European countries (Bulgaria, Czech Republic, Estonia, Poland), where the energy source is dominated by coal (average 40.9%) and oil (average 27.2%). The average share of renewables is 11.2%. This is a relative high emission energy mix. An important element of the EU environmental policy is the European Union Emissions Trading System (EU ETS) to reduce carbon emissions in a cost-effective manner, and to spur the development of new low-carbon technologies [46]. For example, Moessner [47] investigated the impact of carbon prices on inflation. They found that an increase in prices of ETS by USD 10 per ton of CO_2 equivalents increases energy CPI inflation by 0.8 p.p. and headline inflation by 0.08 p.p. The entire cluster is characterized by higher inflation on average than the average within the entire EU, but at the same time less diversification. In this case, it can also be concluded that a similar energy mix favors similar inflation. It is worth noting that in this cluster, there is a subgroup consisting of Bulgaria and the Czech Republic, in which nuclear has a significant share (17.8 and 16.7%, respectively).

Cluster No.4 includes Belgium, Spain, Slovenia and France, where the main source of energy is oil (average 42.74%). It is also worth noting the significant share of nuclear in each of these countries (average 19%). The average share of renewables is 15.5%. The average inflation in this group was lower than in all the surveyed EU countries. The differences in the level of inflation within the cluster were also smaller.

In group 5 (Germany, Romania, Italy, Hungary, Slovakia), the countries are characterized by a relatively even energy mix structure. In each country, oil dominates (average 34%), but gas (average 29.8%), nuclear (in Slovakia and Hungary) and renewables (Romania, Germany, Italy) are also of considerable importance. The average share of renewables for all countries is 15.1%. The average inflation in this cluster was higher than in the EU. In the case of HICP inflation, the diversification was lower, and in the case of core inflation—higher.

The last largest cluster (consisting of Croatia, Ireland, The Netherlands, Greece, Lithuania, Latvia, Portugal and Luxembourg) is definitely characterized by the highest share of oil in the entire EU (average 53.2%). In second place is gas (average 22.9%), then renewables (average 20.1%). The nuclear share is almost zero. In this group of countries, HICP inflation was higher than in the entire EU (there was also higher differentiation within the cluster), and the core inflation was slightly lower (with lower differentiation within the cluster). However, this cluster includes countries that significantly lag behind the rest (Lithuania and Latvia). Without them, the cluster is much more coherent due to the level of inflation, and the dispersion measures are much lower than for the entire EU.

To sum up, it can be said that the countries within a given cluster are characterized by a similar level of inflation, although there are a few outliers. Mostly in clusters with a relatively large share of renewables, the average inflation level was lower than in the entire EU. The lowest inflation rates were recorded in countries with a high share of renewable energy complemented by a high share of nuclear energy. This result may suggest that this energy mix provides the best resilience against external energy shocks.

4. Discussion

The connections between inflation and renewables have already been discussed in the literature on the subject, especially in recent years. A valuable review of the literature in this respect was conducted by Arslan and Yildiz [19]. The results are inconclusive. The lack of a relationship between these two variables in the long term in Brazil was observed by, among others, Deka et al. [30]. However, the authors proved that the use of renewable energy negatively affects the exchange rate (appreciation), so in addition to reducing carbon dioxide emissions, it is a positive phenomenon, which justifies the use of this type of energy. Deka and Dube [48], reach similar conclusions for Mexico. The lack of causality of renewables on inflation was also observed by Malik et al. [49] for Pakistan.

There are also studies indicating that there is a positive relationship between renewables and inflation. Filippidis et al. [31] conducted research on over 200 countries (2000–2019) and pointed to a positive relationship between these variables. The authors mainly argue that the relationship between economic growth and renewable energy is U-shaped, and that the increase in renewable energy consumption reduces income inequality. A positive impact in the short term was also observed in the previously mentioned study by Deka and Dube [48] and Nga [50] for Vietnam.

However, most research indicates a negative relationship between renewables and inflation. Arslan and Yildiz [19] found that there is a cointegration relationship between renewable energy and inflation. It is true that renewable energy sources had no impact on inflation in the short term, but in the long term the impact was negative. The decrease in prices is influenced by the decrease in expenditure resulting from technological development. The authors observed a reduction in inflation in some European countries in the long term, however, they have emphasized the diversity of renewables among EU member states, which means that in some countries, it may take longer to observe a reduction in inflation. A negative relationship between renewables and inflation in the long term among OECD countries was also observed by Yıldırım and Kaya [33]. Research results by Khan et al. [32] show that alternative energy sources and inflation are negatively related in Germany. Economic growth, however, is positively related to environmental quality. The authors formulated recommendations for decision-makers to eliminate subsidies for domestic coal mining and promote profitable environmental policy projects. A negative relationship between prices and renewable energy in Iran was observed by Ghaed et al. [51].

The research results in this study also lead to the conclusion that a higher share of renewables in energy production is accompanied by lower inflation. However, before formulating final conclusions, some annotations should be added to the obtained research results. The purpose of this study was not to describe the entire inflation phenomena in 2022, but one should be aware of the multi-faceted nature of the price shock. It is worth noting that inflation has already been rising compared to previous years. The post-pandemic rebound in demand was of great importance [2,52]. Particular attention should be paid to 2021, when global climatic conditions affected the supply of many products, mainly agricultural ones. Moreover, inflation in 2022 was influenced not only by the degree of dependence on imports of energy products, but above all by the specific countries from which these products were imported. In 2020, Russia was the leading supplier to the EU of all major primary energy raw materials: natural gas, crude oil and hard coal. The different energy mix of individual EU countries causes different energy dependence of individual EU members on Russia [53]. In 2020, the countries with the largest share of energy needs met by Russian imports into the EU were Lithuania, Slovakia and Hungary. The least dependent were Cyprus, Ireland and Luxembourg (Figure 5).

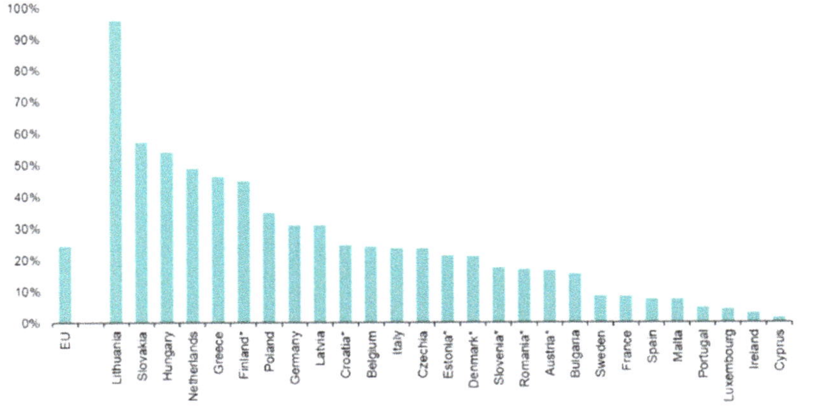

Figure 5. Imports from Russia in gross available energy (EU, 2020). *—estimates for non-reported data. Source: [53].

Despite the above issues, the obtained research results justify the conclusion that a higher share of renewables contributes to increasing the resistance of EU economies to external shocks. Even though European countries rank high in terms of the share of renewables [38], this share in the energy mix is still too small to protect all of the European Union against the inflationary effects of an energy crisis. Therefore, it is advisable to intensify efforts to develop renewable energy sources. Despite the ambitious plan, the difficulty of increasing the resilience of European economies against energy shocks is a complex issue. This is, for example, due to the specificity of the market in Europe. The investigated countries cooperate in the energy sector as members of the European Union. National energy markets are not separated from their neighbours but are interdependent with them. Jankiewicz and Grądzik [54] point out the problem with the rapid development of renewable energy sources concentrated in some regions as well as the difference in infrastructure conditions in respective countries. The unplanned power flows are especially prevalent during windy days. For example, in periods when German electricity is cheap, and especially at times of high wind power production in the northern part of the country, Austria imports this electricity either for consumption or storage purposes. Unplanned power flows cause negative consequences for the transit country because interconnectors are permanently overloaded [54]. Despite technical problems related to insufficient transmission capacity, international electricity trading is developing throughout Europe. European policy is intended to develop territorial cooperation by increasing the diversity of energy sources and transmission routes [55].

It is also important to remember that renewable energy sources are not completely free of fuel costs. Variations in fuel prices can materially affect the total costs of conventional generation technologies, but direct comparisons to "competing" renewable energy generation technologies must take into account issues such as dispatch characteristics (e.g., baseload and/or dispatchable intermediate capacity vs. peaking or intermittent technologies). In addition to fuel costs, capital and Operation and Maintenance (O&M) costs also influence the prices of energy produced [36]. In the short term, this may suppress involvement in investments in renewable energy sources, especially in countries where the technological infrastructure in this area is not yet developed. In recent years, this issue has been of significant importance. Table 4 contains a comparison of the levelized cost of energy in 2021–2023 conducted by Lazard [36].

Table 4. Levelized Cost of Energy Comparison ($/MWh).

Source	2021	2023
Solar PV	36	60
Geothermal	75	82
Wind—Onshore	38	50
Gas Peaking	173	168
Nuclear	167	180
Coal	108	117
Gas Combined Cycle	60	70

Source: [36].

According to the Lazard analysis [36], in the years 2021–2023, the cost of energy from both renewable and non-renewable sources increased. This may be due to the fact that fossil fuels are also needed to run renewable energy sources. The increases themselves may also be the result of the pandemic. However, certain renewable energy generation technologies are already cost-competitive with conventional generation technologies. This is a good forecast for the future, which allows us to assume that along with the growing competitiveness of renewable energy sources, their share in the energy mix will increase. This will undoubtedly contribute to the solution of the so-called energy trilemma: ensuring energy security, environmental sustainability and energy affordability and availability [56, 57]. This is particularly important for EU countries that are poor in fossil fuels and can therefore improve their trade balance with the use of renewables.

5. Conclusions

Based on the obtained research results, it can be concluded that in the face of the energy shock in 2022, a higher share of renewables in energy production was conducive to lower inflation. It follows that increasing renewables should not only be good for the planet, but also good for the economy and good for consumers. It is true that this study only concerns the specific year 2022, but one can also argue that such effects are the result of the accumulation of renewables in the long term, because in clusters which contain countries with a large share of this type of energy (meaning they had been implemented over a long time), average inflation was relatively low. Moreover, the research indicates a potential variant of the energy mix that is most resistant to energy and inflation shocks. A larger share of renewable energy complemented by nuclear energy could increase the country's resistance to external supply shocks.

Therefore, it is justified to formulate recommendations for further efforts to increase the share of renewables in energy production. Any facilitation of investing in renewable energy sources accelerates the reduction in costs related to the operation of these technologies. It is also important to remember the limitations of renewables in the sense that improving renewable energy technologies can reduce energy costs to some extent but cannot change the basic characteristics of energy sources. For example, improving the production of solar panels reduces production costs, but the problem of seasonal intermittency still exists. New technology makes it possible to produce ethanol from switchgrass instead of corn. However, land availability limits switchgrass production. It is indicated that offshore wind energy and solar photovoltaics may be the only renewable sources important enough to replace fossil fuels [18]. It is worth emphasizing that this corresponds to the research results obtained in this study. "Wind" and "solar" had the greatest importance in the multiple regression equation.

Future work will investigate other relationships between the energy mix and macroeconomic indicators, such as economic growth, employment and the economic cycle. The analysis of time series and the study of long-term dependencies, e.g., Granger causality or cointegration, are also extremely important. Such research will allow us to answer the question to what extent an increase in the share of renewables in energy production will change the dynamics of inflation in more economically stable conditions, not only in the face of an energy shock.

Author Contributions: Conceptualization, K.K. and Ł.M.; methodology, Ł.M.; formal analysis, K.K. and Ł.M.; investigation, K.K. and Ł.M.; resources, K.K.; data curation, Ł.M.; writing—original draft preparation, K.K. and Ł.M.; writing—review and editing, K.K. and Ł.M.; visualization, K.K. and Ł.M. All authors have read and agreed to the published version of the manuscript.

Funding: This research received no external funding.

Data Availability Statement: Publicly available datasets were analyzed in this study. This data can be found here: https://ourworldindata.org/ (accessed on 29 September 2023), https://ec.europa.eu/eurostat (accessed on 29 September 2023). The processed data presented in this study are available on request from the corresponding author.

Conflicts of Interest: The authors declare no conflict of interest.

References

1. International Energy Agency. World Energy Outlook 2022 Executive Summary. 2022, p. 4. Available online: https://www.iea.org/reports/world-energy-outlook-2022 (accessed on 25 September 2023).
2. Dagar, V.; Malik, S. Nexus between macroeconomic uncertainty, oil prices, and exports: Evidence from quantile-on-quantile regression approach. *Environ. Sci. Pollut. Res.* **2023**, *30*, 48363–48374. [CrossRef] [PubMed]
3. Svabova, L.; Kramarova, K.; Chabadova, D. Impact of the COVID-19 Pandemic on the Business Environment in Slovakia. *Economies* **2022**, *10*, 244. [CrossRef]
4. Neri, S.; Busetti, F.; Conflitti, C.; Corsello, F.; Monache, D.D.; Tagliabracci, A. Energy Price Shocks and Inflation in the Euro Area. Questioni di Economia e Finanza (Occasional Papers). No. 792. 2023. Available online: https://www.bancaditalia.it/pubblicazioni/qef/2023-0792/QEF_792_23.pdf?language_id=1 (accessed on 15 September 2023).
5. Mirza, N.; Naqvi, B.; Rizvi, S.K.A.; Boubaker, S. Exchange rate pass-through and inflation targeting regime under energy price shocks. *Energy Econ.* **2023**, *124*, 106761. [CrossRef]
6. Peersman, G.; van Robays, I. Oil and the Euro Area Economy. *Econ. Policy* **2009**, *24*, 603–651. [CrossRef]

7. Baumeister, C.; Peersman, G.; Van Robays, I. The Economic Consequences of Oil Shocks: Differences across Countries and Time. In *Inflation in An Era of Relative Price Shocks, Proceedings of the Conference held at the H.C. Coombs Centre for Financial Studies, Kirribilli, Australia, 17–18 August 2009*; Fry, R., Jones, C., Kent, C., Eds.; The Reserve Bank of Australia: Sydney, Australia, 2010; pp. 90–128.
8. Conflitti, C.; Luciani, M. Oil Price Pass-through into Core Inflation. *Energy J.* **2019**, *40*, 221–247. [CrossRef]
9. Grossman, P.Z. Energy shocks, crises and the policy process: A review of theory and application. *Energy Policy* **2015**, *77*, 56–69. [CrossRef]
10. Dutta, A.; Jana, R.K.; Das, D. Do green investments react to oil price shocks? Implications for sustainable development. *J. Clean. Prod.* **2020**, *266*, 121956. [CrossRef]
11. Naeem, M.A.; Peng, Z.; Suleman, M.T.; Nepal, R.; Shahzad, S.J.H. Time and frequency connectedness among oil shocks, electricity and clean energy markets. *Energy Econ.* **2020**, *91*, 104914. [CrossRef]
12. Murshed, M.; Tanha, M.M. Oil price shocks and renewable energy transition: Empirical evidence from net oil-importing South Asian economies. *Energy Ecol. Environ.* **2021**, *6*, 183–203. [CrossRef]
13. Baumeister, C. Pandemic, War, Inflation: Oil Markets at a Crossroads? NBER Working Paper 31496. 2023. Available online: https://www.nber.org/papers/w31496 (accessed on 20 September 2023).
14. Dagher, L.; Hasanov, F.J. Oil market shocks and financial instability in Asian countries. *Int. Rev. Econ. Financ.* **2023**, *84*, 182–195. [CrossRef]
15. Urom, C.; Guesmi, K.; Abid, I.; Dagher, L. Dynamic integration and transmission channels among interest rates and oil price shocks. *Q. Rev. Econ. Financ.* **2023**, *87*, 296–317. [CrossRef]
16. Gros, D.; Shamsfakhr, F. Global Energy Price Inflation with a European Twist. In-Depth Analysis, Monetary Dialogue Papers. 2022. Available online: https://www.europarl.europa.eu/RegData/etudes/IDAN/2022/733997/IPOL_IDA(2022)733997_EN.pdf (accessed on 23 September 2023).
17. Gros, D. Is Putin's War Driving Up Commodity Prices? Project Syndicate. 2022. Available online: https://www.project-syndicate.org/commentary/ukraine-war-minor-effect-on-energy-and-commodity-prices-by-daniel-gros-2022-04 (accessed on 18 September 2023).
18. Timmons, D.; Harris, J.M.; Roach, B. *The Economics of Renewable Energy*; Global Development and Environmental Institute, Tufts University: Medford, MA, USA, 2014. Available online: https://www.bu.edu/eci/files/2019/06/RenewableEnergyEcon.pdf (accessed on 19 September 2023).
19. Arslan, Ü.; Yildiz, T. Does the Transition to Renewable Energy Increase in European Countries? Available online: https://www.researchsquare.com/article/rs-1408222/v1 (accessed on 30 September 2023).
20. European Commission, The European Green Deal, Brussels, 11.12.2019, COM(2019) 640 Final. Available online: https://eur-lex.europa.eu/legal-content/EN/TXT/?uri=COM%253A2019%253A640%253AFIN (accessed on 12 September 2023).
21. Eurostat, Publications Office of the European Union. Sustainable Development in the European Union. Monitoring Report on Progress towards the SDGs in an EU Context. Luxembourg. 2023. Available online: https://ec.europa.eu/eurostat/web/products-flagship-publications/w/ks-04-23-184 (accessed on 25 September 2023).
22. Directive (EU) 2018/2001 of The European Parliament and of the Council of 11 December 2018 on the Promotion of the Use of Energy from Renewable Sources, Official Journal of the European Union L 328/83. Available online: https://eur-lex.europa.eu/legal-content/EN/TXT/PDF/?uri=CELEX:32018L2001 (accessed on 25 September 2023).
23. Borovik, M.R.; Albers, J.D. Participation in the Illinois solar renewable energy market. *Electr. J.* **2018**, *31*, 33–39. [CrossRef]
24. Keramitsoglou, K.M.; Mellon, R.C.; Tsagkaraki, M.I.; Tsagarakis, K.P. Clean not green: The effective representation of renewable energy. *Renew. Sustain. Energy Rev.* **2016**, *59*, 1332–1337. [CrossRef]
25. Bhowmik, C.; Bhowmik, S.; Ray, A.; Pandey, K.M. Optimal green energy planning for sustainable development: A review. *Renew. Sustain. Energy Rev.* **2017**, *71*, 796–813. [CrossRef]
26. Sharvini, S.R.; Noor, Z.Z.; Chong, C.S.; Stringer, L.C.; Yusuf, R.O. Energy consumption trends and their linkages with renewable energy policies in East and Southeast Asian countries: Challenges and opportunities. *Sustain. Environ. Res.* **2018**, *28*, 257–266. [CrossRef]
27. Ziebell, A.C.; Singh, V.K. Energy indicator in sustainable urban energy metabolism and challenges. *Energy* **2018**, *171*, 26–31. [CrossRef]
28. Wierzbicka, W. Activities Undertaken in the Member Cities of the Polish National Cittaslow Network in the Area of "Energy and Environmental Policy". *Energies* **2022**, *15*, 1309. [CrossRef]
29. Ozdarski, W.T.; Witkowska-Dąbrowska, M.; de Jesus, I.M. Feasibility study of the use of renewable energy sources in households. *Olszt. Econ. J.* **2021**, *16*, 251–262. [CrossRef]
30. Deka, A.; Cavusoglu, B.; Dube, S. Does renewable energy use enhance exchange rate appreciation and stable rate of inflation? *Environ. Sci. Pollut. Res.* **2022**, *29*, 14185–14194. [CrossRef]
31. Filippidis, M.; Tzouvanas, P.; Chatziantoniou, I. Energy poverty through the lens of the energy-environmental Kuznets curve hypothesis. *Energy Econ.* **2021**, *100*, 105328. [CrossRef]
32. Khan, I.; Tan, D.; Azam, W.; Hassan, S.T. Alternate energy sources and environmental quality: The impact of inflation dynamics. *Gondwana Res.* **2022**, *106*, 51–63. [CrossRef]
33. Yıldırım, U.; Kaya, M. Seçilmiş OECD Ülkelerinde Yenilenebilir Enerji Tü-ketiminin Makro Ekonomik Belirleyicileri. *Çankırı Karatekin Üniversitesi İktisadi İdari Bilim. Fakültesi Dergisi* **2021**, *11*, 267–289. [CrossRef]

34. Çakmak, N.; Gözen, M. An analysis of systematic risk factors associated with renewable energy supports mechanism applied in Turkey. *J. Bus. Innov. Gov.* **2021**, *4*, 57–81. Available online: https://dergipark.org.tr/en/pub/jobig/issue/63580/971988 (accessed on 19 September 2023).
35. Zhang, M.; Zhang, S.; Lee, C.C.; Zhou, D. Effects of trade openness on renewable energy consumption in OECD countries: New insights from panel smooth transition regression modelling. *Energy Econ.* **2021**, *104*, 105649. [CrossRef]
36. Lazard's Levelized Cost of Energy Analysis—Version 16.0, April 2023. Available online: https://www.lazard.com/media/typdgxmm/lazards-lcoeplus-april-2023.pdf (accessed on 15 September 2023).
37. Eurostat Database. Available online: https://ec.europa.eu/eurostat/databrowser/view/PRC_HICP_AIND__custom_7648658/default/table (accessed on 29 September 2023).
38. Our World in Data. Available online: https://ourworldindata.org/explorers/energy?tab=table&facet=none&country=CYP~MLT&pickerSort=asc&pickerMetric=entityName&Total+or+Breakdown=Select+a+source&Energy+or+Electricity=Primary+energy&Metric=Share+of+total&Select+a+source=Fossil+fuels (accessed on 29 September 2023).
39. Dudek, A.; Walesiak, M. The Choice of Variable Normalization Method in Cluster Analysis. In *Education Excellence and Innovation Management: A 2025 Vision to Sustain Economic Development during Global Challenges*; Soliman Khalid, S., Ed.; International Business Information Management Association: Seville, Spain, 2020.
40. Milligan, G.W.; Cooper, M.C. A Study of Standardization of Variables in Cluster Analysis. *J. Classif.* **1988**, *5*, 181–204. [CrossRef]
41. Ward, J.H., Jr. Hierarchical Grouping to Optimize an Objective Function. *J. Am. Stat. Assoc.* **1963**, *58*, 236–244. [CrossRef]
42. Szymańska, A. Dochody podatkowe w krajach Unii Europejskiej. *Wiadomości Stat.* **2017**, *62*, 79–99. [CrossRef]
43. Gąsiorowski, P.; Kuszewski, T. Analiza ekonometryczna. In *Decyzje Menedżerskie z Excelem*; Szapiro, T., Ed.; PWE: Warsaw, Poland, 2000.
44. Rószkiewicz, M. *Metody Ilościowe w Badaniach Marketingowych*; Wydawnictwo Naukowe PWN: Warszawa, Poland, 2002.
45. Available online: https://www.mapchart.net/europe.html (accessed on 1 October 2023).
46. Dechezleprêtre, A.; Nachtigall, D.; Venmans, F. The joint impact of the European Union emissions trading system on carbon emissions and economic performance. *J. Environ. Econ. Manag.* **2023**, *118*, 102758. [CrossRef]
47. Moessner, R. Effects of Carbon Pricing on Inflation. *CESifo Work. Pap.* **2022**, *9563*, 1–20. [CrossRef]
48. Deka, A.; Dube, S. Analyzing the causal relationship between exchange rate, renewable energy and inflation of Mexico (1990–2019) with ARDL bounds test approach. *Renew. Energy Focus* **2021**, *37*, 78–83. [CrossRef]
49. Malik, I.A.; Siyal, G.; Abdullah, A.B.; Alam, A.; Zaman, K.; Kyophilavong, P.; Shahbaz, M.; Baloch, S.U.; Shams, T. Turn on the lights: Macroeconomic factors affecting renewable energy in Pakistan. *Renew. Sustain. Energy Rev.* **2014**, *38*, 277–284. [CrossRef]
50. Nga, N.T.V. Monetary policy, exchange rate, renewable energy and economic growth: An empirical analysis of Vietnam. *Accounting* **2021**, *7*, 1315–1324. [CrossRef]
51. Ghaed, E.; Naji Meidani, A.A.; Raji Asadabadi, M. Investigation of the role of renewable and new energies on the inflation rate of Iran. *J. Renew. New Energy* **2021**, *8*, 125–131. Available online: https://www.jrenew.ir/article_111157.html (accessed on 12 September 2023).
52. Ascari, G.; Bonomolo, P.; Hoeberichts, M.; Trezzi, R. The Euro Area Great Inflation Surge. SUERF Policy Brief No. 548. 2023. Available online: https://www.suerf.org/suerf-policy-brief/63635/the-euro-area-great-inflation-surge (accessed on 12 September 2023).
53. Eurostat, Statistics Explained. Available online: https://ec.europa.eu/eurostat/statistics-explained/index.php?title=Archive:EU_energy_mix_and_import_dependency (accessed on 15 November 2023).
54. Jankiewicz, S.; Grądzik, P. Renewable Energy sources as a barier to the EU's common energy policy—On example of Poland and Germany. *Res. Pap. Wrocław Univ. Econ.* **2017**, *46*, 111–118. [CrossRef]
55. Commission Unveils Key Energy Infrastructure Projects to Integrate Europe's Energy Markets and Diversify Sources, European Commission. Available online: https://ec.europa.eu/energy/en/news/commission-unveils-list-195-key-energy-infrastructure-projects (accessed on 21 September 2023).
56. Khan, I.; Zakari, A.; Dagar, V.; Singh, S. World energy trilemma and transformative energy developments as determinants of economic growth amid environmental sustainability. *Energy Econ.* **2022**, *108*, 105884. [CrossRef]
57. Khan, I.; Zakari, A.; Zhang, J.; Dagar, V.; Singh, S. A study of trilemma energy balance, clean energy transitions, and economic expansion in the midst of environmental sustainability: New insights from three trilemma leadership. *Energy* **2022**, *248*, 123619. [CrossRef]

energies

Article

Modeling the Consumption of Main Fossil Fuels in Greenhouse Gas Emissions in European Countries, Considering Gross Domestic Product and Population

Alicja Kolasa-Więcek [1,*], Agnieszka A. Pilarska [2,*], Małgorzata Wzorek [3], Dariusz Suszanowicz [1] and Piotr Boniecki [4]

[1] Institute of Environmental Engineering and Biotechnology, Faculty of Natural Sciences and Technology, University of Opole, St. Kominka 6, 46-020 Opole, Poland; d.suszanowicz@uni.opole.pl

[2] Department of Hydraulic and Sanitary Engineering, Poznań University of Life Sciences, St. Piątkowska 94A, 60-649 Poznan, Poland

[3] Department of Process and Environmental Engineering, Opole University of Technology, St. Mikołajczyka 5, 46-271 Opole, Poland; m.wzorek@po.opole.pl

[4] Department of Biosystems Engineering, Poznań University of Life Sciences, Wojska Polskiego 50, 60-627 Poznan, Poland; bonie@up.poznan.pl

* Correspondence: akolasa@uni.opole.pl (A.K.-W.); pilarska@up.poznan.pl (A.A.P.)

Abstract: Poland ranks among the leading European countries in terms of greenhouse gas (GHG) emissions. Many European countries have higher emissions per capita than the EU average. This research aimed to quantify the complex relationships between the consumption variables of the main fossil fuels, accounting for economic indicators such as population and gross domestic product (GDP) in relation to GHG emissions. This research attempted to find similarities in the group of 16 analyzed European countries. The hypothesis of an inverted U-shaped environmental Kuznets curve (EKC) was tested. The resulting multiple regression models showed similarities in one group of countries, namely Poland, Germany, the Czech Republic, Austria and Slovakia, in which most of the variables related to the consumption of fossil fuels, including HC and BC simultaneously, are statistically significant. The HC variable is also significant in Denmark, Estonia, the Netherlands, Finland and Bulgaria, and BC is also significant in Lithuania, Greece and Belgium. Moreover, results from Ireland, the Netherlands, and Belgium indicate a negative impact of population on GHG emissions, and in the case of Germany, the hypothesis of an environmental Kuznets curve can be accepted.

Keywords: GHG emission; fossil fuel; GDP; environmental Kuznets curve; multiple regression; EU; modeling

1. Introduction and Literature Review

Economic progress, increasing consumption and the unsustainable exploitation of the natural environment contribute to the emission of many dangerous pollutants. Changes occurring in the environment affect entire ecosystems, from changes in biodiversity [1]; the extinction of plants [2] and animal species [3]; the poisoning of water, soil and air [4]; and threats to human health and life [5,6]. Broadly understood sustainable development and the rational management of natural resources are particularly important today. The emission of greenhouse gases as a result of human activities increases the greenhouse effect. The rising atmospheric temperature and global warming have been attracting the attention of scientists and researchers in research centers around the world for many years [7]. A variety of modern methods and tools are widely used to assess [8,9], prevent [10], reduce [11,12] and predict the volume of GHG emissions [13,14]. The main sources of greenhouse gases are the combustion of fossil fuels for energy generation and transportation, industry, deforestation, agriculture, water bodies (wetlands) and urban activities [15]. A particularly important source of GHG emissions is the use of fossil fuels by the energy sector.

In some European countries, the consumption of fossil fuels is still high (Figure 1).

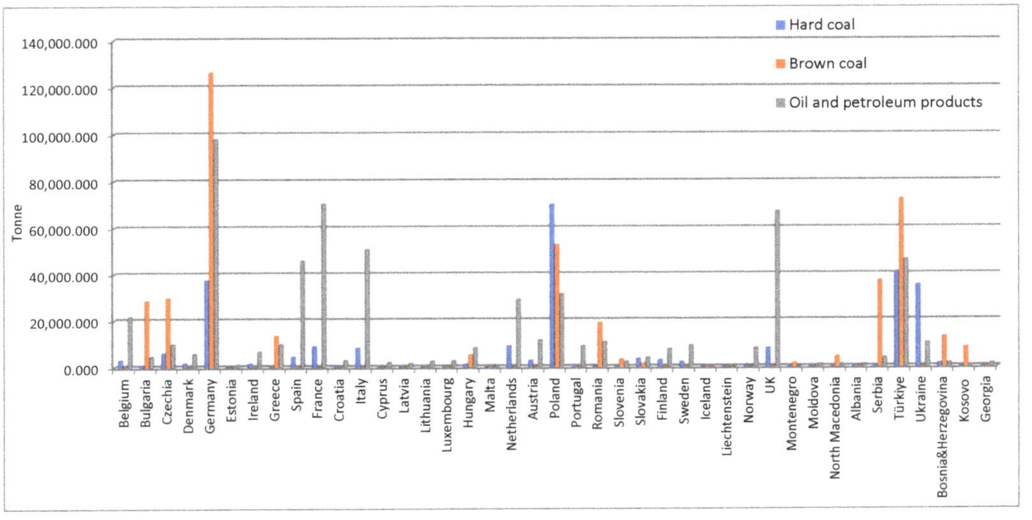

Figure 1. Consumption of basic fossil fuels in geopolitical terms in Europe in 2021 [16].

It is estimated that energy accounts for more than three-quarters of total GHG emissions globally [17]. This problem concerns both developing countries [18–20] and developed, modern economies [21]. The European Union (EU) is a pioneer in the fight against pollution and climate change, having established the United Nations Framework Convention on Climate Change and introducing a range of pollution and energy measures to reduce greenhouse gas emissions. Air pollution remains a major challenge for most EU countries, including those located in Central and Eastern Europe [22]. Many scientific centers undertake research on processes and issues associated with emissions from the energy sector in order to reduce GHG emissions. In the research conducted by Wu, J., et al. [23], the authors analyze the marginal impact of energy price and economic freedom on Europe's CO_2 emissions using the extended Stochastic Impacts by Regression on Population, Affluence and Technology ecology model together with the spatial econometric models. In another study, researchers analyzed a set of selected German and French cities in terms of their relative ecological efficiency, calculated as the ratio of their GDP to CO_2 emissions [24]. The study conducted by Wenlong, Z., et al. [25] attempts to empirically analyze the environmental impact of certain important factors, e.g., energy efficiency, technological innovation, trade openness and institutional quality, in 10 Asian economies over the period 1995–2018. The researchers Forsius, M., et al. [26] integrated the results of three spatially distributed modeling systems (FRES, PREBAS, Zonation) to assess the potential to achieve carbon neutrality by 2035 in Finland at both national and regional scales, by simultaneously considering the conservation goals included in the EU biodiversity strategy. In another study, researchers examine how population growth, wealth and energy technology factors influence CO_2, CH_4 and NO_2 emissions from the power sector in Bangladesh. This study applied extended stochastic effects through regression to a population, wealth and technology model using ridge regression for the analysis. The study also examined the presence of an environmental Kuznets curve (EKC) for greenhouse gas emissions [27]. The research carried out by Fatima, T., et al. [28] aimed to determine the main factors influencing CO_2 emissions in Chinese industry in the period 1991–2016. Based on the logarithmic Divisa index (LMDI) method, the change in industry-related CO_2 emissions was decomposed into the energy structure effect, income effect, energy intensity effect, carbon dioxide emissions effect and labor effect. In the study [29], the authors discuss activities promoting energy efficiency implemented in the socio-economic sectors: electricity, construction, logistics and transport, along with their impact on the carbon footprint. Using the Crystal Ball simulation model, [30] estimates the cumulative increase (35%) in CO_2 emissions over ten years without nuclear power in Ghana's energy mix. After including nuclear power in the country's energy mix, the model estimates a reduction in CO_2 emissions of 12.5% over the same period and 18.2% over 20 years. Of the 21 EU countries that reduced their emissions, only 6 also decreased their GDP (Czech Republic, Estonia, Lithuania, Luxembourg, Hungary and Poland), which means 15 EU countries (Portugal, Croatia, Belgium, Malta, France, Spain, the Netherlands, Germany, Austria, Romania, Italy, Cyprus, Greece, Slovenia and Bulgaria) managed to reduce emissions while increasing their GDP [31].

The availability of modern computational techniques and advanced statistical methods allows the study of correlations on a scale that previously seemed impossible. Ecological Kuznets curves have also been subject to countless verifications. Scientists have analyzed greenhouse gas emission factors and the impact of urbanization on environmental degradation and tested the U-Kuznets curve hypothesis in various configurations. All possible measures of environmental devastation are correlated with GDP per capita indicators. Indeed, it is often possible to match the observed patterns with an inverted letter "U" [32–34]. An example confirming the existence of a correlation is the study of the relationship between income and CO_2 emissions for 26 high-income OECD countries and 52 emerging countries [35]. The results of research conducted for 17 OECD countries confirm the EKC hypothesis and indicate that GDP per capita and GDP per capita squared have a positive and negative impact on CO_2 emissions, respectively, and that renewable energy consumption has a negative impact on CO_2 emissions [36]. The results of research on Croatia also prove the existence of the U-Kuznets curve hypothesis between CO_2 emissions and economic growth in the long run [23]. In other studies, using Turkey as an example, the overall results showed that the level of GDP per capita that could reduce environmental pollution has not been achieved, and the use of renewable energy is not a solution to reducing CO_2 emissions [37]. Another study in Kenya found that fossil fuel energy use, GDP, urbanization and openness to trade increase air pollution in the long- and short-term. However, renewable energy consumption mitigates air pollution in the long- and-short term. Moreover, financial development also reduces air pollution, but only in the long run [38]. There are also critical voices and some studies that question the existence of the effect postulated by the environmental Kuznets curve [39]. The data generally do not support the concept of classical or environmental Kuznets curves, which are the inverse of the U. Rather, they emphasize that there is a trade-off between environment and equity [40].

Compared to European countries, Poland ranks highly in terms of GHG emissions (Figure 2). Countries such as Germany, the UK and Poland, which occupy the leading positions in terms of GHG emissions (Figure 2), are characterized by the highest consumption of hard coal and/or brown coal in the EU.

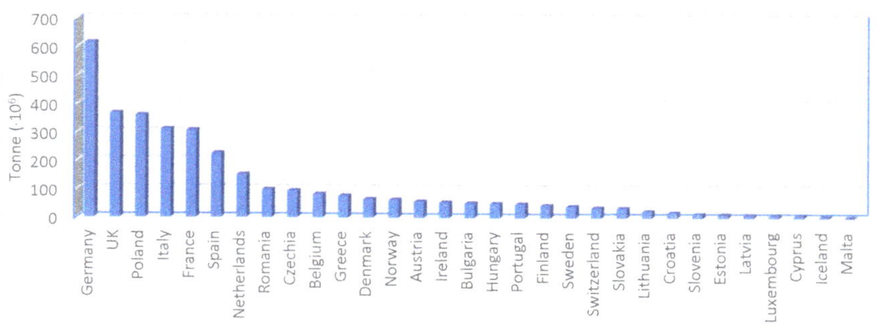

Figure 2. Annual greenhouse gases in CO_2 equivalent 2021 [16,41].

Additional information about the scale of emissions in European countries is provided by data on GHG emissions per capita (Figure 3).

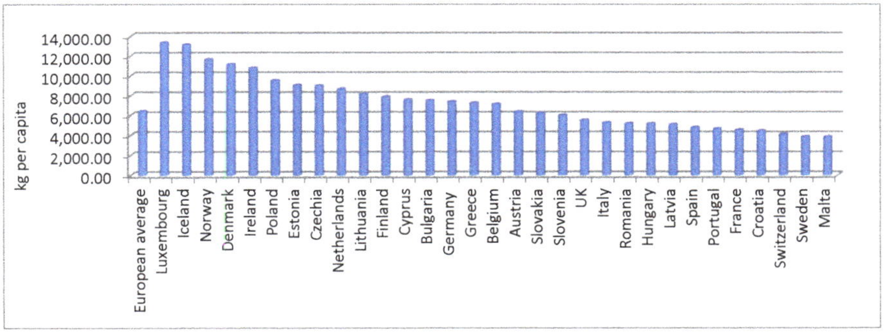

Figure 3. GHG emissions in kg per capita in Europe 2021 [16,41].

Many European countries have emissions that are higher than the EU average. The highest in 2021 were recorded in Luxembourg, Iceland, Norway, Denmark and Ireland. These are highly industrialized countries and high consumerism is probably reflected in GHG emissions per capita.

In Poland, the main sources of air pollution are emissions from fuel combustion, mainly from the energy industry and individual households and transportation [42]. Poland is among the leading countries in Europe in geopolitical terms with the highest inland consumption of hard coal in 2021, with consumption totaling 69,981,742 thousand tonnes, followed by Turkey at 40,926,606 thousand tonnes, Germany at 37,372,647 thousand tons and Ukraine at 35,110.9 thousand tonnes. In the case of the inland consumption of brown coal, Germany dominates with a consumption of 126,182,054 thousand tonnes followed by Turkey with 72,255,627 and Poland with 52,614,807 [16].

The research aimed to quantify the complex relationships between the consumption variables of the main fossil fuels (hard coal, brown coal, crude oil, natural gas), taking into account economic indicators such as population and GDP, in relation to the volume of GHG emissions. The study attempts to group the 16 European countries considered in terms of similarities in the importance of the studied variables. The research was conducted using multiple regression analysis. Based on the obtained models, simulations and forecasts of GHG emissions can be carried out. Most countries achieve reductions in CO_2 emissions, with most maintaining or increasing GDP adequately. Another novelty of the study is obtaining information on the correlation of GDP with GHG emissions in the group of European countries examined in terms of the existence of the environmental Kuznets curve hypothesis. The study of the Kuznets curve aims to confirm the existence of a negative correlation of the above-mentioned variables, while also answering whether the theory of environmental decoupling takes place in EU countries.

2. Data Source and Methodological Framework

The input material was data from international public statistics databases of the Eurostat and OECD [16,41]. The selection of data for research is not accidental. The main source of CO^2 emissions is the burning of fossil fuels, which is still practiced in many European countries. For this reason, the group of explanatory variables includes the consumption of fossil fuels in the analyzed EU countries. However, in order to examine the existence of the environmental Kuznets curve, i.e., to determine whether GHG reduction is negatively correlated with the level of GDP, data on GDP and population size were taken into account.

The research was conducted based on data from the period 1995–2021. The group of explanatory (independent) variables included data on the consumption of the main fossil fuels (hard coal, oil and petroleum products, natural gas, brown coal, gas) and economic data (i.e., gross domestic product and population). The explained (dependent) variable is GHG emissions (GHG). The research group consisted of 16 EU countries where GHG emissions per capita are higher than the EU average: Luxembourg, Denmark, Ireland, Poland, Estonia, Czech Republic, the Netherlands, Lithuania, Finland, Cyprus, Bulgaria, Germany, Greece, Belgium, Austria and Slovakia (Figure 4). This research was conducted using the Statistica v. 13 statistical package.

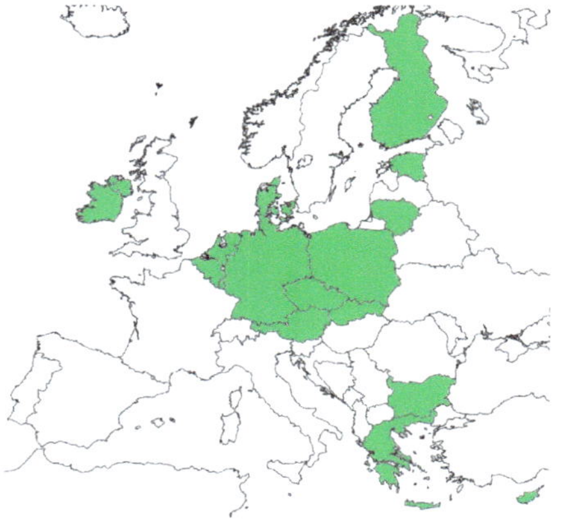

Figure 4. Countries included in the study.

Descriptive statistics of the data set for analyzed European countries are presented in Table 1.

Table 1. Statistical description of the data.

Variable	Mean	St. Dev.	Min.	Max.
Luxemburg				
OPP *	2617.1	392.7	1775.0	3104.5
HC *	117.1	50.0	57.2	242.0
NG *	990.7	259.3	636.0	1402.8
GDP **	32,191.6	12,517.9	13,123.3	55,797.4
Pop ***	505,154.2	71,780.6	408,625.0	640,064.0
GHG ****	10,828.3	1227.4	8590.9	12,993.8
Denmark				
OPP	7595.7	1036.5	5727.8	9445.0
HC	6326.1	3228.1	1237.1	15,009.0
NG	4146.4	813.4	2827.8	5150.0
GDP	168,566.2	42,782.3	99,481.0	256,566.4
Pop	5,516,857.7	187,415.6	5,227,861.0	5,850,189.0
GHG	65,647.1	12,948.4	44,483.8	92,928.7
Ireland				
OPP	7287.9	857.1	5529.0	8486.0
HC	2282.6	655.6	588.2	3002.0
NG	4513.7	747.3	2875.0	5501.8
GDP	163,753.6	79,151.1	57,259.9	362,050.6
Pop	4,320,267.7	455,518.8	3,601,300.0	5,011,460.0
GHG	64,917.7	4360.6	58,852.5	71,814.5

Table 1. *Cont.*

Variable	Mean	St. Dev.	Min.	Max.
Poland				
OPP	22,409.6	4968.4	13,449.0	31,345.3
HC	82,553.3	11,075.0	63,475.6	110,131.0
BC *	59,827.1	4321.0	46,107.0	65,934.0
NG	16,683.4	3099.5	12,196.0	23,541.9
GDP	550,045.3	207,974.3	248,423.3	956,864.9
Pop	38,310,532.6	135,324.3	38,115,909.0	38,533,789.0
GHG	404,308.5	20430.1	371,312.4	460,383.5
Estonia				
OPP	1135.7	83.2	893.0	1309.0
HC	68.7	31.7	3.5	130.0
NG	703.0	175.6	420.7	992.0
GDP	21,109.3	8882.8	7796.6	38,356.9
Pop	1,351,716.3	34,022.8	1,314,545.0	1,436,634.0
GHG	18,818.5	2570.8	11,407.1	22,046.4
Czech Republic				
OPP	8944.9	672.0	7799.0	9913.0
HC	8985.7	2260.0	5288.9	14,180.0
BC	43,757.8	6204.3	29,675.2	52,723.0
NG	8940.5	634.5	7522.0	9895.0
GDP	208,737.0	61,304.4	119,877.0	317,766.8
Pop	10,412,272.3	153,547.5	10,200,774.0	10,700,155.0
GHG	140,656.1	12,640.1	113,072.1	161,709.7
The Netherlands				
OPP	31,706.1	2087.1	27,475.6	34,899.2
HC	13,006.5	2126.0	6492.9	17,977.0
BC	26.9	12.7	6.0	56.6
NG	47,421.9	3711.1	40,073.1	55,951.3
GDP	530,235.0	118,950.4	303,567.7	749,498.1
Pop	16,487,267.3	583,278.3	15,459,004.0	17,533,048.0
GHG	206,412.1	19,310.8	164,367.8	241,662.1
Lithuania				
OPP	2240.2	386.7	1696.0	3213.0
HC	175.2	126.0	1.0	372.0
BC	96.6	126.0	0.0	392.0
NG	2664.4	386.9	2137.0	3553.0
GDP	45,773.6	18,389.4	18,025.7	81,312.8
Pop	3,187,915.6	286,946.3	2,794,137.0	3,629,102.0
GHG	21,277.2	1461.3	19,493.7	24,730.0
Finland				
OPP	9270.3	733.0	7649.0	10,509.0
HC	5574.1	1555.2	2726.0	8862.0
NG	3779.8	855.5	2353.0	5036.0
GDP	145,954.8	32,807.1	83,792.3	201,196.3
Pop	5,326,196.2	144,144.4	5107,787.0	5,541,020.0
GHG	67,762.9	10,595.3	47,756.3	85,490.4
Cyprus				
OPP	2335.1	243.9	1893.0	2822.0
HC	29.1	20.2	0.0	66.4
BC	0.4	0.5	0.0	1.0
GDP	17,780.3	5051.8	9319.5	26,501.9
Pop	781,973.4	81,219.5	650,860.0	900,357.0
GHG	8640.1	748.3	6972.2	10,025.4

Table 1. *Cont.*

Variable	Mean	St. Dev.	Min.	Max.
Bulgaria				
OPP	4217.1	471.9	3308.0	5233.0
HC	2859.3	1347.3	828.0	4916.0
BC	28,976.3	3304.8	22,283.0	36,996.0
NG	3456.7	752.9	2645.0	5794.0
GDP	76,051.6	25,793.3	39,843.9	128,169.2
Pop	7,575,279.5	470,148.8	6,877,742.5	8,406,067.0
GHG	60,642.2	5608.4	47,984.6	72,615.2
Germany				
OPP	116,180.3	9687.8	97,879.0	132,781.0
HC	61,297.3	10,534.1	31,197.7	75,362.0
BC	170,692.3	18,930.8	107,429.6	194,811.0
NG	91,270.0	5193.7	76,897.0	97,389.0
GDP	2,370,100.6	497,805.5	1,612,022.5	3,256,792.8
Pop	82,079,774.9	749,883.4	80,274,981.0	83,196,076.0
GHG	957,161.4	103,495.8	730,922.7	1137,869.0
Greece				
OPP	13,489.6	2451.1	9168.6	16,754.0
HC	677.1	400.7	265.6	1480.0
BC	54,110.4	16,202.7	13,213.7	70,855.0
NG	3199.6	1690.3	36.0	6446.4
GDP	206,312.7	32,851.6	136,520.1	265,211.8
Pop	10,859,677.0	168,322.1	10,562,160.0	11,121,344.0
GHG				
Belgium				
OPP	22,195.8	838.4	19,987.5	23,621.6
HC	6804.3	2909.0	2863.3	11,859.0
BC	189.2	219.9	0.0	730.0
NG	17,146.2	1672.7	12,792.0	20,476.0
GDP	309,968.5	74,820.6	190,570.7	452,167.7
Pop	10,770,320.6	478,660.1	10,136,814.0	11,552,615.0
GHG	113,438.3	17,955.9	75,464.5	136,748.1
Austria				
OPP	12,052.7	768.0	10,383.0	13,522.5
HC	3652.3	428.3	2749.9	4328.0
BC	659.3	676.9	59.6	1743.0
NG	8843.3	547.7	7844.3	9921.8
GDP	256,426.9	60431.8	157,751.2	355,971.0
Pop	8,357,953.1	322485.5	7,948,278.0	8,951,520.0
GHG	82,768.9	4683.3	73,910.8	92,588.6
Slovakia				
OPP	3526.1	265.7	3066.0	4045.0
HC	4359.3	676.4	2594.0	5330.0
BC	3560.0	1512.2	1306.0	7221.0
NG	6052.2	959.5	4535.0	7633.0
GDP	85,709.0	28,715.5	39,059.4	124,676.7
Pop	5,407,179.9	25,951.2	5,363,676.0	5,458,827.0
GHG	46,566.7	4807.4	37,187.9	53,180.0

* Thousand tonnes, ** Euro per capita, *** number, **** Thousands tonnes of CO_2 equivalent.

Linear multiple regression was used for modeling, because it allows for the searching and quantitative description of complex relationships. The built multiple regression model allows for the examination of the influence of many independent variables ($X_1, X_2, \ldots, X_k$) on one dependent variable (Y). The most frequently used type of multiple regression is linear regression. It is an extension of linear regression models based on Pearson's linear correlation coefficient. It assumes a linear relationship between the studied variables. The linear multiple regression model takes the following form:

$$Y = \alpha_0 + \alpha_1 X_1 + \alpha_2 X_2 + \ldots + \alpha_n X_n + \epsilon \quad (1)$$

where:

Y—dependent variable;
$X_1, X_2, \ldots, X_n$—independent variables;
$\alpha_1, \alpha_2, \ldots, \alpha_n$—parameters;
ϵ—random component (the rest of the model).

Multiple regression analysis allowed us to test complex models that examine how multiple explanatory variables predict the values of a single variable. The backward stepwise regression analysis method was used in this research. This method involves, in its first step, the construction of a model that includes all potential explanatory variables, and then gradually eliminates them in order to maintain the model with the highest determination coefficient, while maintaining the significance of the parameters. For example, in Denmark, where Pearson's linear correlation coefficient showed the statistical significance of five independent variables (Table 2) as a result of the stepwise regression analysis, three variables are ultimately included in the regression model (the table in Section 3). Finally, the accuracy of the obtained models is confirmed based on the values of the regression coefficients (R, adj. R^2) and other statistics (F, p). Based on the obtained model, a forecast of the dependent variable—GHG emissions—is made.

Table 2. Pearson's correlation coefficients between studied variables and the dependent variable, GHG emissions.

	Country	OPP	HC	NG	BC	GDP	Pop
1	Luxemburg	**0.707**	−0.165	**0.919**	-	0.048	−0.08
2	Denmark	**0.934**	**0.968**	**0.775**	-	**−0.948**	−0.963
3	Ireland	**0.876**	**0.585**	−0.17	-	−0.371	−0.484
4	Poland	**−0.483**	**0.886**	**−0.507**	**0.424**	**−0.544**	−0.178
5	Estonia	0.343	**0.525**	0.385	-	−0.327	0.068
6	Czech Republic	**−0.412**	**0.912**	**0.451**	**0.979**	**−0.894**	−0.851
7	The Netherlands	−0.039	**0.452**	**0.742**	−0.077	**−0.944**	−0.963
8	Lithuania	**0.455**	−0.172	**0.411**	**0.437**	**−0.439**	0.524
9	Finland	**0.752**	**0.922**	**0.891**	-	**−0.733**	−0.85
10	Cyprus	**0.877**	0.291	-	**0.63**	**0.629**	0.424
11	Bulgaria	**0.536**	**0.667**	**0.778**	0.332	**−0.597**	0.69
12	Germany	**0.941**	**0.927**	0.205	**0.734**	**−0.955**	−0.205
13	Greece	**0.462**	-	**−0.514**	**0.952**	0.161	0.952
14	Belgium	0.214	**0.946**	**−0.566**	**0.576**	**−0.948**	−0.98
15	Austria	**0.831**	**0.877**	**0.481**	**0.438**	**−0.42**	−0.495
16	Slovakia	**−0.305**	**0.945**	**0.917**	**0.834**	**−0.922**	−0.897

Explanation: hard coal—HC; oil and petroleum products—OPP; natural gas—NG; brown coal—BC; gross domestic product—GDP; population—Pop.

3. Results and Discussion

Table 2 contains the results of Pearson's linear correlation with respect to the studied variables. Empty fields in the table ("-") mean that the specified source is not used by the country. The variables statistically significant are for $p < 0.05$ are indicated in bold and these variables were used in further analysis.

Table 3 presents the results of the multiple linear regression analysis. The values of the determined R^2 coefficient refer to the model fitted to the data. Introducing too many variables may result in the model over-fitting the data. Therefore, in addition to the adjusted R^2 coefficient, the result of Fisher's test (F), which is particularly important when comparing models, can also be used. The higher the F test value, the better the fit of the model. The values of the F statistics and the corresponding level of test probability p indicate a significant linear relationship (exception: Estonia).

Table 3. The results of multiple linear regression.

	Country	R	R^2	Adjusted R^2	F (Fischer)	Significance Level p
1	Luxemburg	0.930	0.866	0.855	77.390	0.000
2	Denmark	0.998	0.995	0.994	901.012	0.000
3	Ireland	0.956	0.914	0.903	81.847	0.000
4	Poland	0.986	0.971	0.964	141.782	0.000
5	Estonia	0.529	0.280	0.220	4.668	0.019
6	Czech Republic	0.997	0.995	0.993	629.450	0.000
7	The Netherlands	0.994	0.988	0.986	458.309	0.000
8	Lithuania	0.973	0.947	0.935	75.611	0.000
9	Finland	0.995	0.989	0.987	388.783	0.000
10	Cyprus	0.974	0.949	0.940	102.424	0.000
11	Bulgaria	0.909	0.827	0.785	20.042	0.000
12	Germany	0.994	0.988	0.985	442.695	0.000
13	Greece	0.995	0.990	0.987	405.966	0.000
14	Belgium	0.991	0.982	0.978	227.175	0.000
15	Austria	0.985	0.970	0.961	108.503	0.000
16	Slovakia	0.994	0.987	0.984	328.888	0.000

A rejection of statistically insignificant variables results in a higher coefficient of determination (Adj. R^2). The coefficients of determination indicate a compatibility model fits to the actual data and a proportion of the total variability of the dependent variable is explained by the resulting model. The value of adj. R^2 is usually smaller than the previous R^2. The coefficients of determination in most cases are very high (Adj. $R^2 > 0.9$ for Denmark, Ireland, Poland, Czech Republic, the Netherlands, Lithuania, Finland, Cyprus, Germany, Greece, Belgium, Austria, abd Slovakia) or high (Adj. $R^2 > 0.8$ for Luxembourg) and explain the variance of the GHG emission variable. High values of model fit coefficients to real data seem to be justified, because fuel fossil combustion is the main source of CO_2 emissions. The difference in individual countries lies in the type of fuels. For example, the resulting model for Poland explains over 96% of the variability of GHG emissions. The remaining 4% can be explained by other variables not included in this study. The obtained results indicate the highest proportion of HC. Attention should be paid to the case of Estonia, where the obtained parameter values Adj. $R^2 = 0.22$, F test = 4.67 and $p = 0.019$ indicate that there is no linear relationship between the analyzed variables. It should be assumed that other, perhaps non-linear methods would give a more satisfactory result. In the case of Bulgaria, due to the low value of the F-test result, it can also be assumed that a more satisfactory result could have been obtained using another method.

Table 4 summarizes the obtained equations for GHG emissions in individual countries.

Table 4. The regression equations for GHG emissions.

	Country	Regression Equation
1	Luxemburg	Y = 0.8NG + 5519.8
2	Denmark	Y = 0.23OPP + 0.18NG + 0.58HC
3	Ireland	Y = 0.8OPP − 0.23Pop + 4225.7
4	Poland	Y = 0.16OPP + 1.42HC + 0.5NG + 0.14BC + 0.48GDP + 50,708.8
5	Estonia	Y = 0.47HC + 15,309.4
6	Czech Republic	Y = 0.12OPP + 0.35HC + 0.09NG + 0.64BC + 860,627
7	The Netherlands	Y = −0.54Pop + 0.25HC + 0.33NG + 400,730
8	Lithuania	Y = 0.97OPP + 0.65NG + 0.39BC

Table 4. *Cont.*

	Country	Regression Equation
9	Finland	Y = 0.44Pop + 0.01OPP + 0.48HC + 0.22NG + 188,493.2
10	Cyprus	Y = 0.46Pop + 0.61OPP + 0.88GDP + 5218
11	Bulgaria	Y = 0.53HC + 1.07NG
12	Germany	Y = 0.06OPP + 0.08HC + 0.13BC − 0.34GDP +402,191.8
13	Greece	Y = 0.48OPP + 0.16NG + 0.6BC
14	Belgium	Y = −1.69Pop + 0.14NG + 0.24BC + 0.78DGP + 623,443
15	Austria	Y = 0.4OPP + 0.41HC + 0.26NG + 0.28BC
16	Slovakia	Y = 0.2Pop + 0.59HC + 0.47NG + 0.4BC – 190,009

The obtained results were interpreted for Poland as an example:

- When the consumption of OPP increases by 1 thousand tonnes, GHG emissions will increase by 0.16 thousand tonnes of CO_2 equivalent, if other variables remain unchanged;
- When HC consumption increases by 1 thousand tonnes, GHG emissions will increase by 1.42 thousand tonnes of CO_2 equivalent, if other variables remain unchanged;
- When NG consumption increases by 1 thousand tonnes, GHG emissions will increase by 0.5 thousand tonnes of CO_2 equivalent, if other variables remain unchanged;
- When BC consumption increases by 1 thousand tonnes, GHG emissions will increase by 0.14 thousand tonnes of CO_2 equivalent, if other variables remain unchanged;
- When GDP increases by 1 thousand tonnes, GHG emissions will increase by 0.48 thousand tonnes of CO_2 equivalent, if other variables remain unchanged.

Most variables related to the consumption of fossil fuels were statistically significant in the following countries: Poland, Czech Republic, Finland, Germany, Austria and Slovakia, including HC and BC in most of the countries in the above group except Finland. The HC variable is also significant in Denmark, the Netherlands, Finland and Bulgaria, and BC in Lithuania, Greece and Belgium. The results obtained from the regression analysis can be largely justified by observing the per capita consumption of hard coal and lignite in the studied group of countries (Figure 5).

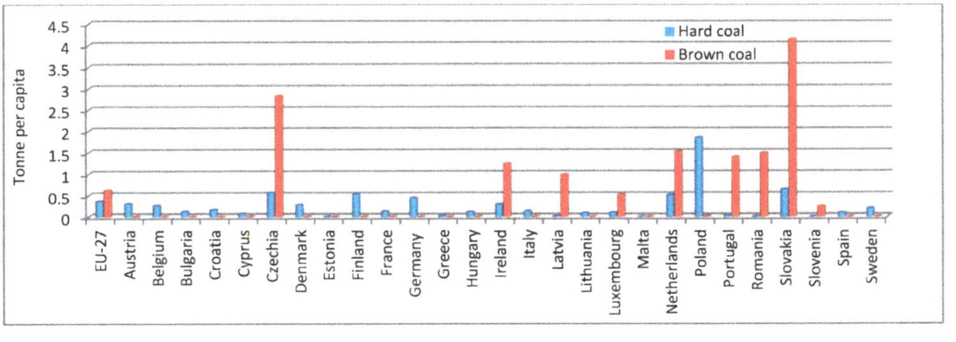

Figure 5. Inland consumption of hard coal and brown coal per capita in 2021; authors' own calculations based on [16,41].

Different characteristics were observed in Ireland, the Netherlands and Belgium, where the variable population negatively correlates with GHG emissions. Other studies also indicate that the impact of population growth on emissions is negligible for certain old EU Member States [43]. The results also showed that in Germany, the GDP variable correlates negatively with GHG emissions, which means that as GDP increases, GHG emissions decrease. On this basis, the hypothesis of an inverted U-shaped environmental Kuznets curve in Germany can be accepted.

In several analyzed countries (Finland, Czech Republic, Germany, the Netherlands, Greece, Belgium) a noticeable trend towards systematic reduction in GHG emissions can be observed (Figure 6). Concurrently, an increase in emissions in 2021 was observed in each of the analyzed countries. The analysis of the data indicates that in all countries in 2021, compared to previous years, there was an increase in the consumption of one or several fossil fuels (for example, the consumption of hard coal increased in Denmark, Luxemburg, Ireland, Poland, the Netherlands, Czech Republic, Lithuania, Finland, Cyprus, Bulgaria, Germany, Greece, Belgium and Slovakia). Generally, GHG emission reductions in EU member states result from the countries' obligations towards EU policy, which involves, among others, the phasing out of fossil fuels, trading in CO_2 emission allowances and promoting and implementing renewable energy sources.

Based on the generated model, a forecast of GHG emissions was made and compared with actual values (Figure 6). In most cases, the obtained models indicate a very good fit to real values (except Estonia). In the case of Bulgaria, although the model deviates slightly from the actual values, it can generally be assumed that it averages and follows the observed tendency.

In that countries, it is highly probable that some explanatory variables that were not included in this study may have a significant impact on GHG emissions. It should also be assumed that there is a relationship between the studied variables and GHG emissions, but it is a non-linear relationship; therefore, the use of multiple linear regression analysis in this case did not provide satisfactory results. Perhaps another reason in the case of Estonia is the impact of the hard coal consumption variable, where a very significant decrease was recorded in the analyzed period of 1995–2021 (Table 5).

Table 5. HC consumption in Estonia.

Year	HC Consump. in Thousand Tonnes
1995	103,000
1996	115,000
1997	98,000
1998	75,000
1999	79,000
2000	87,000
2001	109,000
2002	61,000
2003	44,000
2004	58,000
2005	56,000
2006	70,000
2007	130,000
2008	129,000
2009	87,000
2010	60,000
2011	70,000
2012	64,000
2013	61,000
2014	78,000
2015	29,000
2016	27,000
2017	48,000
2018	50,000
2019	50,345
2020	12,280
2021	3500

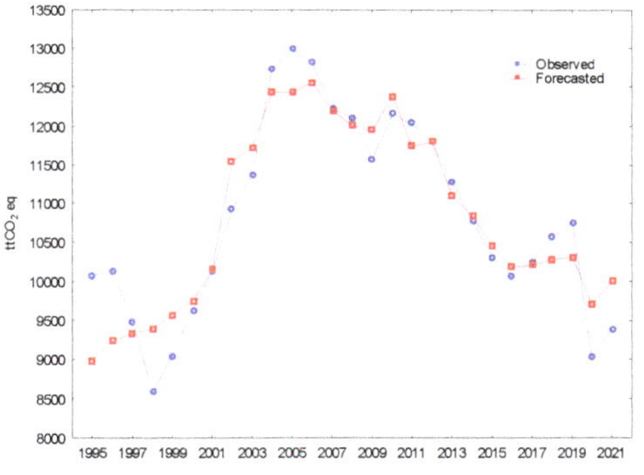

(**a**) Luxemburg

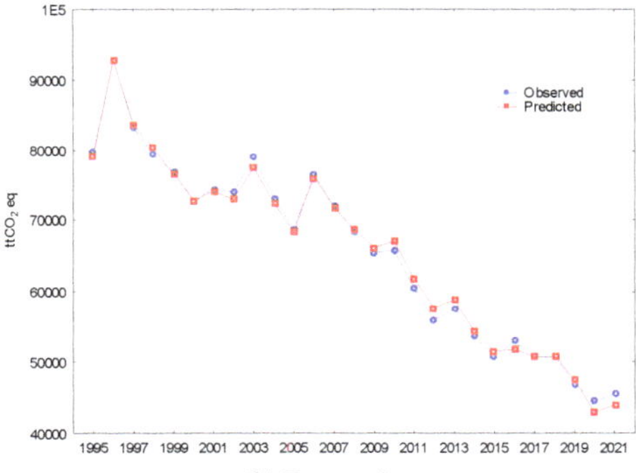

(**b**) Denmark

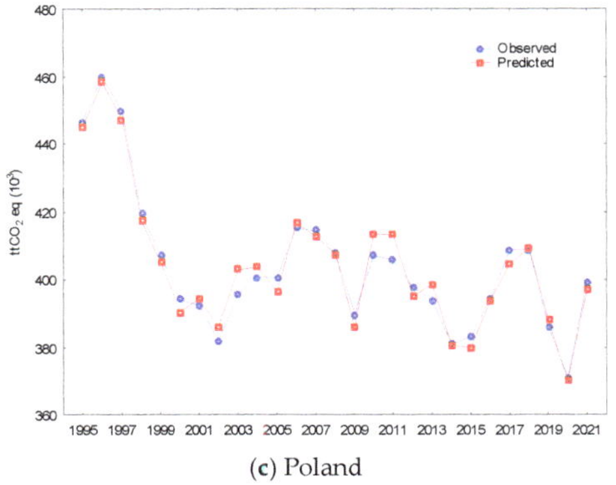

(**c**) Poland

Figure 6. *Cont.*

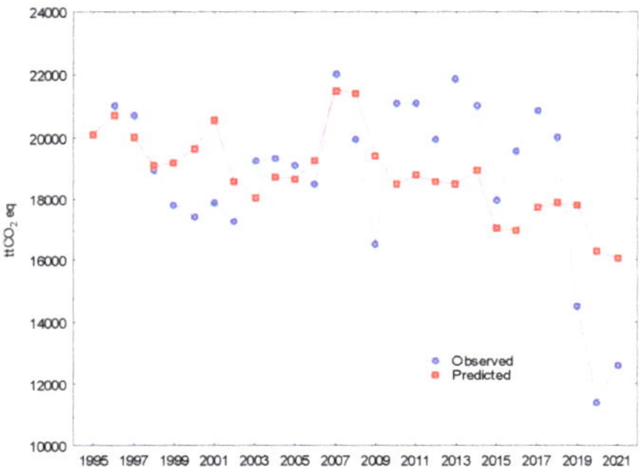

(**d**) Estonia

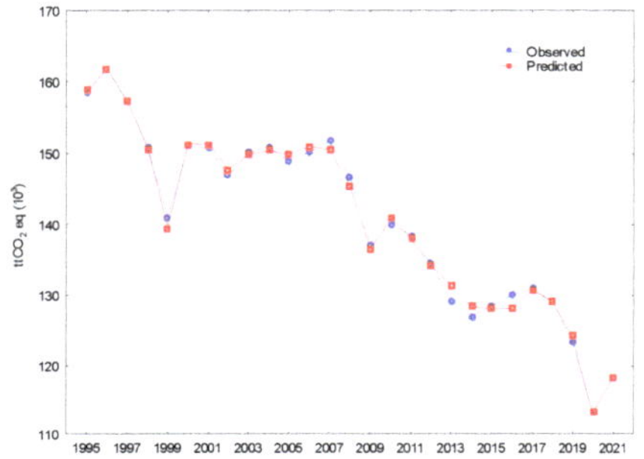

(**e**) Czech Republic

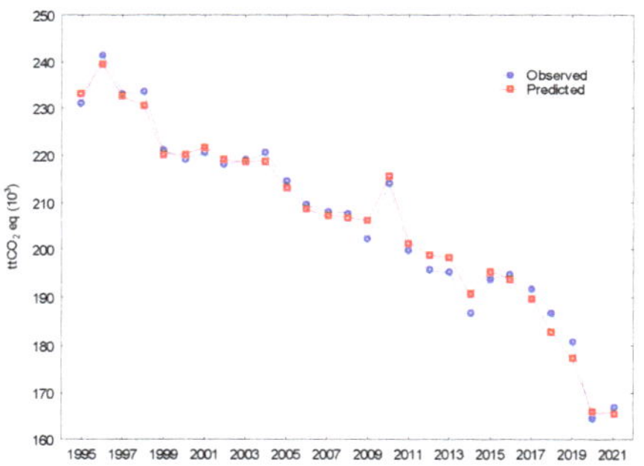

(**f**) The Netherlands

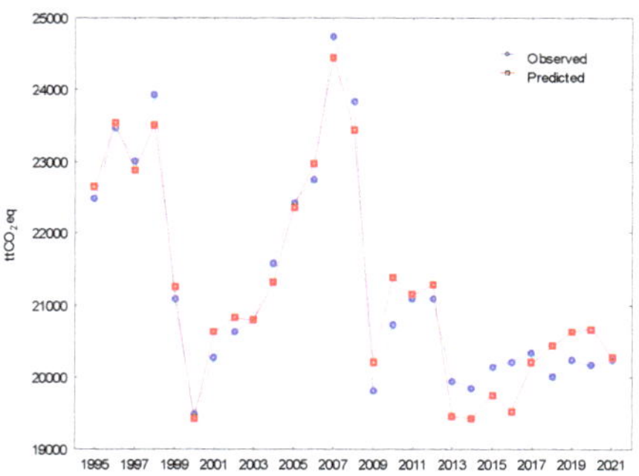

(**g**) Lithuania

Figure 6. *Cont.*

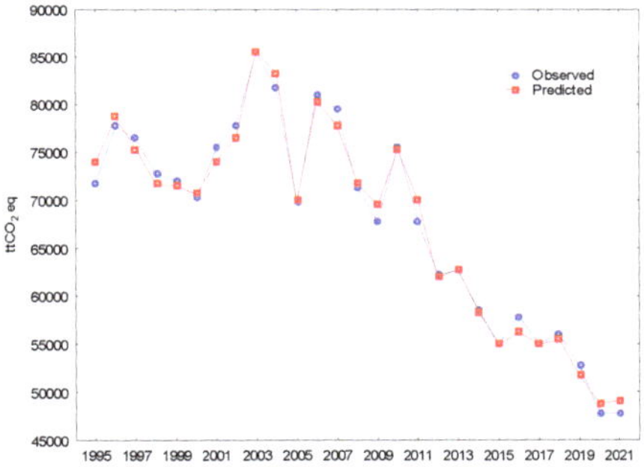

(**h**) Finland

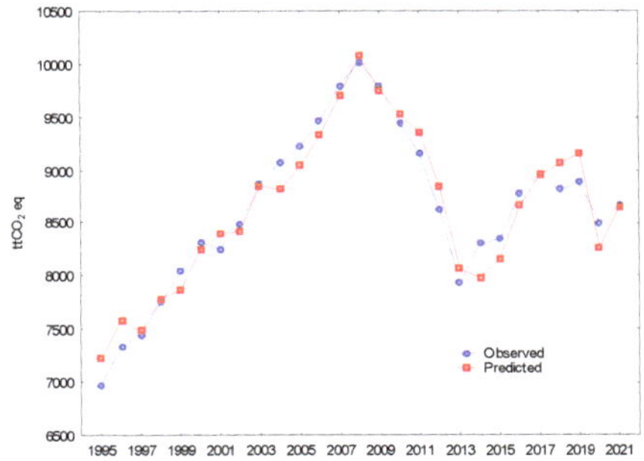

(**i**) Cyprus

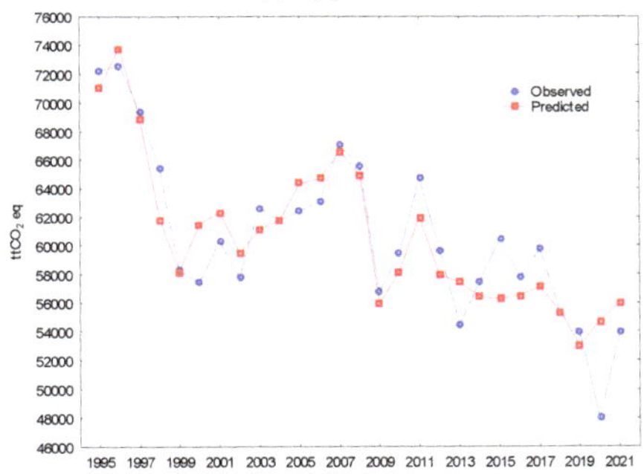

(**j**) Bulgaria

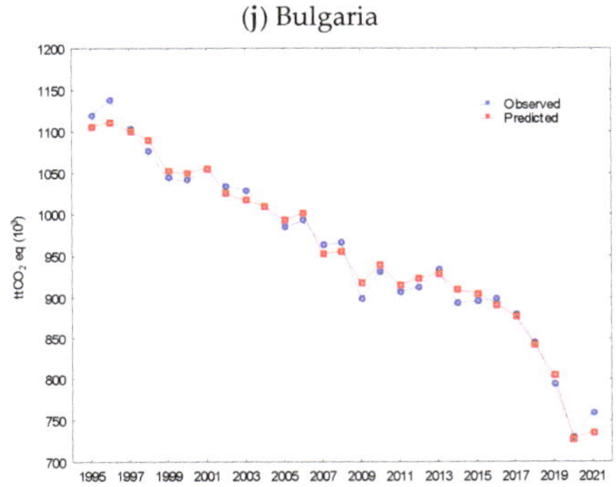

(**k**) Germany

Figure 6. *Cont.*

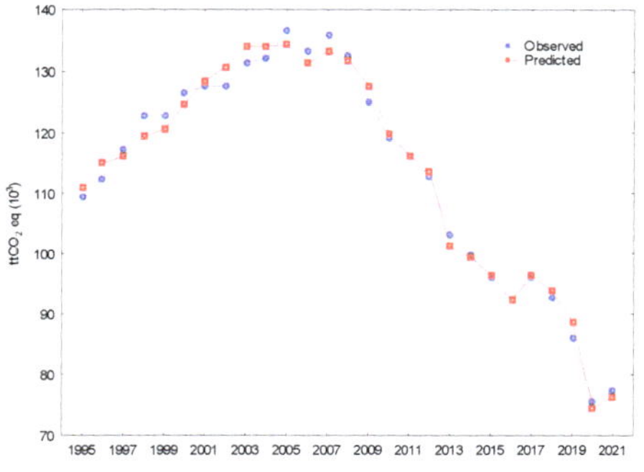

(**l**) Greece

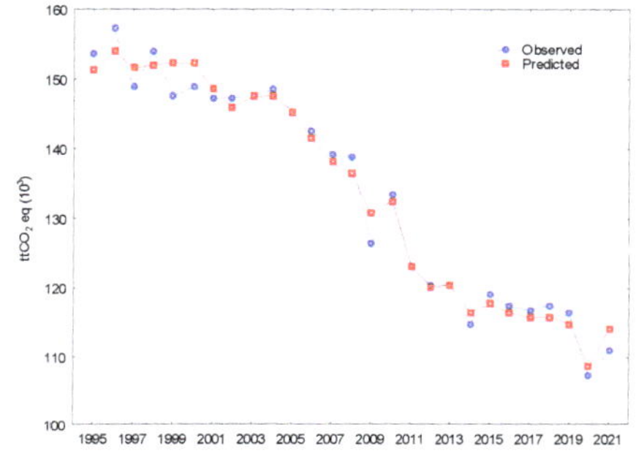

(**m**) Belgium

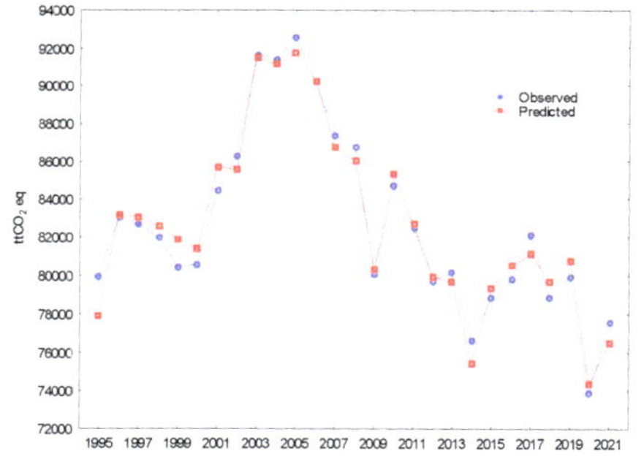

(**n**) Austria

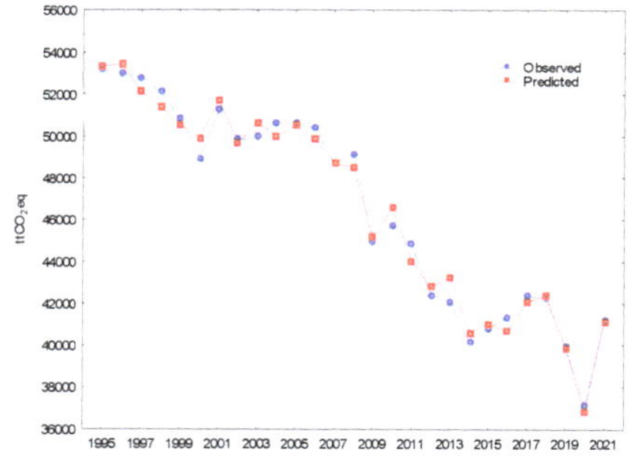

(**o**) Slovakia

Figure 6. Observed and predicted GHG emissions.

4. Conclusions

Any information that complements knowledge about GHG emissions is a valuable source. The multiple regression analysis used in the study is intended to better understand the quantitative relationships between variables. It was used to confirm or refute the formulated hypotheses regarding the phenomena being studied. Multiple regression provides the opportunity to predict the development and simulation of processes. The significant value of regression models lies in the scientific information contained in the resulting equation. The high value of the F statistic and the corresponding level of test probability *p* confirm a statistically significant linear relationship between selected variables in individual countries and GHG emissions (except Estonia). The models obtained as a result of this research demonstrate a very good fit to actual values (exception: Estonia) and can be used to make GHG emission forecasts. The confirmation of a statistically significant linear relationship between the studied independent variables and the dependent variable is the high value of the F statistics (exceptions are Estonia and Bulgaria) and the corresponding level of test probability p ($p < 0.001$). The resulting adjusted R^2 is high (Adj. $R^2 > 0.8$ for Luxembourg) or very high (Adj. $R^2 > 0.9$ for Denmark, Ireland, Poland, Czech Republic, the Netherlands, Lithuania, Finland, Cyprus, Germany, Greece, Belgium, Austria and Slovakia) and largely explains the variance of the GHG emission variable.

Population in several EU Member States, namely Ireland, the Netherlands and Belgium showed negative correlations with GHG emissions. It is noteworthy that these are countries that joined the EU quite early: Ireland in 1973 and the Netherlands and Belgium in 1958. The results of this study open new opportunities for policymakers to design comprehensive financial, economic and energy policies to minimize the harmful effects of environmental pollution. When determining the allocation of GHG emission quotas for individual EU countries, population dynamics, GDP, incomes and the structure and intensity of energy consumption should all be considered, because the study results indicate that these variables significantly affect the volume of CO_2 emissions in the countries examined.

Author Contributions: Conceptualization, A.K.-W.; Methodology, A.K.-W.; Software, A.K.-W., D.S. and P.B.; Validation, A.A.P.; Formal analysis, A.K.-W. and D.S.; Investigation, A.K.-W. and M.W.; Resources, A.A.P. and P.B.; Data curation, M.W. and D.S.; Writing—original draft, A.K.-W.; Writing—review & editing, A.A.P.; Visualization, M.W. and D.S.; Supervision, A.A.P.; Funding acquisition, A.A.P. All authors have read and agreed to the published version of the manuscript.

Funding: This research received no external funding.

Data Availability Statement: Data are contained within the article.

Conflicts of Interest: The authors declare no conflict of interest.

References

1. Kappelle, M.; Van Vuuren, M.M.I.; Baas, P. Effects of climate change on biodiversity: A review and identification of key research issues. *Biodivers. Conserv.* **1999**, *8*, 1383–1397. [CrossRef]
2. Pimm, S.L. What we need to know to prevent a mass extinction of plant species. *Plants People Planet* **2020**, *3*, 7–15. [CrossRef]
3. Courchamp, F.; Ivan Jarić, I.; Albert, C.; Meinard, Y. The paradoxical extinction of the most charismatic animals. *PLoS Biol.* **2018**, *16*, e2003997. [CrossRef] [PubMed]
4. Valipour, M.; Mousavi, S.M.; Valipour, R.; Rezaei, E. Air, water, and soil pollution study in industrial units using environmental flow diagram. *J. Basic Appl. Sci. Res.* **2012**, *2*, 12365–12372.
5. Munzel, T.; Hahad, O.; Daiber, A.; Landrigan, P.J. Soil and water pollution and human health: What should cardiologists worry about? *Cardiovasc. Res.* **2022**, *119*, 440–449. [CrossRef] [PubMed]
6. Kolasa-Więcek, A.; Suszanowicz, D. Air pollution in European countries and life expectancy—Modelling with the use of neural network. *Air Qual. Atmos. Health* **2019**, *12*, 1335–1345. [CrossRef]
7. Antonakakis, N.; Chatziantoniou, I.; Filis, G. Energy consumption, CO_2 emissions, and economic growth: An ethical dilemma. *Renew. Sust. Energ. Rev.* **2017**, *68*, 808–824. [CrossRef]
8. Qureitem, G.; Al-Khatib, I.A.; Anayah, F. Inventory of main greenhouse gas emissions from energy sector in Palestine. *Environ. Monit. Assess.* **2020**, *192*, 63. [CrossRef]
9. Rajbhandari, S.; Winyuchakrit, P.; Pradhan, B.B.; Chaichaloempreecha, A.; Piti Pita, P.; Limmeechokchai, B. Thailand's net-zero emissions by 2050: Analysis of economy-wide impacts. *Sustain. Sci.* **2023**, 1–14. [CrossRef]

10. Chowdhury, M.M.I.; Rahman, S.M.; Abubakar, I.R.; Aina, Y.A.; Hasan, M.A.; Khondaker, A.N. A review of policies and initiatives for climate change mitigation and environmental sustainability in Bangladesh. *Environ. Dev. Sustain.* **2021**, *23*, 1133–1161. [CrossRef]
11. Stošić, M.; Ivezić, V.; Tadić, V. Tillage systems as a function of greenhouse gas (GHG) emission and fuel consumption mitigation. *Environ. Sci. Pollut. Res.* **2021**, *28*, 16492–16503. [CrossRef] [PubMed]
12. Abbas, H.S.M.; Xu, X.; Sun, C. Role of foreign direct investment interaction to energy consumption and institutional governance in sustainable GHG emission reduction. *Environ. Sci. Pollut. Res.* **2021**, *28*, 56808–56821. [CrossRef]
13. Zubir, M.A.; Bong, C.P.C.; Ishak, S.A.; Ho, W.S.; Hashim, H. The trends and projections of greenhouse gas emission by the livestock sector in Malaysia. *Clean. Technal. Environ. Policy* **2022**, *24*, 363–377. [CrossRef]
14. Kolasa-Więcek, A. Use of Artificial Neural Networks in Predicting Direct Nitrous Oxide Emissions from Agricultural Soils. *Ecol. Chem. Eng. S* **2013**, *20*, 419–428. [CrossRef]
15. Ramachandra, T.V.; Aithal, B.H.; Sreejith, K. GHG footprint of major cities in India. *Renew. Sust. Energ. Rev.* **2015**, *44*, 473–495. [CrossRef]
16. Eurostat. Available online: https://ec.europa.eu/eurostat/data/database (accessed on 28 October 2023).
17. IEA. Available online: https://www.iea.org/data-and-statistics/data-tools/greenhouse-gas-emissions-from-energy-data-explorer (accessed on 12 September 2023).
18. Masoomi, M.; Panahi, M.; Samadi, R. Demand side management for electricity in Iran: Cost and emission analysis using LEAP modeling framework. *Environ. Dev. Sustain.* **2022**, *24*, 5667–5693. [CrossRef]
19. Khan, M.T.I.; Yaseen, M.R.; Ali, Q. The dependency analysis between energy consumption, sanitation, forest area, financial development, and greenhouse gas: A continent-wise comparison of lower middle-income countries. *Environ. Sci. Pollut. Res.* **2018**, *25*, 24013–24040. [CrossRef]
20. Dioha, M.O.; Kumar, A. Exploring greenhouse gas mitigation strategies for agriculture in Africa: The case of Nigeria. *Ambio* **2020**, *49*, 1549–1566. [CrossRef]
21. Gamtessa, S.; Çule, M. Marginal abatement costs for GHG emissions in Canada: A shadow cost approach. *Clean. Technal. Environ. Policy* **2023**, *25*, 1323–1337. [CrossRef]
22. Simionescu, M.; Radulescu, M.; Cifuentes-Faura, J. Pollution and electricity price in the EU Central and Eastern European countries: A sectoral approach. *Environ. Sci. Pollut. Res.* **2023**, *30*, 95917–95930. [CrossRef]
23. Ahmad, N.; Du, L.; Lu, J.; Wang, J.; Li, H.-Z.; Hashmi, M.Z. Modelling the CO2 emissions and economic growth in Croatia: Is there any environmental Kuznets curve? *Energy* **2017**, *123*, 164–172. [CrossRef]
24. Moutinho, V.; Madaleno, M.; Robaina, M.; Villar, J. Advanced scoring method of eco-efficiency in European cities. *Environ. Sci. Pollut. Res.* **2018**, *25*, 1637–1654. [CrossRef] [PubMed]
25. Wenlong, Z.; Tien, N.H.; Sibghatullah, A.; Asih, D.; Soelton, M.; Ramli, Y. Impact of energy efficiency, technology innovation, institutional quality, and trade openness on greenhouse gas emissions in ten Asian economies. *Environ. Sci. Pollut. Res.* **2023**, *30*, 43024–43039. [CrossRef] [PubMed]
26. Forsius, M.; Holmberg, M.; Junttila, V.; Kujala, H.; Schulz, T.; Paunu, V.V.; Savolahti, M.; Minunno, F.; Akujärvi, A.; Bäck, J.; et al. Modelling the regional potential for reaching carbon neutrality in Finland: Sustainable forestry, energy use and biodiversity protection. *Ambio* **2023**, *52*, 1757–1776. [CrossRef] [PubMed]
27. Aziz, S.; Chowdhury, S.A. Drivers of greenhouse gas emissions in the electricity sector of Bangladesh. *Clean. Technol. Environ. Policy* **2023**, *25*, 237–252. [CrossRef]
28. Fatima, T.; Xia, E.; Cao, Z.; Khan, D.; Fan, J.-L. Decomposition analysis of energy-related CO_2 emission in the industrial sector of China: Evidence from the LMDI approach. *Environ. Sci. Pollut. Res.* **2019**, *26*, 21736–21749. [CrossRef] [PubMed]
29. Mneimneh, F.; Ghazzawi, H.; Ramakrishna, S. Review Study of Energy Efficiency Measures in Favor of Reducing Carbon Footprint of Electricity and Power, Buildings, and Transportation. *Circ. Econ. Sust.* **2023**, *3*, 447–474. [CrossRef]
30. Dawood, A.M.A.; Darko, E.O.; Glover, E.T. Ghana's nuclear power and its potential to curtail carbon dioxide emission. *Clean. Technol. Environ. Policy* **2021**, *23*, 2883–2893. [CrossRef]
31. Eurostat. Available online: https://ec.europa.eu/eurostat/web/products-eurostat-news/w/ddn-20230816-1 (accessed on 3 September 2023).
32. Kasman, A.; Duman, Y.S. CO_2 emissions, economic growth, energy consumption, trade and urbanization in new EU member and candidate countries: A panel data analysis. *Econ. Model.* **2015**, *44*, 97–103. [CrossRef]
33. Pao, H.T.; Tsai, C.M. CO_2 emissions, energy consumption and economic growth in BRIC countries. *Energy Policy* **2010**, *38*, 7850–7860. [CrossRef]
34. Apergis, N.; Ozturk, I. Testing environmental Kuznets curve hypothesis in Asian countries. *Ecol. Ind.* **2015**, *52*, 16–22. [CrossRef]
35. Ozdemir, O.; Özokcu, S. Economic growth, energy, and environmental Kuznets curve. *Renew. Sustain. Energy Rev.* **2017**, *72*, 639–647. [CrossRef]
36. Bilgili, F.; Koçak, E.; Bulut, Ü. The dynamic impact of renewable energy consumption on CO_2 emissions: A revisited environmental Kuznets curve approach. *Renew. Sustain. Energy Rev.* **2016**, *54*, 838–845. [CrossRef]
37. Pata, U.K. Renewable energy consumption, urbanization, financial development, income and CO_2 emissions in Turkey: Testing EKC hypothesis with structural breaks. *J. Clean. Prod.* **2018**, *187*, 770–779. [CrossRef]

38. Al-Mulali, U.; Solarin, S.A.; Ozturk, I. Investigating the presence of the environmental Kuznets curve (EKC) hypothesis in Kenya: An autoregressive distributed lag (ARDL) approach. *Nat. Hazards* **2016**, *80*, 1729–1747. [CrossRef]
39. Jackson, T. *Wohlstand ohne Wachstum: Leben und Wirtschaften in Einer Endlichen Welt*; Oekom: München, Germany, 2013; p. 206, ISBN 978-3-865-81414-2.
40. Nuroğlu, E.; Kunst, R. Kuznets and Environmental Kuznets Curves for Developing Countries. In *Industrial Policy and Sustainable Growth Part of the Series Sustainable Development*, 1st ed.; Murat, Y., Ed.; Kuznets and Environmental Kuznets Curves for Developing Countries; Springer Nature Singapore Pte Ltd.: Singapore, 2017; pp. 1–14. [CrossRef]
41. OECD. Available online: www.stats.oecd.org (accessed on 30 October 2023).
42. Kolasa-Więcek, A.; Suszanowicz, D.; Pilarska, A.A.; Pilarski, K. Modelling the Interaction between Air Pollutant Emissions and Their Key Sources in Poland. *Energies* **2021**, *14*, 6891. [CrossRef]
43. Martínez-Zarzoso, I.; Bengochea, A.; Morales-Lage, R. The impact of population on CO_2 emissions: Evidence from European countries. *Environ. Resour. Econ.* **2007**, *38*, 497–512. [CrossRef]

energies

MDPI

Review

New Foods as a Factor in Enhancing Energy Security

Eulalia Skawińska [1,*] and Romuald I. Zalewski [2]

[1] Faculty of Economics and Management, University of Zielona Góra, 65-417 Zielona Góra, Poland
[2] Commission of Commodity Science, Polish Academy of Science, 61-772 Poznań, Poland; r.zalewski@wez.uz.zgora.pl
* Correspondence: e.skawinska@wez.uz.zgora.pl

Abstract: Increasing energy security is a crucial component of achieving the Sustainable Development Goals (SDGs). Three main factors influence energy security: (1) the efficiency of resource use in energy production, (2) the extent of energy losses, and (3) the use of new energy sources. Novel food products can impact these factors, and this paper explores whether they are being studied in the context of reducing energy consumption. Specifically, we investigate the role of technical progress and know-how in the creation and development of novel food products and whether novel methods of food production using artificial intelligence aim to reduce energy expenditures while improving product quality, variety, and the use of new energy sources. This paper seeks to examine the impact determinants of novel foods on energy security, considering economic, technological, social, and environmental aspects of knowledge about new food. To implement the study, the relevant international literature published in the past ten years have been reviewed and methods of modeling, visualization, and descriptive statistics applied. The review is structured into three sections: the first section presents ways to save energy and other resources in the food production chain through the intensive use of artificial intelligence tools; the second section presents the development of novel food products; and the last section presents marketing challenges for novel foods. The findings show that the topic addressed by this paper is currently critical, with many authorities, research centers, food producers, and energy producers interested. However, the research problem remains open, as a systematic review of secondary sources revealed little knowledge of the topic under study, and each author's study presents a new solution. The conclusion is that utilizing new foods and innovative production techniques that require less energy not only enhances production diversity but also improves its quality.

Keywords: new foods; energy saving; artificial intelligence (AI)

1. Introduction

As humanity makes social and economic progress and grows in number, the demand for and the price of energy are increasing. Energy is obtained from two types of sources—non-renewable (such as coal, oil, natural gas, and radioactive elements) and renewable (such as solar, wind, hydroelectric, geothermal, and biomass). Currently, the world consumes approximately 170,000 TWh of primary energy, out of which 40,000 TWh is derived from coal and 45,000 TWh from oil [1]. The total energy consumption in Poland, e.g., has increased from 169.4 GWh in 2019 to 173.46 GWh in 2021. Due to the deficit of approximately 1 GHw and rising side costs such as the cost of coal, oil, transport, and external costs (e.g., carbon pollution), there is a growing interest in renewable sources of energy. It is expected that wind and solar electricity will account for more than a quarter of total electricity generation by 2026, while the use of fossil fuels is expected to decrease. The total global energy demand is expected to ease in 2023 and to grow by slightly less than 2% in 2023, down from a rate of 2.3% in 2022 and the average annual growth rate of 2.4% observed over the 2015–2019 period. This trend is mainly driven by declining electricity demand in advanced economies, which face the ongoing effects of the global energy crisis and slower economic growth due to COVID-19 [2,3]. However, it is projected that this trend will end in 2024, and the demand growth will accelerate again.

It is important to note that around 30% of the world's energy is used in the agricultural and food industry. Energy is required at every stage of the food value chain, from the production of agricultural inputs to agricultural production in the field, harvesting, processing, food production, transportation, marketing, and consumption. While primary agriculture consumes only about 20% of the sector's energy, food processing, including transportation, accounts for around 40% of energy usage, significantly contributing to global energy consumption along agricultural value chains [4,5]. In developed countries, the agriculture, livestock, and fishing industries consume approximately 25% of total energy, while 45% of this energy is consumed in food processing and distribution, and 30% is used in retail, preparation, and cooking. Developing countries, on the other hand, consume less energy for production but require more energy for distribution and cooking.

The issue of energy is related to two factors that often compete—energy security through energy access and environmental sustainability. In the current scenario of rapidly transitioning to a decentralized, decarbonized, and digital system, it is essential to maintain a balance between these two factors while achieving equivalent goals. In this context, a compromise should be made to address the necessary challenge. The global energy sector faces an energy trilemma that involves sustainable development, energy decarbonization, and the reliable provision of energy [6]. To better understand the opportunities and disruptions of a decentralized energy system, energy security, affordability, and quality priorities need to be assessed. A sustainable energy transition and reduction in energy consumption through fixed asset investments contribute to economic growth. This growth is directly correlated with the quality of the environment in the short term. Investments in fixed assets and reduction in energy consumption are positively associated with environmental quality. Additionally, energy and biological security are strengthened. The decarbonization process can be accelerated, e.g., through the implementation of general carbon taxes.

Energy security and its economic and political stability must be given long-term priority, as supported by numerous opinions [7]. As energy security is crucial for economic growth and development, the energy development rate will need to grow by 3.2% per year, which is nearly three times more than in 2019 and twice as much as in the past [8]. However, according to the International Energy Agency, if humanity continues at the current pace, by 2050, emissions and pollution will increase by 130% [8]. Thus, a comprehensive energy transition development program is necessary to reevaluate the relationship between economics, nature, and the environment. Critical components include transitioning towards more sustainable models and economic structures that promote social equity, investment flows, and energy patterns [9].

Assuring sustainable growth and development in the long run calls for technical, social, political, economic, ethical, and research integrity solutions to ensure a reliable supply of clean and affordable energy and nutritious food for everyone, which highly impact well-being and health. With the developed countries committed to the achievement of the seventeen Sustainable Development Goals (SDGs), this applies particularly to the following SDGs: good health and well-being (3), affordable and clean energy (7), and responsible consumption and production (12) [10]. Increasing resource use efficiency to create energy and reducing energy losses is the main solution to enhance energy security. New, more sustainable foods offer a potential to help in achieving this goal.

The aim of this paper is to investigate the impact of new foods on energy security, covering economic, technological, and social aspects. For this, several research questions have been explored: (1) whether new foods have been studied in the context of reducing energy consumption; (2) the role that technical progress and knowledge play in the development of new foods; (3) whether new methods of food production using artificial intelligence (AI) aim to reduce energy expenditure and improve product quality and variety; and finally, (4) whether the process of generating new foods is market-sustainable. The research was conducted using the method of standard literature review applied to global literature from the last ten years, including theoretical, empirical, and statistical works found on four academic literature databases: Scopus, Web of Science, Science Direct, and Google Scholar. To ensure systematic results, three criteria have been applied: keywords, scientific nature of publications, and research area. Works selected have been published from 2012 to 2023 and focused on economic, social, technological, and environmental aspects. Both authors independently evaluated the chosen literature. Only highly rated publications were selected based on how helpful the literature was in solving the research questions. The resulting database of 200 works required further analysis, and the findings are presented in individual chapters of the review. In our examination, we applied various methods including deduction, analysis, and literature review. The paper is divided into three logically organized sections. The first section explores new methods to save energy in the food chain that heavily rely on the use of artificial intelligence. The other two sections cover the definition and design of novel food products, as well as the challenges novel food products face in relation to consumer acceptance.

The topic discussed in this paper is highly significant and up-to-date, which is supported by the attention of many authorities, research centers, scientists, as well as food and energy producers, as the bulk of the scientific, white, grey, and other non-scientific literature indicates, e.g., [11]. However, despite the considerable attention it has received, the research problem remains still open and relevant.

2. AI-Aided Ways to Save Energy in Food Production

Energy intensity across all EU countries, measured in kG of oil equivalent per thousand pounds, showed a decline during the period 2011–2021 in all countries. The countries that made the most progress in this regard were Bulgaria, Estonia, and Malta [12]. These achievements were possible in part due to the use of new, more efficient food drying technologies [13], overcoming barriers, applying new ideas [14], waste management, and modernization of production processes, e.g., [15]. Energy savings in the food industry's cold supply chains also contributed to reducing energy consumption [16].

In the European Union, all member states have been required to reduce their industrial energy consumption steadily from 2011 to 2021. As per the regulations, these countries must save an average of 1.5% of their total energy consumption yearly by 2030. The annual energy savings will begin at 1.3% until the end of 2025, and then gradually increase to 1.9% during the final period until the end of 2030 [17]. Meeting these recommendations requires broader energy conservation in the food sector and the economy in general. The food industry encompasses all steps from the farm to the consumer, including agriculture, purchasing, wholesaler processing, commercial units, and the final customer. This process takes a considerable amount of time and involves multiple stages: raw materials are produced, stored, processed, and transported between different units as the product becomes more mature or finished. These activities are influenced by environmental conditions such as temperature, humidity, and light, which can affect the quality of the material or product. Depending on the type of food product, the appropriate physical conditions, especially temperature and humidity, need to be maintained and monitored during transportation of raw materials, semi-finished, or finished food products [18]. Published research shows that 33% of the world's food is wasted along the food chain [19–21], and with it the energy and other resources used for its production and marketing, making food waste and loss a serious challenge. It is estimated that around 14% of the food produced is lost between the time of harvest and delivery to the market. Furthermore, approximately 18% of global food production is wasted, with 11% of this waste happening in households, 5% in restaurants and bars, and 2% in trade [22].

The social and economic significance of the aforementioned issues prompts their management through information technology (IT) such as artificial intelligence (AI) and machine learning (ML). Since the late 1950s, there has been an ongoing a shift from mechanical electronic technology to digital electronics. This transformation has accelerated in the past 50 years, driven by the invention of the Internet and advancements in computer hardware and mobile technology, leading to what is known as the digital revolution [23]. The rapid development of information and communication technology has transformed various aspects of social life, economies, businesses, and management, leading to the emergence of digital economy [24]. In Europe, this process has been particularly intensive in the past decade. Since 2015, digital adoption has increased notably throughout the EU, with the digital economy and society index (DESI), an index tracking indicators on Europe's digital performance, rising from below 40 in 2015 to above 60 in 2020 [25].

AI and computer science can improve agricultural production for consumer needs, according to various authors, e.g., [26], including food processing in terms of its sensory advantages and distribution channels. The literature on the subject covers various scenarios and use cases of machine learning, machine vision, and sustainable global learning for future food production [27]. As the food production chain encompasses various stages, it is essential to involve energy-efficient technologies throughout the whole chain including farming, appropriate packaging, and maintaining proper temperature during transport and storage. The following section explores the potential and the application of these new technologies to save energy along the food chain in more detail.

2.1. Artificial Intelligence and Machine Learning

The definition of artificial intelligence is complex, but for the purpose of clarity, it is limited to artificially intelligent systems, simulating human intelligence processes, especially computer systems. Specific applications of AI include expert systems, natural language processing, speech recognition, and machine vision [28]. There are various definitions available, but most of them fall into one of the following four categories: systems that can think like humans, act like humans, think rationally, and act rationally. Additionally, an AI computer should have the ability to gather knowledge, apply it, and learn. Machine learning is a branch of artificial intelligence (AI) and computer science focusing on the use of data and algorithms to imitate the way humans learn, gradually improving its accuracy [29]. Artificial intelligence models require large amounts of labeled data, which can be expensive and biased. Self-supervised learning provides a solution by allowing models to learn from unlabeled data without overt labeling. This approach enables the model to learn from large datasets without the need for human labeling [30]. With the help of computer technologies and artificial intelligence, it is now possible to use large sets of experimental data from actual production processes to create new equipment and algorithms for real-time intelligent machines. These machines can build models that characterize individual manufacturing processes, which can facilitate production operations and product quality controls. These models are commonly known as artificial intelligence-based computer vision [31]. Investment activity in this area is dominated by start-ups and rising [32], with the global market growth from EUR 0.5 trillion to EUR 4.1 trillion between 2012 and 2015.

During the latter half of 2010, advancements in the fields of nutritional science, food chemistry, food analytical methods, and artificial intelligence started to intersect. For instance, AI has been integrated into nutritional epidemiology by facilitating dietary pattern analysis and identifying foodborne illnesses using large-scale data analysis. Additionally, AI has been useful in food toxicity assessment, image diagnosis, and personalized nutrition [33]. Other technologies based on the so called fourth-generation (4.0 IR) technologies such as the Internet of Things, cyber-physical systems, smart manufacturing, smart factories, cloud computing, and cognitive computing, which make up the Industry 4.0, can also improve food production processes [34]. One such example is 'Food Traceability 4.0' (FT 4.0), i.e., tracking food products using digital technologies in the food chain. Application of FT 4.0 has enormous potential to improve food tracking, reduce food waste, and prevent food fraud, leading to smarter food traceability. This, in turn, creates new opportunities in the food industry and improves customer trust [35].

2.2. Artificial Intelligence in Agriculture

World agriculture is on the cusp of significant improvements as AI technologies rapidly advance and become more widely used in this area. AI has the potential to revolutionize crop management and agricultural productivity by enabling rapid diagnosis of plant diseases, efficient application of agrochemicals, and providing expert agronomic advice to growers. However, the systemic risks of using ML and AI models as expert systems in agriculture are often under-appreciated and poorly understood. Some of the key risk factors include the interoperability, reliability, and relevance of data [36]. The suggested risk-mitigation measures are applying frameworks for responsible and human-centered innovation, setting data cooperatives for improved data transparency and ownership rights, and initial deployment of agricultural AI in digital sandboxes [36].

By analyzing large data sets, AI algorithms can extract valuable information that supports decision-making and management in areas such as crop farming, animal production, automatic control processes, and robotics. These models and the information they provide are essential tools for improving the efficiency and productivity of agri-food systems. The collection of data in various agri-food sectors such as cultivation, field crop production, horticulture, animal production, water management, irrigation, machinery, barriers, and challenges can lead to better management decisions but also create new challenges. The use of AI technology in this domain brings up various risks, ethical issues, and societal implications. Questions must be raised about how to govern the usage of these technologies, and how to incorporate socio-ethical value considerations into the policy and legal frameworks under development [37] as current agricultural practices transition to smart farming, robotic farming, and the use of automated drones [38].

2.3. Artificial Intelligence in Food Processing and Distribution

Numerous papers and books have presented and evaluated AI- and ML-based approaches for various food processing operations, including drying, frying, baking, canning, extrusion, encapsulation, and fermentation, to predict the kinetics of these processes [39]. The development of ML-based models and their practical implementation can be achieved through a step-by-step procedure, as outlined in [40]. Another area of the food production chain in which the integration of artificial intelligence (AI) applications is becoming increasingly popular is transportation. Here, ICT technologies can help optimize transportation routes and reduce losses [41] and costs [42,43]. By applying machine learning (ML) and robotics, it is possible to control transportation routes and ensure the safe transportation of fresh or sensitive food [44]. The goal of these methods is to integrate various elements of the food transport chain to maintain better quality, sustainability, and efficiency.

According to STATISTA, the demand for food robotics equipment is expected to increase by about 5 million units in the next 5 to 7 years, among others due to the implementation of AI and ML in the food industry [45]. The use of these technologies can bring numerous benefits, such as reducing packaging costs, speeding up service, increasing consumer satisfaction, and minimizing human errors [38]. This, in turn, leads to lower losses, increased positive feedback, and increased consumer trust [46]. AI applications also reduce the time taken for placing orders by voice, making the process more personal and advantageous, especially in large processing plants. This brings profits, particularly in the long term. The elimination of irregularities contributes to reducing losses and costs within the company. For instance, errors in quality levels, numerous complaints due to incorrect grammage, poor labeling or marking, and exceeding permitted temperatures or humidity during storage and transportation can be avoided [18].

Artificial intelligence enabled agents, Internet of Things (IoT), sensors, and blockchain technology can be combined to maximize the supply network and increase the revenue of all parties involved along the agri-food value chain [47]. Blockchain is a technology that can record multiple transactions from multiple parties across a complex network. Changing the records inside the blockchain requires the consensus of all parties involved, thus giving a high level of confidence in the data [48]. This way, blockchain technology can support the traceability and transparency of the food supply chain, possibly increasing the trust of consumers, and in combination with AI, intelligent precision farming can be achieved.

The list of scientific papers exploring different applications of new technologies in a variety of food processing operations is long, while more systematic and comprehensive works are few. Among the latter, the following books are especially helpful [49–52].

2.4. Artificial Intelligence in Preservation and Storage of Food

A significant issue in the food chain is the storage and preservation of food. Every year, a large amount of food becomes spoiled during transportation and storage due to the inability to maintain freshness for an extended period of time. However, the new phase change technology for cold energy storage has revolutionized cold chain logistics. This technology offers several advantages, including stable temperatures and high energy storage density [53]. Effective temperature control is crucial for maintaining freshness and preventing spoilage throughout the food supply chain [18,54]. Having materials that can reduce temperature fluctuations and keep perishable produce fresh for a longer time is crucial. According to Chen et al., phase change materials (PCMs) can absorb or release large amounts of latent heat during phase transition, with minimal temperature fluctuations [55,56].

The literature review shows that only a few studies address the use of PCMs to preserve food freshness. Refrigeration systems using cold storage PCMs have been found to maintain the ideal temperature state in refrigerated trucks. Additionally, the cost of this method is half that of traditional refrigeration transportation [57]. Adding a phase change cold storage layer to the reefer enclosure reduced total energy consumption by 4.7% compared to conventional cold storage [58]. There have been several recent reports on insulation PCM boxes that are suitable for refrigerated transportation of fruits and vegetables. These boxes come in various forms, such as trays, multilayer films, foams, and others. Additionally, a new study has provided a novel idea for the design of temperature control materials that can be used in cold chain transportation of food, vegetables, and fruits [53].

2.5. Artificial Intelligence and Energy Issues

Application of AI also offers a vast potential for energy saving. The management of electrical grids is becoming increasingly complex as the world shifts towards renewable energy sources such as solar and wind power. This is where AI can play a crucial role. By predicting fluctuations in demand and adjusting the supply accordingly, AI can help balance the grid, prevent blackouts, and increase energetic efficiency. Additionally, AI can optimize the placement and operation of renewable energy infrastructure, such as wind turbines and solar panels, to maximize their energy output. It can also help reduce energy consumption in frozen food transportation by optimizing routes and traffic, reducing fuel consumption, and CO_2 emissions [59]. In manufacturing, AI can optimize processes, identify inefficiencies, and improve energy efficiency to reduce costs. Harnessing the power of AI can work towards a world where energy is used efficiently, waste is minimized, and the reliance on fossil fuels is reduced [60].

There are three technological gaps that hinder the application of AI in energy saving: (a) difficulty in selecting or combining technologies, (b) various energy-saving effects, and (c) AI effects on control characteristics. AI can optimize manufacturing processes and reduce energy consumption by identifying inefficiencies in production lines and suggesting adjustments. It can help you optimize energy sources and automate energy-saving actions. AI algorithms can also predict energy demand and supply, allowing us to reduce greenhouse gas emissions and mitigate the effects of climate change. A universal workflow combining various AI technologies can achieve consistent energy-saving effects by providing qualitative and quantitative recommendations tailored to different application fields by Squeo [61].

As demonstrated above, AI and robotics are poised to revolutionize every aspect of the food industry, from agriculture and manufacturing to distribution, cooking, and the creation of new products and food sorting options, while the application of ML can help optimize various processes and production kinetics, reduce energy consumption, optimize time, and achieve better-quality products [44]. Moreover, AI can assist businesses in developing sustainable processes, products, and services, thus minimizing their environmental impact. Prioritizing the application of AI technologies in the pursuit of a greener, more sustainable future, offers potential to make a significant impact on reducing carbon emissions and slowing down climate change.

However, developing such technologies requires significant scientific and experimental effort. Also, some experts believe that the ability of AI to create new food products is somewhat exaggerated, as the quality and acceptance of food go beyond the quality and quantity of individual basic ingredients [62]. The type, quality, and quantity of additional ingredients, as well as their characteristics that affect the look, smell, taste, and texture, are equally important. As the demand for better-quality, sustainable food grows, there will be an increasing need for suppliers to meet these expectations. Research laboratories and their employees will play an important role in this transformation, where non-rational factors such as taste, colors, aroma, composition, and other sensory and emotionally connotated properties play a vital role.

3. Design of Novel Food

Various organizations and individuals, directly or indirectly involved in food production, distribution, quality control, sales, and market operations, are constantly seeking new ideas to overcome technical challenges and introduce innovative food products. The interpretation of the concept of a new food product varies across the literature. However, in most countries, the legal definition of a new food product is well-defined. For example, in the United States, the Federal Trade Commission recommends that the term "new" should only be attributed to products that have been available on the market through the usual distribution network for a maximum of six months. Despite the passage of time, Ph. Kotler, in his well-known book [63], suggests that new products can be categorized based on their degree of novelty and innovation. These categories include products that create a completely new market and those that enable companies to enter an existing market for the first time, which can be achieved either through horizontal development (by substitution) or vertical development (by quality). Additional products complementing the existing offer, improved and repositioned products for new markets or market segments, are also considered new product lines [64,65].

In the European Union, novel food is defined as food that has not been consumed to a significant degree by humans in the EU before 15 May 1997, when the first regulation on novel food came into force. Novel food includes newly developed and innovative products, as well as those produced using new technologies, traditional foods from outside the EU, and products not consumed before May 1997 [66]. Novel foods must be safe for consumers and properly labeled to avoid misleading consumers. If a novel food is intended to replace another food, it must not be nutritionally advantageous for the consumer. Pre-market authorization of novel foods is necessary based on an evaluation that adheres to the principles mentioned above. In the European Union, the complexities and risk assessments of novel foods are overseen by the European Food Safety Authority (EFSA). When preparing a novel food application, all available knowledge on the novel food should be considered [67].

New foods can be categorized based on their nature or source. Botanical, animal-based, and microorganism-based foods are significant areas of interest, with consumers, policymakers, and the industry increasingly exploring alternative and sustainable dietary choices [68].

3.1. Development of New Food Products

Developing ideas for new food products or secondary versions of existing ones requires a thorough development process to create a mature design. The journey from the initial idea to the final product can be lengthy due to the need to fulfill several requirements at once. The process of introducing an innovation follows a series of steps outlined below (Table 1).

Table 1. Stages in the development of novel foods. Based on Zarba C. et al. [69].

Action	Scope of Activity
Concept	Exploring market needs and technological opportunities.
Analysis	Literature studies. Consultation and observation of similar activities in the past. Pilot, quarter, or semi-technical studies. Analyzing environmental impact (water, energy consumption). Experiencing how to use, describe, and characterize it.
Preliminary tests	Assessing costs, opportunities for innovation and optimization under conditions of incomplete information and uncertainty.
Economic analysis	Analysis and decision to continue or abandon the project.
Implementation	Two-thirds of products reach the industrial production phase.

New products can be created through original designs, modifications, and improvements to existing products. These changes can be in functional, social, or symbolic dimensions and can be accompanied by new services. In the food market, the number of truly innovative products is limited. There are twelve categories of novel food products, including convenient, light, organic, ethnic, healthy, premium, deluxe, delicatessen, nostalgic, purposeful, variant, and truly new [70–72].

According to recent classifications, there are seven conditions that food producers must consider when creating novel products [68]. These include meeting consumer demands and sensory attributes, considering the seasonality of ingredients, sourcing and distributing local ingredients, ensuring the safety and traceability of ingredients and final products, manufacturing the food products on a large scale, and minimizing the environmental impact.

Another six-point model of product innovation was presented by Hittmar et al. [73]. Its components include the following steps: (1) ensuring the information base; (2) realization of organizational changes; (3) the move to lateral changes; (4) the identification of the innovation process and its management; (5) creating a map of the process; and (6) process optimization. Meanwhile, Bigliardi et al. [74] classify three models of innovation implemented in the food industry: technology push, demand-pull, and stage gate. Another interesting model supported by open innovation has been proposed in the early 2000s by Aiking and de Boer [75]. In this model, food design involves applying design thinking to every step of the food value chain. The aim is to promote innovation centered on humans throughout the whole food value chain, from production to disposal.

The development of new food products requires the use of innovative methods and engineering processes that ensure the product properties are unique and environmentally safe. Proper control of nutrient availability, satiety, and overall well-being can influence consumer health. Sensory profiles can also be adjusted, which is important when introducing new raw materials and technologies that can shape food choices and habits. Poor sensory experiences or consumer biases can affect these decisions [76].

As part of the product development process, the screening and evaluation of numerous product ideas is an essential activity. The objective is to identify and shortlist the most promising product ideas with the greatest potential for success. This process is iterative and involves multiple rounds of evaluation, during which the number of product ideas is reduced (Figure 1). As the product design process continues, the expenditure associated with product development increases.

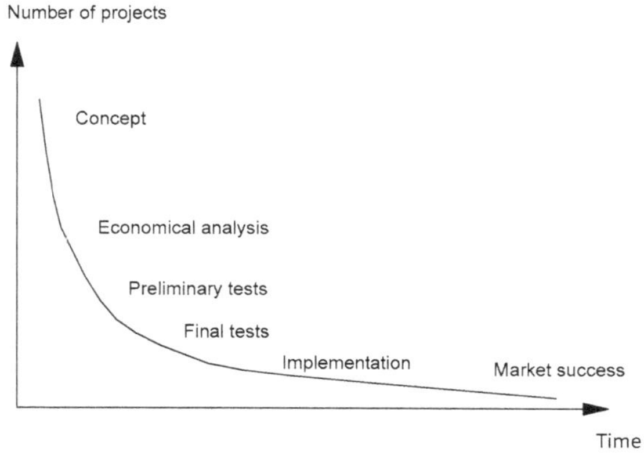

Figure 1. Number of innovative ideas at different stages of the product development process. Source: [77], p. 398.

Establishing connections between the level of achievement in launching new food products and the various factors impacting this level is crucial. To evaluate the success of these launches, one must consider aspects such as market share, sales targets, future profits, technical aspects, company impact and reputation, competitiveness, research and development activities, quality management, implementation speed, and how well the products meet consumer expectations. Steward-Knox and Mitchell argue that "consumer knowledge and retailer involvement are key success factors in food product development" [78]. Costa et al. suggest that both technical knowledge and market information are essential for effective product development [79]. Novel food technologies are important for food security, safety, and sustainability. However, consumers are often hesitant to accept them, as discussed by Siegrist and colleagues [46,80]. It is crucial for food-producing companies to anticipate the inevitable expenses of developing new products, regardless of whether they succeed or fail. The company's top management should play an active role in the product development process by establishing the vision, aligning resources, and rallying support from team members. The most significant methods for finding an idea for a new food product include:

- developing a list of the product's current attributes and looking for possible improvements;
- analyzing relationships between attributes in similar products;
- identifying structural elements of the product (morphological method);
- brainstorming in combination with the Ishikawa method.

New ideas can originate from various sources within and without the organization, such as R&D, distributors, consulting firms, universities, and industry institutes. These ideas help shape the new product and ensure that it is competitive and meets the approval of potential buyers. A combination of intellectual and practical efforts is required to define the characteristics of the new product and ensure it is successful in the market. Thus, the role of R&D representatives has been growing in recent years due to two reasons. Firstly, they are experts in product and technological requirements for its creation, and secondly, they are needed to create a prototype or a model of a new product, select appropriate technologies, raw materials, and compositions to ensure its economic implementation, and make sure the product is attractive and stable for the consumer during distribution and use, and resistant to the variability of conditions [81].

Many companies rely on analyzing their competitors' products to generate new ideas. This can involve purchasing their products and studying their quality features or gathering information from published sources [82]. Such activities are often part of the company's imitation strategy, especially when targeting sales to segments with low. In large companies, the process of developing new food product concepts is typically formalized and includes the following: idea generation, evaluation and screening, market research, product specification, feasibility study, production process development, and prototype development and testing [83]. This is performed to better organize activities, improve management efficiency, and comply with quality management systems such as ISO 9001 and ISO 26000 [84]. Despite the diversity of approaches and positions, the following elements can be distinguished: (1) identifying opportunities and needs, (2) creating an idea to meet these needs, (3) creating a product that materializes the idea, (4) involvement in a variety of activities prior to the start of production, making up the whole.

Several strategies informing the creation process of new food can be adopted by the team responsible. The type of new product strategy chosen is determined by three continuous variables: (1) the degree of innovation in the product (continuous or radical innovation); (2) the type of marketing orientation (demand, customer); and (3) the type of technological orientation (current, completely new, etc.) [77]. The following figure places the variables and the strategies graphically in a 3D space of innovation, marketing, and technology, with final states as corners of a cube (Figure 2).

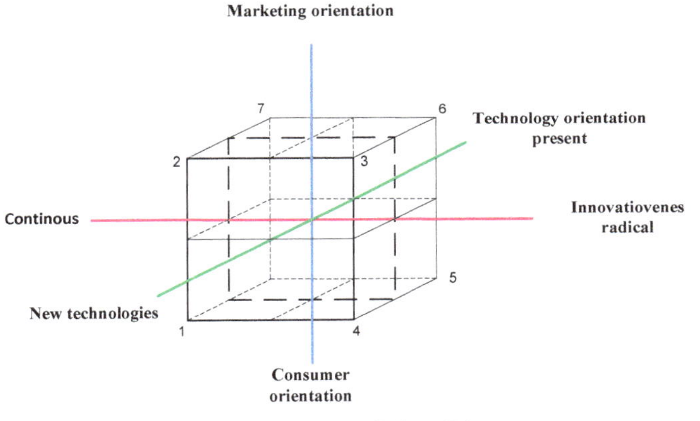

Figure 2. Product strategy space. Source: [77], p. 404.

Strategies 1 to 4 focus on the use of new technology and research for development, while strategies 5 to 8 are more conservative in terms of technology. However, for specific products, these strategies may not be used in their pure form. For example, strategy 4 is consumer-focused and open to radical innovation and new technology. Strategy 6 involves concentric product diversification, which aims to meet the needs of a new group of consumers through radical innovations while utilizing current technology. On the other hand, strategy 8 is about segmental product differentiation. It aims to expand the scope of satisfying the needs of current and new consumers through continuous and dynamic innovations to master new market segments, without the need to change technology.

A selection process for new food product ideas is often conducted using the 'weighted index' method [85–87]. The process of evaluating new product proposals involves assigning scores to each main factor based on their importance: idea generation, concept testing, feasibility (regulations, technology, formulation, ingredients, processing, facilities, packaging, distribution, shelf life, safety, and finance), test marketing, commercialization, or product life cycle. The sum of these scores should be close to 100, and the scores of the sub-factors within each main factor cannot exceed the assigned value. The proposals with the highest scores have the best chance of being implemented. It is possible to modify the list of main and sub-factors based on specific needs.

When selecting a new food product from several options, there are various general factors that need to be considered. These include assessing the legal conditions, determining whether the innovation idea aligns with the product objectives, gauging compatibility with the company's marketing mix strategy (including product, pricing, distribution, and promotion policies), evaluating the idea in relation to the resources available to the company (such as research and development, production, human resources, and financial and managerial capabilities), and ensuring that it aligns with the company's goals, such as fitting in with the current product portfolio and maintaining competitiveness in the market. Factors are scored with weights as percentages, summing to 100%. Proposals scoring below 55% are disqualified, while scores between 70 and 85% are excellent [86].

3.2. *Artificial Intelligence in Development of Novel Food*

The role of information technology is becoming increasingly crucial in the design processes of new foods. Initially, ICT was used primarily for information and education purposes such as weather reports, crop prices, and monitoring. It also played an important role in improving the food supply chain [88,89]. Over the past decade, AI has brought about significant changes in the food industry. It has helped in devising effective marketing strategies, boosting food sales, and analyzing eating habits and preferences. AI has also played a crucial role in food design, new product development, and predicting health problems associated with food consumption [90]. Furthermore, AI has emerged as a major solution to food waste problems by estimating food demand quantity, predicting waste volumes, and supporting effective cleaning methods through smart waste management [91,92].

Information and computer technology applications can be used at different stages of food product development, including process management, R&D, consumer research (needs, preferences, testing, predictions), market analytics, product design, and post-product marketing. A report issued by the European Food Safety Authority indicates that food safety regulations are expected to undergo a transition towards the utilization of artificial intelligence (AI), with a primary focus on machine learning (ML), for the purpose of real-time big data analysis to assess risks. Based on the intensified cooperation with society, and an increase in the amount of information provided by society, AI will facilitate a much more effective risk assessment [93]. The aim of these methods is to reduce the need for costly and time-consuming physical experiments on foods by using laboratory-designed digital equivalents [94]. Extrapolating big data enables also taking into consideration individual customer feedback and thus creating individualized niche products [95].

However, when it comes to food, naturalness is often viewed positively, while technological applications are seen negatively [80]. In fact, according to multiple sources, approximately 80% of new products developed by large food and beverage companies end up failing [96]. Improving consumer satisfaction is crucial for the success of novel food products, and the challenges and ways to achieve this will be discussed in the following section.

4. Acceptance of Novel Food

It is becoming increasingly important to cater to the demand for sustainable and healthy food products, as well as keep up with societal changes. However, many adults tend to stick to familiar foods due to their past experiences. Recent studies reveal that people's individual preferences have a significant impact on how they react to new and unfamiliar foods [97]. The acceptance or rejection of these foods by consumers depends on several factors, such as their mental traits, values, attitudes, expectations, and dietary preferences. Therefore, product developers must consider these factors during the development process.

The success of a new product in the market depends on its quality, which is determined by the degree of healthiness, attractiveness, and availability of food, considering the raw materials, technology, and price. The most significant characteristics of food that consumers consider during purchase are freshness, appearance, taste, smell, and durability. These characteristics are related to sensory attributes and can be evaluated based on personal experience. Health effects, ease of preparation, fat content, and the absence of harmful substances are also important factors to be considered.

The success factors for new products on the market, besides consumer needs and preferences, also include new trends. As a result, food companies should pay close attention to existing and future trends and may even partner with consumers to develop new products. Additionally, the marketing activities used by the company play a significant role in determining the market success of new food products. Companies must keep in mind that the final consumers ultimately decide whether new products will succeed in the market [98]. Many studies show that specific reasons for product failure include food neophobia, rejection of novel technology, suboptimal stimulation levels, sensory/consumer research limitations, and inadequate marketing support [99,100]. Specifically, Tuorila and Hartmann [97] have identified that traits such as disgust, food neophobia, and related factors can significantly impact a person's willingness to try new foods and become a major barrier to accepting novel food alternatives. In their research, they discuss two new food trends: meat alternatives and products for health and well-being. They conclude that launching new foods successfully requires a thorough understanding of how consumers perceive the product and the traits that determine whether they will accept or reject it. It is noteworthy that packaging design and linguistics have been shown to create an emotional attachment to the product surpassing the effect of taste [101]. This information has enabled innovative technologies to be applied along the agri-food chain in optimization of packaging and marketing materials, including graphics and advertisements. Emerging opportunities, such as new raw materials and technologies that enable prolonged shelf-life, also promote the development of new food products (Table 2).

Table 2. Motives for developing novel foods from the perspectives of consumers and the industry.

Perspective of Consumers		Perspective of Industry
Organic	Personalized nutrition	New resources
Sensory properties	Food variety	Extension of shelf-life
Well-being	Ethical concerns	Business profit
Allergies intolerances	Animal welfare	Technological advances
		Energy consumption

Source: own compilation based on [97].

Meat and meat products have a significant impact on the environment, particularly in terms of water, carbon, and energy footprints. Therefore, it is essential for the food industry to educate consumers about meat substitutes and cultured meat. However, studies show that consumers prioritize sensory properties over environmental concerns [102–104]. To encourage sustainable and healthy diets, food designers can enhance product acceptability using molecular gastronomy principles. In doing so, new food concepts aim to meet consumer expectations in terms of taste, appearance, nutritional value, and health benefits. Product positioning and marketing play a significant role in the development of these concepts, with consulting agencies driving the process. One such start-up, Innogusto, established in 2018, aims to create gastronomic dishes using meat alternatives such as insects, algae, fungi, and microorganisms, to increase the acceptance of these protein sources in the future [105]. Similar strategies have been developed and implemented to tackle food production problems. These include methods to improve agricultural productivity, optimize resource use (such as water, energy, and packaging materials), enhance food nutrition through genetic engineering, produce new alternatives to food and food ingredients (such as cellular cultures, insects, algae, and dietary fibers), and preserve biodiversity. These solutions aim to meet current and future food demands sustainably by optimizing natural resources (water and energy) and restructuring the food industry models [106].

Gathering information from potential and existing clients is crucial for any business to develop new products. There are various ways of obtaining such information, e.g., categorized interviews like survey interviews and uncategorized interviews that can be conducted by sales representatives, interviewers, or distributors, such as wholesalers or retailers. In addition, companies can directly receive feedback from consumers through comments, opinions, and suggestions, which can be sent via post or telephone, such as a dedicated customer hotline. Additionally, it is important to analyze customer complaints using the Deming PDCA cycle and the model proposed by Kano in 1984 [107]. In his book, Kano investigated the correlation between the level of expectation, performance, and the resulting satisfaction when it comes to food products (Figure 3). These expectations vary over time and depend on many internal and external factors. New products appearing on the food market are located in a three-dimensional space of satisfaction, expectation, and performance.

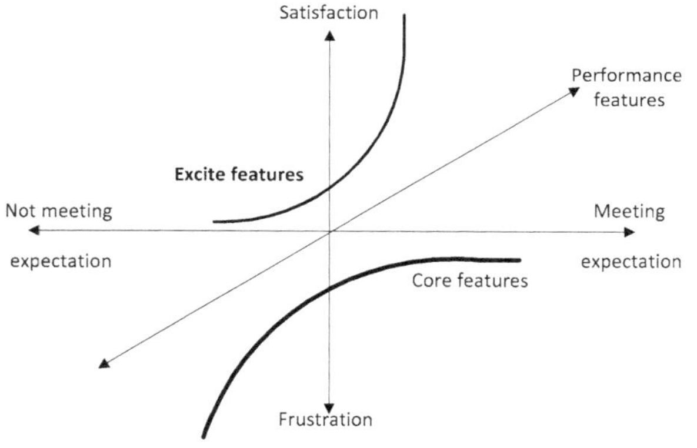

Figure 3. Kano model of the relationship between customer satisfaction, expectation, and frustration. Source: own drawing based on [107].

The classic PDCA cycle can be used to improve and better tailor a new product to the needs of customers [108]. The product idea and strategy undergo both informal and formal verification processes. Expert panels, consumer tastings, and sensory analysis specialists test the best variants, which are then subject to further studies. The evaluation includes measures of consumer acceptance, uniqueness, and expected sales frequency defined by the manufacturer, as well as sensory qualities such as taste, smell, and overall experience.

Sensory attributes of food have been and still are [109,110], intrinsically linked to its attractiveness and, to a large extent, to its healthiness and are absolutely necessary at all times when evaluating the concept of a new product and in the different phases of its implementation. According to Moskowitz [87], the sensory analysis should be used in the following instances: in preliminary work to identify the sensory attributes that create the image of a product; to demonstrate the imperfection or sensory mismatch between currently produced food, the preferences, and needs of the consumer in a given market; to create a product formulation that corresponds to the concept or intended image; or to capture the combination of conceptual elements that correspond to the product. According to Schifferstein [111], in order to attract customers in markets with high competition, products should not only be of good quality and visually attractive, but they should also offer interesting and engaging experiences. Therefore, it is important that research on sensory evaluation is connected to research on product aesthetics, meaning, and emotions. Last but not least, trust is a crucial element when it comes to accepting new products, especially food. A strong level of trust in the producer or brand, along with established systems for tracking the ingredients and products (traceability), is essential for increasing the acceptance of the product. According to research, higher levels of trust are directly proportional to higher levels of product acceptance [62].

During the development phase, it is crucial to experiment with the product on a small scale. This involves formulating the product to ensure optimal nutritional value, shelf life, and packaging proposal. Sensory tests are conducted during this phase, using trained individuals, as well as consumer tests, to ensure the product meets their expectations [109,110]. The objective is to develop a product that satisfies consumer expectations in a particular market. During this phase, various aspects such as taste, appearance, consistency, and smell are fine-tuned by making small adjustments to the composition, concentration of components, and production and technological conditions. Before finalizing mass production, a prototype of the product is put through a marketing test to gauge consumer response to factors such as price, packaging, advertising, promotion, distribution, and to assess its shelf life [112]. The response of the test markets is a good indicator of how the market as a whole will receive the new product. Completing all the above stages successfully is necessary before deciding to move forward with production. Depending on the previously chosen strategy, the manufacturer will need to spend money on technology and marketing, prepare for production, and launch the product on the market. This marks the beginning of the new product's life cycle.

5. Conclusions

Energy security and conservation are crucial topics for sustainable development. With the growing world population and the food sector already accounting for a significant amount of energy usage, there is a clear need for a deeper understanding of how new foods can enhance energy security. The review delves into the potential impact of new food products on enhancing energy security in line with the Sustainable Development Goals (SDGs). Specifically, the current state of research on the interplay between technical progress and know-how in creating and developing novel food products has been examined and factors for consumer acceptance and market success for new foods presented.

It has been found that novel foods and innovative production techniques can reduce the need for energy at various stages of the food production chain while also leading to an increased diversity of production and better product quality. In the last ten years, AI has significantly impacted the agri-food industry by promoting crop sustainability, optimizing processing, devising effective marketing strategies, boosting food sales, and analyzing eating habits and preferences. It has also played a crucial role in food design, new product development, maintaining health and safety systems during transportation, managing food and energy waste, and predicting health problems associated with food consumption.

Although some authors believe AI can or will soon be able to design new food products, we present a more comprehensive and human-oriented overview of design and development methods for novel foods and reducing energy consumption and propose a model for stimulating the development of novel foods, along with scenarios for practical application.

It is important to note that the vast majority of new food products do not succeed on the market, with consumer acceptance being the critical factor. Low acceptance can be attributed to various aspects, including food neophobia, rejection of new technology, suboptimal stimulation levels, poor sensory/consumer research, and inadequate marketing support. Therefore, the food industry must consider these factors and invest in trustworthy food information systems and brand-building to gain consumer favor and trust. Another challenge in designing and launching novel energy-efficient food products is the need for collaboration between academia, laboratories, technology experts, marketing specialists, food agencies, and various governmental and non-governmental organizations involved in legal and promotional aspects of food production.

The findings support that novel food products have the potential to improve food and energy security. Upon conducting a systematic review of secondary sources, little research was found regarding the use of AI to lower energy consumption in novel food. Each author's study proposes a unique solution, leaving the research problem open. With many authorities, research centers, and food and energy producers interested in the subject of novel food, there is a need for further research on the impact determinants of these new food products on energy security and their marketability. This research should consider economic, technological, social, and environmental aspects in order to gain a more comprehensive understanding of the potential offered by novel food.

Author Contributions: Conceptualization, E.S. and R.I.Z.; methodology, E.S. and R.I.Z.; software, R.I.Z.; validation, E.S. and R.I.Z.; formal analysis, E.S. and R.I.Z.; investigation, E.S. and R.I.Z.;

resources, E.S.; Data curation, R.I.Z.; writing—original draft preparation, E.S. and R.I.Z.; writing—review and editing, E.S.; visualization, R.I.Z.; supervision, E.S. All authors have read and agreed to the published version of the manuscript.

Funding: This research received no external funding.

Data Availability Statement: Not applicable.

Conflicts of Interest: The authors declare no conflicts of interest.

References

1. Ritchie, H.; Rose Rosado, P. Energy. Published online in OurWorldInData.org. 2022. Available online: https://ourworldindata.org/energy (accessed on 8 October 2023).
2. Electricity Market Report—Update 2023. Executive Summary. Available online: https://www.iea.org/reports/electricity-market-report-update-2023/executive-summary (accessed on 8 October 2023).
3. Dong, L. Toward Resilient Agriculture Value Chains: Challenges and Opportunities. *Prod. Oper. Manag.* **2021**, *30*, 666–675. [CrossRef]
4. FAO. "Energy-Smart" Food for People and Climate—Issue Paper. 2011. Available online: http://www.fao.org/docrep/014/i2454e/i2454e00.pdf (accessed on 28 October 2023).
5. Chilling Prospects. Promoting Sustainable Agricultural Food Chain through the Energy Smart Food Programs. 2022. Available online: https://www.seforall.org/data-stories/promoting-sustainable-agricultural-food-chains (accessed on 8 October 2023).
6. Liu, H.; Khan, I.; Zakari, A.; Alharthi, M. Roles of trilemma in the world energy sector and transition towards sustainable energy: A study of economic growth and the environment. *Energy Policy* **2022**, *170*, 113238. [CrossRef]
7. Wiatros-Motyka, M. Global Electricity Review 2023. Ember. Available online: https://ember-climate.org/insights/research/global-electricity-review-2023/ (accessed on 8 October 2023).
8. Söderholm, P. The green economy transition: The challenges of technological change for sustainability. *Sustain. Earth* **2020**, *3*, 6. [CrossRef]
9. IEA. An Updated Roadmap to Net Zero Emissions by 2050. In *World Energy Outlook 2022*; International Energy Agency (IEA): Paris, France, 2022. Available online: https://www.iea.org/reports/world-energy-outlook-2022/an-updated-roadmap-to-net-zero-emissions-by-2050 (accessed on 10 October 2023).
10. United Nations; Department of Economic and Social Affairs. Sustainable Development—17 Goals. Available online: https://sdgs.un.org/goals (accessed on 10 October 2023).
11. European Commission. Food 2030—The EU's Research and Innovation Policy to Support the Transition towards Sustainable, Healthy and Inclusive Food Systems. Available online: https://research-and-innovation.ec.europa.eu/research-area/environment/bioeconomy/food-systems/food-2030_en (accessed on 10 October 2023).
12. Energy Statistics—An Overview. Eurostat. Available online: https://ec.europa.eu/eurostat/statistics-explained/index.php?title=Energy_statistics_-_an_overview (accessed on 10 October 2023).
13. Hnin, K.K.; Zhang, M.; Mujumdar, A.S.; Zhu, Y. Emerging food drying technologies with energy-saving characteristics: A review. *Dry. Technol.* **2018**, *37*, 1465–1480. [CrossRef]
14. Clairand, J.-M.; Briceno-Leon, M.; Escriva-Escriva, G.; Pantaleo, A.M. Review of Energy Efficiency Technologies in the Food Industry: *Trends Barriers Oppor. IEEE* **2020**, *8*, 48015–48029. [CrossRef]
15. Hall, G.M.; Howe, J. Energy from waste and the food processing industry. *Process Saf. Environ. Prot.* **2012**, *90*, 203–212. [CrossRef]
16. Marchi, B.; Zanoni, S. Energy efficiency in cold supply chains of the food and beverage sector. *Transp. Res. Procedia* **2022**, *67*, 56–62. [CrossRef]
17. European Parliament, Press Release from 11.07.2023, Ref.: 20230707IPR02421 "Parliament Adopts New Rules to Boost Energy Savings". Available online: https://www.europarl.europa.eu/news/en/press-room/20230707IPR02421/parliament-adopts-new-rules-to-boost-energy-savings (accessed on 10 October 2023).
18. Skawińska, E.; Zalewski, R.I. Economic impact of temperature control during food transportation a Covid perspective. *Foods* **2022**, *11*, 467. [CrossRef]
19. Hegnsholt, E.; Unnikrishnan, S.; Pollmann-Larsen, M.; Askelsdottir, B.; Gerard, M. Tackling the 1.6 Billion-Ton Food Loss and Waste Crisis. 2018. Available online: https://www.bcg.com/publications/2018/tackling-1.6-billion-ton-food-loss-and-waste-crisis.aspx (accessed on 11 October 2023).
20. Corrado, S.; Caldeira, C.; Eriksson, M.; Hanssen, O.J.; Hauser, H.-E.; van Holsteijn, F.; Liu, G.; Östergren, K.; Parry, A.; Secondi, L.; et al. Food waste accounting methodologies: Challenges, opportunities, and further advancements. *Glob. Food Secur.* **2019**, *20*, 93–100. [CrossRef]
21. Chaboud, G.; Daviron, B. Food losses and waste: Navigating the inconsistencies. *Glob. Food Secur.* **2017**, *12*, 1–7. [CrossRef]
22. United Nations. Reducing Food Loss and Waste: Taking Action to Transform Food Systems. Available online: https://www.un.org/en/observances/end-food-waste-day (accessed on 11 October 2023).
23. Evangelista, R.; Guerrieri, P.; Meliciani, V. The economic impact of digital technologies in Europe. *Econ. Innov. New Technol.* **2014**, *23*, 802–824. [CrossRef]

24. Williams, L.D. Concepts of Digital Economy and Industry 4.0 in Intelligent and Information Systems. *Int. J. Intell. Netw.* **2021**, *2*, 122–129. [CrossRef]
25. Anderton, R.; Jarvis, V.; Labhard, V.; Petroulakis, F.; Rubene, I.; Vivian, L. ECB Economic Bulletin, Issue 8/2020. Available online: https://www.ecb.europa.eu/pub/economic-bulletin/articles/2021/html/ecb.ebart202008_03~da0f5f792a.en.html (accessed on 11 October 2023).
26. Patrício, D.I.; Rieder, R. Computer vision and artificial intelligence in precision agriculture for grain crops: A systematic review. *Comput. Electron. Agric.* **2018**, *153*, 69–81. [CrossRef]
27. Kok, J.N.; Boers, E.J.W.; Kosters, W.A.; van der Putten, P. Artificial Intelligence: Definitions, Trends, Techniques and Cases. UNESCO Encyclopedia of Life Support Systems (EOLSS). Available online: https://www.eolss.net/Sample-Chapters/C15/E6-44.pdf (accessed on 11 October 2023).
28. TechTarget Editorial. Available online: https://www.techtarget.com/searchenterpriseai/definition/AI-Artificial-Intelligence (accessed on 12 October 2023).
29. IBM. What Is Machine Learning. Available online: https://www.ibm.com/topics/machine-learning (accessed on 12 October 2023).
30. Lang, H.; Poon, H. Self-Supervised Self-Supervision by Combining Deep Learning and Probabilistic Logic. *Proc. AAAI Conf. Artif. Intell.* **2021**, *35*, 4978–4986. [CrossRef]
31. Kakani, V.; Nguyen, V.H.; Kumar, B.P.; Kim, H.; Pasupuleti, V.R. A critical review on computer vision and artificial intelligence in food industry. *J. Agric. Food Res.* **2020**, *2*, 100033. [CrossRef]
32. Bryant, J. Infographic to Share: Ag-Tech on the Rise. Available online: https://myfarmlife.com/equipment/infographic-to-share-ag-tech-on-the-rise/ (accessed on 12 October 2023).
33. Miyazawa, T.; Hiratsuka, Y.; Toda, M.; Hatakeyama, N.; Ozawa, H.; Abe, C.; Cheng, T.-Y.; Matsushima, Y.; Miyawaki, Y.; Ashida, K.; et al. Artificial intelligence in food science and nutrition: A narrative review. *Nutr. Rev.* **2022**, *80*, 2288–2300. [CrossRef]
34. Dalenogare, L.S.; Benitez, G.B.; Ayala, N.F.; Frank, A.G. The expected contribution of Industry 4.0 technologies for industrial performance. *Int. J. Prod. Econ.* **2018**, *204*, 383–394. [CrossRef]
35. Hassoun, A.; Abdullah, N.A.; Aït-Kaddour, A.; Ghellam, M.; Beşir, A.; Zannou, O.; Önal, B.; Aadil, R.M.; Lorenzo, J.M.; Khaneghah, A.M.; et al. Food traceability 4.0 as part of the fourth industrial revolution: Key enabling technologies. *Crit. Rev. Food Sci. Nutr.* **2022**, 1–17. [CrossRef]
36. Tzachor, A.; Devare, M.; King, B.; Avin, S.; Héigeartaigh, S. Responsible artificial intelligence in agriculture requires systemic understanding of risks and externalities. *Nat. Mach. Intell.* **2022**, *4*, 104–109. [CrossRef]
37. *PE 734.711*; European Parliamentary Research Service Scientific Foresight Unit (STOA). EPRS: Oklahoma City, OK, USA, 2023.
38. Kumar, I.; Rawat, J.; Mohd, N.; Husain, S. Opportunities of Artificial Intelligence and Machine Learning in the Food Industry. *J. Food Qual.* **2021**, *2021*, 4535567. [CrossRef]
39. Khan, I.H.; Sablani, S.S.; Nayak, R.; Gu, Y. Machine learning-based modeling in food processing applications: State of the art. *Compr. Rev. Food Sci. Food Saf.* **2022**, *21*, 1409–1438. [CrossRef]
40. Zalewski, R.I. Rola planowania eksperymentów w poprawie jakości i innowacyjności. In *Nowe Metody Statystyczne Wspomagające Zarządzanie Jakością*; Zalewski, R.I., Ed.; Commission of Commodity Science, Polish Academy of Science: Poznań, PL, USA, 2007; pp. 99–143.
41. Mavani, N.R.; Ali, J.M.; Othman, S.; Hussain, M.A.; Hashim, H.; Rahman, N.A. Application of Artificial Intelligence in Food Industry—A Guideline. *Food Eng. Rev.* **2022**, *14*, 134–175. [CrossRef]
42. Katiyar, S.; Khan, R.; Kumar, S. Artificial Bee Colony Algorithm for Fresh Food Distribution without Quality Loss by Delivery Route Optimization. *Hindawi J. Food Qual.* **2021**, *2021*, 4881289. [CrossRef]
43. Karaboga, D.; Akay, D. A comparative study of Artificial Bee Colony algorithm. *Appl. Math. Comput.* **2009**, *214*, 108–132. [CrossRef]
44. Sudeep Srivastava. Food Automation—How AI and Robotics Are Transforming the Future of Industry. Available online: https://appinventiv.com/blog/automation-in-food-industry/ (accessed on 12 October 2023).
45. STATISTA. Food Robotics: Global Market Unit Volume 2020–2030. 2022. Available online: https://www.statista.com/statistics/1290022/food-robotics-global-market-unit-volume (accessed on 12 October 2023).
46. Siegrist, M.; Hartmann, C. Consumer acceptance of novel food technologies. *Nat. Food* **2020**, *1*, 343–350. [CrossRef]
47. Chesbrough, H. Open Innovation. In *The New Imperative for Creating and Profiting from Technology*; Harvard Business School Press: Boston, MA, USA, 2003.
48. Olsen, P.; Borit, M.; Sayd, S. Applications, Limitations, Costs, and Benefits Related to the Use of Blockchain Technology in the Food Industry, Report 4/2019, Nofima. Available online: https://nofima.brage.unit.no/nofima-xmlui/bitstream/handle/11250/2586121/Report+04-2019.pdf?sequence=1 (accessed on 14 October 2023).
49. Birwal, P.; Goyal, M.R.; Sharma, M. *Handbook of Research on Food Processing and Preservation Technologies: Volume 3: Computer-Aided Food Processing and Quality Evaluation Techniques*; Apple Academic Press: Palm Bay, FL, USA, 2021. [CrossRef]
50. Hassanien, A.E.; Soliman, M. (Eds.) *Artificial Intelligence: A Real Opportunity in the Food Industry*; Springer International Publishing: Berlin/Heidelberg, Germany, 2022; ISBN 978-3-031-13701-3.
51. Heft, D.; Oluwasenn, A.C. *Sensing and Artificial Intelligence for Food Manufacturing*; Taylor & Francis Ltd.: London, UK, 2023. [CrossRef]

52. Khadir, M.T. *Artificial Neural Networks in Food Processing: Modeling and Predictive Control*; Walter de Gruyter GmbH: Berlin, Germany, 2021. [CrossRef]
53. Zhang, Y.; Xu, Y.; Lu, R.; Zhang, S.; Hai, A.M.; Tang, B. Form-stable cold storage phase change materials with durable cold insulation for cold chain logistics of food. *Postharvest Biol. Technol.* **2023**, *203*, 112409. [CrossRef]
54. Ndraha, N.; Hsiao, H.-I.; Vlajic, J.; Yang, M.-F.; Lin, H.-T.V. Time-temperature abuse in the food cold chain: Review of issues, challenges and recommendation. *Food Control* **2018**, *89*, 12–21. [CrossRef]
55. Chen, C.; Ding, R.; Yang, S.; Wang, J.; Chen, W.; Zong, L.; Xie, J. Development of thermal insulation packaging film based on poly (vinyl alcohol) incorporated with silica aerogel for food packaging application. *LWT-Food Sci. Technol.* **2020**, *129*, 109568. [CrossRef]
56. Chen, C.; Chen, R.; Zhao, T.; Wang, L. A comparative study of linear polyurea and crosslinked polyurea as supports to stabilize polyethylene glycol for thermal energy storage. *Renew. Energy* **2022**, *183*, 535–547. [CrossRef]
57. Liu, M.; Saman, W.; Bruno, F. Development of a novel refrigeration system for refrigerated trucks incorporating phase change material. *Appl. Energy* **2012**, *92*, 336–342. [CrossRef]
58. Fioretti, R.; Principi, R.; Corpetaro, B. Refrigerated container envelope with a PCM (Phase Change Material) layer: Experimental and theoretical investigation in a representative town in Central Italy. *Energy Convers. Manag.* **2016**, *122*, 131–146. [CrossRef]
59. Meneghetti, A.; Ceschia, S. Energy-efficient frozen food transports: The Refrigerated Routing Problem. *Int. J. Prod. Res.* **2020**, *58*, 4164–4181. [CrossRef]
60. CARILEC. Green AI: The Role of Artificial Intelligence in Energy Conservation. Available online: https://www.carilec.org/green-ai-the-role-of-artificial-intelligence-in-energy-conservation/ (accessed on 14 October 2023).
61. Donaldson, D.L.; Jayaweera, D. Applications and Issues of the Experimental Designs for Mixture in the Food Sector. *Foods* **2021**, *10*, 1128. [CrossRef]
62. Siddiqui, S.A.; Zannou, O.; Karim, I.; Kasmiati; Awad, N.M.H.; Gołaszewski, J.; Heinz, V.; Smetana, S. Avoiding Food Neophobia and Increasing Consumer Acceptance of New Food Trends—A Decade of Research. *Sustainability* **2022**, *14*, 10391. [CrossRef]
63. Kotler, P. *Marketing Management*; Prentice Hall: Upper Saddle River, NJ, USA, 1991; p. 342.
64. Herman, R. The distribution of product innovation in food industry. *Agribusiness* **1997**, *13*, 319–334. [CrossRef]
65. Azanedo, L.; Garcia-Garcia, G.; Stone, J.; Shahin Rahimifard, S. An Overview of Current Challenges in New Food Product Development. *Sustainability* **2020**, *12*, 3364. [CrossRef]
66. European Commission. What Is Novel Food? Available online: https://ec.europa.eu/food/safety/novel-food_en (accessed on 14 October 2023).
67. Ververis, E.; Ackerl, R.; Azzollini, D.; Colombo, P.A.; de Sesmaisons, A.; Dumas, C.; Fernandez-Dumont, A.; da Costa, L.F.; Germini, A.; Goumperis, T.; et al. Novel foods in the European Union: Scientific requirements and challenges of the risk assessment process by the European Food Safety Authority. *Food Res. Int.* **2020**, *137*, 109515. [CrossRef]
68. European Parliament, Novel Foods. Complementary Impact Assessment Reviewing and Updating the European Commission's 2008 Impact Assessment for a Regulation on Novel Foods. European Parliamentary Research Service. Available online: https://ec.europa.eu/food/system/files/2018-07/novel-food_leg_impact-assessment-2013.pdf (accessed on 14 October 2023).
69. Zarbà, C.; Chinnici, G.; Hamam, M.; Bracco, S.; Pecorino, B.; D'Amico, M. Driving Management of Novel Foods: A Network Analysis Approach. *Front. Sustain. Food Syst.* **2022**, *5*, 2021. [CrossRef]
70. SavorEat Ltd. Product Development of Food: Strategy, Innovations, Trends, and Examples. 2022. Available online: https://savoreat.com/product-development-of-food-strategy-innovations-trends-and-examples (accessed on 15 October 2023).
71. Newton, A. The Roadmap to New Product Development for the Food Industry Columbus Global Columbus Global UK Ltd. 2022. Available online: https://www.columbusglobal.com/en-gb/blog/the-roadmap-to-new-product-development-in-the-food-industry (accessed on 15 October 2023).
72. Aramouni, F.; Deschenes, K. *Methods for Developing New Food Products: An Instructional Guide*; DEStech Publications, Inc.: Lancaster, PA, USA, 2015; ISBN 978-1-60595-432-5. Available online: https://www.destechpub.com/wp-content/uploads/2015/01/Methods-for-Developing-New-Food-Products-preview.pdf (accessed on 15 October 2023).
73. Hittmar, Š.; Varmus, M.; Lendel, V. Proposal of model for effective implementation of innovation strategy to business. *Procedia Soc. Behav. Sci.* **2014**, *109*, 1194–1198. [CrossRef]
74. Bigliardi, B.; Ferraro, G.; Filippelli, S.; Galati, F. Innovation models in food industry: A review of the literature. *J. Technol. Manag.* **2020**, *15*, 97–107. [CrossRef]
75. Aiking, H.; de Boer, J. The next protein transition. *Trends Food Sci. Technol.* **2020**, *105*, 515–522. [CrossRef]
76. Krishna, A. An integrative review of sensory marketing: Engaging the senses to affect perception, judgment and behavior. *J. Consum. Psychol.* **2012**, *22*, 332–351. [CrossRef]
77. Zalewski, R.I. *Zarządzanie Jakością w Produkcji, Żywności*, 3rd ed.; Wydawnictwo Academy of Economics: Poznań, Poland, 2018.
78. Stewart-Knox, B.; Mitchell, P. What separates the winners from the losers in new food product development? *Trends Food Sci. Technol.* **2003**, *14*, 58–64. [CrossRef]
79. Costa, A.I.A.; Jongen, W.M.F. New insights into consumer-led food product development. *Trends Food Sci. Technol.* **2006**, *17*, 457–465. [CrossRef]
80. Roman, S.; Sanchez-Siles, L.M.; Siegrist, M. The importance of food naturalness for consumers: Results of a systematic review. *Trends Food Sci. Technol.* **2017**, *67*, 44–57. [CrossRef]

81. Earle, M.D.; Earle, R.L. Food Industry Research and Development. In *Government and the Food Industry: Economic and Political Effects of Conflict and Co-Operation*; Wallace, L.T., Schroder, W.R., Eds.; Springer: Boston, MA, USA, 1997. [CrossRef]
82. Mirtalaie, M.A.; Hussain, O.K.; Chang, E.; Hussain, F.K. A decision support framework for identifying novel ideas in new product development from cross-domain analysis. *Inf. Syst.* **2017**, *69*, 59–80. [CrossRef]
83. Rudder, A.; Ainsworth, P.; Holgate, D. New food product development: Strategies for success? *Br. Food J.* **2001**, *103*, 657–671. [CrossRef]
84. ISO and Food. International Organisation for Standardization (ISO). 2018. Available online: https://www.iso.org/files/live/sites/isoorg/files/store/en/PUB100297.pdf (accessed on 17 October 2023).
85. Relich, M.; Pawlewski, M. A fuzzy weighted average approach for selecting portfolio of new product development projects. *Neurocomputing* **2017**, *231*, 19–27. [CrossRef]
86. Lambin, J.-J.; Schuling, I. Marketing Driven Management. In *Strategic and Operational Marketing*, 3rd ed.; Palgrave Macmillan: New York, NY, USA, 2012.
87. Moskowitz, H.R. On the intersection of products and Concepts: Opportunities for Sensory Analysis to Improve Commercial Development Process. *Food Qual. Prefer.* **1999**, *10*, 333–342. [CrossRef]
88. Carmela Annosi, M.; Brunetta, F.; Capo, F.; Heideveld, L. Digitalization in the agri-food industry: The relationship between technology and sustainable development. *Manag. Decis.* **2020**, *58*, 1737–1757. [CrossRef]
89. Ramirez-Asis, E.; Vilchez-Carcamo, J.; Thakar, C.M.; Phasinam, K.; Kassanuk, T.; Naved, M. A review on role of artificial intelligence in food processing and manufacturing industry. *Mater. Today Proc.* **2021**, *51*, 2462–2465. [CrossRef]
90. European Parliament. Artificial Intelligence in the Agri-Food Sector. Applications, Risks and Impacts. Panel for the Future of Science and Technology. European Parliamentary Research Service. Available online: https://www.europarl.europa.eu/RegData/etudes/STUD/2023/734711/EPRS_STU(2023)734711_EN.pdf (accessed on 15 October 2023).
91. Adeogba, E.; Barty, P.; O'Dwyer, E.; Guo, M. Waste-to resource transformation: Gradient boosting modeling for organic fraction municipal solid waste projection. *ACS Sustain. Chem. Eng.* **2019**, *7*, 10460–10466. [CrossRef]
92. Calp, M. Estimation of personnel food demand quantity for businesses by using artificial neural networks. *J. Polytech.* **2019**, *22*, 675–686. [CrossRef]
93. Bronzwaer, S.; Kass, G.; Robinson, T.; Tarazona, J.; Verhagen, H.; Verloo, D.; Vrbos, D.; Hugas, M.; European Food Safety Authority (EFSA). *Food Safety Regulatory Research Needs 2030*; EFSA: Parma, Italy, 2019. [CrossRef]
94. Mengucci, C.; Ferranti, P.; Romano, A.; Masi, P.; Picone, G.; Capozzi, F. Food structure, function and artificial intelligence. *Trends Food Sci. Technol.* **2022**, *123*, 251–263. [CrossRef]
95. Tao, Q.; Ding, H.; Wang, H.; Cui, X. Application Research: Big Data in Food Industry. *Foods* **2021**, *10*, 2203. [CrossRef] [PubMed]
96. Stanton, J. Food and Beverage Manufacturers Increasing Number of New Products Developed. Food Processing, Endeavor Business Media, LLC. 2013. Available online: https://www.foodprocessing.com/business-of-food-beverage/business-strategies/article/11331678/food-and-beverage-manufacturers-increasing-number-of-new-products-developed (accessed on 16 October 2023).
97. Tuorila, H.; Hartmann, C. Consumer responses to novel and unfamiliar foods. *Curr. Opin. Food Sci.* **2020**, *33*, 1–8. [CrossRef]
98. Matysiak-Pejas, R. Success factors in new product development in the food sector. In *Consumer Trends and New Product Opportunities in the Food Sector*; Grunert, K.G., Ed.; Wageningen Academic Publisher: Wageningen, The Netherlands, 2017; pp. 221–232.
99. Costa, A.I.A.; Monteiro, M.J.P. Food Product Introduction Failure: Reasons and Remedies. In *Reference Module in Food Science*; Elsevier: Amsterdam, The Netherlands, 2018. [CrossRef]
100. Ryynänen, T.; Olsen, A. We must have the wrong consumers—A case study on new food product development failure. *Br. Food J.* **2014**, *116*, 707–722. [CrossRef]
101. Gunaratne, N.M.; Fuentes, S.; Gunaratne, T.M.; Torrico, D.D.; Francis, C.; Ashman, H.; Viejo, C.G.; Dunshea, F.R. Effects of packaging design on sensory liking and willingness to purchase: A study using novel chocolate packaging. *Heliyon* **2019**, *5*, e01696. [CrossRef]
102. Forde, C.G.; de Graaf, K. Influence of Sensory Properties in Moderating Eating Behaviors and Food Intake. *Front Nutr.* **2022**, *9*, 841444. [CrossRef]
103. Hoek, A.C.; Luning, P.A.; Weijzen, P.; Engels, W.; Kok, F.J.; de Graaf, C. Replacement of meat by meat substitutes. A survey on person and product-related factors in consumer acceptance. *Appetite* **2011**, *56*, 662–673. [CrossRef]
104. Weinrich, R. Opportunities for the adoption of health-based sustainable dietary patterns: A review on consumer research of meat substitutes. *Sustainability* **2019**, *11*, 4028. [CrossRef]
105. Innogusto. Available online: www.innogusto.com (accessed on 17 October 2023).
106. Valoppi, F.; Agustin, M.; Abik, F.; de Carvalho, D.M.; Sithole, J.; Bhattarai, M.; Varis, J.J.; Arzami, A.N.A.B.; Pulkkinen, E.; Mikkonen, K.S. Insight on Current Advances in Food Science and Technology for Feeding the World Population. *Front. Sustain. Food Syst.* **2021**, *5*, 626227. [CrossRef]
107. Kano, W. *The Management and Control of Quality*; Ewans, J.K., Lindsay, W.M., Eds.; West Publishing Co. Minneapolis: St. Paul, MN, USA, 1989; p. 149.
108. Deming, W.E. *Quality, Productivity and Competitive Position*; Cambridge Mass. MIT Center for Advanced Engineering Study: Cambridge, MA, USA, 1982.

109. Van Trijp, H.C.M.; Schiferstein, H.N.J. Sensory analysis in marketing practice; Comparison and Integration. *J. Sens. Stud.* **1995**, *10*, 127–148. [CrossRef]
110. Ruiz-Capillas, C.; Herrero-Herranz, A. *Sensory Analysis and Consumer Research in New Product Development*; MDPI: Basel, Switzerland, 2021. [CrossRef]
111. Schifferstein, H.N.J. From salad to bowl: The role of sensory analysis in product experience research. *Food Qual. Prefer.* **2010**, *21*, 1059–1067. [CrossRef]
112. Van Herpen, E.; Koen, J. Less waste versus higher quality; how to stimulate consumer demand for frozen food. *Br. Food J.* **2021**, *13*, 340–358.

energies MDPI

Article

Energy Management of Hydrogen Hybrid Electric Vehicles—Online-Capable Control

David Machacek [1,*], Nazim Yasar [1], Fabio Widmer [1], Thomas Huber [2] and Christopher Onder [1,*]

[1] IDSC, ETH Zurich, Sonneggstrasse 3, 8092 Zurich, Switzerland
[2] Robert Bosch GmbH, 71701 Schwieberdingen, Germany
* Correspondence: davidm@ethz.ch (D.M.); onder@idsc.mavt.ethz.ch (C.O.)

Abstract: The results shown in this paper extend our research group's previous work, which presents the theoretically achievable hydrogen engine-out NO_x^{eo} (H_2-NO_x^{eo}) Pareto front of a hydrogen hybrid electric vehicle (H_2-HEV). While the Pareto front is calculated offline, which requires significant computing power and time, this work presents an online-capable algorithm to tackle the energy management of a H_2-HEV with explicit consideration of the H_2-NO_x^{eo} trade-off. Through the inclusion of realistic predictive data on the upcoming driving mission, a model predictive control algorithm (MPC) is utilized to effectively tackle the conflicting goal of achieving low hydrogen consumption while simultaneously minimizing NO_x^{eo}. In a case study, it is shown that MPC is able to satisfy user-defined NO_x^{eo} limits over the course of various driving missions. Moreover, a comparison with the optimal Pareto front highlights MPC's ability to achieve close-to-optimal fuel performance for any desired cumulated NO_x^{eo} target on four realistic routes for passenger cars.

Keywords: hydrogen internal combustion engine; hybrid electric vehicles; H_2-NO_x^{eo} trade-off; extremely low NO_x^{eo}; energy management

1. Introduction

A major challenge of today's society is to decrease the worldwide CO_2 emissions, of which a major part stems from the transportation sector. In 2022, it was accountable for almost one-fourth of the global CO_2 emissions [1]. Hence, a major part of global decarbonization politics includes ever more stringent emission legislation in this sector [2,3]. This signifies an extension of technological choices for the light-duty market in the EU. However, currently, the traditional combustion engine is still the predominant propulsion technology. In 2023, the global light-electric-vehicle sales accounted for roughly 16% of new-car sales [4]. Therefore, the need to find alternatives to the conventional combustion engine incentivizes investments into advanced propulsion technologies that can comply with the current tank-to-wheel CO_2 goal and offer the potential for well-to-wheel and life cycle net zero CO_2 emissions.

1.1. Context and Motivation

Hydrogen is the simplest energy carrier for vehicles, produced from electricity generated by renewable power plants [5]. It can be utilized in two ways: First, hydrogen can be used in fuel cells to generate an electric current, which is used to power an electric motor or stored in a battery. Second, hydrogen can be burned in a hydrogen internal combustion engine (H_2ICE), which directly provides mechanical power. Both technologies have advantages and disadvantages. The fuel cell is known to emit only water and reaches high efficiency [6] but requires a high degree of hydrogen purity [7]. The H_2ICE can burn hydrogen, even of low purity, but produces NO_x emissions in the process. In this text, the focus is put on the H_2ICE and how to tackle engine-out NO_x^{eo} emissions.

Strategies to operate H_2ICEs with extremely low NO_x^{eo} emissions usually rely on ultra-lean hydrogen combustion [8,9], which is achieved by using adequate turbocharging [10,11]. However, ultra-lean combustion limits the achievable power density of H_2ICEs, thus limiting their applicability in mobile applications. One possible solution for this issue is the use of a secondary torque source, such as an electric motor, to assist the H_2ICE. The result is a H_2-HEV with two onboard energy storage devices, i.e., hydrogen and a battery. This also introduces an additional degree of freedom, which is typically managed by an energy management system (EMS) algorithm. In the following, a literature review on such algorithms is provided. Particular emphasis is placed on EMS algorithms that explicitly consider NO_x emissions.

1.2. Literature Review

The EMS algorithm is concerned with the distribution of the driver's power request between the vehicle's different propulsion components. The presented algorithms can be divided into two categories: offline algorithms, which require a large amount of computational power and are used to evaluate the theoretical potential of a system, and online-capable controllers, which are computationally much leaner and can thus be implemented on a real vehicle. In this text, the focus is placed on online-capable EMS controllers.

The earliest EMS controllers reported in the literature are based on simple heuristics (e.g., [12]). Their main goal is to minimize the vehicle's fuel consumption. However, the power distribution not only affects fuel consumption but also has a widespread impact on a vehicle's powertrain components. While more and more control objectives can be incorporated into an appropriately designed EMS, this naturally leads to an increase in complexity. Also, the inclusion of predictive data to plan the vehicle's battery usage for the upcoming driving mission has been studied [13]. The integration of thermal effects in the power distribution decision process is investigated in [14,15]. The authors of [16] show that in the case of a diesel HEV, the decision of the EMS also has a large influence on its NO_x emissions. Compared with HEVs powered by traditional fuels, the literature on EMS algorithms for H_2-HEVs is scarce, especially with a focus on NO_x emissions.

The authors of [17] investigate the NO_x emissions resulting from simulations of a parallel H_2-HEV on the worldwide harmonized light-duty test cycle. However, they lack the use of an advanced EMS control algorithm, as their main focus is the investigation of the propulsion architecture. The only study which discusses the connection between the EMS and the H_2-NO_x^{eo} trade-off of a H_2-HEV is presented in [18]. Here, a strong emphasis is laid on the potential of H_2-HEVs to reduce NO_x^{eo} emissions via the dedicated use of the system's degrees of freedom to satisfy the driver's power request. The discussed degrees of freedom are managed by the EMS and include, among others, the power distribution between the H_2ICE and the electric motor, and the driving mode. The presented study shows that given an optimal use of the EMS algorithm, reduced NO_x^{eo} emissions can only be achieved with an increased amount of H_2 consumption. Such a trade-off is described by using a Pareto curve.

Figure 1 shows a Pareto curve depicting the optimal normalized H_2-NO_x^{eo} trade-off of a mixed H_2-HEV on a specific driving mission. Each point on this curve represents one optimal EMS calibration. This means that for each point on this curve, accumulated NO_x^{eo} emissions cannot be decreased without increasing H_2 consumption, and vice versa. Points above the curve are called Pareto sub-optimal, points below the curve cannot be reached. The H_2-optimal EMS calibration is depicted by the point farthest to the right on the Pareto curve. The NO_x^{eo}-optimal EMS calibration is depicted by the point farthest to the left on the Pareto curve. As the DP algorithm is used, the results represent the global optimum. However, DP cannot be practically implemented on a real vehicle due to its immense requirements for computational power and computation time.

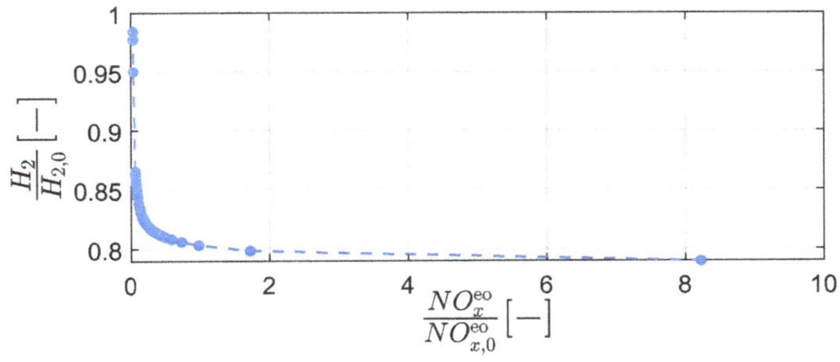

Figure 1. A Pareto curve representing the H_2-NOx^{eo} trade-off of a series H_2-HEV.

The following three conclusions are drawn in the presented study:

(a) The potential for NO_x reduction is much larger for H_2-HEVs than for conventional diesel-powered HEVs, as well as standard H_2 vehicles.

(b) Although ultra-lean combustion of hydrogen–air mixtures allows H_2ICEs to emit near-zero NO_x^{eo} emissions, this is a highly delicate process. Small deviations from the chosen operating point of the H_2ICE can increase the instantaneous NO_x^{eo} emissions by over two orders of magnitude.

(c) The mixed hybrid drivetrain architecture is required to achieve consistent NO_x^{eo} reductions across a wide range of challenging driving missions. However, it is more complex than the standard parallel or series hybrid architectures.

Point **(b)** suggests the use of an online optimization-based control algorithm to balance the delicate H_2-NO_x^{eo} trade-off. Point **(c)**, however, complicates the use of online optimization-based control algorithms, as the mixed hybrid drivetrain architecture introduces integer control variables, resulting in a mixed-integer optimization problem. Currently, there is no online-capable controller in the literature that achieves similar performance to what is predicted by the offline computed potential analysis presented in [18].

1.3. Research Statement

This paper aims at closing this gap. An MPC algorithm is used for the EMS algorithm, since the incorporation of conflicting control goals, such as the delicate H_2-NO_x^{eo} trade-off, is straightforward in optimization problems. A careful separation of the integer and the continuous control variables allows for the circumvention of solving a mixed-integer optimization problem online. To summarize, the contribution of this publication is twofold:

1. To the authors' best knowledge, this publication presents the first online-capable EMS controller for a H_2-HEV, explicitly accounting for the H_2-NO_x^{eo} trade-off.
2. A case study, using the same mixed H_2-HEV as discussed in [18], allows for a comparison between the proposed online-capable EMS controller and the full theoretically reachable Pareto front obtained by the DP algorithm. The results show that the proposed online-capable controller reaches close-to-optimal performance on all investigated driving missions, covering a broad range of driving scenarios.

1.4. Paper Structure

The remainder of this paper is organized as follows: In Section 2, the investigated mixed H_2-HEV is introduced. Moreover, a simplified mathematical model, which is later leveraged to formulate an optimization problem suitable for online-capable control, is presented. In Section 3, a method to avoid solving a mixed-integer program is presented. Moreover, a convex optimization problem which can be solved quickly is presented. In Section 4, the proposed EMS controller structure is presented. In Section 5, the EMS algorithm is compared with the DP-optimal solution in a case study encompassing in total four different driving missions. In Section 6, the main contributions are summarized, and an outlook on future research is presented.

2. Modeling

In this section, first, the investigated vehicle is outlined, and a sketch of the EMS optimization problem including NO_x^{eo} is presented. Second, a motivation for the MPC-based controller structure is given; this then requires some model simplifications to ensure the online-capable use of the MPC's optimization problem.

2.1. Map-Based Powertrain Model

Figure 2 depicts a schematic representation of the vehicle's powertrain discussed in this text. The component models are characterized by using maps created from testbench measurements, which is why they will be together referred to as the "map-based" model from here on. The maps for the H_2ICE and the motor and generator are given in the following; however, a detailed description of the map-based model is provided in [18].

$$\dot{m}_{\mathrm{f}}^{\mathrm{map}} = f(\omega_e, T_e), \tag{1}$$
$$\dot{m}_{\mathrm{NO_x}}^{\mathrm{map}} = f(\omega_e, T_e), \tag{2}$$
$$P_{\mathrm{l_{mot/gen}}}^{\mathrm{map}} = f(\omega_{mot/gen}, T_{mot/gen}). \tag{3}$$

where ω_e denotes the engine speed; T_e denotes the engine torque; and $\omega_{mot/gen}$ and $T_{mot/gen}$ denote the speed and torque of the electric motor or the generator, respectively. The battery state of charge is described by the variable SoC and is modeled by using a Thevenin equivalent circuit model.

The vehicle under investigation is a so-called mixed H_2-HEV. It is equipped with one electric motor; which provides propulsive power and is used for regenerative braking; a H_2ICE, which is connected to the wheels via a clutch and a fixed transmission ratio; and a generator. The H_2ICE and the generator together form the engine–generator unit. A mixed hybrid electric vehicle can be operated in three different driving modes, which are introduced in the following.

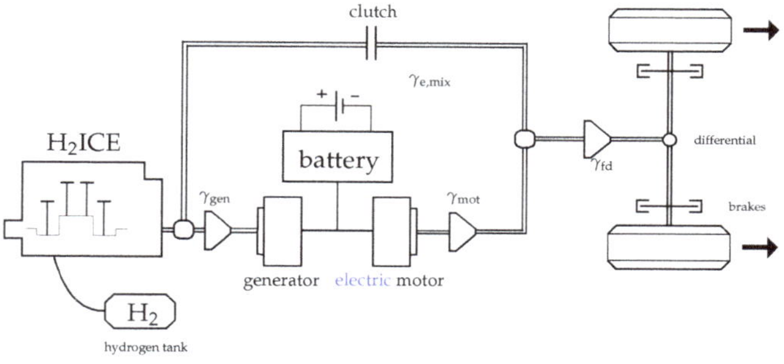

Figure 2. A schematic of the mixed H_2-HEV drivetrain considered in this paper. The double-bar lines indicate a mechanical connection. The single-bar lines indicate electrical connections.

2.1.1. Driving Modes

The driving modes are denoted by $M \in \{1, 2, 3\}$ in this work and are listed in Table 1. In series mode, the H_2ICE is operating, but it is mechanically decoupled from the wheels. Any requested power is fully delivered by the electric motor, and the H_2ICE is only used to generate electric power through the generator. In parallel mode, the H_2ICE is operating and is mechanically coupled to the wheels. In this mode, both propulsion systems (H_2ICE and electric motor) can provide power to the wheels. In EV mode, the H_2ICE is off and mechanically decoupled from the wheels. The entire torque request is met by the electric motor.

Table 1. Driving modes of the mixed H_2-HEV investigated in this paper.

Series mode	H_2ICE = ON	clutch = OPEN	$M = 1$
Parallel mode	H_2ICE = ON	clutch = CLOSED	$M = 2$
EV mode	H_2ICE = OFF	clutch = OPEN	$M = 3$

2.1.2. Rotational Speeds

Depending on the driving mode, the rotational speeds of the individual powertrain components can be calculated. From the vehicle speed (v) and the wheel radius (r_w), the wheel speed is calculated as

$$\omega_w = \frac{v}{r_w}. \tag{4}$$

As the electric motor is always mechanically coupled to the wheels, its speed is directly defined by

$$\omega_{\text{mot}} = \omega_w \cdot \gamma_{\text{fd}} \cdot \gamma_{\text{mot}}, \tag{5}$$

where γ_{fd} and γ_{mot} are the fixed gear ratios of the final drive and the electric motor, respectively. Although the engine is not always mechanically coupled to the wheels, it is always mechanically coupled to the generator. Therefore, ω_e is defined by the generator speed (ω_{gen}):

$$\omega_e = \frac{\omega_{\text{gen}}}{\gamma_{\text{gen}}}, \tag{6}$$

where γ_{gen} is the fixed gear ratio of the generator.

The generator speed depends on the driving mode. In parallel mode, ω_{gen} is fixed by the vehicle speed. In EV mode, the generator is not used, and its speed is zero. In series mode, the generator is mechanically decoupled from the wheels, but its speed can be chosen freely, as it does not depend on the vehicle speed.

$$\omega_{\text{gen}} \begin{cases} \in [\omega_{\text{gen}}^{\min}, \omega_{\text{gen}}^{\max}], & \text{if } M = 1, \\ = \omega_w \cdot \gamma_{\text{fd}} \cdot \gamma_{\text{gen}}, & \text{if } M = 2, \\ = 0, & \text{if } M = 3. \end{cases} \tag{7}$$

where $\omega_{\text{gen}}^{\min}$ and $\omega_{\text{gen}}^{\max}$ represent physical limits imposed by the engine idle speed and the maximum engine speed.

2.1.3. EMS including NO_x^{eo}

By using the map-based model and the description of the rotational speeds, the following sketch of the EMS optimization problem including NO_x^{eo} is stated:

$$\min_{T_e, T_{\text{mot}}, T_{\text{gen}}, \omega_{\text{gen}}, M} \int_{t_0}^{t_f} \dot{m}_f^{\text{map}} dt \tag{8a}$$

$$\text{s.t.} \quad \frac{d\text{SoC}}{dt} = f(\text{SoC}, I_b) \tag{8b}$$

$$\text{SoC}(t_0) = \text{SoC}_0 \tag{8c}$$

$$\text{SoC}(t_f) = \text{SoC}_0 \tag{8d}$$

$$\frac{dm_{\text{NO}_x}}{dt} = \dot{m}_{\text{NO}_x}^{\text{map}} \tag{8e}$$

$$m_{\text{NO}_x}(t_0) = 0 \tag{8f}$$

$$m_{\text{NO}_x}(t_f) = \bar{m}_{\text{NO}_x}^{\text{target}} \tag{8g}$$

$$T_{\text{req}_{\text{GB}}} = \begin{cases} T_{\text{mot}_{\text{GB}}}, & \text{if } M = 1, \\ T_{\text{e}_{\text{GB}}} + T_{\text{mot}_{\text{GB}}} + T_{\text{gen}_{\text{GB}}}, & \text{if } M \in \{2,3\} \end{cases} \tag{8h}$$

$$T_{\text{gen}_{\text{GB}}} = T_{\text{e}_{\text{GB}}}, \quad \text{if } M = 1, \tag{8i}$$

where I_b denotes the electrical current flowing from the battery and the subscript $_{\text{GB}}$ denotes the evaluation of the corresponding torques at the level of the gearbox (introduced further down in the text). The goal of this optimization problem is to minimize the cumulated hydrogen consumption over the driving mission, which starts at time instance t_0 and ends at time instance t_f. Additionally, a cumulated $\text{NO}_\text{x}^\text{eo}$ target has to be satisfied at the end of the driving mission. Constraint (8h) ensures that the torque request (T_{req}) of the driver must be met by the sum of all the drivetrain's components. Note that this optimization problem is a multi-objective, nonlinear, mixed-integer program and as such, in general, notoriously hard to solve. One time- and resource-consuming solution approach based on the DP algorithm is presented in [18]. However, the aim of this publication is to derive an online-capable controller, and the DP algorithm cannot be used for energy management due to its low computational efficiency.

One control method which is well known to cope with multi-objective goals and nonlinear system dynamics is based on the model predictive control scheme. Therefore, such a controller architecture is aimed for. However, the technique of model predictive control relies on solving an optimization problem. Therefore, the formulation of a simplified version of the sketch presented in Equation (8), which allows for fast and efficient numerical optimization, is needed. One possible ally is the family of convex optimization problems, as there exist highly efficient solvers to compute their solution quickly. In order to formulate a convex optimization problem, two steps must be taken: First, a simplified model needs to be introduced, which allows for the formulation of a convex optimization problem. Second, the mixed-integer nature of the optimal control problem in Equation (8) needs to be circumvented by dropping the driving mode from the control space.

2.2. Simplified Powertrain Model

The goal with the simplified model structure is to approximate the map-based model with the use of convex functions. Figure 3 depicts the power flows in the powertrain. The vehicle's differential and the final drive are lumped together, forming the gearbox, which has constant efficiency (η_{GB}). The power request after the gearbox is defined as

$$P_{\text{req}_{\text{GB}}} = \begin{cases} P_{\text{req}} \div \eta_{\text{GB}}, & \text{if } P_{\text{req}} \geq 0 \\ P_{\text{req}} \cdot \eta_{\text{GB}}, & \text{if } P_{\text{req}} < 0. \end{cases} \tag{9}$$

The gearboxes of the electric motor and the generator have constant efficiencies η_{mot} and η_{gen}. Based on these, the electric motor power and the generator power at the gearbox level are calculated as follows:

$$P_{\text{mot}_{\text{GB}}} = \begin{cases} P_{\text{mot}} \cdot \eta_{\text{mot}}, & \text{if } P_{\text{mot}} \geq 0, \\ P_{\text{mot}} \div \eta_{\text{mot}}, & \text{if } P_{\text{mot}} < 0, \end{cases} \tag{10}$$

$$P_{\text{gen}_{\text{GB}}} = P_{\text{gen}} \cdot \eta_{\text{gen}}, \tag{11}$$

where P_{mot} and P_{gen} are the motor's and generator's mechanical power. The electric motor losses and generator losses are denoted by $P_{\text{l}_{\text{mot}}}$ and $P_{\text{l}_{\text{gen}}}$. The sum of the electrical components' mechanical power and their losses is referred to as their source power:

$$P_{\text{s}_{\text{mot}}} = P_{\text{mot}} + P_{\text{l}_{\text{mot}}}, \tag{12}$$

$$P_{\text{s}_{\text{gen}}} = P_{\text{gen}} + P_{\text{l}_{\text{gen}}}. \tag{13}$$

The battery source power (P_{s_b}) includes the battery losses and is modeled by using a piece-wise quadratic approximation:

$$P_{\text{s}_\text{b}} = \begin{cases} r_1^+ \cdot P_b + r_2^+ \cdot P_b^2, \text{ if } P_b \geq 0 \\ r_1^- \cdot P_b + r_2^- \cdot P_b^2, \text{ if } P_b < 0. \end{cases} \tag{14}$$

where the model parameters r_1^+, r_2^+ and r_1^-, r_2^- are used to distinguish between battery charging and discharging losses. The value P_b is defined as the summed power, which is required by the electric motor, the generator, and the vehicle's auxiliaries:

$$P_b = P_{s_{\mathrm{mot}}} + P_{s_{\mathrm{gen}}} + P_{\mathrm{aux}}. \tag{15}$$

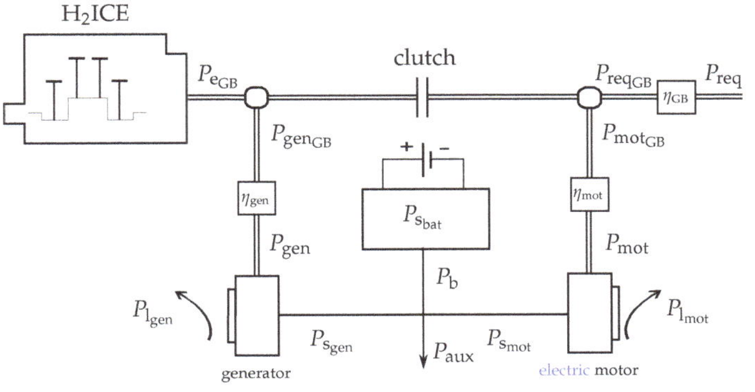

Figure 3. A schematic overview of the power flows in the mixed H_2-HEV. The double-bar lines indicate a mechanical connection. The single-bar lines indicate electrical connections.

The rest of the simplified model is divided into two different modeling structures, depending on the driving mode. In parallel mode, the powertrain components' rotational speeds are predefined by the vehicle speed, which means that speed-dependent model parameters can be utilized. In series mode, additional speed-independent model parameters need to be identified for the generator and the H_2ICE.

2.2.1. Parallel Mode

The hydrogen consumption map is approximated by using a speed-dependent quadratic Willans approach:

$$\dot{m}_{\mathrm{f}} = f_0(\omega_{\mathrm{e}}) + f_1(\omega_{\mathrm{e}}) \cdot P_{\mathrm{e_{GB}}} + f_2(\omega_{\mathrm{e}}) \cdot P_{\mathrm{e_{GB}}}^2. \tag{16}$$

Figure 4 shows the model fit for the entire hydrogen map.

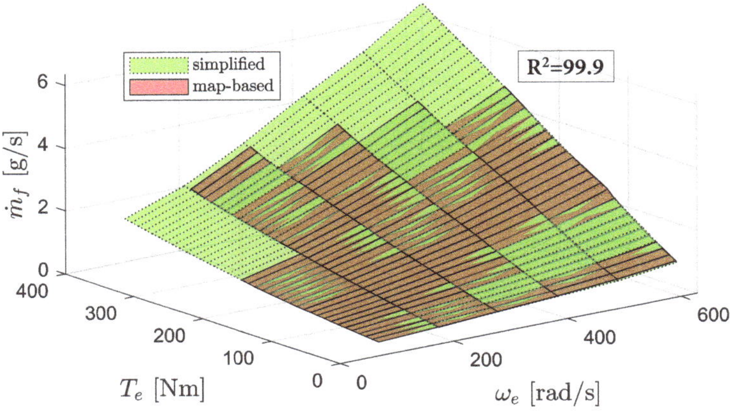

Figure 4. Comparison between the map-based model and the simplified model for the hydrogen consumption in parallel mode.

The electric motor and generator losses are both approximated by using a speed-dependent piece-wise affine model:

$$P_{\text{l}_{\text{mot}}} = \max_j\{a_{0,j}(\omega_{\text{mot}}) + a_{1,j}(\omega_{\text{mot}}) \cdot P_{\text{mot}}\}, \text{ for j} \in \{1, ..., 5\},$$
$$P_{\text{l}_{\text{gen}}} = \max_j\{b_{0,j}(\omega_{\text{gen}}) + b_{1,j}(\omega_{\text{gen}}) \cdot P_{\text{gen}}\}, \text{ for j} \in \{1, ..., 5\}. \quad (17)$$

The model accuracy is found to be sufficient when using five affine functions. Figure 5 shows the model fit for the electric motor and generator evaluated for one rotational speed of the corresponding component.

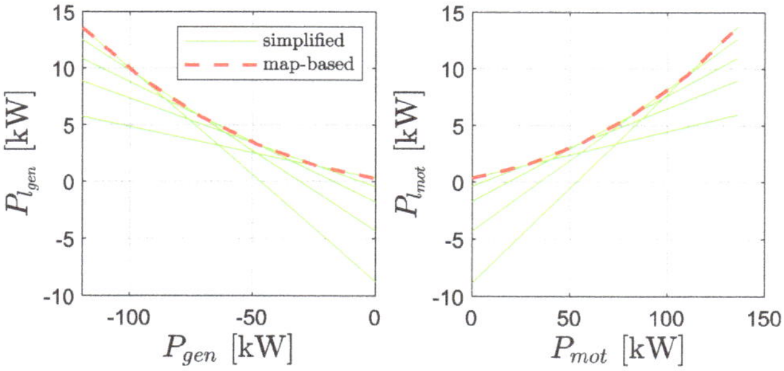

Figure 5. Comparison between the map-based model and the simplified model for the electric motor losses and the generator losses in parallel mode evaluated at $\omega_{\text{gen}} = 4000$ rpm and $\omega_{\text{mot}} = 4000$ rpm.

The $\text{NO}_\text{x}^{\text{eo}}$ emission map is approximated by using a speed-dependent piece-wise affine model:

$$\dot{m}_{\text{NO}_\text{x}} = \max_j\{d_{0,j}(\omega_e) + d_{1,j}(\omega_e) \cdot P_{e_{\text{GB}}}\}, \text{ for j} \in \{1, ..., 10\}. \quad (18)$$

The model accuracy is found to be sufficient when using 10 affine functions. Figure 6 shows the model fit for the entire $\text{NO}_\text{x}^{\text{eo}}$ emission map.

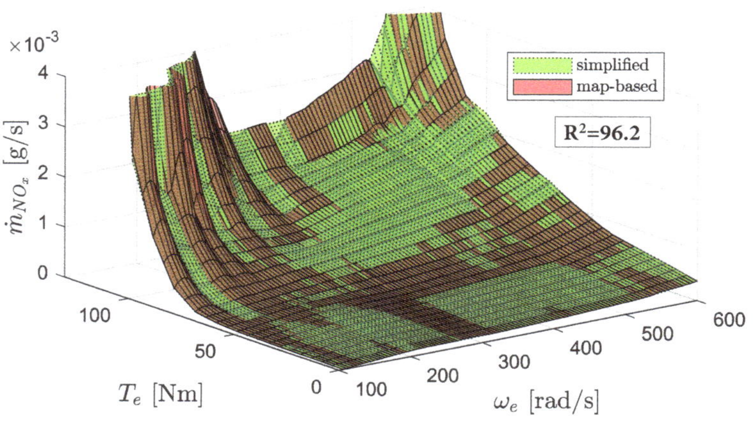

Figure 6. Comparison between the map-based model and the simplified model for the $\text{NO}_\text{x}^{\text{eo}}$ emissions in parallel mode.

2.2.2. Series Mode

Operating the vehicle in series mode detaches the engine–generator unit from the rest of the drivetrain, which allows one to choose its operating point independently of the vehicle speed. For example, to produce fixed generator power P_{gen}, the selection of $(\omega_{\mathrm{gen}}, T_{\mathrm{gen}})$ can be optimized. From a control perspective, the series mode increases the flexibility of using the individual propulsion systems to satisfy EMS controller goals. From a system modeling perspective, however, the additional degree of freedom impedes the formulation of convex functions. This challenge is tackled by performing the pre-optimization of the engine–generator unit, covering all its operating conditions. The goal is to predefine, for each obtainable generator power, which combinations of generator speed and generator torque are optimal candidates with respect to hydrogen consumption and/or NO_x^{eo} emissions. The pre-optimization is based on the following extended cost function:

$$J = \dot{m}_f + \psi \cdot \dot{m}_{\mathrm{NO_x}}. \tag{19}$$

where the trade-off factor, ψ, is used as a weight factor to reflect the H_2-NO_x^{eo} trade-off. Algorithm 1 is used to perform the pre-optimization. In the following, it is explained in detail:

- Step 1: The generator power and the trade-off factor (ψ) are discretized.
- For each realizable P_{gen}, all possible combinations (ω_{e}, T_{e}) that result in $P_{\mathrm{gen}}(j)$ are identified. By using the map-based model, the corresponding hydrogen consumption $\dot{m}_f^{\mathrm{map}}$ and the NO_x^{eo} emissions $\dot{m}_{\mathrm{NO_x}}^{\mathrm{map}}$ are calculated (steps 3–5).
- By looping over all ψ, Equation (19) is used to formulate the extended cost for all identified pairs of (ω_{e}, T_{e}) and the corresponding trade-off weight ψ (step 7).
- Minimizing the extended cost function over all previously identified operating points (ω_{e}, T_{e}) yields the optimal engine operating point (ω_{e}^*, T_{e}^*) for the corresponding ψ (step 8).
- Finally, for the generator power ($P_{\mathrm{gen}}(j)$) and the trade-off parameter (ψ), the following optimal values are stored for later use: optimal engine power (P_{e}^*), optimal hydrogen consumption ($\dot{m}_f^*$), and optimal NO_x^{eo} emissions ($\dot{m}_{\mathrm{NO_x}}^*$) (steps 9–11).

Algorithm 1 Pre-optimization for series mode.

1: $\vec{P}_{\mathrm{s_{gen}}} \leftarrow$ Uniformly distributed vector $\in [-80\ \mathrm{kW}, 0\ \mathrm{kW}]$
2: $\vec{\psi} \leftarrow$ Uniformly distributed vector $\in [0, 150]$
3: $\vec{\omega}_{\mathrm{e}} \leftarrow$ Uniformly distributed vector $\in [\omega_e^{\mathrm{min}}, \omega_e^{\mathrm{max}}]$
4: **for** $j \in \{1, 2, ..., |\vec{P}_{\mathrm{s_{gen}}}|\}$ **do**
5: $\quad\vec{T}_{\mathrm{e}} \leftarrow$ for each entry in $\vec{\omega}_{\mathrm{e}}$, find T_{e} that results in $\vec{P}_{\mathrm{s_{gen}}}[j]$ according to Equations (3), (11) and (13)
6: $\quad\vec{\dot{m}}_f \leftarrow f(\vec{\omega}_{\mathrm{e}}, \vec{T}_{\mathrm{e}})$ ▷ Equation (1)
7: $\quad\vec{\dot{m}}_{\mathrm{NO_x}} \leftarrow f(\vec{\omega}_{\mathrm{e}}, \vec{T}_{\mathrm{e}})$ ▷ Equation (2)
8: $\quad$**for** $k \in \{1, 2, ..., |\vec{\psi}|\}$ **do**
9: $\quad\quad\vec{J} \leftarrow \vec{\dot{m}}_f + \psi[k] \cdot \vec{\dot{m}}_{\mathrm{NO_x}}$ ▷ Equation (19)
10: $\quad\quad i^* \leftarrow \arg\min \vec{J}$ ▷ index of minimizer in $\vec{J}$
11: $\quad\quad\dot{m}_f^* \leftarrow \vec{\dot{m}}_f[i^*]$
12: $\quad\quad\dot{m}_{\mathrm{NO_x}}^* \leftarrow \vec{\dot{m}}_{\mathrm{NO_x}}[i^*]$
13: $\quad\quad P_{\mathrm{e_{GB}}}^* \leftarrow \vec{\omega}_{\mathrm{e}}[i^*] \cdot \vec{T}_{\mathrm{e}}[i^*]$
14: $\quad\quad P_{\mathrm{s_{gen}}}^* \leftarrow \vec{P}_{\mathrm{s_{gen}}}[j]$
15: $\quad\quad\psi^* \leftarrow \psi[k]$
16: $\quad\quad$Store $(P_{\mathrm{e_{GB}}}^*, \psi^*, \dot{m}_f^*, P_{\mathrm{s_{gen}}}^*, \dot{m}_{\mathrm{NO_x}}^*)$
17: $\quad$**end for**
18: **end for**
19: **Return:** $\dot{m}_f = f(\psi, P_{\mathrm{e_{GB}}})$ ▷ based on stored variables
20: **Return:** $P_{\mathrm{s_{gen}}} = f(\psi, P_{\mathrm{e_{GB}}})$ ▷ based on stored variables
21: **Return:** $\dot{m}_{\mathrm{NO_x}} = f(\psi, P_{\mathrm{e_{GB}}})$ ▷ based on stored variables

Here, the lower bound $\underline{P}_{s_{gen}} = -80$ kW is sufficiently large to cover the entire relevant range of engine–generator usage. The upper generator bound of 0 kW is set, as the generator is not used to provide additional mechanical propulsion power at the wheel. Setting the trade-off factor to its lower bound of $\psi = 0$ results in an objective solely concerned about hydrogen consumption. Setting $\psi = 150$, on the other hand, results in an objective that is solely concerned about NO_x^{eo} emissions. Further increasing ψ does not change the result of step 8. Thus, selecting $\psi \in [0, 150]$ represents the complete achievable H_2-NO_x^{eo} trade-off.

Based on the results of the pre-optimization, the following model parameters are fitted. Notably, compared with the modeling approach for the parallel mode, these mappings have no explicit dependency on rotational speeds.

Figure 7 shows the approximation for the generator source power and the hydrogen consumption based on above pre-optimized values. The generator source power is approximated by using a hyperplane:

$$P_{s_{gen}} = \beta_0 + \beta_1 \cdot P_{e_{GB}} + \beta_2 \cdot \psi. \quad (20)$$

The hydrogen consumption is approximated by using multiple hyperplanes:

$$\dot{m}_f = \max_j \{\phi_{0,j} + \phi_{1,j} \cdot P_e + \phi_{2,j} \cdot \psi\}, \text{ for j} \in \{1, ..., 5\}. \quad (21)$$

The model accuracy is found to be sufficient when using 5 hyperplanes.

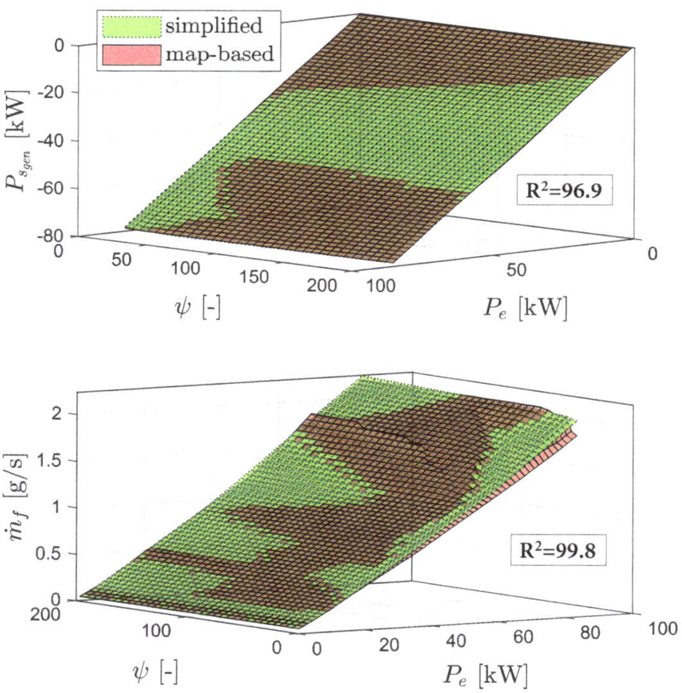

Figure 7. (**Upper plot**) Comparison between the map-based model and the simplified model for the hydrogen consumption in series mode. (**Lower plot**) Comparison between the map-based model and the simplified model for the generator source power in series mode.

Figure 8 depicts the approximation of the NO_x^{eo} emissions in series mode, consisting of multiple hyperplanes.

$$\dot{m}_{NO_x} = \max_j\{\delta_{0,j} + \delta_{1,j} \cdot P_e + \delta_{2,j} \cdot \psi\}, \text{ for j} \in \{1, ..., 17\}. \tag{22}$$

The fitting process is performed by using a slightly adapted version of the convex adaptive partitioning algorithm [19].

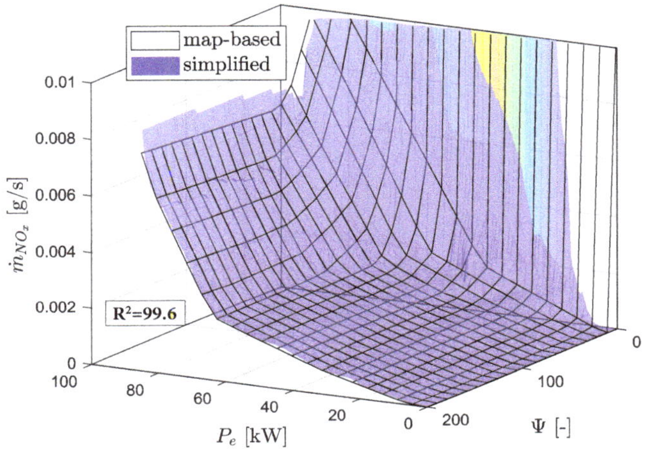

Figure 8. Comparison between the map-based model and the simplified model for the NO_x^{eo} emissions in series mode.

The convex adaptive partitioning algorithm is free to decide the number of hyperplanes and converges at the number of 17.

2.3. Optimization Parameters

The simplified model provides an approximation of the drivetrain components. Next, to arrive at the formulation of an optimization problem, the set of dynamic states and the set of control inputs are described.

The dynamic states are the accumulated NO_x^{eo} and the SoC:

$$\mathbf{x} = \begin{bmatrix} m_{NO_x} \\ \text{SoC} \end{bmatrix}. \tag{23}$$

The differential equation for the accumulated NO_x^{eo} is defined by Equations (18) and (22) and reads as

$$\frac{\mathrm{d}m_{NO_x}}{\mathrm{d}t} = \begin{cases} \delta_{1,j} \cdot P_e + \delta_{2,j} \cdot \psi + \delta_{0,j}, & \text{if } M = 1 \\ d_{1,j} \cdot P_e + d_{0,j}, & \text{if } M = 2. \end{cases} \tag{24}$$

The differential equation for the SoC is defined as follows:

$$\frac{\mathrm{dSoC}}{\mathrm{d}t} = -\frac{P_{s_b}}{Q_0}, \tag{25}$$

where Q_0 denotes the battery capacity and P_{s_b} denotes the battery source power.

The control inputs for the simplified powertrain model are given in the vector

$$\mathbf{u} = \begin{bmatrix} P_{\mathrm{e_{GB}}} \\ P_{\mathrm{mot_{GB}}} \\ P_{\mathrm{gen_{GB}}} \\ \psi \end{bmatrix}. \tag{26}$$

A validation of the simplified model description is presented in Figure 9. Both models are simulated by using the input trajectories depicted by the black lines. The resulting trajectories for SoC and $\mathrm{NO_x^{eo}}$ and the resulting hydrogen consumption are depicted by using the red lines and the green lines.

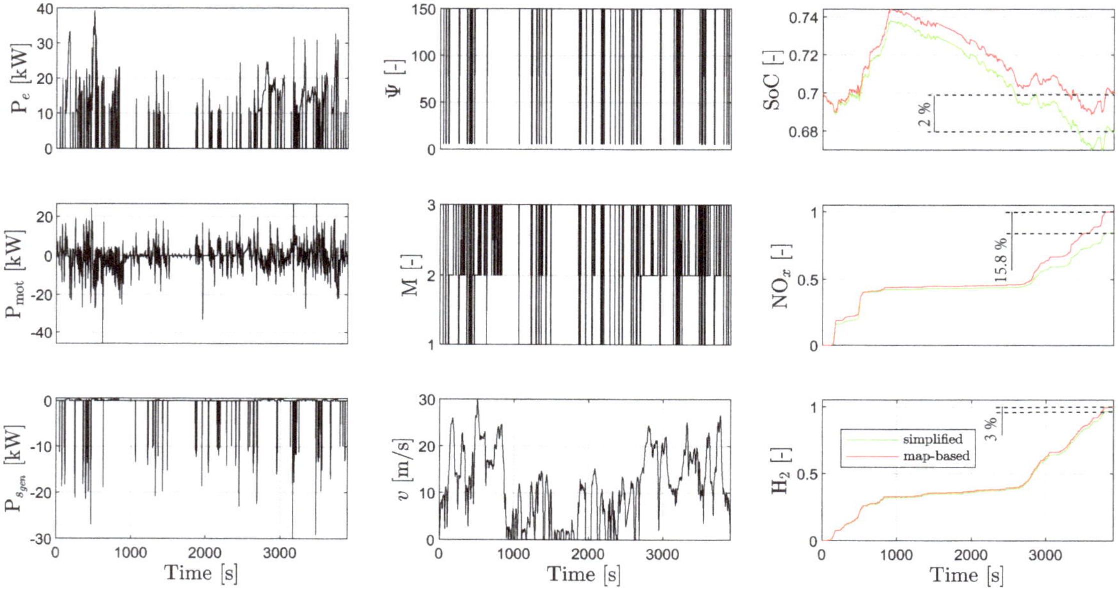

Figure 9. Comparison between the map-based and the simplified models. The black trajectories are used as inputs. The resulting state trajectories of the map-based and the simplified models are visualized with red and green lines, respectively.

An offset results when the input trajectories are evaluated on the simplified model. The offset of 3% in hydrogen consumption and the offset of 2% in the final SoC are limited. However, the offset of 15.8% in cumulated $\mathrm{NO_x^{eo}}$ emissions is larger. Nevertheless, the accuracy of the simplified model was found to be good enough for its use as a control-oriented model.

3. Control-Oriented Optimization Problem

In this section, the simplified model description is leveraged to formulate a convex optimization problem. However, in order to omit the formulation of a mixed-integer program (see Equation (8)), a time-resolved trajectory of integer driving mode M has to be found in advance.

3.1. Driving Mode Estimation

The estimation of the optimal driving mode, M, is based on the work published in [20]. Here, the authors propose a map which is defined on the $(P_{\mathrm{req}} - v)$ plane to estimate the optimal driving mode. If the map is chosen carefully, a close-to-optimal estimation of the driving mode is possible based on the vehicle velocity and the power request. An exemplary map used in this work is depicted in Figure 10. It is obtained by solving Equation (8) by using the DP algorithm for a specific driving mission. The parameters P_{crit}, v_{crit}, and P_{par} describe regions to estimate the optimal driving mode. By assuming predictive knowledge of the vehicle's upcoming time-resolved velocity and power request, a time-resolved estimation of the future optimal driving mode can also be made.

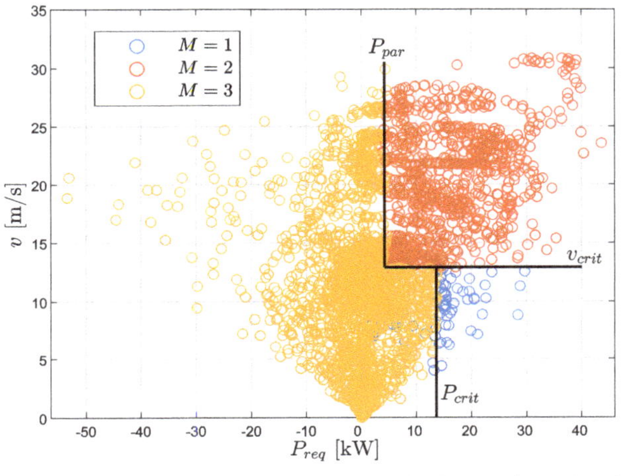

Figure 10. Optimal driving mode distribution for the mixed H_2-HEV, obtained with the DP algorithm.

3.2. Convex Optimization Problem

Based on an estimation of the time-resolved trajectory for the integer driving mode (M), the convex optimization problem, which is from here on referred to as the $\mathcal{COP}$, is explained step-by-step in the following.

3.2.1. Cost Function and Dynamics

The cost function and dynamic state equations of the $\mathcal{COP}$ read as follows:

$$\min_{\mathbf{u}} \int_{t_0}^{t_f} (\dot{m}_f + \kappa_{\mathrm{P_e}} \cdot \epsilon_{\mathrm{P_e}})\, dt + \kappa_{\mathrm{NO_x}} \cdot \epsilon_{\mathrm{NO_x}} \tag{27a}$$

$$\text{s.t.} \quad \frac{\mathrm{dSoC}}{\mathrm{d}t} = \text{Equation (25)} \tag{27b}$$

$$\mathrm{SoC}(t_0) = \mathrm{SoC}_0 \tag{27c}$$

$$\mathrm{SoC}(t_f) \geq \bar{\mathrm{SoC}}^{\mathrm{RTG}} \tag{27d}$$

$$\frac{\mathrm{d}m_{\mathrm{NO_x}}}{\mathrm{d}t} = \text{Equation (24)} \tag{27e}$$

$$m_{\mathrm{NO_x}}(t_0) = m_{\mathrm{NO_{x,0}}} \tag{27f}$$

$$m_{\mathrm{NO_x}}(t_f) \leq \bar{m}_{\mathrm{NO_x}}^{\mathrm{RTG}} + \epsilon_{\mathrm{NO_x}} \tag{27g}$$

$$P_{\mathrm{eGB}} \leq P_{\mathrm{eGB}}^{\max} + \epsilon_{\mathrm{P_e}} \tag{27h}$$

$$\epsilon_{\mathrm{P_e}} \geq 0 \tag{27i}$$

$$\epsilon_{\mathrm{NO_x}} \geq 0 \tag{27j}$$

The cost function (27a) consists of three parts: First, there is the integral of the hydrogen consumption, $\dot{m}_f$, just as in the original formulation in (8a). Second, there is the integral of the quantity $\kappa_{\mathrm{P_e}} \cdot \epsilon_{\mathrm{P_e}}$, which, together with Equation (27h,i), represents a soft constraint on engine power. It is used to ensure that the optimization problem remains feasible, even if an unrealistically high power request is predicted. A detailed explanation is provided in [20]. Third, there is the scalar term $\kappa_{NO_x} \cdot \epsilon_{NO_x}$, which, together with Equation (27g,j), represents a soft constraint on accumulated NO_x^{eo}. It is used to ensure that disturbances leading to increased NO_x^{eo} cannot lead to an infeasible optimization problem. It is important to ensure that constraints are only softened if no feasible solution with strict constraint compliance exists. Thus, the weights κ_{P_e} and κ_{NO_x} have to be at least as high as the corresponding optimal Lagrange multipliers for the original problem [21]. Since these Lagrange multipliers are not known a priori, the weighting is chosen very large, i.e., 10^5, here.

The definition of the dynamic states requires three ingredients: First, both dynamic states are initialized with the values of SoC_0 and $m_{NO_{x,0}}$ (Equation (27c,f)). Second, the evolution of the dynamic states is defined by Equation (27b,e). Third, the terminal conditions for both dynamic states are given by Equation (27d,g).

3.2.2. Power Split

The goal of the $\mathcal{COP}$ is to satisfy the time-resolved power request trajectory optimally by using the optimal control inputs $\mathbf{u}$. In the following, the power balance is described depending on the driving mode:

$$\underline{M = 1:} \quad \begin{cases} P_{\mathrm{req_{GB}}} & = P_{\mathrm{mot_{GB}}} \\ P_{\mathrm{s_{gen}}} & = \beta_1 \cdot P_{\mathrm{e_{GB}}} + \beta_2 \cdot \psi + \beta_0 \end{cases} \tag{28}$$

$$\underline{M = 2:} \quad \begin{cases} P_{\mathrm{req_{GB}}} & = P_{\mathrm{gen_{GB}}} + P_{\mathrm{mot_{GB}}} + P_{\mathrm{e_{GB}}} \end{cases} \tag{29}$$

$$\underline{M = 3:} \quad \begin{cases} P_{\mathrm{req_{GB}}} & = P_{\mathrm{mot_{GB}}} \\ P_{\mathrm{e_{GB}}} & = 0 \\ P_{\mathrm{gen_{GB}}} & = 0 \end{cases} \tag{30}$$

Based on the optimal power distribution, the hydrogen consumption, NO_x^{eo} emissions, and electric losses result are formulated by using constraint relaxations.

3.2.3. Constraint Relaxations

In the following, equality constraints (16)–(18), (21), and (22) are relaxed, which is necessary to obtain a convex feasible set for the optimization parameters:

$$\dot{m}_f \geq \begin{cases} \alpha_i P_{\mathrm{e_{GB}}} + \beta_i s + \gamma_i, i \in \{1,..,5\}, & \text{if } M = 1 \\ f_0 + f_1 P_{\mathrm{e_{GB}}} + f_2 P_{\mathrm{e_{GB}}}^2, & \text{if } M = 2 \\ 0, & \text{if } M = 3, \end{cases} \tag{31a}$$

$$\dot{m}_{\mathrm{NO_x}} \geq \begin{cases} \alpha_i P_{\mathrm{e_{GB}}} + \beta_i s + \gamma_i, i \in \{1,..,17\}, & \text{if } M = 1 \\ \alpha_i P_{\mathrm{e_{GB}}} + \gamma_i, i \in \{1,..,10\}, & \text{if } M = 2 \\ 0, & \text{if } M = 3, \end{cases} \tag{31b}$$

$$P_{\mathrm{l_{gen}}} \geq \begin{cases} 0, & \text{if } M \in \{1, 3\} \\ \alpha_i P_{\mathrm{gen}} + \gamma_i, i \in \{1,..,5\}, & \text{if } M = 2, \end{cases} \tag{31c}$$

$$P_{\mathrm{l_{mot}}} \geq \alpha_i P_{\mathrm{mot}} + \gamma_i, i \in \{1,..,5\}. \tag{31d}$$

An in-depth explanation of constraint relaxations is presented in [22]. However, intuitively, each parameter subject to the relaxed Equation (31a–d) should be as small as possible to accumulate as little hydrogen consumption, NO_x^{eo} emissions, or electrical losses as possible. Therefore, the equations hold with equality if used within an optimization problem.

3.2.4. Battery

The following set of constraints reformulates the convex battery model introduced by Equations (14) and (15):

$$P_b^{\text{tot}} = P_{\text{s}_{\text{mot}}} + P_{\text{s}_{\text{gen}}} + P_{\text{aux}} \tag{32a}$$
$$P_b^{\text{tot}} = P_b^+ + P_b^- \tag{32b}$$
$$P_{\text{s}_\text{b},1} \geq r_1^+ P_b^+ + r_2^+ (P_b^+)^2 \tag{32c}$$
$$P_{\text{s}_\text{b},2} \geq r_1^- P_b^- + r_2^- (P_b^-)^2 \tag{32d}$$
$$P_{\text{s}_\text{b}} \geq P_{\text{s}_\text{b},1} + P_{\text{s}_\text{b},2} \tag{32e}$$
$$P_b^+ \geq 0 \tag{32f}$$
$$P_b^- \leq 0. \tag{32g}$$

The auxiliary variables P_b^{tot}, P_b^+, P_b^-, $P_{\text{s}_\text{b},1}$, and $P_{\text{s}_\text{b},2}$ are introduced to formulate the piecewise quadratic battery model using convex set representations. Equation (32c,d) are relaxed equality constraints.

3.2.5. Input and State Domains

The feasible control input domain and state domain are given in the following:

$$\mathbf{u} \in \mathcal{U} \tag{33a}$$
$$\mathbf{x} \in \mathcal{X}, \tag{33b}$$

where $\mathcal{U}$ includes the physical limitations of the propulsion systems. Feasible set $\mathcal{X}$ consists of constant upper and lower battery bounds, i.e., SoC $\in [0.5, 0.9]$. It is important to note that the $\mathcal{COP}$ is a convex optimization problem only if the driving mode (M) is known in advance. The $\mathcal{COP}$ is parsed by using the software package YALMIP, R2023 [23] and solved by using the solver MOSEK, v10.1.31 [24]. Euler forward integration is used, and the time discretization is set to 1 s.

4. Controller Structure

In this section, the proposed control architecture to calculate the vehicle control inputs is presented. The overall controller architecture is presented in Figure 11 and is based on the work presented in [20]. The controller consists of three controller levels: the *lower-level controller*; the so-called *MPC*; and the reference trajectory generator, which is from here on referred to as the *RTG*. Each controller level receives a power request signal, i.e., P_{req}, $\hat{P}_{\text{req}}$, and $\bar{P}_{\text{req}}$, respectively. The corresponding power request signals are either determined by the driver or calculated by using the elevation profile (Γ) of the driving mission and a corresponding predicted velocity signal ($\bar{v}$ or $\hat{v}$). As predictive data are likely to be imperfect and of lower accuracy for long-term predictions than for short-term predictions, the accents $\bar{\{\}}$ and $\hat{\{\}}$ are used to denote signals of differing prediction accuracy. The corresponding power request signals are calculated in the block labeled PR by using a set of equations of the longitudinal vehicle dynamics [25].

In the following, a bottom-up approach is used to explain the overall controller structure in detail. The focus is on the communication between the individual controller levels.

4.1. Lower-Level Controller

The *lower-level controller* calculates the powertrain control inputs T_e, T_{mot}, T_{gen}, ω_{gen}, and M in a real-time-feasible fashion. It is a non-predictive controller and obtains the current value P_{req}. The control inputs are passed on to the vehicle plant model, which can be interpreted as a digital twin of the mixed H_2-HEV. The battery SoC and the accumulated $\text{NO}_\text{x}^{\text{eo}}$ emissions can be accessed from the vehicle plant model in a causal fashion. They are denoted with SoC^{meas} and $m_{\text{NO}_\text{x}}^{\text{meas}}$. The *lower-level controller* is based on the concept of Pontryagin's minimum principle (PMP) and requires knowledge of the so-called costates, which can be interpreted as equivalent cost terms in a local Hamiltonian minimization problem [26]. The theory of PMP states that for an optimization problem, there exists one costate variable for each dynamic state variable. Therefore, two costates are required for the EMS at hand: λ_{SoC} and λ_{NO_x}. Given an estimation of the costates, the calculation of controller inputs based on the application of PMP has already been shown multiple times (see, e.g., [27]) and is thus not further discussed in this text.

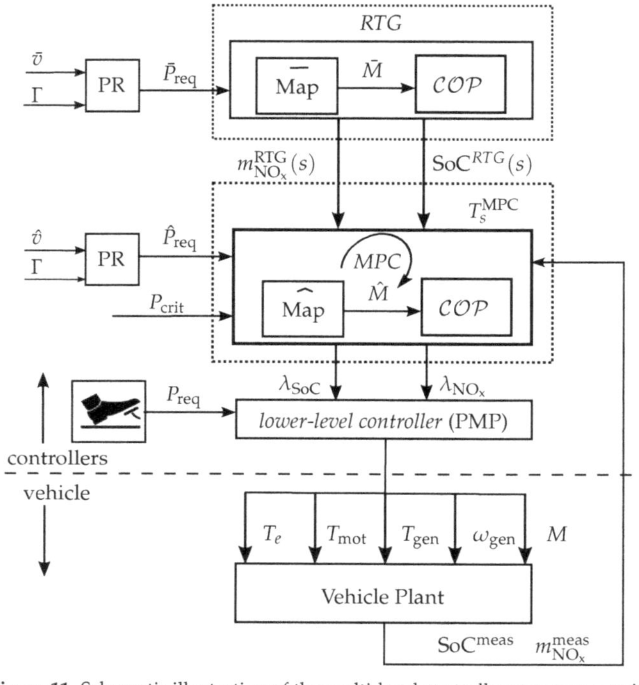

Figure 11. Schematic illustration of the multi-level controller structure consisting of three controller levels: *RTG*, *MPC*, and *lower-level controller*.

4.2. MPC

The *MPC* is used to estimate the value of the optimal costates, which are required by the *lower-level controller*. It solves the $\mathcal{COP}$ based on medium-term time-resolved predictive information for the power request of the upcoming driving mission. In Equation (27), the parameter t_f is replaced by $t_0 + H_p$, where H_p denotes the length of the predictive information and is called the prediction horizon. The *MPC* also consists of a map, denoted in Figure 11 by the block $\widehat{\text{Map}}$, which is defined on the $(\hat{P}_{\text{req}} - \hat{v})$ plane, and the $\mathcal{COP}$. This map is not static but is adapted over time by using a learning algorithm, which is explained in detail in [20]. The controller is updated every T_s^{MPC} seconds, where each update includes the following steps: The map is used to estimate the time-resolved optimal driving mode $\hat{M}$ based on the predicted power request. The dynamic states of the $\mathcal{COP}$ (Equation (27c,f)) are initialized by using the signals SoC^{meas} and $m_{\text{NO}_x}^{\text{meas}}$. The terminal conditions for the dynamic states (Equation (27d,g)) are set. This is performed by using interpolation of the spatially defined vectors $\text{SoC}^{\text{RTG}}(s)$ and $m_{\text{NO}_x}^{\text{RTG}}(s)$ and the predicted vehicle position $s(t_0 + H_p)$ at the end of the prediction horizon. The two costates λ_{SoC} and λ_{NO_x} are obtained by solving the $\mathcal{COP}$ [20].

4.3. *Reference Trajectory Generator*

The reference trajectory generator is used to calculate suitable reference state trajectories for the *MPC*. The use of reference trajectories for online-capable EMS algorithms is a well-documented technique (see, e.g., [28]). Typically, these approaches only encompass a reference for the battery. In order to ensure the possibility to aim at a user-defined long-term NO_x^{eo} target $\bar{m}^t_{NO_x}$, a NO_x^{eo} trajectory also needs to be included in this work. The *RTG* is a predictive algorithm, receiving long-term time-resolved predictive information for the power request of the entire driving mission. It consists of a map, denoted in Figure 11 by the block $\bar{\text{Map}}$, which is defined on the $(\bar{P}_{\text{req}} - \bar{v})$ plane, and the $\mathcal{COP}$. The map is used to estimate the time-resolved optimal driving mode $\bar{M}$ based on the predicted power request. The $\mathcal{COP}$ in the *RTG* is set up only once and solved prior to departure. The initialization of the dynamic states is performed by using the current battery SoC and $m_{NO_{x,0}} = 0$ (Equation (27c,f)). The terminal condition for the battery is set to be charge-sustaining (Equation (27d)), and the terminal condition for the cumulated NO_x^{eo} emissions is set to the user-defined long-term target $\bar{m}^t_{NO_x}$.

Algorithm 2 explains the procedure for obtaining reference trajectories. First, the algorithm is initialized with a user-defined NO_x^{eo} target and a critical power request ($P^{\text{RTG}}_{\text{crit}}$) (see Figure 10). After calculating the predicted power request for the entire driving mission, the $\mathcal{COP}$ is solved iteratively by using the estimated optimal driving mode, resulting from the current $P^{\text{RTG}}_{\text{crit}}$. If the optimal control problem can be solved while respecting the user-defined NO_x^{eo}, i.e., $\epsilon^{\text{RTG}}_{NO_x} = 0$, the resulting optimal state trajectories are used as reference trajectories for the *MPC*. In most cases, only one iteration suffices. However, Equation (27g) is checked for slacking first. If this constraint requires slacking, then the resulting NO_x^{eo} trajectory does not satisfy the user-defined accumulated NO_x^{eo} target. If this trajectory were used as a reference for the *MPC*, the controller would track a NO_x^{eo} target, which is different from the user-defined target by design. If slacking is needed, $P^{\text{RTG}}_{\text{crit}}$ is decreased for the next iteration. This allows one to use the series mode more often, allowing for more ultra-low NO_x^{eo} operating points of the H_2ICE. During simulation, it was seen that decreasing $P^{\text{RTG}}_{\text{crit}}$ below -40 kW did not lead to noticeable changes in the solution of the $\mathcal{COP}$. It was, therefore, used as a termination criterion. If the termination criterion is met but slacking is still required, the resulting reference state trajectories are used. However, the NO_x^{eo} state trajectory is linearly scaled to $\bar{m}^t_{NO_x}$. Although such a post-processing results is an infeasible NO_x^{eo} state trajectory for the *RTG*, it can be feasible for the *MPC*, which calculates optimal solutions based on different predictive data.

Algorithm 2 *RTG* iterations.

Initialize:
  $\bar{m}^t_{NO_x}$
  $P^{\text{RTG}}_{\text{crit}} = 12{,}000$ W
  $\epsilon^{\text{RTG}}_{NO_x} = 1$
$\bar{P}^{\text{RTG}}_{\text{req}} \leftarrow \bar{\Gamma}, \bar{v}$
while $\epsilon^{\text{RTG}}_{NO_x} \neq 0 \mid P^{\text{RTG}}_{\text{crit}} > -40$ kW **do**
  $M \leftarrow$ map $\{\bar{P}_{\text{req}}, \bar{v}\}$
  $\{\text{SoC}^{\text{RTG}}, m^{\text{RTG}}_{NO_x}, \epsilon^{\text{RTG}}_{NO_x}\} \leftarrow$ solve $\mathcal{COP}(\bar{P}_{\text{req}}, M)$
  if $\epsilon^{\text{RTG}}_{NO_x} \neq 0$ **then**
    $P^{\text{RTG}}_{\text{crit}} = P^{\text{RTG}}_{\text{crit}} - 500$ W
  end if
end while

5. Case Study

This section provides a case study on the performance of the *MPC*. To quantify the performance of controllers, the following relative performance criteria are used:

$$e_{\mathrm{NO_x}} = 1 - \frac{m_{\mathrm{NO_x}}(t_f)}{\bar{m}^t_{\mathrm{NO_x}}}, \tag{34}$$

$$e_{\mathrm{SoC}} = 1 - \frac{\mathrm{SoC}(t_f)}{\mathrm{SoC}(t_0)}, \tag{35}$$

$$e_{\mathrm{H_2}} = \left(m_{\mathrm{H_2}}(t_f) + \frac{Q_0 \cdot (\mathrm{SoC}(t_f) - \mathrm{SoC}(t_0))}{\eta^{\mathrm{const}}_{\mathrm{H_2ICE}} \cdot \eta^{\mathrm{const}}_{\mathrm{gen}} \cdot H_l} \right) / m^{\mathrm{DP}}_{\mathrm{H_2}} - 1. \tag{36}$$

The first two criteria, i.e., the $\mathrm{NO_x^{eo}}$ error ($e_{\mathrm{NO_x}}$) and the SoC error (e_{SoC}), quantify the violation of the user-defined $\mathrm{NO_x^{eo}}$ target and the charge sustainability constraint at the end of the driving mission, respectively. The corrected hydrogen error ($e_{\mathrm{H_2}}$) denotes the additional fuel spent with respect to the DP-optimal result. To account for the additional energy stored in the battery at the end of the driving mission, the fuel consumption is corrected by using the lower heating value of hydrogen (H_l), constant engine efficiency $\eta^{\mathrm{const}}_{\mathrm{H_2ICE}} = 35$ %, and constant generator efficiency $\eta^{\mathrm{const}}_{\mathrm{gen}} = 90$ %.

For all results, the *MPC*'s prediction horizon is set to 450 s. The *MPC*'s update time is $T_s = 2$ s. The map-based model is used as the vehicle plant.

Overall, after introducing the driving missions, the subsequent case study is built in three steps, which together form an inductive approach to analyze the proposed controller's performance. First, the controller's ability to adhere to a single $\mathrm{NO_x^{eo}}$ target is investigated on one specific driving mission. Second, on the same driving mission, the full range of achievable $\mathrm{NO_x^{eo}}$ targets is investigated. Third, the controller's performance is analyzed for a broad range of different realistic driving missions.

5.1. Driving Missions

Figure 12 depicts the four driving missions that are used in the remainder of this text to test the performance of the suggested EMS algorithm. The first driving mission is extracted from the open-source software SUMO, v1.20.0 [29] and is referred to as *urban cycle*. The second driving mission is based on recorded sensor data of a real vehicle and features a mix of urban driving, rural driving, and highway driving and is referred to as *real driving cycle*. The third and fourth driving missions are generated from combinations of measured and simulated velocity profiles. They are used to represent extreme driving conditions and are referred to as *mountain driving cycle* and *highway driving cycle*.

The driving missions are visualized in Figure 12 over the travel distance s. The visualization includes three velocity trajectories for each mission in the upper plot and the elevation profile in the lower plot. The blue lines depict the actual vehicle speed v, which is interpreted as the driver's request and is not known in advance by the *MPC* and the *RTG*. The red lines depict a predicted velocity ($\hat{v}$), which is available to the *MPC*. The dashed black lines depict a crude velocity estimation ($\bar{v}$), which could represent, for instance, the legal speed limits.

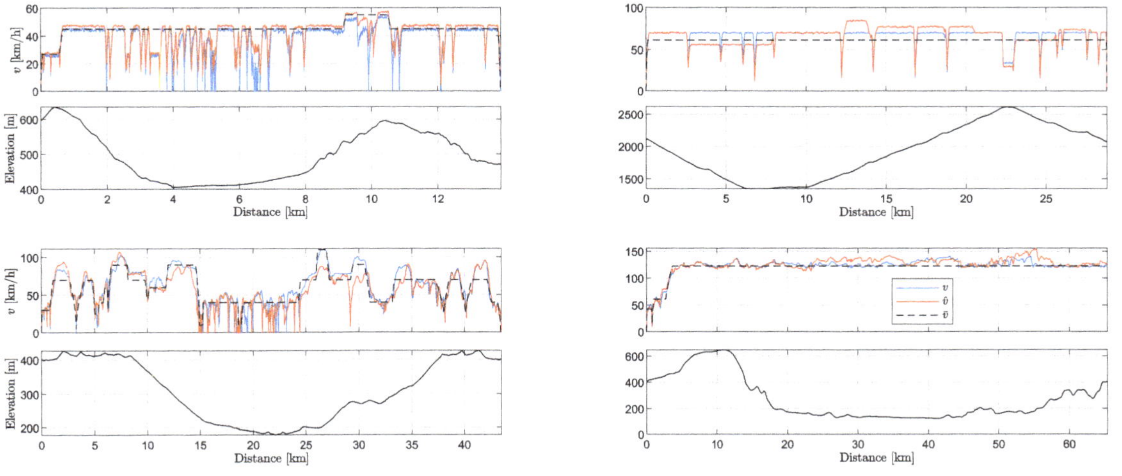

Figure 12. Driving missions investigated in the case study. The blue velocity lines are the actual driving missions. The red lines are predicted signals, which are available to the *MPC*. The black lines are crude velocity predictions, which are available to the *RTG*.

5.2. Single-NO_x^{eo}-Target Adherence

First of all, the *MPC*'s ability to comply with a user-defined NO_x^{eo} target is analyzed. Here, two sources of disturbance are investigated. First, the model mismatch between the map-based model and the simplified model is always present. Second, the *MPC* relies on inaccurate predictive data. To distinguish between the performance losses resulting from either of the two disturbances, a *Prescient MPC*, which has access to the entire trajectory of the actual vehicle speed for the entire driving mission, is introduced. Together with the full elevation profile, the *Prescient MPC*, therefore, has perfect knowledge of the entire upcoming driving mission, which means that the only disturbance to this controller remains the model mismatch.

Figure 13 shows a comparison between the results obtained by the DP algorithm, the *Prescient MPC*, and the *MPC* on the *real driving cycle*. For all three methods, the same NO_x^{eo} target is chosen. The upper plot shows the accumulated NO_x^{eo} emissions, normalized by the amount achieved by the DP algorithm. Both MPC controllers are able to satisfy the NO_x^{eo} target. When looking at the three trajectories, it can be observed that all three methods have a similar way of distributing the NO_x^{eo} emissions across the driving mission: In time window **A**, the NO_x^{eo} emissions are comparably high. In time window **B**, all algorithms choose to emit less NO_x^{eo} emissions than in the first part of the driving mission. Finally, in time window **C**, all three algorithms show increased NO_x^{eo} emissions. The *Prescient MPC* yields a NO_x^{eo} error of 0.01%, and the *MPC* yields a NO_x^{eo} error of 0.4%. It is, therefore, concluded that realistic velocity mispredictions have a very low impact on satisfying the NO_x^{eo} target on this driving mission and for this specific NO_x^{eo} target.

The middle plot depicts the SoC trajectories. In time window **D**, the *MPC* chooses to divert from the SOC trajectory chosen by the DP algorithm and the *Prescient MPC*. This is a result of the mispredictions and its reduced prediction horizon. Even though, in time window **E**, both model predictive controllers choose a less aggressive use of the battery than the DP algorithm, both algorithms are able to operate the vehicle in charge-sustaining mode.

The bottom plot depicts the corrected hydrogen error of the three algorithms. Time windows **D** and **E** highlight the reflection of the algorithms' different choices of battery usage in their corresponding hydrogen allocation strategy. If an algorithm chooses to charge the battery more than the DP algorithm, its hydrogen consumption is increased as a result of transferring energy from the hydrogen storage into the battery. Ultimately, both model predictive controllers result in close-to-optimal hydrogen consumption: the *Prescient MPC* reaches e_{H_2} = 1.3%, and the *MPC* reaches e_{H_2} = 2.3%. The slightly lower hydrogen consumption of the *Prescient MPC* is due to its access to perfect predictive data, which are leveraged to calculate a better estimation of the costates.

In this subsection, it was shown that the *MPC* is able to comply with one user-defined NO_x^{eo} target and satisfy charge-sustaining vehicle operation in the *real driving cycle* while reaching close-to-optimal hydrogen consumption. Moreover, the additional loss of optimality by using realistic predictive data instead of perfect predictive data is found to be marginal. The next step is to show that the proposed *MPC* is able to satisfy all physically possible choices of NO_x^{eo} targets, without the need to adjust or re-tune the controller.

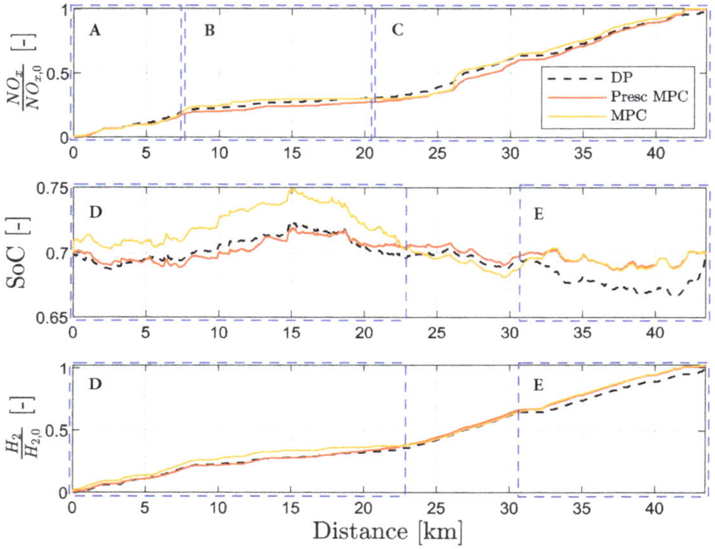

Figure 13. The performance analysis of the *MPC* for one NO_x^{eo} target for the *real driving cycle*.

5.3. NO_x^{eo}-Target Expansion

To investigate the ability of the *MPC* to reach different NO_x^{eo} targets, a simulation study is performed on the *real driving cycle* for varying NO_x^{eo} targets. Figure 14 includes a total of three plots, which are explained in detail in the following paragraphs.

The top plot shows the entire physically achievable H_2-NO_x^{eo} Pareto front in gray. This is calculated offline, using the DP algorithm with perfect predictive data and the full map-based model. The gray arrow depicts the maximum amount of NO_x^{eo} reduction over the H_2 optimal solution calculated by the DP algorithm for this driving mission. It is defined as 100%. The blue curve depicts the performance achieved by the online-capable *MPC*, which is subject to model mismatch and mispredictions, as mentioned above. This curve will from now on be referred to as the *MPC*'s performance front. Its results are obtained by running twelve simulations on the *real driving cycle* with twelve different NO_x^{eo} targets spanning the entire Pareto front. The *MPC* is able to reproduce a large part of the offline achievable Pareto front, including extremely low NO_x^{eo} targets, recovering more than 99% of the NO_x^{eo} reduction potential over the H_2-optimal solution. The dashed blue line is a horizontal extension of the point that is marked by a blue X. It serves as a visual aid and depicts a part of the Pareto front that cannot be reached by the *MPC*, even if the NO_x^{eo} target is set to a higher value. This is a result of the model mismatch which is introduced by using the simplified emission models, which underestimate NO_x^{eo}. However, as indicated by the DP-optimal Pareto front, there is an extremely flat trade-off between H_2 and NO_x^{eo}

in this region. Consequently, the benefits in H_2 efficiency are offset by a significant surplus of NO_x^{eo} emissions.

The middle plot depicts the *MPC*'s NO_x^{eo} error and SoC error for all twelve iterations. The blue line is very steep when approaching the lowest NO_x^{eo} targets of the theoretically reachable Pareto front. This shows that more NO_x^{eo} is emitted by the *MPC*; thus, these targets cannot be achieved by the online-capable controller. However, already for the NO_x^{eo} target, which is only 1.4% of the maximum NO_x^{eo} emissions on this driving mission, the *MPC*'s NO_x^{eo} error is 5% and is well contained below 1% for higher NO_x^{eo} targets. The region of very high NO_x^{eo} targets is here visualized by the steeply declining error towards negative values, highlighting that the *MPC* cannot achieve extremely high NO_x^{eo} targets. This part corresponds to the region represented by the dashed blue line in the top plot.

The bottom plot shows the *MPC*'s corrected hydrogen consumption error for the different NO_x^{eo} targets. It shows that for the entire achievable Pareto front, the *MPC* is able to operate the vehicle with close-to-optimal hydrogen consumption.

In this subsection, it was shown that the *MPC* is able to achieve close-to-optimal hydrogen consumption consistently for a wide variety of NO_x^{eo} targets. Furthermore, the *MPC* can achieve almost the entire possible NO_x^{eo} reduction potential that was predicted by DP. This shows that the *MPC* is able to reproduce almost the entire H_2-NO_x^{eo} Pareto front on the *real driving cycle*. To conclude the case study, it remains to be shown whether these results can be generalized to other driving missions.

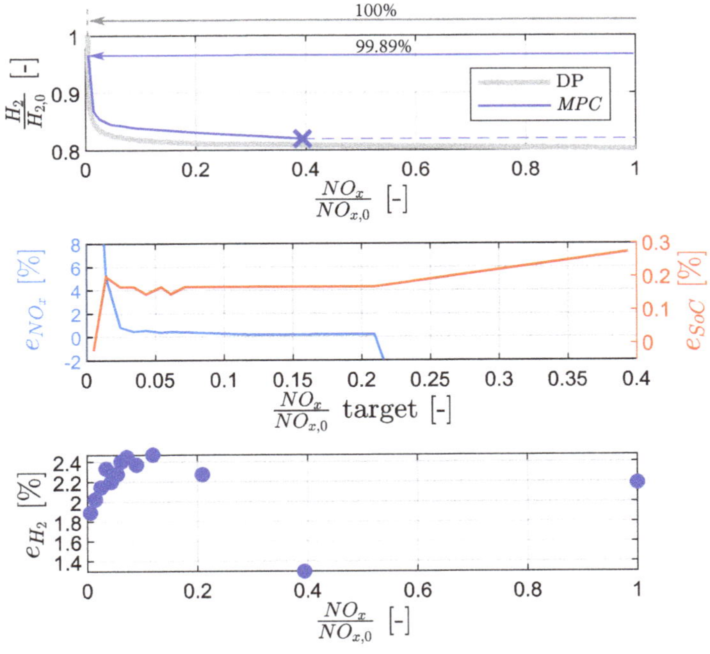

Figure 14. The performance analysis of the *MPC* for the entire reachable H_2-NO_x^{eo} performance front for the *real driving cycle*.

5.4. Driving Mission Generalization

In the following, the *MPC*'s ability to reproduce the H_2-NO_x^{eo} Pareto front is presented for a broad range of driving missions. Figure 15 depicts the comparison between the offline calculated DP-optimal Pareto fronts and the *MPC*'s online calculated performance fronts obtained for the *urban driving mission*, the *mountain driving mission*, and the *highway driving mission*. For the sake of comparison, the results obtained for the *real driving mission* are also included. The gray lines depict the entire, physically achievable H_2-NO_x^{eo} Pareto front, calculated by the DP algorithm with perfect predictive data and the full map-based model. The blue curves represent the performance fronts achieved by the online-capable *MPC*, which is subject to model mismatch, as well as mispredictions. Its results are obtained by running multiple simulations on the corresponding drive mission but with different NO_x^{eo} targets.

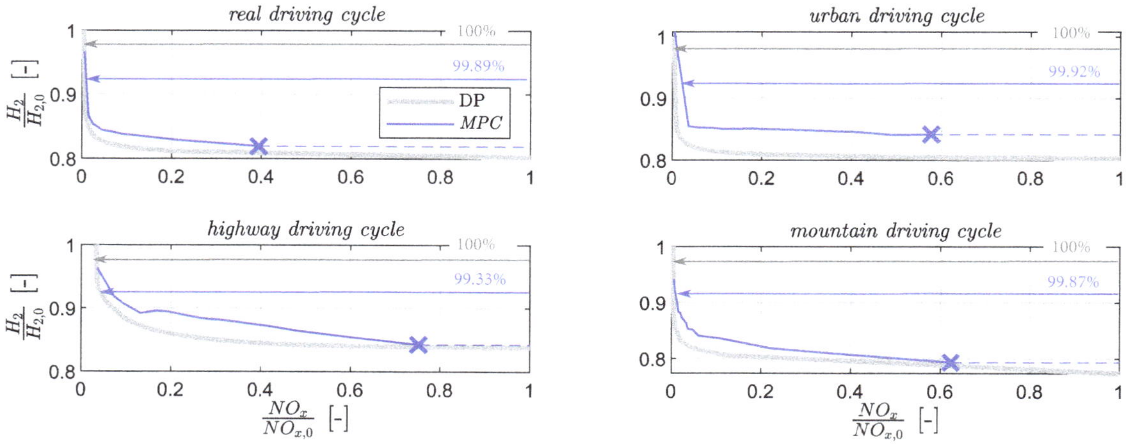

Figure 15. The performance analysis of the *MPC* with a focus on the H_2-NO_x^{eo} performance front on all four driving missions.

On all four driving missions, the *MPC* is able to reach extremely low NO_x^{eo} emissions, recovering more than 99% of the NO_x^{eo} reduction potential over the H_2-optimal solution. It cannot capture the theoretically possible EMS calibrations to the far right-hand side of the Pareto front which result in the lowest possible H_2 consumption. However, as indicated by the DP-optimal Pareto fronts, there is an extremely flat trade-off between H_2 and NO_x^{eo} for all investigated driving missions in this region. Consequently, the benefits in H_2 efficiency are offset by a significant surplus of NO_x^{eo} emissions.

Table 2 summarizes the *MPC*'s performance fronts for all investigated driving missions. The two columns $e_{\mathrm{H_2}}^{\mathrm{mean}}$ and $e_{\mathrm{H_2}}^{\mathrm{max}}$ represent the mean and the maximum values, respectively, of all the $e_{\mathrm{H_2}}$ resulting from the simulation on the corresponding driving mission and the individual NO_x^{eo} targets. Two things can be noted here: First, in all four driving missions, the maximum hydrogen error is close to the mean hydrogen error. This suggests that variance in the controller performance can be characterized as low. Second, across all four driving missions, the mean hydrogen error is similar. This suggests that the controller performance is similarly consistent across different driving missions. Overall, these results show that the proposed *MPC* is able to provide predictable, close-to-optimal hydrogen consumption for a wide variety of driving missions and for almost the entire physically reachable range of H_2-NO_x^{eo} trade-off.

Table 2. Key characteristics regarding the hydrogen consumption of the *MPC* over its entire performance front.

	$e_{\mathrm{H_2}}^{\mathrm{mean}}$	$e_{\mathrm{H_2}}^{\mathrm{max}}$
Real driving mission	2.18%	2.47%
Urban driving mission	4.66%	5.13%
Mountain driving mission	3.79%	6.62%
Highway driving mission	4.15%	6.91%

6. Conclusions

To the authors' best knowledge, this paper provides the first results of an online-capable control algorithm for the EMS of a H_2-HEV explicitly including NO_x^{eo} emissions. The algorithm is based on a multi-level control approach and features a supervisory MPC algorithm, which solves a convex optimization problem. The proposed controller is evaluated in a case study, comprising three levels of complexity: First, the controller is shown to be able to satisfy a user-defined NO_x^{eo} target while operating the vehicle in charge-sustaining mode and achieving close-to-optimal hydrogen consumption. Second, the controller is shown to be able to extend its close-to-optimal hydrogen consumption to a wide range of different NO_x^{eo} targets. Third, it is shown that the proposed controller exhibits predictable performance for a wide variety of driving missions. In all tested scenarios, an increase of 4.66% at most in hydrogen consumption is observed compared with the globally optimal solution. More importantly, for each driving mission, more than 99% of the possible NO_x^{eo} reduction potential can be reached by the online-capable MPC controller. In summary, this means that the proposed controller can almost fully exploit the theoretical potential of an H_2-HEV regarding the H_2-NO_x^{eo} trade-off.

It is important to emphasize that the obtained NO_x^{eo} stem from a map, which is derived from steady-state measurements. While these maps can predict the NO_x^{eo} emissions of an H_2ICE well during steady-state operation, higher NO_x^{eo} emissions may occur during transient operating point switches of the H_2ICE. Therefore, no statements regarding compliance with legislative limits should be made.

In future research, the development and inclusion of an additional dynamic NO_x^{eo} emission model and its inclusion in the MPC could be looked into. Also, the possibility to adjust the H_2-NO_x^{eo} trade-off online could lead to changed requirements for the exhaust gas aftertreatment system. This could create an avenue for exploring the optimization of EMS calibration in conjunction with the design and operation of the exhaust gas aftertreatment system.

Author Contributions: Conceptualization, D.M., N.Y., T.H. and C.O.; methodology, D.M., N.Y. and T.H.; software, D.M. and N.Y.; validation, D.M.; writing—original draft preparation, D.M.; writing—review and editing, D.M. and F.W.; visualization, D.M. and F.W.; supervision, C.O. All authors have read and agreed to the published version of the manuscript.

Funding: This research received no external funding.

Data Availability Statement: The original contributions presented in the study are included in the article, further inquiries can be directed to the corresponding authors.

Conflicts of Interest: Author Thomas Huber was employed by the company Robert Bosch GmbH. The remaining authors declare that the research was conducted in the absence of any commercial or financial relationships that could be construed as a potential conflict of interest.

Abbreviations

The following abbreviations are used in this manuscript:

CO_2	carbon dioxide
$\mathcal{COP}$	convex optimization problem
DP	dynamic programming
EMS	energy management system
EU	European Union
EV	electric vehicle
GB	gearbox
H_2	hydrogen
H_2ICE	hydrogen combustion engine
HEV	hybrid electric vehicle
MPC	model predictive control
NO_x	nitrogen oxides

NO_x^{eo}	engine-out nitrogen oxides
OCP	optimal control problem
PMP	Pontryagin's minimum principle
PR	power request (block diagram schematic)
RTG	reference trajectory generator
SoC	state of charge

References

1. IEA. *CO_2 Emissions in 2022*; International Energy Agency: Paris, France, 2022.
2. EPA. epa.gov. 2021. Available online: https://www.epa.gov/regulations-emissions-vehicles-and-engines/final-rule-revise-existing-national-ghg-emissions (accessed on 8 June 2023).
3. EU Regulation. Regulation (EU) 2019/631 of the European Parliament and of the Council of 17 April 2019 Setting CO_2 Emission Performance Standards for New Passenger Cars and for New Light Commercial Vehicles, and Repealing Regulations (EC) No 443/2009 and (EU) No 510/2011. 2019. Available online: https://eur-lex.europa.eu/legalcontent/EN/TXT (accessed on 10 January 2024).
4. IEA. *Global EV Outlook 2023*; International Energy Agency: Paris, France, 2024.
5. Hänggi, S.; Elbert, P.; Bütler, T.; Cabalzar, U.; Teske, S.; Bach, C.; Onder, C. A review of synthetic fuels for passenger vehicles. *Energy Rep.* **2019**, *5*, 555–569. [CrossRef]
6. Hassan, Q.; Azzawi, I.D.; Sameen, A.Z.; Salman, H.M. Hydrogen Fuel Cell Vehicles: Opportunities and Challenges. *Sustainability* **2023**, *15*, 11501. [CrossRef]
7. Lider, A.; Kudiiarov, V.; Kurdyumov, N.; Lyu, J.; Koptsev, M.; Travitzky, N.; Hotza, D. Materials and techniques for hydrogen separation from methane-containing gas mixtures. *Int. J. Hydrogen Energy* **2023**, *48*, 28390–28411. [CrossRef]
8. Sementa, P.; de Vargas Antolini, J.B.; Tornatore, C.; Catapano, F.; Vaglieco, B.M.; Sánchez, J.J.L. Exploring the potentials of lean-burn hydrogen SI engine compared to methane operation. *Int. J. Hydrogen Energy* **2022**, *47*, 25044–25056. [CrossRef]
9. Zhao, F.C.; Sun, B.G.; Yuan, S.; Bao, L.Z.; Wei, H.; Luo, Q.H. Experimental and modeling investigations to improve the performance of the near-zero NOx emissions direct-injection hydrogen engine by injection optimization. *Int. J. Hydrogen Energy* **2023**, *49*, 713–724. [CrossRef]
10. Güler, İ.; Kılıçaslan, A.; Küçük, T.; Corsini, D. Transient and altitude performance analysis of hydrogen fuelled internal combustion engines with different charging concepts. *Int. J. Hydrogen Energy* **2023**, *49*, 1112–1122. [CrossRef]
11. Bao, L.Z.; Sun, B.G.; Luo, Q.H. Experimental investigation of the achieving methods and the working characteristics of a near-zero NOx emission turbocharged direct-injection hydrogen engine. *Fuel* **2022**, *319*, 123746. [CrossRef]
12. Sciarretta, A.; Back, M.; Guzzella, L. Optimal control of parallel hybrid electric vehicles. *IEEE Trans. Control. Syst. Technol.* **2004**, *12*, 352–363. [CrossRef]
13. Ambühl, D. Energy Management Strategies for Hybrid Electric Vehicles. Ph.D. Thesis, ETH Zurich, Zürich, Switzerland, 2009.
14. Machacek, D.T.; Barhoumi, K.; Ritzmann, J.M.; Huber, T.; Onder, C.H. Multi-level model predictive control for the energy management of hybrid electric vehicles including thermal derating. *IEEE Trans. Veh. Technol.* **2022**, *71*, 10400–10414. [CrossRef]
15. Hu, Q.; Amini, M.R.; Kolmanovsky, I.; Sun, J.; Wiese, A.; Seeds, J.B. Multihorizon model predictive control: An application to integrated power and thermal management of connected hybrid electric vehicles. *IEEE Trans. Control. Syst. Technol.* **2021**, *30*, 1052–1064. [CrossRef]
16. Ritzmann, J.; Peterhans, C.; Chinellato, O.; Gehlen, M.; Onder, C. Model Predictive Supervisory Control for Integrated Emission Management of Diesel Engines. *Energies* **2022**, *15*, 2755. [CrossRef]
17. Kyjovský, Š.; Vávra, J.; Bortel, I.; Toman, R. Drive cycle simulation of light duty mild hybrid vehicles powered by hydrogen engine. *Int. J. Hydrogen Energy* **2023**, *48*, 16885–16896. [CrossRef]
18. Machacek, D.T.; Ozan, N.; Huber, T.; Onder, C.H. Energy Management of Hydrogen Hybrid Electric Vehicles—A Potential Study. *arXiv* **2023**, arXiv:2309.09804.
19. Hannah, L.A.; Dunson, D.B. Multivariate convex regression with adaptive partitioning. *J. Mach. Learn. Res.* **2013**, *14*, 3261–3294.
20. Machacek, D.T.; van Dooren, S.; Huber, T.; Onder, C.H. Learning-Based Model Predictive Control for the Energy Management of Hybrid Electric Vehicles Including Driving Mode Decisions. *IEEE Trans. Veh. Technol.* **2023**, *73*, 5113–5127. [CrossRef]
21. Kerrigan, E.C.; Maciejowski, J.M. Soft Constraints and Exact Penalty Functions in Model Predictive Control. In Proceedings of the United Kingdom Automatic Control Council (UKACC) International Conference on Control, Cambridge, UK, 6–8 July 2000.
22. Murgovski, N.; Johannesson, L.; Hu, X.; Egardt, B.; Sjöberg, J. Convex relaxations in the optimal control of electrified vehicles. In Proceedings of the 2015 American control conference (ACC), Chicago, IL, USA, 1–3 July 2015; pp. 2292–2298.
23. Lofberg, J. YALMIP: A toolbox for modeling and optimization in MATLAB. In Proceedings of the 2004 IEEE International Conference on Robotics and Automation (IEEE Cat. No. 04CH37508), Taipei, Taiwan, 2–4 September 2004; pp. 284–289.
24. MOSEK ApS. *The MOSEK Optimization Toolbox for MATLAB Manual*; Version 9.0.; MOSEK ApS: Copenhagen, Denmark, 2019.
25. Guzzella, L.; Sciarretta, A. *Vehicle Propulsion Systems*; Springer: Berlin/Heidelberg, Germany, 2013; Volume 3.
26. Boltyanskiy, V.; Gamkrelidze, R.; Mishchenko, Y.; Pontryagin, L. *Mathematical Theory of Optimal Processes*; John Wiley & Sons: Hoboken, NJ, USA, 1962.

27. Kim, N.; Cha, S.; Peng, H. Optimal control of hybrid electric vehicles based on Pontryagin's minimum principle. *IEEE Trans. Control. Syst. Technol.* **2010**, *19*, 1279–1287.
28. Ambuhl, D.; Guzzella, L. Predictive reference signal generator for hybrid electric vehicles. *IEEE Trans. Veh. Technol.* **2009**, *58*, 4730–4740. [CrossRef]
29. Behrisch, M.; Bieker, L.; Erdmann, J.; Krajzewicz, D. SUMO–simulation of urban mobility: An overview. In Proceedings of the SIMUL 2011, The Third International Conference on Advances in System Simulation. ThinkMind, Barcelona, Spain, 23–29 October 2011.

energies

Article

Socioeconomic Factors Driving the Transition to a Low-Carbon Energy System

Evangelia Karasmanaki *, Spyros Galatsidas * and Georgios Tsantopoulos

Department of Forestry and Management of the Environment and Natural Resources, Democritus University of Thrace, Pantazidou 193, 68200 Orestiada, Greece; tsantopo@fmenr.duth.gr
* Correspondence: evkarasm@fmenr.duth.gr (E.K.); sgalatsi@fmenr.duth.gr (S.G.)

Abstract: Citizen participation via different investment schemes may be a promising solution to the financing barriers inhibiting energy transition. In this regard, citizens may be approached as potential investors in renewables, but, to mobilize their capital, strategies need to be developed. Much like other services or products seeking to improve their market position, renewable energy investments by citizens also require dedicated efforts to acquire a strong market position. Using a large sample of Greek citizens, this study investigated whether it is possible to identify distinct and addressable citizen clusters which can enable energy developers and marketers to effectively address the preferences and needs of potential investor segments. The performance of k-means cluster analysis identified four clusters: Indifferent Investors were neither driven by economic or social factors, Enthusiastic Investors were motivated both by economic and social factors, Pro-environmental Investors were driven by the environmental benefits, and Social Investors were motivated by the social aspects of the investment. Moreover, each cluster demonstrated different levels of willingness-to-invest in renewable energy and were knowledge about renewable energy investments. It was concluded that citizens should not be approached as a homogeneous target group by marketing experts and policymakers, while novel strategies should be followed.

Keywords: energy transition; renewable energy; socioeconomic drivers; citizen investment; cluster analysis; attitudes to renewable energy

1. Introduction

The European Union (EU) is widely recognized as a leader in renewable energy due to its continuous policy efforts to increase the penetration of renewable energy sources (RES) in member states' energy mixes. The energy policy of the EU is characterized by setting increasingly ambitious targets, and more recently, it has set an EU-wide emission reduction target of a minimum 55% net emissions by year 2030 as part of the "FiT for 55" package [1]. The outcome of these efforts seems more critical than ever given the recent energy crisis due to the armed conflicts around the globe. As future development is at risk, the deployment of renewable energy sources is a necessary step for the continent. As an EU member state, Greece presents an interesting case for the study of energy transition because it has introduced a multitude of regulations and policies to promote the penetration of renewable energy in its energy mix, and at the same time, it possesses a remarkable renewable potential [2,3]. The country exerts great efforts to lower its greenhouse gas emissions and, according to the latest national report, greenhouse gas emissions (without measuring the levels of LULUCF) amounted to 77.50 Mt CO_2 eq in 2021, showing a notable decrease of 25.48% compared to reference year 1990, in which the emissions were equal to 101.74 Mt CO_2 eq (including all emissions sources) [4,5]. To contextualize the latter, the corresponding emission reduction rate achieved across EU was 30% [6]. The impressive decrease achieved by Greece can be ascribed, to a significant extent, to the policy efforts to increase the installed capacity of renewables. More recently, Greece has revised its

National Energy and Climate Plan and has set even more ambitious national targets to be met by 2030 compared to EU-wide targets [7]. The new targets involve a minimum of a 35% percentage of renewable energy in the final gross energy consumption, a 42% decrease in greenhouse gas emissions compared to 1990 levels, and an at least 38% rate of energy efficiency in final energy consumption. Moreover, the Greek parliament has also passed the Climate Law, which proposes additional mitigation measures, such as higher penetration of electric vehicles (EVs) [8]. The country has also been implementing a highly radical lignite phase-out schedule, which aims at the total decommissioning of all lignite plants. Despite these efforts and the progress towards emission reduction targets, the country still needs increase the install capacities of renewables. Yet, the transition to a renewable-based energy system is facing significant cost barriers, and attention is required to closing investment gaps because public finance alone cannot suffice to ensure continuous progress towards the decarbonization of the energy sector [9,10].

As alternative financing concepts are required, investments in renewable energy sources can be a straightforward way for citizens to participate in energy transition [11–14]. Citizens now have the option to participate in different investment schemes and, in this way, to provide a considerable part of the capital necessary for the deployment of renewable energy [15]. In many EU countries, citizens have made substantial investments in renewable energy, thereby contributing to their country's energy transition by allocating financial resources to energy policy implementation [16]. It is important to note, however, that in these countries, there were established mechanisms for citizen investment in renewables [17].

The promising role of citizens in financing energy transition brings forward the need to conduct more research to understand the intricacies of citizen decision-making. So far, many studies on the subject have focused on the sociodemographic, financial, and non-financial variables that affect investment decisions [11–14]. From a broader perspective, it has been observed that all investments, including those made by citizens, are resulting from rational and well-calculated decision-making. In this regard, marketers and energy developers should not be surprised that citizens tend to express rational and well-rounded preferences for renewable energy investments and, most notably, for return on investment [18,19]. Moreover, citizens, as potential RES investors, have been found to prefer investments where the certainty of financial returns is considerable [20,21]. Return on investment has also been found to be more influential to citizens' investment decision compared to the estimated investment depreciation [20].

In addition to the factors described above, other economic factors that have a positive effect on investment decisions involve incentives, such as subsidies, tax credits, and financing aid for house improvements (in cases where the incentives are intended to induce investments in solar energy). More analytically, Vasseur and Kemp [15] indicated that incentives, along with aid for house improvements, act as strong motives for citizens to invest in residential photovoltaic systems. Conversely, the risk of the investment can act as an inhibiting factor for citizens that are principally interested in the certainty of the investment [21]. At the same time, citizens express an acute aversion to long-term investments, and it is due to this aversion that citizens feel more comfortable to invest along with energy communities or public electricity companies [21]. A recent study has confirmed that citizens as investors are very sensitive to the risk of losing the entire investment [22].

Another critical economic factor in citizens' decision-making is the cost of the investment, which, if deemed too high, acts as a barrier per se to investments in renewables [23]. The study by Vasseur and Kemp [15] indicated that the investment cost was seen as the primary obstacle to investing by the majority of respondents. In addition, around half of respondents in the same study thought that the cost must be significantly reduced for them to invest, and many of them would invest only under a highly attractive subsidy scheme [15]. The crucial role of cost in investment decision-making also accounts for high-incomers' higher willingness to invest compared to those with fewer financial resources [24], but it also places attention on the adverse influence of insufficient savings and restricted loan access on RES investment decisions [21]. In contrast to the cost, lower anticipated household electricity costs due to RES investment act as a strong motivation for householders, especially with regard to investments in residential renewable systems. In such cases, achieving electricity cost reductions through investments in RES seems to be a strategic choice for citizens with an excess of disposable income that try to ensure lower household expenses in the future [25].

In addition to the economic factors discussed above, citizens' investment decision is also affected by social factors. Specifically, individuals' perceptions about their role in society and their ability to influence other society members can be a motivating factor. Schelley [26] noticed that a considerable part of citizens who had adopted residential renewable systems viewed themselves as advocates for the feasibility of residential renewable systems and believed they should set an example within their community. At the same time, the social circle of potential investors is often a strong driver because if adopters are included in it, potential investors are able not only to become acquainted with new technologies and explore their feasibility, but also discuss their concerns [27]. In addition, individuals may also be persuaded to invest if community members that are viewed as significant by the community have invested or if the investment is highly valued by most community members [28]. The strong effect of community attitudes was recently corroborated by Knauf and le Maitre [29].

The technique of market segmentation may prove useful in providing a new angle of the role of socioeconomic factors in citizens' investment decisions, as it is a procedure in which the potential market is divided into distinct subsets of individuals who express the same needs, preferences, and priorities [15,30]. There are many advantages to this approach, and most importantly, it enables marketers to address the problem of consumer heterogeneity by segmenting them in order to identify and address different clusters. Marketers are also able to choose one or more clusters as a target market and reach them with the use of tailored marketing strategies and tools. To put this differently, this kind of approach enables marketers to capitalize on the preferences and needs of consumers in each cluster in order to enhance the probability of as many of them as possible choosing the product or service in question. The fundamental idea underlying market segmentation is that consumers are different, and thus one strategy cannot be effective for all of them [15,30].

Much like every kind of service or product that seeks to improve its position in the market, renewable energy investments can also be seen as a product that requires dedicated efforts in order to acquire a strong market position. To that end, it is necessary to better understand citizens' decision-making and try to segment them in order to develop strategies to mobilize their capital. Despite the potential benefits of citizen segmentation, there has so far been scant research efforts on developing citizen RES investor clusters. Of the existing research works, Vasseur and Kemp [15] conducted a web-based survey in 2011 on householders in the Netherlands and developed a segmentation model for technological innovations. The clusters were based on whether the adoption was voluntary or not, as well as on whether the household could be characterized as a potential adopter or rejecter. Petrovich et al. [31] examined Swiss homeowners' preferences for building retrofits and developed two clusters of potential solar adopters, with the 'premium' segment expressing preference for colored and building integrated solar systems, and the 'value' segment being more sensitive to the price of the systems. Another segmentation study was performed by Ebers, Broughel, and Hampl [32] in Austria and Switzerland, and focused on the socio-demographic and the socio-psychological characteristics of potential investors. This study managed to identify four clusters of potential investors based on their place of residence (urban and rural dwellers), as well as on their acceptance of wind energy projects in the vicinity of their residence. The groups were described as rural and urban wind energy enthusiasts and as urban and rural wind energy skeptics. More recently, Gáspár et al. [33] examined investment preferences based on sociodemographic characteristics, and identified a group preferring green investment aspects to financial returns and a group that can be affected by suitable marketing strategies.

Another stream of the literature has developed clusters of residents according to their willingness to participate in energy communities. Specifically, Wang et al. [34] classified residents in three clusters according to innovation diffusion groups in view of their attitude to participate in energy communities: Earlier Adopters, Mid-term Adopters, and Later Adopters. Heiskanen et al. [35] examined investor heterogeneity in Finland by focusing on investment drivers and available investment options, and identified investors whose investment was induced by profit seeking motives through the feed-in-tariff system and those investors who were mostly real estate owners and invested in heat pumps in order to benefit from the low interest rates.

It can be seen that numerous previous studies on citizen willingness-to-invest in renewable energy have primarily examined the economic and social factors affecting citizen decision to invest, while the true extent of this influence remains somewhat vague. That is, most studies have noted the role of socioeconomic factors, but very few of them have tried to examine whether and how economic in conjunction with social factors affect citizen decision-making. Moreover, most relevant studies often depend on datasets derived from small sample sizes, while much research focuses on citizens that have already invested in specific investment schemes rather than on non-investors whose willingness-to-invest is much more relevant to examine if we are to devise new strategies to mobilize citizen RES investment. Furthermore, the ability to classify citizens based on how they evaluate the effect of socioeconomic factors is underexplored, even though it may hold the potential to introduce a novel approach to policymaking design. There is thus a remarkable gap in the literature which is crucial to fill for various reasons. Understanding how citizens can be classified according to their evaluation of socioeconomic investment drivers is necessary in order to inform targeted interventions and support mechanisms. A precise idea of citizen decision-making may also be the missing link to designing investment mechanisms that correspond better to citizens' needs and preferences. It is, therefore, worthwhile to examine if such segmentation is possible because this may reveal novel nuances that may have been missed in marketing strategies and policies. In addition, as attitudes to renewables constantly change, it is necessary to conduct research regularly so that marketers and policymakers have a precise idea on the factors that affect citizen investment [27]. Using a large citizen sample that allows for a better differentiation among clusters, this study aims to investigate whether it is possible to identify distinct and addressable citizen clusters which can enable energy developers and marketers to effectively attract and address the preferences and needs of potential investor segments.

The remainder of this paper is structured as follows: The second Section describes in great detail the methods that were followed to carry out the study, and then, in the third Section, the developed clusters are presented. Afterward, the results reported in this paper are discussed, and in the last Section, conclusions are reached and study limitations are stated.

2. Research Methodology

2.1. Research Area

The analysis presented in this paper is part of a broad research that focused on citizens' attitudes towards renewable energy investments. The study area involved all households in Greece, a member state of the EU located in southeastern Europe. Greece was chosen as the study area because it has a notable renewable energy potential and, if energy transition is successfully implemented, the country can become a remarkable renewable energy producer in the EU [2,3].

2.2. Questionnaire Design and Sampling Method

The research was performed using a structured questionnaire which was constructed drawing on the literature on citizen investors' socioeconomic profile, motives, barriers, and preferences [11,15,18–21,23–25,32]. In addition, the design of the questionnaire took into account the policy framework in place for investments in renewables in the study area ensuring that investment parameters were considered in the content of the items. The questionnaire involved eight thematic sections gathering information about (i) respondents' perceptions of energy topics, (ii) their knowledge on RES investments, (iii) their willingness to invest, (iv) their evaluation of economic, environmental, and social factors affecting the investment decision, (v) their perceptions of barriers to RES investments, (vi) their satisfaction levels with various stakeholders' actions to enable RES investments, (vii) their environmental attitudes, and (viii) their sociodemographic characteristics. Results reported in this paper are based on the data from the fourth Section, which examines respondents' evaluation of factors affecting their decision to invest in renewables. It should be noted that all items were measured with a five-point Likert scales to enable a higher level of precision and a wide range of statistical analyses.

The efficacy and accuracy of the questionnaire was ensured by performing a pilot study on a limited sample of 30 Greek citizens from diverse socioeconomic backgrounds. In view of the results of the pilot study, some changes were made. Specifically, a few phrases in the items had to be reworded to ensure that they would be understood by respondents, the response scale for one item had to be changed to facilitate responses, and the order of three items was re-arranged to improve the coherence of the questionnaire and to avoid confusion among respondents. In accordance with Law 4521/2018 and, in particular, the provisions stated in Article 23, this research applied for and was granted a permit from the Research Ethics Committee of the Democritus University of Thrace (Decision No. 3/9 December 2019), the entity responsible for examining and approving every research conducted by researchers affiliated with the university.

To recruit respondents and to attain a representative citizen sample, the simple random sampling method was used [36]. Compared to other sampling methods, this method has also some distinct benefits, and most importantly, it does not require a very high level of knowledge about the population under study [37,38]. The population under study consisted of all households in Greece. It should be noted that the use of households is a classic case of utilizing a group of people as a sampling unit, while this method is significantly more convenient and affordable [39]. The process of choosing one member from each randomly selected household ensured that the same member would not be always chosen [36,39].

To calculate the sample size, the formula of simple random sampling (1) was used, and in particular, simple random sampling without replacement was employed. For this reason, the factor of finite population correction could be excluded, as the size of the sample (n) was somewhat 'limited' in comparison to the population size (N) [40]. According to our estimation, 1536 citizens had to participate in the study.

$$n = \frac{t^2 \cdot \bar{p} \cdot (1 - \bar{p})}{e^2} = \frac{1.96^2 \cdot 0.50 \cdot (1 - 0.50)}{0.025^2} \cong 1536 \tag{1}$$

In this formula, t expresses the value of Student's t-distribution with $(1 - \alpha) = 95.0\%$. In this calculation, $n - 1$ degrees of freedom are used, while e expresses the level of precision (that is, the margin of error). Because the size of pre-sampling is somewhat large (greater than 50), the value of t is retrieved from standard normal distribution tables, and in this way, for a probability of 95%, the value is 1.96 [40]. In addition, the p-bar estimates the proportion, and e expresses the highest acceptable difference between the sampling mean and the mean of the unknown population. It may be accepted that it is 0.025, i.e., 2.5%.

Moreover, pre-sampling using a sample of 50 subjects had to be conducted in order to estimate the real proportion of the population for each variable. This was a necessary step because the questionnaire must be able to accurately calculate a substantial number of different variables. The variable 'Gender' was the one that gave the highest sample size, and when the calculated sample sizes are similar but may be afforded, then the highest sample size must be selected [40]. All questionnaires were filled in through personal interviews with the respondents in order to ensure that questionnaires were fully understood and completed. Finally, 1536 respondents participated in the study, and to ensure their anonymity, the questionnaires were placed in envelopes and were destroyed after data coding.

2.3. Data Processing

After questionnaire collection, the data were coded and entered into the Statistical Package for Social Sciences (SPSS, version 22). Prior to cluster analysis, factor analysis had to be performed to decrease the number of observed variables into fewer latent variables (factors) which would retain the most information of the initial dataset [41]. As for the significance of the principal components, we used the criterion proposed by Guttman and Kaiser, according to which, the limit for acquiring the necessary number of principal components is determined using the eigenvalues that are ≥ 1 [42,43]. The analysis was completed with a varimax rotation of the principal components to optimize results [44]. A factor analysis was used in this study to obtain the factors from the following multivariates: 'economic reasons for investing in renewable energy (Q7)' and 'social reasons for investing in renewable energy (Q9)'.

K-means cluster analysis was chosen for this study because it classifies observations into a predefined number of clusters (k), optimizing inter-cluster differences and minimizing intra-cluster variance on specified variables [45]. In addition, in the k-means method, the first observations can be reclassified into another cluster if it is a more suitable fit. This type of clustering provides a series of solutions which correspond to different numbers of clusters, that is, k-means [46]. In k-means clustering, the algorithm used for defining the contribution of observations is based on the closest classified centroid. An observation falls into the cluster which presents the smallest distance between the observation and the cluster centroid. If centroids are already known, they can be defined in the start. If, however, they are not known, they need to be estimated through an iteration process. The application of k-means clustering means that the classification of new observations into clusters is based on the proximity of each observation to these centers [45]. The center of each cluster is merely the numerical mean of all variables for the variables falling into the respective cluster. The initial values of the new observations are then standardized in z values, and afterward, the Euclidean distances from cluster centers are estimated. The observation falls into the cluster that presents the smallest distance [47].

Following the integration of new observations into clusters, the estimation of the clusters' real centers becomes feasible. These values are the final centers (centroids), which are slightly different from initial ones, especially in cases where the observations had not fallen into any cluster previously. The distances are able to indicate the degree of the differentiation among clusters, and this is achieved through the estimation of Euclidean distances for every two centers [47].

To build the citizen typology, a series of steps were followed. First, the variables to establish the typology had to be selected. The selected variables involved the economic and social factors affecting the respondents' decisions to invest in renewable energy. As this was a high number of variables, a factor analysis had to be performed in order to discover the relationships among the variables and underlying dimensions. Then, to identify the most appropriate number of clusters, various clustering combinations were examined. Afterwards, respondents were classified based on the selected criterion using cluster analysis and, finally, results were evaluated [46]. The description of citizen profiles also took into account respondents' demographic characteristics as well as investment-related variables. It is important to note that the chi-squared test was performed to verify that the clusters differ in terms of sociodemographic variables. Finally, a typology of citizens was established, and the types were named by the researchers in a way that reflected the citizens' characteristics in relation to their evaluation of socioeconomic factors related to RES investments.

3. Results

3.1. Sample Characteristics

In the sample, female respondents outnumbered (by 51.6%) their male counterparts, and with regard to age, most respondents were aged 41 to 50 years old (by 27.9%). With regard to occupation, the respondent percentages who reported being employed in the private (by 21.2%) and public (by 19.9%) sectors were considerably higher in comparison to the other occupations. Regarding education level, most respondents were university graduates (by 22.3%), as well as upper secondary school graduates (by 20.8%). Regarding respondents' family status, around half of them were married (by 51%), while a considerable percentage reported having two children (by 28.3%). Most respondents reported residing in urban centers (by 64.1%). As for their annual income, 28.5% of reported earning between EUR 10,001 and EUR 20,000, and 20.1% reported earning between 5001 and 10,000 Euros a year.

3.2. Profiling Citizens as Potential Investors in Renewable Energy

In order to investigate the underlying structure of respondents' views on the economic and social reasons for investing in renewable energy, a principal components analysis with Varimax rotation was carried out. Regarding the multivariate 'economic reasons for investing in renewable energy (Q7)', Cronbach's alpha value was equal to 0.883, the Keiser-Meyer-Olkin index scored 0.888, and a Bartlett's test of sphericity rejected the null hypothesis (Chi-Square = 5932.216, df = 6, $p < 0.001$). The analysis yielded two factors, 'Subsidies, low investment taxation and improved income and electricity cost (PC1_1)' and 'Optimal investment opportunity and protection from oil price fluctuations (PC1_2)' (Table 1).

Then, to examine the structure of responses for the multivariate 'social reasons for investing in renewable energy Q9', a principal components analysis with Varimax rotation was again performed. Cronbach's alpha value was 0.802, the Keiser-Meyer-Olkin index scored 0.712, and a Bartlett's test of sphericity rejected the null hypothesis (Chi-Square = 3283.840, df = 10, $p < 0.001$). The analysis gave two factors: 'Boosting social profile (PC2_1)' and 'Pro-environmental behavior and example (PC2_2)' (Table 1).

In order to identify the structure underlying citizens' views on the economic and social reasons for investing in renewable energy sources, second-order factor analysis was performed. Before the performance of the second-order factor analysis, the following necessary tests were carried out: the Keiser–Meyer–Olkin index had a value of 0.448, and Bartlett's sphericity test rejected the null hypothesis (Chi-Square = 350.703, df = 6, $p < 0.001$). The analysis gave two factors. The first factor (P1) is named 'Pro-environmentalism, favorable investment conditions and benefits for income and electricity cost' and involves the variables 'Pro-environmental behavior and example (PC2_2)' and 'Subsidies, low investment taxation and improved income and electricity cost (PC1_1)'. The second factor (P2) is named 'Social profile, sound investment and independence from oil prices (P2)' and includes the variables Optimal investment opportunity and protection from oil price fluctuations (PC1_2)' and 'Boosting social profile (PC2_1)' (Table 2).

Table 1. Factor analysis results with varimax rotation of the reasons for investing in RES.

Principal Components—Variables Loaded in Each Factor	Loadings	Eigenvalue
Subsidies, low investment taxation, and improved income and electricity cost (PC1_1)		**4.664**
Income increase through selling produced energy	0.772	
High subsidies granted for RES investments	0.752	
Reduction in electricity costs	0.737	
Low taxation on RES investments	0.693	
Acquisition of stable income	0.600	

Table 1. *Cont.*

Principal Components—Variables Loaded in Each Factor	Loadings	Eigenvalue
Optimal investment opportunity and protection from oil price fluctuations (PC1_2)		**0.915**
The relatively low investment cost compared to other investment types (e.g., asset purchase)	0.811	
Short depreciation time	0.769	
Capital investment	0.690	
Protection from oil price fluctuations due to geopolitical crises	0.690	
Total variance (%): 61.995		
Keiser–Meyer–Olkin = 0.888, Bartlett χ^2 = 5932.216, df = 6, $p < 0.001$		
Boosting social profile (PC2_1)		2.828
Increasing respect from friends and acquaintances	0.895	
Social prestige through entrepreneurial activity	0.885	
Pro-environmental behavior and example (PC2_2)		1.032
The desire to adopt pro-environmental behavior	0.817	
The desire to set a good example for my family	0.785	
The desire to set a good example for society	0.759	
Total variance (%): 77.206		
Keiser–Meyer–Olkin = 0.712, Bartlett χ^2 = 3283.840, df = 10, $p < 0.001$		

Table 2. Factor loadings of the data after Varimax rotation for all reasons for investing in renewable energy.

Principal Components—Results of Second-Order Factor Analysis	Loadings	Eigenvalue	Variance (%)
Pro-environmentalism, favorable investment conditions and benefits for income and electricity cost (P1)		1.452	36.292
Pro-environmental behavior and example (PC2_2)	0.816		
Subsidies, low investment taxation, and improved income and electricity cost (PC1_1)	0.653		
Social profile, sound investment and independence from oil prices (P2)		1.025	25.621
Optimal investment opportunity and protection from oil price fluctuations (PC1_2)	0.744		
Boosting social profile (PC2_1)	0.722		
Total variance (%): 61.913			
Kaiser–Meyer–Olkin = 0.448, Bartlett χ^2 = 350.703, df = 6, $p < 0.001$			

3.3. Description of Citizen Clusters

The application of k-means cluster analysis served to indicate the association of each member to the cluster. The ANOVA was performed to test if respondents can be clustered according to socioeconomic reasons for investing in renewable energy sources. Results pointed to the existence of important differences, and thus the proposed clustering could be accepted.

Two-, three-, and four-cluster solutions were explored using the k-means procedure, which was used in order to identify natural groupings of respondents based on their evaluation of socio-economic reasons to invest in renewable energy sources. The best solution was given by the solution of four-clusters, as it provided an optimal balance between within-cluster homogeneity and between-cluster heterogeneity.

Table 3 shows dimensions' average scores for the characterization of clusters. Cluster 1 can be labeled as 'Indifferent Investors' and represents only 9.4% of the sample (it is the smallest cluster). Its members were not affected positively by both factors, and when compared to the other clusters, this cluster stood out as giving the lowest importance to "Pro-environmentalism, favorable investment conditions and benefits for income and electricity cost". The second cluster is the second largest cluster, as it represents 30.5% of the sample. This cluster can be named 'Enthusiastic Investors' because its members classify both factors positively. It is also the cluster that ascribed higher importance to the factor concerning "Social profile, sound investment and independence from oil prices". Cluster 3 can be designated as 'Pro-environmental Investors' and includes 28.9% of the sample. Its members attached greater value to factor 'Pro-environmentalism, favorable investment conditions and benefits for income and electricity cost', but lesser value to 'Social profile, sound investment and independence from oil prices'. Finally, the fourth cluster can be labeled as 'Social Investors', and is the largest cluster (with a small difference from Clusters 2 and 3), including 31% of the sample. Members of this cluster are characterized by the higher level of importance they ascribed to 'Social profile, sound investment and independence from oil price fluctuations' and the lower importance they attributed to 'Pro-environmentalism, favorable investment conditions and benefits for income and electricity cost'.

Table 3. Mean factor loadings of the four clusters.

	CL1 (9.4%) Indifferent Investors	CL2 (30.5%) Enthusiastic Investors	CL3 (28.9%) Pro-Environmental Investors	CL4 (31%) Social Investors	Cluster		Error		F	*p*-Value
					Mean Square	df	Mean Square	df		
(P1)	−2.10465	0.55281	0.54963	−0.41651	334.265	3	0.347	1532	962.208	0.000
(P2)	−0.13794	0.95132	−1.04727	0.08358	306.200	3	0.402	1532	761.028	0.000

The analysis of each cluster's profile is presented in Table 4. The analysis is based on statistical significant differences using chi-square tests between the four clusters. The clusters are thus described based on the level of differentiation that they entail and the overall positive or negative effect of the examined socioeconomic factors. More specifically, the analysis has indicated four clusters: 'Indifferent Investors', 'Enthusiastic Investors', 'Pro-environmental Investors', and 'Social Investors'.

Cluster 1 'Indifferent Investors':

Cluster 1 (smallest cluster), whose members were indifferent to socioeconomic reasons for investing in RES, consists of citizens who are likely to be public and private employees. Most are primary school and university graduates. In addition, most members in this cluster are married and earn between EUR 10,001 and EUR 20,000 per year. A high proportion from this cluster reported having little knowledge about investments in renewables, and was significantly less willing to invest compared to citizens in the other clusters.

Cluster 2 'Enthusiastic Investors':

Cluster 2 (second largest cluster) was labeled 'Enthusiastic Investors' because the members of this cluster classified socioeconomic factors positively, while higher importance was given to factor concerning "Social profile, sound investment and independence from oil prices". Members of this cluster are likely to be public and private employees, and compared to the other clusters, this one has the highest proportion of crop farmers. Most citizens are upper secondary school and university graduates. This cluster has also more vocational training school graduates compared to the other clusters. Much like the previous cluster, the majority is married and earns between EUR 10,001 and EUR 20,000. More than half citizens in this cluster reported knowing much and very much about renewable energy investments. This cluster reported significantly higher willingness to invest in renewables, and only as few as 10% from this cluster would not invest.

Cluster 3 'Pro-environmental Investors':

Citizens in Cluster 3, whose members attached greater value to the factor 'Pro-environmentalism, favorable investment conditions and benefits for income and electricity cost', are likely to be public and private employees. Meanwhile, this cluster presents the highest percentages of pensioners and holders of Bachelor's and Master's degrees. Most citizens in this cluster are married and earn between EUR 10,001 and EUR 20,000 a year and have a moderate knowledge about RES investments. Citizens from this cluster would either invest lower sums of money (EUR 500–2000) or not invest at all.

Table 4. Chi-squared tests results of differences between clusters.

Variable	Scale	CL1 (9.4%)	CL2 (30.5%)	CL3 (28.9%)	CL4 (31%)
Socio-demographic variables					
Occupation	Public employee	17.2%	21.3%	23.8%	15.7%
	Private employee	17.9%	20.5%	23.6%	20.5%
	Freelancer	14.5%	12.4%	9.9%	13%
	Entrepreneur	8.3%	5.5%	2%	4%
	Homemaker	6.2%	5.8%	4.7%	7.3%
	Crop farmer	6.2%	7.5%	3.8%	5.9%
	Livestock farmer	2.1%	2.1%	1.3%	1.9%
	Retired	15.2%	14.7%	18.7%	14.5%
	Unemployed	12.4%	10.2%	12.1%	17.2%
Education level	Primary school	22.1%	15.1%	17.3%	18.9%
	Lower secondary school	6.2%	9.0%	4.9%	8.2%
	Technical school	3.4%	2.3%	1.6%	4.2%
	Vocational training school	4.8%	8.7%	7.2%	6.9%
	Upper secondary school	17.2%	22.0%	16.9%	24.3%
	Vocational education and training (VET)	12.4%	10.0%	10.3%	9.9%
	University	20.7%	21.5%	26.1%	19.9%
	Master's degree	10.3%	10.0%	13.3%	6.9%
	Doctoral degree	2.8%	1.3%	2.5%	0.8%
Family status	Unmarried	29.0%	30.9%	42.5%	40.0%
	Married	57.9%	55.7%	48.3%	46.8%
	Divorced	6.2%	5.8%	2.7%	5.9%
	Widowed	6.9%	7.7%	6.5%	7.3%
Income	Less than EUR 5000	13.8%	9.6%	10.8%	11.7%
	EUR 5001–10,000	19.3%	19.6%	17.5%	23.1%
	EUR 10,001–20,000	24.1%	31.1%	33.9%	22.0%
	EUR 20,001–30,000	4.8%	10.7%	6.7%	6.9%
	More than EUR 30,000	6.2%	3.0%	3.4%	4.4%
Information about RES investments	Not at all—it is the first time I have heard that there is such a possibility	11.7%	3.2%	5.2%	6.9%
	Slightly—I know very few things about RES investments	34.5%	28.1%	32.4%	32.1%
	Moderately—I am familiar with the basic types of RES investments	26.9%	39.9%	46.7%	41.1%
	Much—I have a good knowledge about different investments in RES, risks, and return on investments	10.7%	37.0%	22.5%	29.8%
	Very much—I have a detailed knowledge about RES investments	19.3%	20.7%	13.3%	16.4%
Willingness to invest in RES	Unwilling to invest	37.9%	10.0%	23.8%	25.2%
	Less than EUR 500	16.6%	16.4%	17.5%	12.6%
	EUR 500–1000	5.8%	17.5%	17.8%	20.5%
	EUR 1000–2000	17.9%	15.4%	18.7%	17.4%
	EUR 2000–5000	8.3%	13.0%	9.7%	11.3%
	EUR 5000–10,000	4.1%	12.8%	7.4%	6.1%
	EUR 10,000–20,000	0.0%	7.9%	2.5%	2.7%
	>EUR 20,000	4.1%	7.0%	2.7%	4.2%

Cluster 4 'Social Investors':

Cluster 4, the largest cluster, involves citizens characterized by the higher importance they ascribe to 'Social profile, sound investment and independence from oil prices' and the lower importance they attributed to 'Pro-environmentalism, favorable investment conditions and benefits for income and electricity cost'. Most citizens are private employees, while this cluster involves the highest number of unemployed citizens. Most members from this cluster have completed upper secondary school. This cluster has the lowest proportion of university graduates compared to the other clusters. Much like the other clusters, the highest percentage of citizens is married and earns between EUR 5001 and EUR 10,000. In addition, most citizens in this cluster have a moderate knowledge about investments in renewable energy sources. Finally, the majority would not invest in renewables or would invest between EUR 500 and EUR 1000.

4. Discussion

Citizen investment in renewable energy holds promising potential as an alternative means of financing energy transition [11–14]. As public financing may not suffice, citizens can become investors, thereby supplying a part of the necessary capital and contributing to significant energy targets. Much like any other investor group, citizens will proceed to investments only if they perceive the investment as one that is suitable for their own needs, expectations, and preferences [11,16]. Citizens are thus evolving into a new target investor group for which novel marketing approaches must be devised [20,21]. Our study sought to examine whether there is homogeneity in citizens in relation to investments in renewable energy, and indicates remarkable differences that raise several implications. As a consequence, marketing strategies and investment policies need to be tailored according to the differences observed among the four citizen segments.

This study classified citizens based on how they evaluated a set of economic and social factors in terms of their influence on the decision to invest in renewable energy. The segments that resulted from our analysis exhibited considerable heterogeneity, and it is important to observe that the four clusters demonstrated different levels of appreciation of the examined socioeconomic factors. Enthusiastic Investors were the only segment whose investment decision was positively affected by all examined economic and social factors, and thus seemed to equally value the economic benefits as well as the social deference associated with investments made by citizens. On the other hand, Indifferent Investors did value neither the economic nor the social factors. Regarding the remaining two clusters, Pro-environmental Investors valued the environmental benefits and the investment conditions, while Social Investors valued the social aspects of the investment. The observed differentiation is a significant contribution of this study because it points at the different motivations underlying citizens' decision to invest in renewable energy.

An implication of the observed heterogeneity is that it points to the possibility that universal policies and market strategies based on "one-size-fits-all" approaches are not successful in attracting citizen investment in renewables. It would be thus highly meaningful to take into account the observed differences in the design of strategies and policies for mobilizing citizen investments. As an example, marketing strategies emphasizing the positive effect of renewable energy investments on the effort to reduce energy production-related greenhouse gas emissions may be successful in attracting investments from citizens identified in the cluster Pro-environmental Investors. As this cluster represents 28.9% in our representative sample, such strategies can be expected to appeal to a substantial part of Greek citizens. Following the same line of thought, it is also necessary to appeal to Social Investors by designing suitable marketing strategies which will feature the ability of RES investments to turn citizens into social advocates of renewable energy in their community and to enhance their social standing as active citizens that proceed with exemplary deeds in favor of the environment. As Enthusiastic Investors valued both economic and social factors, it may not be necessary to design strategies explicitly for this group as strategies targeted towards Pro-environmental and Social Investors will be effective for them. It is also important for policymakers to observe that all clusters exhibited a clear preference for low investment amounts, highlighting the need to establish investment mechanisms that may be more suitable for citizens, but, at the moment, are not diffused in the study area, such as equity co-investments, revenue-sharing dividends, and crowdfunding options.

Our study also indicates discrepancies among the four clusters with regard to the reported levels of willingness to invest in renewable energy. Specifically, most members of the Enthusiastic Investors cluster expressed pronounced investment willingness, whereas Pro-environmental Investors and Social Investors would either invest very low sums of money or would make no investment at all. In essence, this means that a considerable part of citizens would prefer to invest very low sums of money, confirming previous research showing that citizens who are willing to invest generally prefer to invest low sums of money [14]. Moreover, Indifferent Investors exhibited the lowest level of willingness to invest, which can be associated with their low evaluation of the examined socioeconomic reasons. It therefore seems that the negative evaluation of the socioeconomic RES investment factors is related to low willingness to invest, and, therefore, the understanding of citizens' perceptions to socioeconomic factors is highly meaningful in any effort to understand how to mobilize citizen investment in renewables.

The reported levels of willingness to invest may also be associated with each cluster's reported level of knowledge about RES investments. Enthusiastic Investors, for instance, reported knowing much, whereas Indifferent Investors reported knowing very little about RES investments. In this regard, it is likely that increasing the provision of information about RES investments may be the missing link to turn willingness to invest into actual investment for clusters that reported having moderate knowledge about RES investments [20,24].

5. Conclusions

This study sought to examine, for the first time, whether citizens as potential investors in renewable energy can be classified on the basis of how they evaluate socioeconomic reasons affecting their investment decision. Our analysis has managed to provide a comprehensive socioeconomic factor-based classification of the identified citizen investor segments, named as Indifferent Investors, Enthusiastic Investors, Pro-environmental Investors, and Social Investors (representing 9.4%, 30.5%, 28.9%, and 31% of the sample, respectively). These four identified segments exhibit notable differentiation in terms of their evaluation of socioeconomic reasons related to RES investment, suggesting that there is significant differentiation in the way the examined socioeconomic factors affect citizen investment decision. Therefore, citizens should not be approached as a homogeneous target group by marketing experts and policymakers that seek to mobilize citizen investment in RES. Marketing experts and policymakers should also pay attention to the fact that each cluster has demonstrated a different level of willingness to invest in renewable energy, and has also demonstrated knowledge about renewable energy investments. Specifically, high levels willingness-to-invest were also recorded for Enthusiastic Investors (90%), Pro-environmental Investors (76.3%), and Social Investors (74.8%), whereas only 56.8% of Indifferent Investors would invest in renewables. However, all clusters exhibited a clear preference for small investment sums, bringing forward the need to establish investment mechanisms that are suitable for citizens interested in investing small amounts of money. In terms of their knowledge about RES investments, Indifferent Investors (by 61.4%), Pro-environmental Investors (by 79.1%), and Social Investors (by 73.2%) have a slight or moderate knowledge, whereas Enthusiastic Investors (by 57.7%) reported good levels of knowledge. The evaluation of socioeconomic reasons made by each cluster may enable a high level of precision in the design of different marketing strategies for each cluster. This kind of design can also be expected to be more effective in comparison to the one currently applied, because it will be in line with the expressed preferences of citizens.

Finally, the results reported in this paper should be seen in the light of some limitations which also point to new directions for future research. Most importantly, even though this research used a broad and representative citizen sample, it is limited only to Greece. For this reason, the reported results should be interpreted in view of the energy policy framework and the investment environment in the study area. To achieve a level of EU-wide segmentation, it would be both interesting and meaningful to examine how EU citizens evaluate socioeconomic factors in future research. Another limitation was that this study measured socioeconomic factors among citizens that have not invested, and not actual citizen investors. However, the rationale was that through the right policies and a more careful design of investment schemes, citizens can be induced to invest in renewable energy and, in this way, to facilitate the pursued energy transition. That being said, a future cluster analysis using a sample of actual citizen investors may perhaps help design the marketing strategies and policies with even greater precision. In addition, future research dedicated to RES investments should replicate this socioeconomic factor-based cluster analysis in relation to specific investment schemes, such as the feed-in-tariff and the net-metering options currently available to citizens. Finally, this kind of segmentation analysis should be repeated after some time because it is essential to examine the possibility that citizen investors evolve and the ways in which they evolve. In this way, the design of marketing strategies and policies for citizen investment can be updated according to how citizen preferences and perceptions change.

Author Contributions: Conceptualization, E.K., G.T. and S.G.; methodology, G.T., S.G. and E.K.; software, E.K.; validation, G.T., S.G. and E.K.; investigation, E.K.; data curation, G.T., S.G. and E.K.; writing—original draft preparation, E.K. and G.T.; writing—review and editing, E.K. All authors have read and agreed to the published version of the manuscript.

Funding: This research received no external funding.

Data Availability Statement: Dataset available on request from the authors.

Conflicts of Interest: The authors declare no conflicts of interest.

References

1. European Council Fit for 55. Available online: https://www.consilium.europa.eu/en/policies/green-deal/fit-for-55/ (accessed on 5 May 2024).
2. Chatzizacharia, K.; Benekis, V.; Hatziavramidis, D. A blueprint for an energy policy in Greece with considerations of climate change. *Appl. Energy* **2016**, *162*, 382–389. [CrossRef]
3. Spyridaki, N.-A.; Banaka, S.; Flamos, A. Evaluating public policy instruments in the Greek building sector. *Energy Policy* **2016**, *88*, 528–543. [CrossRef]
4. Ministry of Environment and Energy National Inventory Report of Greece for Greenhouse Gases and other Gases for the Years 1990–2021. Available online: https://ypen.gov.gr/wp-content/uploads/2021/06/2021_NIR_Greece.pdf (accessed on 5 May 2024).
5. Karasmanaki, E. Understanding willingness to pay for renewable energy among citizens of the European Union during the period 2010–20. In *Low Carbon Energy Technologies in Sustainable Energy Systems*; Academic Press: Cambridge, MA, USA, 2021; ISBN 9780128228975.
6. European Commission Climate Change-Driving Forces. Eurostat Statistics Explained. Available online: https://ec.europa.eu/eurostat/statistics-explained/index.php?title=Climate_change_-_driving_forces (accessed on 9 July 2024).
7. Karamaneas, A.; Koasidis, K.; Frilingou, N.; Xexakis, G.; Nikas, A.; Doukas, H. A stakeholder-informed modelling study of Greece's energy transition amidst an energy crisis: The role of natural gas and climate ambition. *Renew. Sustain. Energy Transit.* **2023**, *3*, 100049. [CrossRef]
8. Hellenic Parliament Law 4936/2022–National Climate Law. Available online: https://www.hellenicparliament.gr/Nomothetiko-Ergo/Anazitisi-Nomothetikou-Ergou?law_id=0b7f36df-2e5b-4d4b-b5f3-ae9900a07542 (accessed on 15 April 2014).
9. Karasmanaki, E.; Galatsidas, S.; Ioannou, K.; Tsantopoulos, G. Investigating Willingness to Invest in Renewable Energy to Achieve Energy Targets and Lower Carbon Emissions. *Atmosphere* **2023**, *14*, 1471. [CrossRef]
10. Zafeiriou, E.; Spinthiropoulos, K.; Tsanaktsidis, C.; Garefalakis, S.; Panitsidis, K.; Garefalakis, A.; Arabatzis, G. Energy and Mineral Resources Exploitation in the Delignitization Era: The Case of Greek Peripheries. *Energies* **2022**, *15*, 4732. [CrossRef]
11. Willis, K.; Scarpa, R.; Gilroy, R.; Hamza, N. Renewable energy adoption in an ageing population: Heterogeneity in preferences for micro-generation technology adoption. *Energy Policy* **2011**, *39*, 6021–6029. [CrossRef]
12. Masini, A.; Menichetti, E. Investment decisions in the renewable energy sector: An analysis of non-financial drivers. *Technol. Forecast. Soc. Change* **2013**, *80*, 510–524. [CrossRef]
13. Collier, S.H.C.; House, J.I.; Connor, P.M.; Harris, R. Distributed local energy: Assessing the determinants of domestic-scale solar photovoltaic uptake at the local level across England and Wales. *Renew. Sustain. Energy Rev.* **2023**, *17*, 113036. [CrossRef]
14. Sirr, G.; Power, B.; Ryan, G.; Eakins, J.; O'Connor, E.; le Maitre, J. An analysis of the factors affecting Irish citizens' willingness to invest in wind energy projects. *Energy Policy* **2023**, *173*, 113364. [CrossRef]
15. Vasseur, V.; Kemp, R. A segmentation analysis: The case of photovoltaic in the Netherlands. *Energy Effic.* **2015**, *8*, 1105–1123. [CrossRef]

16. Salm, S.; Hille, S.L.; Wüstenhagen, R. What are retail investors' risk-return preferences towards renewable energy projects? A choice experiment in Germany. *Energy Policy* **2016**, *97*, 310–320. [CrossRef]
17. le Maitre, J.; Ryan, G.; Power, B.; Sirr, G. Mechanisms to promote household investment in wind energy: A national experimental survey. *Renew. Energy* **2024**, *220*, 119557. [CrossRef]
18. Wilson, C.; Dowlatabadi, H. Models of Decision Making and Residential Energy Use. *Annu. Rev. Environ. Resour.* **2007**, *32*, 169–203. [CrossRef]
19. Shwom, R.; Lorenzen, J.A. Changing household consumption to address climate change: Social scientific insights and challenges. *Wiley Interdiscip. Rev. Clim. Change* **2012**, *3*, 379–395. [CrossRef]
20. Aguilar, F.X.; Cai, Z. Exploratory analysis of prospects for renewable energy private investment in the U.S. *Energy Econ.* **2010**, *32*, 1245–1252. [CrossRef]
21. Curtin, J.; McInerney, C.; Gallachóir, B.Ó.; Salm, S. Energizing local communities—What motivates Irish citizens to invest in distributed renewables? *Energy Res. Soc. Sci.* **2019**, *48*, 177–188. [CrossRef]
22. Guetlein, M.C.; Schleich, J. Understanding citizen investment in renewable energy communities. *Ecol. Econ.* **2023**, *211*, 107895. [CrossRef]
23. Vasseur, V.; Kemp, R. The adoption of PV in the Netherlands: A statistical analysis of adoption factors. *Renew. Sustain. Energy Rev.* **2015**, *41*, 483–494. [CrossRef]
24. Gamel, J.; Menrad, K.; Decker, T. Is it really all about the return on investment? Exploring private wind energy investors' preferences. *Energy Res. Soc. Sci.* **2016**, *14*, 22–32. [CrossRef]
25. Faiers, A.; Neame, C. Consumer attitudes towards domestic solar power systems. *Energy Policy* **2006**, *34*, 1797–1806. [CrossRef]
26. Schelly, C. Residential solar electricity adoption: What motivates, and what matters? A case study of early adopters. *Energy Res. Soc. Sci.* **2014**, *2*, 183–191. [CrossRef]
27. Jager, W. Stimulating the diffusion of photovoltaic systems: A behavioural perspective. *Energy Policy* **2006**, *34*, 1935–1943. [CrossRef]
28. Yee, C.H.; Al-Mulali, U.; Ling, G.M. Intention towards renewable energy investments in Malaysia: Extending theory of planned behaviour. *Environ. Sci. Pollut. Res.* **2022**, *29*, 1021–1036. [CrossRef] [PubMed]
29. Knauf, J.; le Maitre, J. A matter of acceptability? Understanding citizen investment schemes in the context of onshore wind farm development. *Renew. Sustain. Energy Rev.* **2023**, *175*, 113158. [CrossRef]
30. Schiffman, L.G.; Wisenblit, J.; Kumar, S.R. *Consumer Behavior*; Pearson Prentice Hall: Upper Saddle River, NJ, USA, 2010.
31. Petrovich, B.; Hille, S.L.; Wüstenhagen, R. Beauty and the budget: A segmentation of residential solar adopters. *Ecol. Econ.* **2019**, *164*, 106353. [CrossRef]
32. Ebers Broughel, A.; Hampl, N. Community financing of renewable energy projects in Austria and Switzerland: Profiles of potential investors. *Energy Policy* **2018**, *123*, 722–736. [CrossRef]
33. Gáspár, S.; Pataki, L.; Barta, Á.; Thalmeiner, G.; Zéman, Z. Consumer Segmentation of Green Financial Products Based on Sociodemographic Characteristics. *J. Risk Financ. Manag.* **2023**, *16*, 98. [CrossRef]
34. Wang, J.; Liu, F.; Li, L.; Zhang, J. More than innovativeness: Comparing residents' motivations for participating renewable energy communities in different innovation segments. *Renew. Energy* **2022**, *197*, 552–563. [CrossRef]
35. Heiskanen, E.; Jalas, M.; Juntunen, J.K.; Nissilä, H. Small streams, diverse sources: Who invests in renewable energy in Finland during the financial downturn? *Energy Policy* **2017**, *106*, 191–200. [CrossRef]
36. Bryman, A. *Social Research Methods—Oxford University Press*; Oxford University Press: Oxford, UK, 2016.
37. Singh, S. *Advanced Sampling Theory With Applications: How Michael "Selected" Amy*; Kluwer Academic Publishers: Alphen aan den Rijn, The Netherlands, 2003.
38. Gupta, S.; Shabbir, J. On improvement in estimating the population mean in simple random sampling. *J. Appl. Stat.* **2008**, *35*, 559–566. [CrossRef]
39. Matis, K. *Forest Sampling*, 2nd ed.; Democritus University of Thrace: Xanthi, Greece, 2001.
40. Pagano, M.; Gauvreau, K.; Mattie, H. *Principles of Biostatistics*; Chapman and Hall/CRC: Boca Raton, FL, USA, 2022; ISBN 9780429340512.
41. Jolliffe, P.M. *Principal Component Analysis*, 2nd ed.; Springer: New York, NY, USA, 2012.
42. Cattell, R.B. *The Scientific Use of Factor Analysis in Behavioral and Life Sciences*; Springer: Boston, MA, USA, 1978; ISBN 978-1-4684-2264-1.
43. Fragos, C.K. *Methodology of Market Research and Data Analysis with the Use of the Statistical Package SPSS for Windows*; Interbooks Publications: Athens, Greece, 2004.
44. Maxwell, A.E.; Harman, H.H. Modern Factor Analysis. *J. R. Stat. Soc. Ser. A* **1968**, *131*, 615. [CrossRef]
45. Ikotun, A.M.; Ezugwu, A.E.; Abualigah, L.; Abuhaija, B.; Heming, J. K-means clustering algorithms: A comprehensive review, variants analysis, and advances in the era of big data. *Inf. Sci.* **2023**, *622*, 178–210. [CrossRef]

46. Dalmaijer, E.S.; Nord, C.L.; Astle, D.E. Statistical power for cluster analysis. *BMC Bioinform.* **2022**, *23*, 205. [CrossRef] [PubMed]
47. Wu, J. Cluster Analysis and K-means Clustering: An Introduction. In *Advances in K-Means Clustering: A Data Mining Thinking*; Springer: Boston, MA, USA, 2012; pp. 1–16.

energies

MDPI

Article

A Novel Method to Integrate Hydropower Plants into Resource Adequacy Assessment Studies

Christiana I. Kostaki, Pantelis A. Dratsas *, Georgios N. Psarros *, Evangelos S. Chatzistylianos and Stavros A. Papathanassiou

School of Electrical and Computer Engineering, Zografou Campus, National Technical University of Athens (NTUA), 9, Iroon Polytechniou Str., Zografou, 15780 Athens, Greece; christi.kostaki@gmail.com (C.I.K.); vxatzistylianos@mail.ntua.gr (E.S.C.); st@power.ece.ntua.gr (S.A.P.)

* Correspondence: pantelisdratsas@mail.ntua.gr (P.A.D.); gpsarros@mail.ntua.gr (G.N.P.)

Abstract: This paper presents a novel methodology for modeling hydropower plants (HPPs) with and without pumping capability in resource adequacy assessment studies. The proposed method is based on the premise that HPPs should maximize their contribution to system adequacy within their technical constraints by using the energy reserves in their upper reservoirs without significantly deviating from their market schedule. The approach of this paper differs from the conventional operating policies for incorporating HPPs into resource adequacy assessment studies, which either adhere to a fixed market schedule or perform peak shaving, and are inelastic to real-time events or do not resort to realistic temporal correlations between natural water inflows on upper reservoirs and the water discharge needs to cover demand peaks, respectively. The modeling approach focuses on large-reservoir HPPs with natural inflows and is generic enough to deal with both stations incorporating pumping capabilities and those without. It utilizes the state-of-the-art Monte Carlo simulation technique to form the availability of system assets and determine the loss of load incidents. The market schedule and level of reservoir fulfillment for the HPPs are retrieved from a cost-optimal power system simulation algorithm executed offline before the application of the resource adequacy assessment. The effectiveness of the proposed methodology is demonstrated through its implementation in a case study of a power system experiencing different levels of adequacy, comparing the obtained results with various traditional HPP modeling methods from the literature.

Keywords: resource adequacy; hydropower plants; Monte Carlo; real-time redispatch; capacity value

1. Introduction

The ongoing energy transition towards CO_2 emissions' elimination calls for a fundamental transformation of existing power systems, including the gradual decommissioning of thermal power plants in favor of increasing the penetration of renewable energy sources (RES) in the energy mix [1–3]. In such an RES-dominated landscape, flexible assets such as hydropower and battery systems will be required to complement intermittent renewable generation and mitigate any loss of load events, thus allowing the system to maintain acceptable levels of resource adequacy [4–6]. In this regard, HPPs, comprising mature low-carbon generation technologies in existing power systems, are expected to perform a key role as dispatchable power sources, provided that their full potential is exploited and their contribution to system adequacy is properly assessed [7].

The contribution of an asset to the adequacy of a system is assessed based on its impact on the Loss of Load Expectation (LOLE) and Expectation of Energy not Supplied (EENS) reliability indices as a result of its presence in the power system or by using the capacity value metric. Conceptually, the capacity value aims to establish a common basis for comparing the abilities of different generation and storage resources to improve the system's reliability. It is measured using several metrics, such as the equivalent firm capacity (EFC), the equivalent conventional capacity (ECC), and the effective load carrying capability (ELCC) [8–12].

Interestingly, HPPs' capacity values vary significantly across various resource adequacy assessment studies, reflecting the different operating policies applied in each case. Higher capacity values usually indicate that HPPs are simulated according to reliability-driven operating policies, having as their sole objective the contribution of the hydro plants to system adequacy and the reduction in loss of load events. An extensively used reliability-based method is peak shaving, in which the scheduling of the HPP operating profile is adjusted to supply energy to the system exclusively during peak load periods. This is based on the assumption that these periods represent the most crucial time spans for potential energy shortfall occurrences. Although this HPP simulation technique overlooks critical operating constraints, such as the plant's reservoir capacity and the temporal correlation between the peak shaving needs and the availability of natural inflows, which are eventually considered to all be known and available for use at the beginning of every time interval, several recent adequacy studies are still adopting this technique, with variations regarding the application period (e.g., week, month, and year) and the number of hydrological scenarios examined [13–16]. Another major drawback of reliability-based simulation methods, as noted in [17], is their lack of consideration of economic aspects during system operation, leading to unrealistic modeling of HPPs and, thus, to an overoptimistic assessment of their ability to contribute to system adequacy.

In this context, some adequacy studies, such as those of ENTSO-e [18], Pentalateral Energy Forum [19], National Grid ESO [20,21], and IESO [22], take into consideration the economic aspects of power system operation, extracting the HPPs' operating profiles by applying market-oriented unit commitment and economic dispatch models, which can feasibly determine the hourly dispatch of the plants while adhering to their fundamental technical and operational constraints. Apparently, the level of detail of each model leads to a more accurate HPP dispatch plan. For instance, the analysis of ENTSO-e in [18] includes modeling the mandatory water discharge of the plants to shape their final operating profiles. Although these market-based operating policies appear to better reflect the actual operating profiles of HPPs, aspects related to HPPs' realistic operation during the loss of load events for evaluating system adequacy levels are not fully addressed, as, in the respective studies, the market schedule of HPPs remains fixed and inelastic to adjustments to any capacity shortfall incidents.

All adequacy studies in the literature [13–16,18–22] use probabilistic techniques to measure system reliability, as they can capture the inherent uncertainty of stochastic variables, such as unexpected generation outages and variations in RES generation [23]. However, none of them account for unplanned generation outages of HPPs, leading to an overestimation of their ability to contribute to power supply.

A further issue not fully addressed in the literature and the studies carried out by competent institutions is the incorporation and management of real-time outages in adequacy studies. More specifically, available resource adequacy assessment (RAA) studies do not account for the impacts of unplanned outages of system-generating assets on HPP operation. Hence, in RAA, hydro plants adhere to a fixed and unyielding operating profile built upon the maximum system asset availability hypothesis obtained by market-based simulation modeling. This profile persists unchanged, even in the event of unforeseen capacity shortages, restricting the ability of the HPP to contribute to adequacy from a study-level perspective and limiting its capacity value. In reviewing the present methodologies for evaluating system adequacy and potential future improvements that can be implemented by stakeholders, the authors of [17] highlight the need to include real-time outages in resource adequacy assessments and accordingly adjust the operating profiles of the available generation assets. In [24], a real-time redispatch methodology was developed concerning energy storage systems (ESSs), pointing towards real-time operational adjustments of dispatchable generation components and serving as a basis for the proposed HPP redispatch methodology.

Apart from resource adequacy studies carried out by system operators and relevant organizations, the assessment of HPPs' contribution to system adequacy has received limited attention in the relevant literature. The authors of [25] performed a sensitivity analysis regarding HPPs' different storage capacity levels and their impacts on system adequacy employing a probabilistic approach, while in [26], the contribution of HPPs and intermittent RES portfolios was investigated using historical operating data and employing a peak shaving technique. An interesting analysis by Karki R., Hu P., and Billinton R. [27] addressed the coordination of HPPs and wind farms, where HPPs were used either as balancing units for wind power fluctuations or as peaking units to satisfy the system's demand. Reliability indices reached their optimal values when one-third of the available hydro generation was used to regulate wind power fluctuations, while the rest were assigned as peaking units.

Despite hydroelectric plants constituting the principal large-scale energy storage technology in existing power systems, as they fall into the extended storage class due to their substantial reservoir capacity that enables daily and seasonal energy arbitrage [28], there have been no significant improvements in the modeling of their operation and the evaluation of their capacity value in recent RAA studies. Evidently, most of the relevant literature depicts the operation of HPPs in a rather one-dimensional and simplistic manner, applying either a market-driven or a reliability-driven policy, thus failing to capture a balanced real-world operating profile. Key technical and operational aspects, such as their capacity limitations and obligation for mandatory water discharges, are omitted or partially addressed, while the likelihood of unplanned generation outages is completely ignored.

Identifying the aforementioned gaps in the literature, this paper proposes a novel methodology for modeling hydroelectric plants in the context of RAA studies. The underlying principle of the proposed methodology is that HPPs should contribute to mitigating loss of load events in real time to the maximum possible extent while adhering to their technical constraints and avoiding severe deviations from their market schedule to maintain the system's economic operation, as identified by the cost minimization dispatch model that is applied offline prior to the execution of the RAA. To this end, a redispatch algorithm is developed to enforce modifications to HPPs' scheduled operations in real time according to the actual system capacity needs in order to enhance the system adequacy. At the end of a disturbance event, the algorithm ensures the continuity of the market-based scheduled operations of HPPs by applying corrective redispatch orders to restore the reservoir levels to their scheduled values. Thus, HPPs' contribution to system adequacy is captured by maintaining a reasonable operating profile that is capable of satisfying both market-and reliability-driven objectives. This approach provides a more comprehensive perspective than the prevailing simulation approaches, which show a more one-dimensional operating profile. The proposed methodology incorporates the unscheduled outages of hydropower units that occur during real-time operation and implicitly considers the operation of the hydropower fleet for purposes beyond electrification as part of the initial market dispatch plan, including mandatory outflows to cover irrigation and water supply needs.

Overall, a realistic simulation of HPPs' operation is achieved, taking into account their full potential to contribute to adequacy while respecting their inherent constraints. In addition to securing an operational profile that closely resembles the actual operation of HPPs, the adequacy metrics derived from the proposed methodology consistently exhibit a higher level of reliability compared to those obtained from alternative operating policies. This outcome demonstrates the crucial role of the proposed methodology, which involves a real-time assessment of system adequacy and the adjustment of HPPs' dispatch from their market schedule, when required and to the greatest extent technically possible, to ensure the mitigation of inadequacy events without severely compromising market decisions and negatively impacting system operating costs.

The stochastic assessment of system adequacy is performed using the state-of-the art sequential Monte Carlo simulation (SMCS) approach, yielding the LOLE and EENS adequacy indices, as well as the capacity value of the HPP assets, calculated using the EFC metric. The market schedule and scheduled level of reservoir fulfillment of the HPPs are retrieved from a cost-optimal power system simulation algorithm performed before applying the resource adequacy assessment method. The present study examines two HPP types—HPP reservoirs without pumping capabilities and open-loop pumped hydro stations (PHSs). The redispatch methodology is adapted to the operational characteristics of each technology and applied both separately to each HPP type, as well as in aggregate by examining the joint contribution of the two HPP technologies to system adequacy.

The rest of this paper is organized as follows. Section 2 describes, in detail, the principles of the real-time redispatch methodology. Section 3 provides a brief overview of the case study system and the principal outcomes of the proposed methodology. Section 4 outlines a widely used reliability-driven simulation technique that serves as a comparative approach. Section 5 presents a sensitivity analysis regarding HPPs' nameplate capacity and output power. The key findings are summarized in Section 6.

2. Methodology

2.1. Resource Adequacy Assessment Framework

The RAA study of this paper uses the SMCS technique, as this is the most appropriate for dealing with stochastic uncertainties [29]. Multiple states of the modeled system are generated by sampling a sequence of random outages of conventional power plants to determine the available power output of the generation system. A two-state model is assumed to express the annual availability profile of each power plant [30,31]. This profile consists of the two following time intervals: the time period during which a unit is fully available until its next outage event (Time to Failure—TTF) and the time period required to repair the failure and become available again (Time to Repair—TTR). TTF and TTR variables have randomly generated values following exponential distributions with mean values of $\lambda = 1/\text{MTTF}$ and $\mu = 1/\text{MTTR}$, respectively, where MTTF is the mean TTF and MTTR is the mean TTR. TTF and TTR are calculated as shown in Equations (1) and (2), where U and U′ are sequences of uniformly generated numbers in [0,1].

$$\text{TTF} = -\ \text{MTTF} \cdot \ln \text{U} \tag{1}$$

$$\text{TTR} = -\ \text{MTTR} \cdot \ln \text{U}' \tag{2}$$

The MTTR and MTTF values are used to define the Forced Outage Rate (FOR) of each plant, a key parameter that expresses the probability of the plant being out of service.

$$\text{FOR}\ [\%] = \frac{\text{MTTR}}{\text{MTTR} + \text{MTTF}} \cdot 100\% \tag{3}$$

For every SMCS sample year s, the annual availability profile of each thermal and hydropower unit is produced and then aggregated to calculate the total available conventional capacity ($\text{ACC}_{(t,s)}$) of the system, (4). $\text{ACC}_{(t,s)}$ is subsequently compared to the residual load of the system ($\text{P}_{\text{L}_{(t)}}$), defined as the demand load ($\text{P}_{\text{R}_{(t)}}$) plus the ESS charging power ($\text{P}_{\text{ESS-c}_{(t)}}$) minus the RES production ($\text{P}_{\text{RES}_{(t)}}$) and the ESS discharging power ($\text{P}_{\text{ESS-d}_{(t)}}$), as shown in (5). Comparisons are carried out hourly and indicate whether the available generation is sufficient to meet the system's needs or whether there is a capacity deficit. In the case of the latter, inadequacy events are quantified using the Loss of Load Duration ($\text{LLD}_{(s)}$) and Energy Not Supplied ($\text{ENS}_{(s)}$) metrics, which represent the total number of hours per year that load curtailments must be applied and the cumulative amount of energy not supplied, respectively, for every sample year s of the SMCS process, as calculated by (6) and (7).

$$\text{ACC}_{(t,\ s)} = \text{ATC}^{\text{RT}}_{(t,\ s)} + \text{P}^{\text{RT}}_{\text{HPP}_{(t,\ s)}} \tag{4}$$

$$P_{L_{(t)}} = P_{R_{(t)}} + P_{ESS\text{-}c_{(t)}} - P_{RES_{(t)}} - P_{ESS\text{-}d_{(t)}} \quad (5)$$

$$LLD_{(s)} = \sum_{t=1}^{8760} (ACC_{(t,\,s)} < P_{L_{(t)}}) \quad (6)$$

$$ENS_{(s)} = \sum_{t=1}^{8760} \max(0,\ (P_{L_{(t)}} - ACC_{(t,s)})) \quad (7)$$

The final reliability indices reflecting the system's adequacy level as a result of the implementation of the SMCS technique, namely, LOLE (8) and EENS (9), correspond to the mean values of LLD and ENS, respectively, over N SMCS sample years and are calculated as follows.

$$LOLE = \frac{\sum_{s=1}^{N} LLD_{(s)}}{N} \quad (8)$$

$$EENS = \frac{\sum_{s=1}^{N} ENS_{(s)}}{N} \quad (9)$$

The target reliability indices converge to their final value as the number of sample years in the SMCS process rises. In this study, the convergence condition adopted to terminate the SMCS model is set at 1% and is applied to the EENS metric, defined by (10) and (11), as it has a lower convergence rate compared to the LOLE index. The termination criterion is stated by the accuracy metric, calculated using the standard deviation of the EENS after N SMCS sample years.

$$\alpha = \frac{\sigma}{EENS} \quad (10)$$

$$\sigma^2 = \frac{1}{N\cdot(N-1)} \cdot \sum_{s=1}^{N} (ENS_{(s)} - EENS)^2 \quad (11)$$

2.2. HPP Real-Time Redispatch Modeling

As previously stated, the HPP operation is scheduled in accordance with a cost-driven operating policy and developed outside of the SMCS process. More specifically, the operating profiles of ESS and HPPs are derived from applying a market-based generation and storage dispatch algorithm with a cost minimization objective. The generation scheduling algorithm is built upon the linear programming mathematical optimization technique, as in [28]. Optimization has an annual look-ahead horizon to capture the seasonal operations of HPPs and their hourly granularity, while assuming the maximum availability for system assets.

The proposed method takes into account the market schedules of HPPs and real-time capacity outages to determine the final position of the system's HPP fleet. The main feature of the described methodology is that the market schedule of the HPPs, as defined by the market-based cost minimization algorithm of [28], must always be followed to minimize system costs unless a loss of load incident occurs. In such cases, the power output of hydro plants is adjusted (*real-time redispatch*) to mitigate inadequacy incidents as much as technically feasible, deviating from the dispatch orders imposed by the market scheduling processes. Once the HPPs' corrective action fully or partially resolves the loss of load event, HPPs undergo another stage of power output adjustment to restore the stored energy in their upper reservoirs to the levels denoted by their initial market schedule. This is a crucial part of the proposed methodology, as it ensures that the upper reservoir of the hydro plants will always be filled with an adequate amount of water to meet the stations' market obligations, as well as other needs beyond power supply, such as irrigation and water supply, which are explicitly considered in market dispatch processes. Note that the mandatory water discharge of HPPs cannot be compromised to enhance the system adequacy, as this water usage is considered to cover the needs of a higher order of merit.

In brief, the following three stages can be identified regarding the modeling of hydropower plants in resource adequacy assessment study:

- **Normal operating stage:** When the power system is not experiencing any adequacy risk, HPPs operate according to their market schedule.
- **Urgent redispatch stage:** In the occurrence of a shortfall event, HPPs are redispatched from their scheduled operation to contribute to system adequacy, to the greatest extent technically and operationally feasible.
- **Restoration redispatch stage:** Upon the resolution of the shortfall event, HPPs' operating profile isrestored according to their scheduled reservoir levels.

These guidelines apply to the redispatch algorithm developed for each type of hydropower technology, namely HPP reservoirs and open-loop PHSs. Appropriate adjustments are made to ensure that the specific characteristics of each HPP type are properly accounted for in the modeling.

The methodology architecture considers three cases to accurately and comprehensively represent the redispatch of different types of HPPs in the resource adequacy assessment study. These cases address the contribution to adequacy of the following:

(a) HPP reservoirs with natural inflows and without pumping capabilities,
(b) HPP reservoirs with natural inflows and with pumping capabilities (open-loop PHSs), and
(c) the coordinated operation of the above assets in the presence of generation outages to ensure mutual contributions to system adequacy.

The two first cases, (a) and (b), perform redispatch actions only to the HPP examined, while the remaining technology is incorporated into the RAA study with its scheduled operating profile. The third case, (c), addresses the cooperation of the two HPP technologies in mitigating any loss of load events and securing the system adequacy.

A representative redispatch example referring to open-loop PHS operation is shown in Figure 1, together with a corresponding flowchart regarding the individual redispatch algorithm of each HPP technology (cases (a) and (b) above). The four graphs in Figure 1b depict the system operation (Figure 1(b1)), the amount of energy not supplied (Figure 1(b2)), the open-loop PHS's real-time operation (Figure 1(b3)), and the deviation from its scheduled operating profile (Figure 1(b4)). The blue color in the upper graphs corresponds to the total available conventional capacity, comprising the available thermal production and HPPs' generation, according to their initial market schedule ($ATC^{RT}_{(t,\,s)} + P^{SCH}_{HPP_{(t,\,s)}}$). The dark red line indicates the residual load curve (P_L), including RES and ESS operation. The dashed light blue line represents the total available generation after the HPP is redispatched ($ATC^{RT}_{(t,\,s)} + P^{RT}_{HPP_{(t,\,s)}}$).

Figure 1b illustrates the resolution of a shortage event due to an open-loop PHS redispatch that commenced in the second hour of the day, when the demand load requirements exceeded the available output power of the system. This loss of load incident persisted until the sixth hour of the day, prompting the redispatch of the plant to enhance its output and contribute to system adequacy. The open-loop PHS was scheduled to idle during the period of the incident (hours 2 to 6). However, it was eventually switched to production mode to mitigate the loss of load event, operating under the urgent redispatch stage. As described in the corresponding flowchart in Figure 1a, the amount of additional energy supplied by the HPP in each hour from 2 to 6 was equal to the amount of energy not supplied, with a limit up to the hourly maximum available output of the HPP. During the 7th hour of the day, since there was no risk of inadequacy in the system, the hydroelectric plants switched to pumping mode, operating under the restoration redispatch stage, to restore their scheduled reservoir levels (here referred to as the State of Charge—SoC), as indicated by the first step of the redispatch flowchart, which aims to achieve compliance with the scheduled SoC. It is worth noting that, during the first two hours of the restoration (7th–8th), the amount of pumping was being limited by the residual load curve to avoid the occurrence of a further loss of load event. In the 10th hour of the day, the necessity to restore the reduced reservoir level exceeded the scheduled output of the HPP (purple bar in the open-loop PHS operation graph). However, the amount of mandatory water discharge (yellow bar in the open-loop PHS operation graph) was provided to the system as a fundamental function regulating the operating profile of the HPPs, as indicated in the first step of the redispatch flowchart. The 12th hour indicated the conclusion of the HPP's redispatch period. At this point, the $\mathrm{ATC}^{\mathrm{RT}}_{(\mathrm{t,\,s})} + \mathrm{P}^{\mathrm{RT}}_{\mathrm{HPP}_{(\mathrm{t,\,s})}}$ curve aligned with the $\mathrm{ATC}^{\mathrm{RT}}_{(\mathrm{t,\,s})} + \mathrm{P}^{\mathrm{SCH}}_{\mathrm{HPP}_{(\mathrm{t,\,s})}}$ line, and the HPP SoC returned to its market schedule, allowing the HPPs to operate under the normal operating stage. The example provided pertains to a fully resolved loss of load event resulting from the open-loop PHS being redispatched while respecting its technical and operational restrictions.

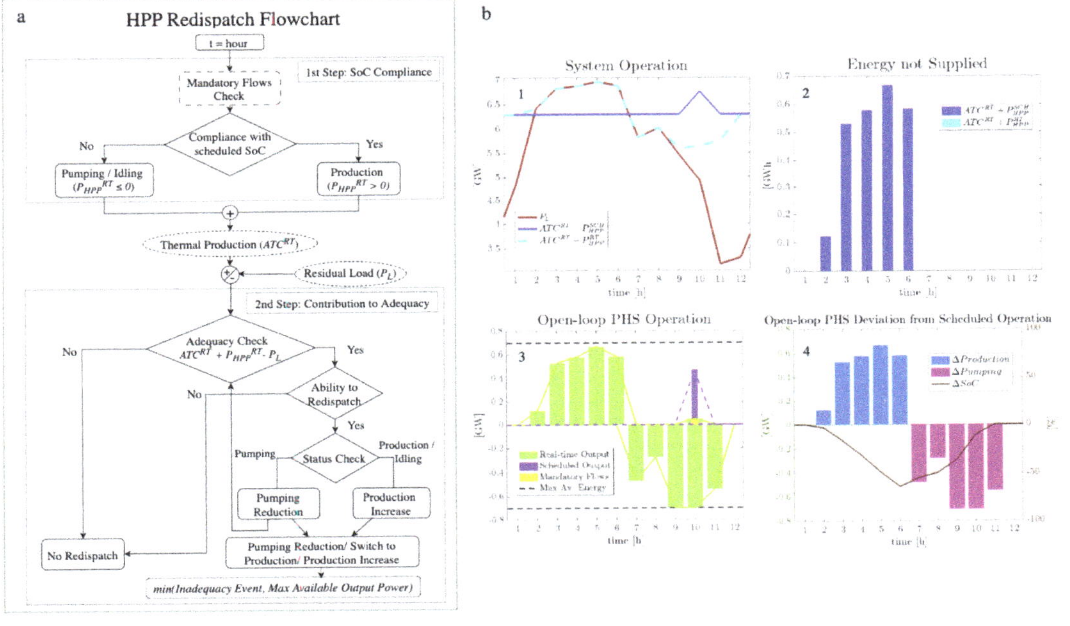

Figure 1. Description of the real-time redispatch algorithm when accounting for each HPP technology separately: (**a**) algorithm flowchart for each hour of a single SMCS sample and (**b**) redispatch example illustrating: (**b1**) system operation; (**b2**) energy not supplied; (**b3**) open-loop PHS scheduled and real-time operation and (**b4**) its production, pumping and SoC deviations between scheduled and real-time operation.

The HPP redispatch flowchart, illustrated in Figure 1a, applies to each HPP technology type being redispatched separately. The main difference between the two HPP technologies concerns the open-loop PHS's pumping mode. In the event of an inadequacy, when hydro plants are forced to operate in the urgent redispatch stage, the reduction in energy that is scheduled for pumping by the open-loop PHS is the first measure to be applied. Conversely, pumping increases in the open-loop PHS when the restoration redispatch stage is applied, to assist the plants in rapidly restoring their lost energy and returning to the normal operating stage as soon as possible. In the case of the HPP reservoir, compliance with the scheduled operation is achieved through idling, which enables the replenishment of the upper reservoir with natural inflows without discharging as scheduled after the clearance of the inadequacy incident.

It is worth noting that the two steps in the flowchart represent the real-time redispatch algorithm that operates on an hourly basis for each sample year s within the SMCS process. The first step ensures compliance with the scheduled operating profile by ensuring the supply of mandatory water discharge and adjusting the HPP's operations to comply with the planned reservoir levels. This step determines the extent to which the HPP's production should be limited after a shortage event, during the restoration redispatch stage. The second step assesses whether the available hourly generation satisfies the system demand load and involves the redispatch actions that are performed in the case of a loss of load event within the power system. The hourly real-time output power of the plant is determined in the last step of the flowchart.

The third case of the proposed methodology entails the redispatch of both HPP technologies and their co-ordination, with the objective of eliminating loss of load events. Thus, it constitutes the most comprehensive framework for simulating the operation and evaluating the contribution of HPPs to resource adequacy assessment. Figure 2a depicts the combined HPP redispatch flowchart, which is conceptually similar to Figure 1a. It outlines the initial step of the algorithm in a concise manner and notes that SoC compliance is implemented independently for each HPP technology. It then proceeds to discuss the sequence of steps to be taken in the event of an inadequacy occurrence, referring to the urgent redispatch stage, which are as follows:

1. Pumping Reduction: This is the initial action to be applied and refers to the hours where the loss of load event is synchronized with the pumping mode of an open-loop PHS. The pumping reduction is equal to the desired energy amount to be supplied or to the maximum energy that has been pumped in this hour. The $P_{L_{(t)}}$ curve is modified, and the energy balance is tested again.
2. Proportional Production Increase: This action pertains to both HPP types and signifies that each HPP type increases its output power in accordance with its installed capacity, contingent upon its energy capacity and the maximum available output power for the given hour. The $ATC^{RT}_{(t,\,s)} + P^{RT}_{HPP_{(t,\,s)}}$ is altered, and the adequacy status is checked again.
3. Extra Production Increase: The final step is initiated if, during the second step, one of the HPP aggregated units is unable to deliver the requested power amount due to an energy deficiency or output power limitation. In such an instance, the remaining one continues to produce either until the depletion of its capacity or until the inadequacy event is entirely alleviated, circumventing the proportionality of a production increase in favor of improving the system adequacy levels.

Figure 2b shows a partially resolved shortfall event arising from the combined contribution of the two HPP technologies to system adequacy. The redispatch performed by the two HPP units decreases the amount of unsupplied energy, as shown in Figure 2(b2). However, the demand cannot be fully covered, as both HPP typologies have reached their maximum output power (hours 4–6). During the remaining two hours of the inadequacy event (hours 7–8), the HPP reservoir (Figure 2(b3)) raises its power output to its hourly maximum, while the production of the open-loop PHS (Figure 2(b4)) drops as its stored energy levels in the upper reservoir deplete (Figure 2(b6)). This aligns with the Extra Production Increase step of the common redispatch algorithm, whereby HPPs without pumping capabilities contribute further to system adequacy due to the energy deficit of the open-loop PHS.

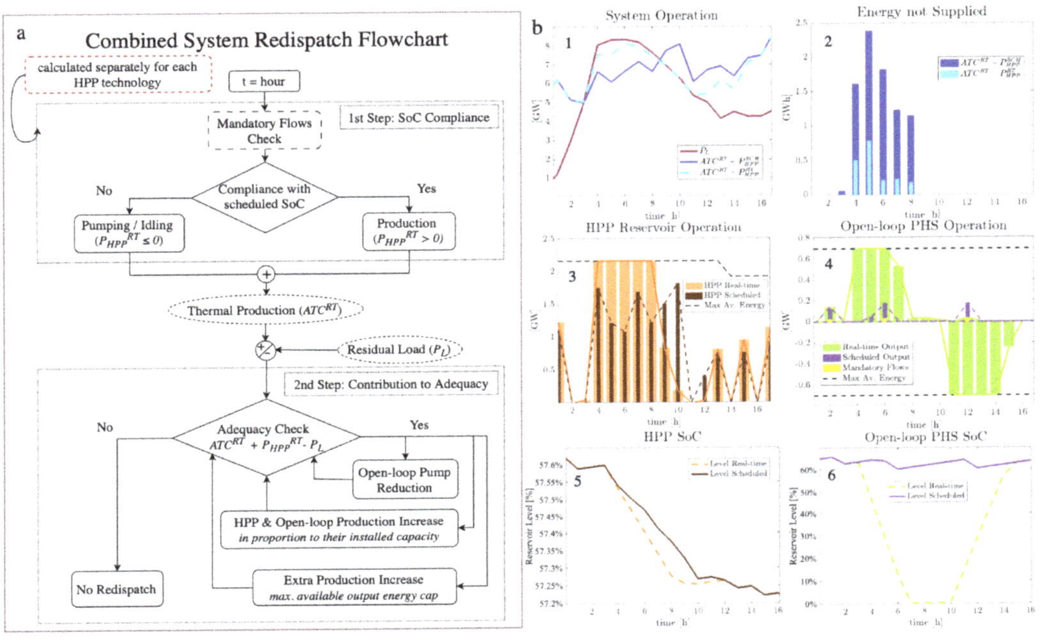

Figure 2. Description of the real-time redispatch algorithm when accounting for the joint contribution of the two HPP technologies in RAA: (**a**) algorithm flowchart for each hour of a single SMCS sample and (**b**) combined redispatch example illustrating: (**b1**) system operation; (**b2**) energy not supplied; scheduled and real-time operation of (**b3**) the HPP reservoir and (**b4**) the open-loop PHS; scheduled and real-time SoC of (**b5**) the HPP reservoir and (**b6**) the open-loop PHS.

2.3. Capacity Value Estimation

The contribution of the HPPs to system adequacy is assessed by applying the capacity value metric, which reflects the additional reliability provided to the power system by integrating the examined production asset [32]. In the present study, the capacity values of HPPs are quantified using the EFC metric, which refers to a perfectly reliable generating unit that would deliver the same system adequacy outcome as the unit under consideration [33,34]. The EFC is calculated through the iterative process shown in Figure 3. The principle is to gradually increase the generation capacity until the target reliability index is reached. At each step, the adequacy metrics obtained by adding firm capacity are recalculated. The smaller the iteration step, the more accurate the output. The final EFC value is equal to the firm capacity that provides the same level of adequacy to the power system as the unit under consideration. It can be expressed either in MW or as a fraction of the installed capacity of the HPP.

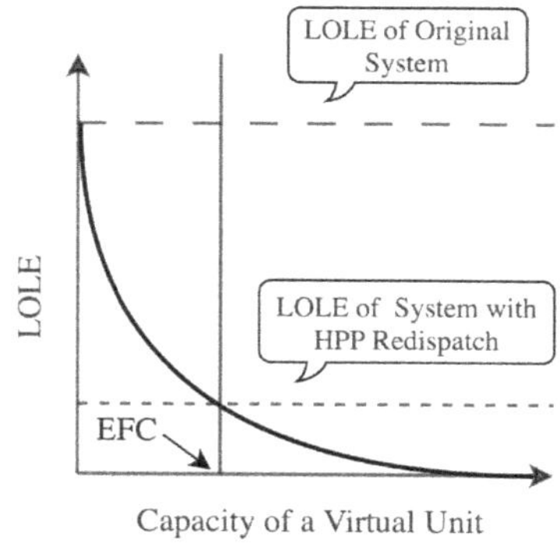

Figure 3. Evaluation methodology for EFC metric.

3. Case Study and Main Results

3.1. Case Study

The proposed methodology is applied to the Greek power system in its stage of expansion in 2030 [35], with some variations in the composition of the thermal fleet to serve the purpose of the analysis. The annual load of the system is approximately 62 TWh, with an hourly peak of 11.6 GW. The system includes 8.3 GW of combined cycle gas turbines (CCGTs), 3.1 GW of hydropower plants (HPPs), 13 GW of photovoltaic (PV) systems, 9 GW of wind farms (both onshore and offshore), and 0.4 GW of small run-of-river hydro. In addition, ~3 GW of energy storage (including both battery systems and closed-loop PHSs) completes the generation mix. As mentioned above, the HPPs are divided into 2.4 GW of aggregated HPP reservoirs without pumping capacity and 0.7 GW of aggregated open-loop PHSs. Their reservoir capacities are 3600 GWh and 4.5 GWh, respectively, which implies tighter constraints for open-loop PHSs in terms of deviation from their scheduled operation due to their limited reservoirs. The operating profile of the RES, ESS, and HPPs, together with the demand load time series, are derived from the CEP model outside the SMCS process.

The reliability indices applied to the conventional units are obtained from the latest ENTSO-e resource adequacy studies [36]. The FOR is 5% for newer and 8% for older thermal units, while the MTTR is 24 h for all conventional power plants. The aforementioned assets refer to the base case power system. The subsequent sensitivity analysis presents modifications to the thermal capacity and nameplate characteristics of the HPPs. The RAA model, the redispatch algorithm, and the power system simulations are conducted in MATLAB [37].

3.2. Main Findings for the Base Case Power System

The proposed real-time redispatch methodology is implemented for 35 different climate years (CYs) for each of the three modeling cases (HPP reservoirs, open-loop PHSs, and their joint redispatch modeling). The objective of this study is to investigate the impact of various wind and solar data on RES production, the influence of different hydrological conditions on HPPs' operating profiles (Figure 4), and variations in the demand load curve, as well as the interaction of the aforementioned parameters of the studied power system. This analysis evaluated the adaptability and effectiveness of the developed methodology by applying it to a wide range of hydrological conditions (Table 1) and operational states for the given generation system. The required climate data are procured from the Paneuropean Climate Database and are those employed by the ENTSO-e in their most recent resource adequacy assessments [36]. Figure 4 presents the annual aggregated reservoir levels of HPP reservoirs, as indicated by the operating profile derived from the cost minimization algorithm.

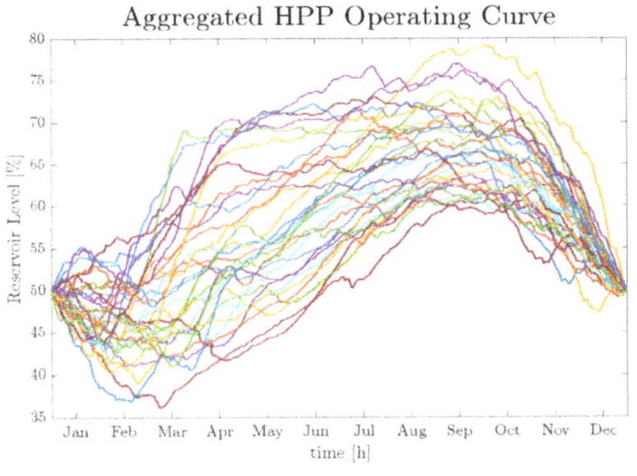

Figure 4. Annual operating curve of HPP reservoir for 35 climate years (with different colors representing each climatic year).

Table 1. HPPs' hydrological data for 35 climate years.

Natural Inflows	Mean Value (TWh)	Standard Deviation (TWh)
HPP Reservoirs	4.2	0.7
Open-loop PHSs	0.6	0.15

The 35 CYs are investigated for both the scheduled market-driven operation policy for HPPs and the redispatch methodology for all three modeling cases. Figure 5 presents a comparison of the retrieved reliability indices and capacity values for the two operational policies. Figure 5(a1–a3) concern the HPP reservoir redispatch, Figure 5(b1–b3) the open-loop PHS redispatch, and Figure 5(c1–c3) their joint redispatch methodology. The CYs refer to the years 1982 to 2016 and are presented in chronological order on the x-axis (i.e., 1 = 1982, 2 = 1983, etc.).

As illustrated in all diagrams of Figure 5, the LOLE and EENS indices resulting from the developed method are consistently lower than those observed when the HPPs adhere to their market schedule. In particular, the proposed methodology achieves a more than 98% improvement in the combined redispatch modeling of both HPP technologies compared to their cost-driven operating profiles (Figure 5(c1,c2)). This is evidenced by the elimination of the initial 7.6 h/year LOLE to only 0.1 h/year of load not satisfied.

Each independently redispatched HPP technology demonstrates enhanced adequacy metrics compared to the LOLE and EENS values derived from its market schedule. The HPP reservoirs exhibit superior performances relative to the open-loop PHSs, as evidenced by their smaller total output power (0.7 GW) versus the output power of the HPP reservoirs (2.4 GW). The joint redispatch of both HPP technologies achieves the highest performance among the three modeling cases, allowing for the availability of 3.1 GW of cumulative output power to perform redispatches from their scheduled operation in the event of inadequacy.

In terms of evaluating their capacity value, normalized to the installed capacity of the asset under consideration, the results are significantly higher when the asset is redispatched (87–93%) than when it operates according to its market schedule (15–46%). It is noteworthy that the open-loop PHS redispatch achieves slightly lower capacity values than the other two modeling cases, due to its comparatively smaller storage capacity (4.5 GWh). This storage volume corresponds to approximately seven hours of full charge/discharge modes to achieve the fill-up/empty-out of their tanks. Consequently, the open-loop PHS is not able to perform frequent redispatch for consecutive hours, as required during severe loss of load events, since this would rapidly deplete their reservoirs.

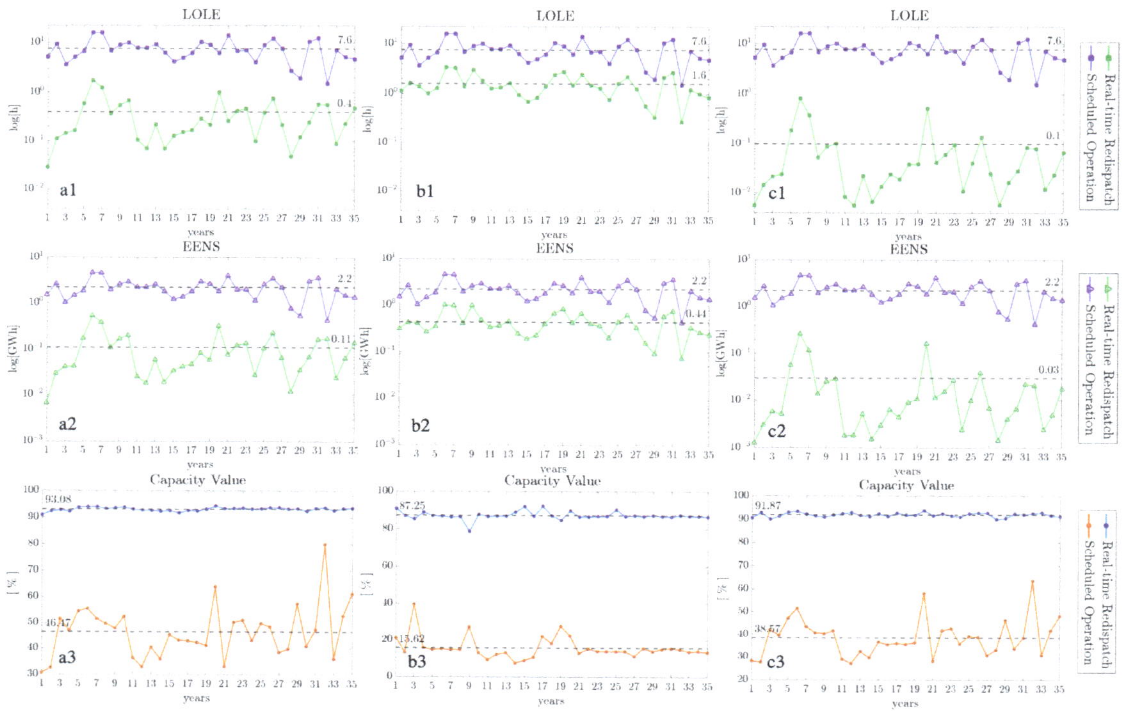

Figure 5. Reliability indices of (**a1**,**b1**,**c1**) LOLE and (**a2**,**b2**,**c2**) EENS and (**a3**,**b3**,**c3**) capacity value results for 35 climate years of the base case power system: (**a1–a3**) HPP reservoir modeling; (**b1–b3**) pen-loop PHS modeling; and (**c1–c3**) both HPP technologies' modeling.

The capacity values retrieved from the cost-optimal operating policy in all three modeling cases are comparatively low, which is consistent with the findings in the relevant literature. This is due to the initial objective of the applied market schedule, which is to minimize system costs rather than to ensure a high level of system reliability.

4. Comparison with a Reliability-Driven Operation Policy

In order to achieve a comprehensive evaluation of the proposed real-time redispatch methodology and its obtained results, a reliability-driven operating policy is chosen as a further comparison method. The most commonly used reliability-oriented approach for HPP modeling is the peak shaving technique, which is briefly explained and subsequently applied to the same base case power system of the 35 CYs analysis to perform a direct comparison between the two HPP modeling approaches.

4.1. Peak Shaving Principles

The fundamental concept of peak shaving is the scheduling of HPPs to operate during peak demand periods [38]. This operation policy considers peaks as the most critical hours for the occurrence of a shortfall event and, therefore, allocates the natural inflows of HPPs to the hours with the highest load values. In this way, the operation of HPPs is integrated into the residual load curve, resulting in a smoothed load curve where the demand peaks have been shaved off. The present modeling approach assumes that the amount of natural inflows is known and fully available to be allocated on demand for the selected time interval where peak shaving is performed. The most common time periods for implementing peak shaving in resource adequacy studies are weekly, monthly, or yearly.

The flowchart diagram of the peak shaving technique is shown in Figure 6a. First, the implementation interval is determined, and then the natural inflows are allocated via an iterative, stepwise process. This process considers the maximum available output power of the given HPP plant and the remaining amount of inflows to be dispatched. Once the iterative calculation is complete, a threshold is established that defines the peak load hours during which the HPP inflows are allocated and the remaining load hours during which the HPPs are not scheduled to operate. Subsequently, the shaved load curve is extracted (Figure 6b).

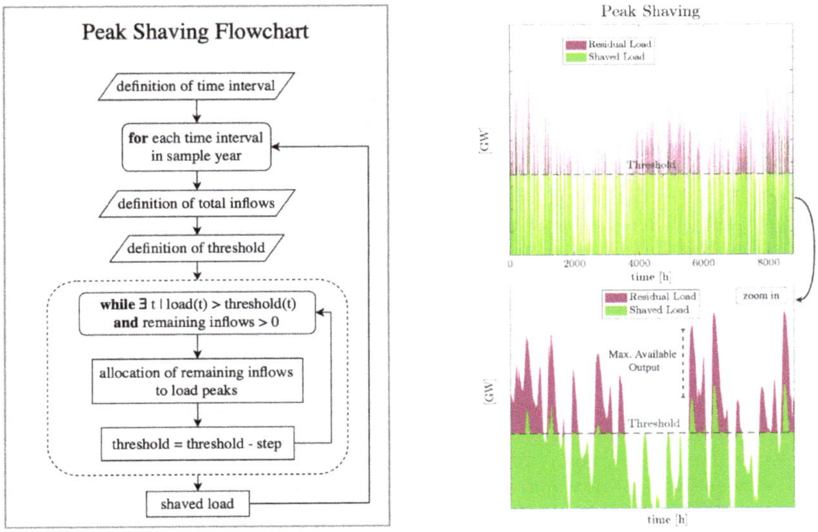

Figure 6. Peak shaving technique: (**a**) shaved load calculation process and (**b**) shaved load curve.

4.2. Adequacy Results

The peak shaving technique is applied to the 35 CYs of the base case study for the most commonly used implementation time intervals (week, month, and year) in all three HPP modeling cases. The results of the joint HPP modeling are shown in Figure 7. For comparison purposes, the corresponding reliability results achieved by the proposed methodology are also shown.

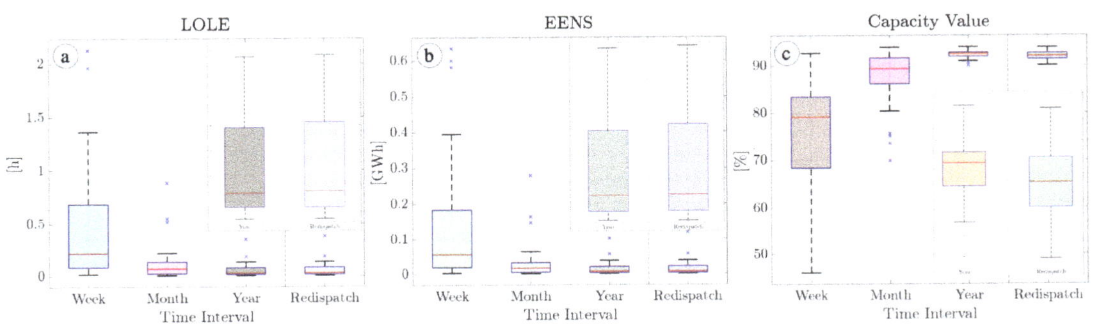

Figure 7. Peak shaving adequacy results for different time intervals: (**a**) LOLE; (**b**) EENS; and (**c**) capacity value.

It can be observed that a shorter application interval is associated with a deterioration in the LOLE (Figure 7a) and EENS (Figure 7b) indices, as well as the capacity value metric (Figure 7c). This is due to the fact that peak shaving dispatches smaller amounts of inflows in shorter implementation periods. In addition, the algorithm only has access to a limited sample of the load curve, while being forced to allocate all available inflows in that particular time frame. As the application period increases, the available inflows can be dispatched in a more optimal manner, as peak load hours can be addressed more effectively.

A further point of interest illustrated in Figure 7 is that the annual peak shaving provides slightly superior reliability results compared to the joint redispatch modeling. This can be attributed to the main assumptions of the peak shaving technique, one of which is neglecting the reservoir capacity of the plant. In this regard, the total amount of inflows is deemed to be known and available at the beginning of each implementation interval and dispatched on demand, without factoring in the reservoir capacity of the plant. In particular, the peak shaving method allows for each plant to be scheduled to either provide energy that is not contained in its tank, resulting in negative reservoir level values, or to store large amounts of energy without having the capacity to support it, exceeding 100% of the reservoir level. Peak shaving's particular drawback arises from its failure to account for the timing of natural inflows, which is partially addressed by shortening the implementation period. However, this is particularly noticeable when applied to plants with a relatively limited capacity, such as open-loop PHSs in the given power system.

Figure 8 depicts the duration curves and operating curves of the reservoir level for each HPP technology, as determined by the scheduled market-driven operating policy (purple line) and annual peak shaving modeling (grey line). As demonstrated in the preceding analysis, the curve of the latter approach in the open-loop PHS modeling (Figure 8(b2)) exceeds the accepted storage range almost throughout the year, indicating a completely infeasible operating profile.

In the case of HPP reservoirs, peak shaving also results in a deviation from their planned operating curves (Figure 8(a2)), albeit modest and within the designed boundaries of their reservoir levels. This is due to the vast aggregated reservoir capacity of the HPP reservoir within the existing power system. This feature allows for redispatch whenever necessary, with the sole constraint being the maximum available output power. Consequently, in the modeling case where only the HPP reservoir is redispatched, the redispatch reliability indices are never exceeded by the corresponding indices obtained by annual peak shaving. This indicates that the annual peak shaving and the redispatch algorithms yield comparable outcomes in the context of HPP reservoir modeling variations, as neither approach is constrained by the plant reservoir capacity at the given adequacy level of the studied power system.

To gain further insight into the performance of peak shaving, different adequacy levels of the considered power system were examined by gradually removing thermal units, leading to lower levels of system adequacy. It was demonstrated that, as the system became increasingly inadequate, the results demonstrated a decline in the performance of the peak shaving technique. This tendency aligns with the core idea of the peak shaving approach, which is to shave off the load peaks, as these periods are thought to be the most pivotal for loss of load events. However, systems with lower levels of reliability face inadequacy events also during non-peak load periods, which the peak shaving approach is not designed to resolve. Nevertheless, these off-peak inadequacy events are effectively addressed by the proposed methodology in a manner analogous to any other loss of load event. This is due to the real-time adequacy evaluation within the SMCS process, which prompts HPPs to perform redispatch whenever necessary. Thus, the capacity values of the HPPs obtained by the redispatch approach maintain their consistently high values across different reliability levels of the studied power system (Figure 9).

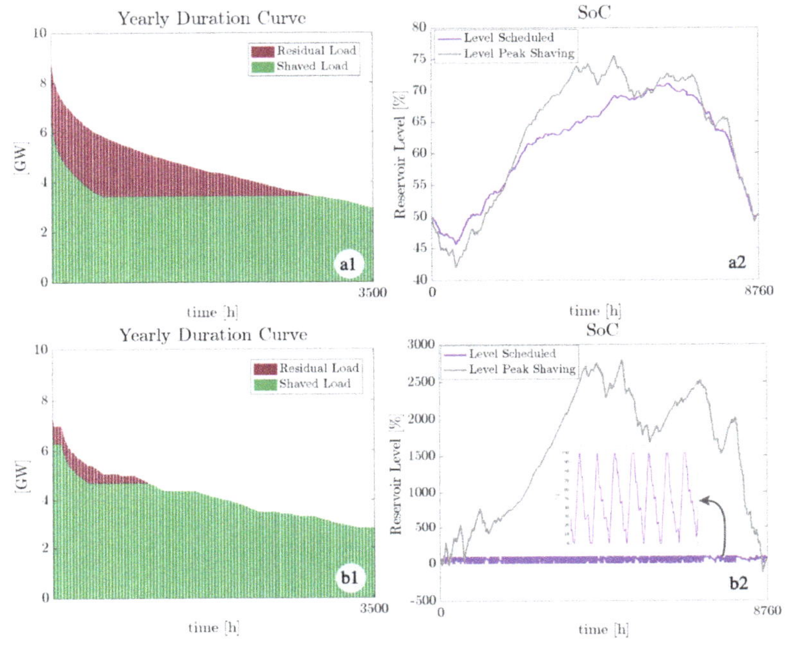

Figure 8. Peak shaving operating profile: (**a1**,**a2**) HPP reservoir and (**b1**,**b2**) open-loop PHS; (**a1**,**b1**) residual and shaved load duration curves and (**a2**,**b2**) annual SoC profile based on cost minimization algorithm and peak shaving policy.

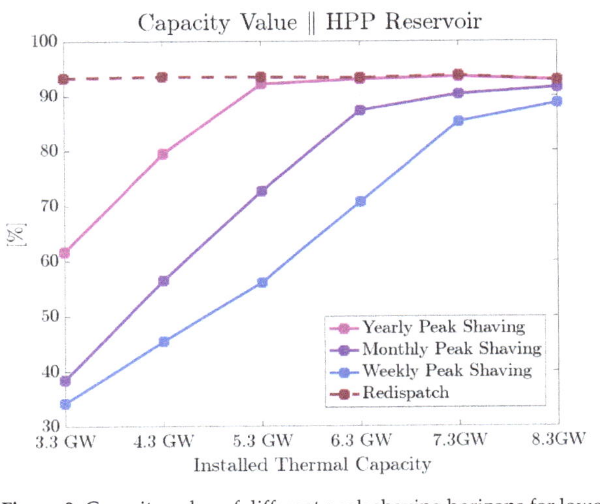

Figure 9. Capacity value of different peak shaving horizons for lower levels of system adequacy in the HPP reservoir modeling variation.

This point of discussion reveals another limitation of the peak shaving technique, namely that, rather than dynamically analyzing the operating state of the system within the HPP simulation process, the technique adheres to a rigid assessment of upcoming adequacy risks within the power system and remains completely unresponsive to real-time generation outages.

The preceding analysis of the peak shaving approach serves to reinforce the view that it fails to account for some of the most important operating parameters of the assets under consideration, resulting in an overestimation of their ability to contribute to system adequacy. In addition, it treats loss of load events as predetermined events, which only applies to significantly adequate power systems where loss of load events tend to be infrequent and minor during peak load hours. The foregoing analysis demonstrated the inability of the peak shaving approach to adequately address shortfall events in power systems with lower levels of adequacy.

5. Sensitivity Analysis

This section performs a sensitivity analysis to investigate the impact of HPP nameplate parameters on system adequacy when operating under the proposed real-time redispatch methodology. This analysis is conducted by gradually increasing both the output power and the storage capacity values of each HPP technology separately and examining their performances at different initial system adequacy levels. A representative CY, namely 2004, is used as a reference case in which the HPP parameters and the time series of the power system exhibit mainly median values.

Figure 10 illustrates the reliability indices and capacity value results for the HPP reservoir and open-loop PHS of the base case power system (8.3 GW CCGTs), where additional hydro units of each type are added in increments of +175 MW. The storage capacity of these additional units is increased in proportion to the output power addition. As extra hydropower units of one technology type are added, the remaining HPP technology remains unaffected. The base case in both configurations (Figure 10(a1–a3,b1–b3)) refers to the initial HPP characteristics described in Section 3, namely 2.4 GW of the HPP reservoir and 0.7 GW of the open-loop PHS. In each of the following steps, the total amount of the hydro generation is identical between the two configurations, with a discrepancy in the hydro generation mix. In particular, the final step of each configuration, where an extra +700 MW of hydropower is added, corresponds to a total of 3.8 GW of hydropower generation and an energy mix of 82% HPP reservoir and 18% open-loop PHS in the case of adding an HPP reservoir. Conversely if open-loop PHS units are added, the mix shifts to 63% and 37%, respectively.

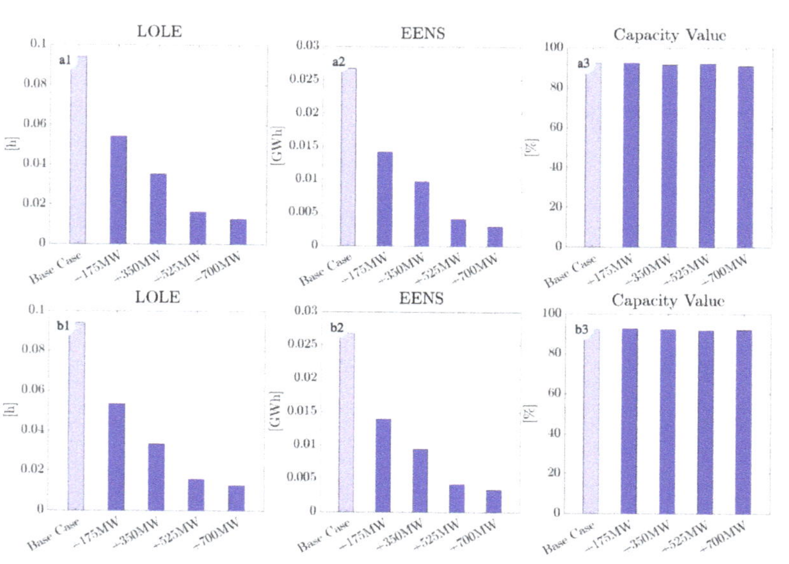

Figure 10. Base Case power system (8.3 GW CCGTs)—reliability results: (**a1**,**b1**) LOLE, (**a2**,**b2**) EENS and (**a3**,**b3**) capacity value for additional (**a1–a3**) HPP reservoirs and (**b1–b3**) open-loop PHSs.

The initial adequacy level of the power system is significantly enhanced by the redispatch of the base case HPP units, with a less than 0.1 h/year LOLE and ~0.025 GWh/year. As anticipated, the introduction of additional hydro units results in a further reduction in the reliability indices, with the final step of generation increase yielding the most favorable adequacy results. It is noted that the incorporation of additional HPP units of both technologies achieves comparable outcomes, with the final increment of +700 MW approaching a value of 0.01 h/year LOLE and ~0.005 GWh/year EENS when either additional HPP reservoir or open-loop PHS units are added. This is explained by the extremely high adequacy level of the studied power system, even in its initial stage (referred to as Base Case in the diagrams of Figure 10), which requires infrequent and undemanding HPP redispatches. Consequently, both technology types can contribute almost equally, in proportion to their installed capacity, since they are not subject to significant operational limitations. The corresponding capacity value results demonstrate minimal variation, remaining stable at approximately 91% to 92%. These values are consistent with the primary objective of the proposed methodology, namely the mitigation of loss of load events.

Figure 11 depicts the same two configurations as in Figure 10, but with a significantly lower level of system adequacy, starting from over 70 h/year LOLE and over 40 GWh/year EENS. Note that, in this case, CCGTs with a total capacity of 3 GW are removed from the system. Once again, the additional hydro units gradually improve the system's reliability level, although it is apparent that the integration of HPP reservoirs provides better results compared to the addition of the same amount of open-loop PHSs. The discrepancy in the final LOLE index between the two configurations is almost 8 h per year, with the extra HPP reservoir achieving a final LOLE of ~21 h/year, while the open-loop PHS achieves ~29 h/year.

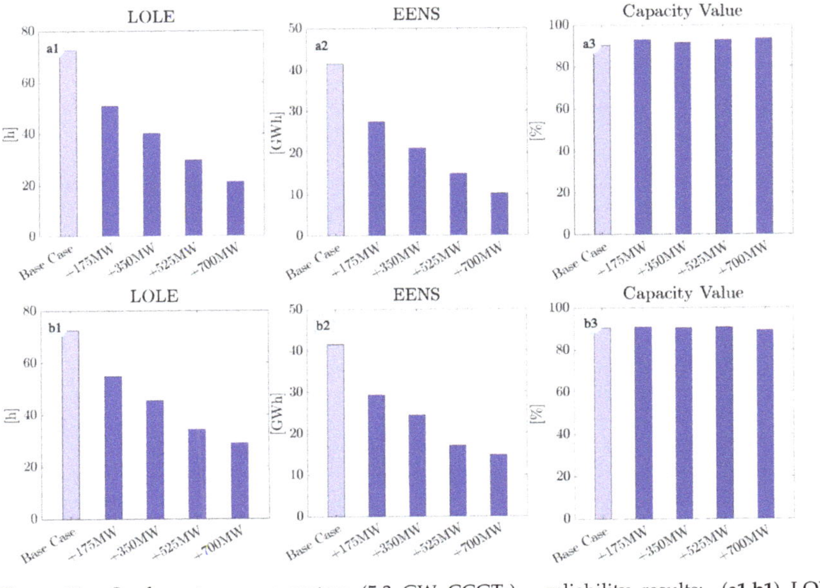

Figure 11. Inadequate power system (5.3 GW CCGTs)— reliability results: (**a1**,**b1**) LOLE, (**a2**,**b2**) EENS and (**a3**,**b3**) capacity value for additional (**a1–a3**) HPP reservoirs and (**b1–b3**) open-loop PHSs.

The reason for this outcome can be attributed to the constrained storage capacity, which influences the open-loop PHS redispatch. Even with a storage capacity of +700 MW for the open-loop PHS, which is also doubled to 9 GWh, the redispatch is still constrained by energy limits. In the case of a relatively inadequate power system, as shown in Figure 11, the HPPs' contribution to system reliability is both extensive and highly demanded. This often leads to the depletion of the open-loop PHS reservoir and, thus, an inability to efficiently contribute to upcoming loss of load events.

To further investigate the performances of the two configurations, intermediate levels of system adequacy are examined between the two "extreme" scenarios of 8.3 GW and 5.3 GW of installed thermal capacity. Figure 12 illustrates the final reliability indices values achieved after the addition of +700 MW of each HPP type for different system adequacy levels. As indicated, lower adequacy levels lead to significant differences in the LOLE and EENS values between the two configurations, however, as the thermal capacity increases, this difference diminishes. This finding lends support to the proposition that there is a robust correlation between the adequacy level of the system and the storage capacity of the unit being redispatched.

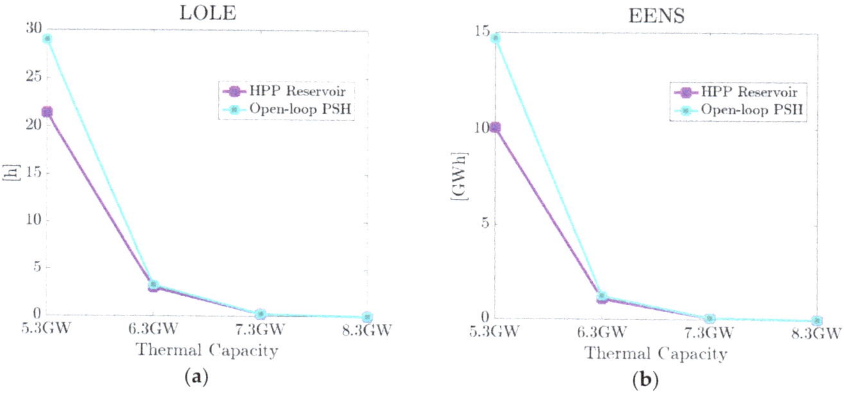

Figure 12. Reliability indices for +700 MW of each HPP technology for different adequacy levels of the same power system: (**a**) LOLE and (**b**) EENS.

It is worth noting that the reason for the varying reliability results between the two configurations lies in the different hydro generation mixes. As the proportion of HPP technology constrained by energy limitations rises, the resulting reliability indices decline, as a greater quantity of generating capacity is unable to fulfil the desired redispatch from its scheduled operation. In both configurations, the dominant HPP technology is the HPP reservoir, which has a high energy capacity due to its large reservoirs and is, therefore, capable of performing the necessary redispatch required by the system's needs.

Despite representing a relatively minor proportion of the total hydro capacity, with capacity shares ranging from 18% to 37% depending on the increment, open-loop PHSs still impact the resulting reliability indices derived from the joint redispatch modeling of the two HPP technologies due to their energy capacity constraints. This can be observed by examining the reliability level enhancement achieved by incorporating each technology type across varying system adequacy levels, as illustrated in Figure 13. Both graphs in Figure 13 demonstrate the reduction in each adequacy index from the base scenario with no additional HPP units to the final step of the analysis, where an additional 700 MW of hydro capacity is added. The same amount of supplementary hydro capacity produces rather moderate outcomes for both the LOLE and EENS metrics in the case of a relatively inadequate power system, compared to the results achieved at higher adequacy levels of the same power system. This verifies the assertion that, as the level of system adequacy declines and the proportion of energy-constrained HPP units in the total hydro capacity mix increases, the final adequacy indices of the examined power system deteriorate.

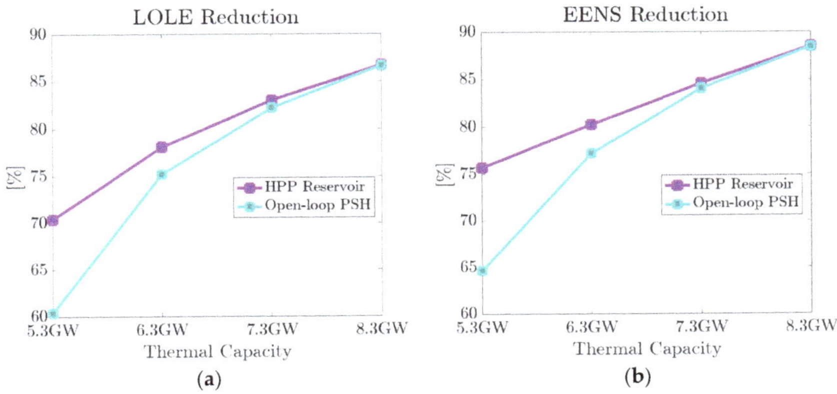

Figure 13. Reliability indices' improvement by the addition of +700 MW of each HPP technology for different adequacy levels of the same power system: (**a**) LOLE reduction and (**b**) EENS reduction.

6. Conclusions

This paper introduces a novel real-time redispatch algorithm for integrating HPPs with and without pumping capabilities into resource adequacy assessment studies. The proposed modeling approach posits that HPPs are subject to a market-driven operating profile, yet are redispatched in real time to ensure that the adequacy requirements of the power system are satisfied, reflecting realistic operating conditions. A reliability evaluation is performed using the Monte Carlo simulation technique, and the Greek power system in its future development stage serves as a case study. The proposed methodology is compared with a market-driven modeling technique, yielding significantly higher reliability indices and capacity value results in all modeling variations. In particular, the improvement in the LOLE index achieved ranges from 78% to 98% when compared to an HPP operating profile following the market schedule alone and being inelastic to inadequacy incident needs. At the same time, the corresponding HPP capacity values are notably high, attaining values between 78.7% and 94.3%. Furthermore, the proposed method is evaluated against the peak shaving technique, a reliability-driven method, achieving balanced results and demonstrating the inability of the latter to properly account for the inherent technical and operational constraints of HPPs. The adaptability and effectiveness of the developed methodology are proven by investigating different reliability levels of the studied power system, where the capacity value metric exhibits a robust performance, consistently exceeding 91%. A sensitivity analysis is also conducted, considering different HPP nameplate parameters and assessing their impacts on the resulting metrics of the proposed modeling algorithm. The analysis reveals that the reservoir capacity of a hydropower plant has a significant impact on its ability to perform redispatches and, thus, its ability to further contribute to system adequacy. Especially in power systems with lower levels of adequacy, the addition of a 700 MW HPP reservoir provided a 41% improvement in LOLE, while the addition of an open-loop PSH achieved only a 30% enhancement.

Author Contributions: Conceptualization, P.A.D., G.N.P. and S.A.P.; methodology, C.I.K., P.A.D. and G.N.P.; software C.I.K., P.A.D. and E.S.C.; validation, P.A.D., G.N.P. and S.A.P.; formal analysis, C.I.K.; investigation, C.I.K., P.A.D. and G.N.P.; resources, S.A.P.; data curation, P.A.D., G.N.P., E.S.C. and S.A.P.; writing—original draft preparation, C.I.K.; writing—review and editing, C.I.K., P.A.D. and G.N.P.; visualization, C.I.K. and P.A.D.; supervision, S.A.P.; project administration, S.A.P. All authors have read and agreed to the published version of the manuscript.

Funding: This research received no external funding.

Data Availability Statement: Data is contained within the article.

Conflicts of Interest: The authors declare no conflicts of interest.

Abbreviations

ACC	Available Conventional Capacity
ATC	Available Thermal Capacity
CCGT	Combined Cycle Gas Turbine
CEP	Capacity Expansion Planning
CY	Climate Year
ECC	Equivalent Conventional Capacity
EENS	Expectation of Energy not Supplied
EFC	Equivalent Firm Capacity
ELCC	Effective Load Carrying Capability
ENS	Energy Not Supplied
ESS	Energy Storage Systems
FOR	Forced Outage Rate
HPP	Hydropower Plant
LLD	Loss of Load Duration
LOLE	Loss of Load Expectation
MTTF	Mean Time To Failure
MTTR	Mean Time To Repair
PHS	Pumped Hydro Station
PV	Photovoltaic
RAA	Resource Adequacy Assessment
RES	Renewable Energy Sources
SMCS	Sequential Monte Carlo Simulation
SoC	State of Charge
TTF	Time To Failure
TTR	Time To Repair
U	Sequences of uniformly generated numbers in [0,1]
s	Monte Carlo sample year
t	Time period
N	Number of sample years of Monte Carlo simulation
$P_{L_{(t)}}$	Residual Load in period t
$P_{R_{(t)}}$	Demand Load in period t
$P_{ESS\text{-}c_{(t)}}$	ESS charging power in period t
$P_{ESS\text{-}d_{(t)}}$	ESS discharging power in period t
$P_{RES_{(t)}}$	RES production in period t
$LLD_{(s)}$	Loss of Load Duration in sample year s
$ENS_{(s)}$	Energy Not Supplied in sample year s
α	Accuracy of EENS value over the Monte Carlo simulation
σ^2	Standard deviation of EENS value over the Monte Carlo simulation

References

1. Mohamad, F.; Teh, J.; Lai, C.M.; Chen, L.R. Development of Energy Storage Systems for Power Network Reliability: A Review. *Energies* **2018**, *11*, 2278. [CrossRef]
2. Psarros, G.N.; Papathanassiou, S.A. Electricity Storage Requirements to Support the Transition towards High Renewable Penetration Levels—Application to the Greek Power System. *J. Energy Storage* **2022**, *55*, 105748. [CrossRef]
3. Mirzapour, O.; Rui, X.; Sahraei-Ardakani, M. Transmission Impedance Control Impacts on Carbon Emissions and Renewable Energy Curtailment. *Energy* **2023**, *278*, 127741. [CrossRef]
4. Yüksel, I. Hydropower for Sustainable Water and Energy Development. *Renew. Sustain. Energy Rev.* **2010**, *14*, 462–469. [CrossRef]
5. Dratsas, P.A.; Psarros, G.N.; Papathanassiou, S.A. Capacity Value of Pumped-Hydro Energy Storage. In Proceedings of the 2022 2nd International Conference on Energy Transition in the Mediterranean Area (SyNERGY MED), Thessaloniki, Greece, 17–19 October 2022; pp. 8–13. [CrossRef]
6. Heidari, M.; Heidari, M.; Soleimani, A.; Khorrami, B.M.; Pinnarelli, A.; Vizza, P.; Dzikuć, M. Techno-Economic Optimization and Strategic Assessment of Sustainable Energy Solutions for Powering Remote Communities. *Results Eng.* **2024**, *23*, 102521. [CrossRef]
7. IEA. *Hydropower Special Market Report*; IEA: Paris, France, 2021. Available online: https://www.iea.org/reports/hydropower-special-market-report (accessed on 10 May 2024).

8. Amelin, M. Comparison of Capacity Credit Calculation Methods for Conventional Power Plants and Wind Power. *IEEE Trans. Power Syst.* **2009**, *24*, 685–691. [CrossRef]
9. Keane, A.; Milligan, M.; Dent, C.J.; Hasche, B.; D'Annunzio, C.; Dragoon, K.; Holttinen, H.; Samaan, N.; Söder, L.; O'Malley, M. Capacity Value of Wind Power. *IEEE Trans. Power Syst.* **2011**, *26*, 564–572. [CrossRef]
10. Dratsas, P.A.; Psarros, G.N.; Papathanassiou, S.A. Capacity Value of Wind Powered Pumped Hydro Storage Virtual Power Plants. In Proceedings of the 2023 IEEE 2nd Industrial Electronics Society Annual On-Line Conference (ONCON), Virtual, SC, USA, 8–10 December 2023. [CrossRef]
11. Liu, G.; Vrakopoulou, M.; Mancarella, P. Assessment of the Capacity Credit of Renewables and Storage in Multi-Area Power Systems. *IEEE Trans. Power Syst.* **2021**, *36*, 2334–2344. [CrossRef]
12. Zhou, Y.; Mancarella, P.; Mutale, J. Modelling and Assessment of the Contribution of Demand Response and Electrical Energy Storage to Adequacy of Supply. *Sustain. Energy Grids Netw.* **2015**, *3*, 12–23. [CrossRef]
13. Ming, Z.; Olson, A.; Jiang, H.; Mogadali, M.; Schlag, N. *Resource Adequacy in the Pacific Northwest*; Energy and Environmental Economics, Inc.: San Francisco, CA, USA, 2019.
14. PJM Manual 21A: Determination of Accredited UCAP Using Effective Load Carrying Capability Analysis. Available online: https://www.pjm.com/-/media/documents/manuals/archive/m21a/m21av0-determination-of-accredited-ucap-using-effective-load-carrying-capability-analysis-08-01-2021.ashx (accessed on 20 April 2024).
15. IPTO Resource Adequacy Study 2020–2030. Available online: https://www.admie.gr/sites/default/files/users/dssas/meleti-eparkeias-ishyos-2020-2030.pdf (accessed on 30 January 2024).
16. Uría-Martínez, R.; Johnson, M.M. *U.S. Hydropower Market Report 2023 Edition*; Oak Ridge National Laboratory (ORNL): Oak Ridge, TN, USA, 2023; pp. 28–32.
17. Carvallo, J.P.; Zhang, N.; Leibowicz, B.D.; Carr, T.; Baik, S.; Larsen, P.H. *A Guide for Improved Resource Adequacy Assessments in Evolving Power Systems Institutional and Technical Dimensions*; Ernest Orlando Lawrence Berkeley National Laboratory: Berkeley, CA, USA, 2023.
18. ENTSO-E. *European Resource Adequacy Assessment/2023 Edition*; ENTSO-E: Brussels, Belgium, 2023.
19. Chowdhury, A.A.; Koval, D.O. Generation Adequacy Assessment. *IEEE Ind. Appl. Mag.* **2009**, *15*, 69–74. [CrossRef]
20. National Grid ESO. *Resource Adequacy in the 2030s*; National Grid ESO: Warwick, UK, 2022.
21. National Grid ESO. *Electricity Capacity Report 2021*; National Grid ESO: Warwick, UK, 2021; Volume 2021, pp. 1–116.
22. Independent Electricity System Operator (IESO). *Methodology to Perform the Reliability Outlook*; IESO: Toronto, ON, Canada, 2020; pp. 1–50.
23. Blanco, M.P.; Spisto, A.; Hrelja, N.; Fulli, G. Generation Adequacy Methodologies Review. In *JRC Science For Policy Report*; European Union: Brussels, Belgium, 2016; p. 106.
24. Dratsas, P.A.; Psarros, G.N.; Papathanassiou, S.A. A Real-Time Redispatch Method to Evaluate the Contribution of Storage to Capacity Adequacy. *IEEE Trans. Power Syst.* **2024**, *39*, 1274–1286. [CrossRef]
25. Cunha, S.H.F.; Borges, C.L.T.; Simon, P.S.; Dias, J.A.S. Probabilistic Capacity Value of a Hydroelectric Plant for Different Storage Capacity Levels. In Proceedings of the 2018 IEEE PES Innovative Smart Grid Technologies Conference Europe (ISGT-Europe), Sarajevo, Bosnia and Herzegovina, 21–25 October 2018; pp. 1–6. [CrossRef]
26. Vehviläinen, I. Joint Assessment of Generation Adequacy with Intermittent Renewables and Hydro Storage: A Case Study in Finland. *Electr. Power Syst. Res.* **2021**, *199*, 107385. [CrossRef]
27. Karki, R.; Hu, P.; Billinton, R. Reliability Evaluation Considering Wind and Hydro Power Coordination. *IEEE Trans. Power Syst.* **2010**, *25*, 685–693. [CrossRef]
28. Chatzistylianos, E.S.; Psarros, G.N.; Papathanassiou, S.A. Insights from a Comprehensive Capacity Expansion Planning Modeling on the Operation and Value of Hydropower Plants under High Renewable Penetrations. *Energies* **2024**, *17*, 1723. [CrossRef]
29. Billinton, R.; Li, W. *Reliability Assessment of Electric Power Systems Using Monte Carlo Methods*; Springer Science & Business Media: Berlin/Heidelberg, Germany, 1994; ISBN 9781489913487.
30. Reddy Atla, C.S.; Balaraman, K.; Patil, A. Estimating Capacity Credit of Wind Energy Sources Using Probabilistic Methods. In Proceedings of the 2013 15th International Conference on Advanced Computing Technologies (ICACT), Rajampet, India, 21–22 September 2013. [CrossRef]
31. Dratsas, P.A.; Psarros, G.N.; Papathanassiou, S.A. Battery Energy Storage Contribution to System Adequacy. *Energies* **2021**, *14*, 5146. [CrossRef]
32. Zhou, Y.; Mancarella, P.; Mutale, J. Framework for Capacity Credit Assessment of Electrical Energy Storage and Demand Response. *IET Gener. Transm. Distrib.* **2016**, *10*, 2267–2276. [CrossRef]
33. Milligan, M.; Porter, K. Determining the Capacity Value of Wind: A Survey of Methods and Implementation. In Proceedings of the Wind Power 2008, Houston, TX, USA, 1–4 June 2008.
34. Dratsas, P.A.; Psarros, G.N.; Papathanassiou, S.A. Feasibility of Behind-the-Meter Battery Storage in Wind Farms Operating on Small Islands. *Batteries* **2022**, *8*, 275. [CrossRef]
35. 4893/2019 B Greek National Plan for Energy and Climate of the Greek Government. *Gov. Gaz. 4893 B* **2019**. [CrossRef]
36. ENTSO-e ERAA Downloads. Available online: https://www.entsoe.eu/outlooks/eraa/2023/eraa-downloads/ (accessed on 20 December 2023).

37. Mathworks MATLAB and Simulink for Technical Computing. Available online: https://www.mathworks.com (accessed on 10 January 2023).
38. Chua, K.H.; Bong, H.L.; Lim, Y.S.; Wong, J.; Wang, L. The State-of-The-Arts of Peak Shaving Technologies: A Review. In Proceedings of the 2020 International Conference on Smart Grid and Clean Energy Technologies (ICSGCE), Kuching, Malaysia, 4–7 October 2020; Volume 2020, pp. 162–166. [CrossRef]

energies

MDPI

Article

How can Fossil-Energy-Free Technologies and Strategies (FEFTS) be adopted in European Farming?

Athanasios T. Balafoutis [1,*] and Bas Paris [1,2]

[1] Institute of Bio-Economy & Agro-Technology, Centre for Research & Technology Hellas, Dimarchou Georgiadou 118, 38333 Volos, Greece; bparis@aua.gr

[2] Department of Natural Resources and Agricultural Engineering, Agricultural University of Athens, 75 Iera Odos Str., 11855 Athens, Greece

* Correspondence: a.balafoutis@certh.gr

Abstract: This paper provides policy recommendations for accelerating the adoption of Fossil-Energy-Free Technologies and Strategies (FEFTS) in the EU agricultural sector. Faster adoption of these technologies and strategies is crucial to achieving the medium- and long-term sustainability targets laid out in EU policy. The prepared policy recommendations originate out of the key outputs and findings of the Horizon 2020 project "AgroFossilFree", including an assessment and evaluation of the current energy use status in EU agriculture, survey results on farmers' needs, ideas and interests on the adoption of FEFTS, FEFTS categories identified through an online inventory of FEFTS called the AgEnergy platform, and key innovative processes through national and transnational workshops that combine expertise from hundreds of keys stakeholders (researchers, innovation brokers, policymakers, farmers, and industry representatives). The policy recommendations are synthesized and presented in the form of 19 policy briefs split into three main categories: those that are related to energy issues in farming and can be applied to any farm and FEFTS type; those that are specific to certain agricultural production systems; and those that are necessary for FEFTS integration in agriculture in general.

Keywords: fossil-energy-free technologies and strategies; policy recommendations; renewable energy; energy efficiency; carbon sequestration; sustainable agriculture; common agricultural policy

1. Introduction

The agricultural sector in the European Union (EU) is at a critical juncture, facing the challenges of transitioning towards a sustainable food and agriculture sector, ensuring food security and developing resilience against climatic change [1]. The agricultural sector remains heavily reliant on fossil energy and contributes significantly to greenhouse gas emissions and environmental degradation. As the urgency to combat climate change intensifies, there is a growing consensus on the need to transition towards fossil-energy-free technologies and strategies (FEFTS) within the agricultural sector. This transition is not only crucial for environmental sustainability, but also for enhancing the resilience and economic viability of agriculture in the face of fluctuating energy prices and depleting fossil fuel reserves.

The EU has set ambitious targets for reducing greenhouse gas emissions, increasing the use of renewable energy, and improving energy efficiency, as outlined in the European Green Deal [2], the Common Agricultural Policy (CAP) [3], and the Farm to Fork Strategy [1]. However, the pace of adoption of FEFTS in agriculture remains slow especially when compared to other economic sectors [4]. In addition, in recent years, considerable attention has been placed on adopting sustainable agricultural practices and decarbonizing the agricultural sector. Still, the role and steps required to effectively transition away from fossil energy in agriculture is not well explored. This paper addresses these critical gaps by providing comprehensive policy recommendations aimed at accelerating the adoption of such technologies and strategies across the European agricultural sector.

Drawing on a thorough analysis of the outputs of the Horizon 2020 project Agro-FossilFree (www.agrofossilfree.eu), based on an interactive innovation model, this paper synthetizes key policy recommendations and proposes targeted policy interventions to accelerate the adoption of FEFTS in the EU agricultural system. The proposed policies emphasize the integration of renewable energy sources (RESs), energy-efficient technologies and practices, and innovative carbon sequestration practices that reduce dependence on fossil fuels. Furthermore, the paper highlights the importance of financial incentives, regulatory frameworks, and capacity-building measures to support farmers in this transition. By fostering a collaborative approach among policymakers, industry stakeholders, and the farming community, the proposed policy recommendations aim to create a conducive environment for the widespread adoption of FEFTS. This paper contributes to the broader discourse on sustainable agriculture by offering actionable recommendations that align with the EU's climate goals and support the sustainable development of the agricultural sector.

Such a study is particularly relevant in the present context, as the EU has established several overarching policies to promote sustainability and the adoption of renewable energy and energy efficient technologies in agriculture. The CAP, the European Green Deal, and the Farm to Fork Strategy are central to these efforts, setting ambitious targets for reducing greenhouse gas emissions, increasing the use of renewable energy and applying more energy efficient technologies and practices. These strategies aim to create a sustainable food system and encourage the adoption of environmentally friendly practices in agriculture [1–3]. However, it is often noted that these policies often lack specific, actionable measures tailored to the agricultural sector. For instance, while the Renewable Energy Directive (RED II) mandates a significant increase in the share of renewable energy in the EU's energy mix, it does not provide detailed guidance on how agricultural stakeholders can contribute to this target [5].

At the national level, several EU member states have implemented framework policies to support the adoption of FEFTS in agriculture. For example, Germany's Renewable Energy Sources Act (EEG) has been instrumental in promoting the use of biogas and solar energy in agriculture. Similarly, Denmark's integrated approach to renewable energy, which combines wind, solar, and bioenergy, has provided a robust framework for sustainable agricultural practices [6]. In addition, there are multiple specific policy initiatives within EU member states that promote FEFTS adoption in agriculture. For renewable energy generation, notable examples include the following: France's "Plan Énergie Méthanisation Autonomie Azote" (EMAA) that promotes biogas from agricultural waste [7], Italy's "Conto Energia" that supports photovoltaic installations on farms [8], and the Netherlands' SDE++ that provides support for RES generation by famers [9]. For energy efficiency measures, notable examples include Spain's "PAREER-CRECE", focused on the energy efficiency of agricultural buildings [10], Poland's "Program Agroenergia" [11], and Germany's "Bundesprogramm Energieeffizienz in der Landwirtschaft" (Federal Program for Energy Efficiency in Agriculture) [12]. There are significant disparities in the implementation of national policies across the EU and some national initiatives are not fully aligned with EU-level goals, leading to a fragmented policy landscape [13].

To address these challenges, several scholars and policy analysts have proposed more coherent and inclusive policy frameworks. Bojnec and Ferto (2022) show that targeted financial incentives, such as subsidies and low-interest loans, lower the barriers to entry for small and medium-sized farms [14]. Jenner et al. (2013) have also argued that regulatory measures, such as feed-in tariffs and renewable energy mandates, should be harmonized across the EU to create a stable and predictable market for renewable energy penetration [15].

Knowledge sharing and capacity-building initiatives are also shown as crucial policy mechanisms in promoting FEFTS in agriculture [16]. Herrera et al. (2019) show the importance of effective advisory services in the sustainability profiles of farms and highlight the importance of training programs and knowledge-sharing platforms to equip farmers with the necessary skills and information [17]. These programs should be designed to reflect the diverse agricultural contexts within the EU and be accessible to farmers in all member states.

While existing policies provide a foundation for the promotion of FEFTS in agriculture, there is a clear need for a comprehensive study that spans the entire EU. Such a study should be representative of the diverse agricultural practices and conditions across member states and capture grassroots-level opinions. This would ensure that policy recommendations are evidence-based and reflect the realities faced by farmers on the ground. Grassroots perspectives are particularly important for designing effective policies. According to recent research, many farmers are hesitant to adopt new technologies due to a lack of awareness and technical knowledge. Engaging with farmers directly can help policymakers understand these barriers and design interventions that are both practical and acceptable to the agricultural community [18]. In this context, the present study that is based on an interactive innovation model that captures the opinions and needs of hundreds of relevant stakeholders across the EU is particularly crucial in developing policies for promoting FEFTS acceleration in the EU agricultural sector.

2. Materials and Methods

2.1. Process for the Integration of Results

The policy recommendations for accelerating the adoption of FEFTS were developed as part of and based on the results of the Horizon 2020 Project AgroFossilFree that followed the EIP-AGRI "multi-actor approach", engaging agricultural and energy sectors' stakeholders to support agricultural defossilization. A bottom-up approach was used to gather information at the grassroots level. First, the policy gaps were identified and then specific policy recommendations were developed using the results of the main activities of the project, which were as follows:

- Literature review to assess and evaluate the energy use state of EU agriculture.
- Identification and detailed description of the successful innovation processes of specific marker-ready FEFTS that defossilize specific agricultural practices.
- Survey execution to 470 farmers and 41 experts from eight countries (Denmark, Germany, Greece, Ireland, Italy, the Netherlands, Poland, and Spain) to primarily identify the main factors affecting FEFTS adoption in EU farms and then extract the needs and interests to shift into a greener farm energy profile.
- Detection and recording of FEFTS derived from international peer-reviewed journal publications, EU research projects' results, and commercial products combined with important training information on how to install and use such FEFTS and financing mechanisms available today to support agricultural defossilization practices.
- Organization of 24 national (three agricultural systems (namely open-field, livestock, greenhouses) workshops in each of the eight countries) and 3 transnational workshops to collaborate directly and interactively with agricultural and energy stakeholders (998 actors in national and 140 in transnational events) to gather ideas on how FEFTS could be integrated in agricultural production systems in a national and EU level.
- Development of an online toolkit, named AgEnergy Platform, where an easily searched inventory of the main FEFTS types would be available for all users, while giving the registered members the ability for FEFTS assessment.

Figure 1 shows explicitly the above mentioned activities, the respective results, and their liaison for producing the proposed policy guidelines to support agricultural defossilization. More specifically, the sequence of activities (in green boxes) and the respected results (in orange bubbles) are given on a national and EU level, respectively. The three main activities on a national level were (i) the survey to farmers and experts, (ii) the successful innovation processes identification of specific FEFTS that are in action and contribute to farming defossilization, and (iii) the organization of national (regional) workshops.

The survey expressed the experiences and thoughts of the 511 interviewed stakeholders and after a thorough qualitative analysis it was possible to extract their knowledge about farm energy related needs, the obstacles that delay FEFTS adoption, and the respective incentives to promote defossilization of EU farming [19,20]. The successful innovation processes analysis identified the ways that the selected use cases overcame the existing barriers (recognized from the survey) and managed to apply their FEFTS in practice [21]. The 24 national workshops were focused on extracting additional information about the needs, obstacles, and motivations that all types of actors in the agricultural and energy sectors would require to promote FEFTS integration in EU agriculture. In addition, the participating stakeholders were asked to point out the most important FEFTS for the defossilization of their sector. A shared methodology was used in all workshops and after an analysis of the workshops' outcomes in different countries, the gaps in existing policies and incentives were retrieved [22,23]. As for the EU level, three more activities were executed, starting with (i) the review of the EU energy use in the three main agricultural systems (open-field, livestock, and greenhouses), following up with (ii) the AgEnergy Platform creation and filling it with inventoried FEFTS of all categories, and (iii) the three transnational workshops (one per agricultural system). The literature review detected the tendencies of energy consumption in all types of farming systems and acknowledged the main conventional technologies and FEFTS that are in use today [24–26]. This way, the starting point for agricultural defossilization was set as the cornerstone of this work methodology. The AgEnergy Platform was filled based on the methodology and standards set from the beginning of the project [27] and it contains low-to-medium-TRL technology solutions as results of scientific publications and research projects and high-TRL commercial technologies ready to use in EU farming. In addition, the platform provides access to relative training material about different types of FEFTS and financing tools to support FEFTS application. Finally, the three transnational workshops were used to validate the needs, obstacles, and motivations extracted from the regional workshops and provide extra ideas on further policies and incentives to increase FEFTS adoption [23].

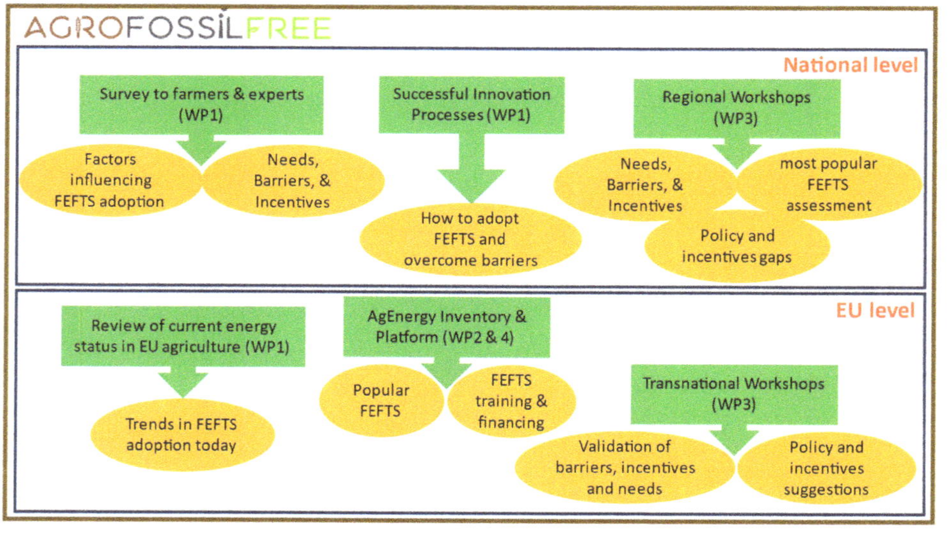

Figure 1. Methodology for the development of policy recommendations to assist the defossilization of EU farming. Each activity (green boxes) produces one or more results (orange bubbles). The above frame considers the work carried out at a national level, while the frame below refers to the EU level.

After the main results for policy recommendations were gathered, an extensive critical analysis process by experts of all the organizations that are a part of the AgroFossilFree consortium was conducted to gather the main relevant outputs. This was largely a qualitative analysis of the results and their importance, as considered by the experts, as well as a detailed investigation into the most common aspects originating from the workshops. This analytical process then allowed for a synthesis of the main results leading to the formulation of a series of policy recommendations presented into 19 specific policy briefs. Experts from the following organizations took part in this process: Agricultural University of Athens (AUA), Agricultural & Environmental Solutions (AGENSO), Aarhus Universitet (AU), the Comite Europeen Des Groupements De Constructeurs Du Machinisme Agricole (CEMA), the Centre for Research & Technology Hellas (CERTH), the Confederazione Generale Dell Agricoltura Italiana (CONFAGRICOLTURA), DELPHY European Conservation Agriculture Federation (ECAF), Iniciativas Innovadoras Sal (INI), Innovationscenter for Økologisk Landbrug (ICOEL), the Instytut Uprawy Nawozenia I Gleboznawstwa, Panstwowy Instytut Badawczy (IUNG-PIB), Lubelski Osrodek Doradztwa Rolniczego W Konskowoli (LODR), RESCOOP EU ASBL, the Agriculture and Food Development Authority (TEAGASC), Trama Tecnoambiental S.L. (TTA), and Wirtschaft Und Infrastruktur Gmbh & Co Planungs Kg (WIP).

In order to validate the results of this methodology, an online consultation meeting with Policy Officers from selected DGs of the European Commission was conducted. A total of 41 people participated and the selected 19 policy recommendations/briefs were presented. After a thorough discussion, the feedback was used to assist in the optimization of the policy recommendations/briefs. Their revised versions were presented at a brokerage event in Brussels, Belgium, on 22 June 2023. After the critical review of the participants at the brokerage event and an external reviewer of the project, the policy recommendations were optimized and finalized.

2.2. Limitations

There are a number of methodological limitations relevant to the present study. The synthesizing of the project results to policy recommendations, particularly derived from workshops, was predominantly a qualitative process. This reliance on qualitative data introduces potential biases, as the experts and stakeholders involved are active participants in the agricultural sector who may have vested interests or specific perspectives that could influence their contributions. These biases can arise from the individual priorities, experiences, or sectoral positions of the participants, potentially skewing the findings or recommendations. However, efforts were made to mitigate such biases through cross-validation and expert reviews. The use of cross-validation involved comparing inputs from different workshops or stakeholder groups to identify consistent themes and minimize the influence of outlier opinions. Furthermore, additional expert reviews were used to provide an external assessment of the data synthesis, ensuring that conclusions drawn were robust and reflected a balanced interpretation of the broader context. Despite these efforts, it is important to acknowledge that any qualitative analysis inherently carries a degree of subjectivity, and the interpretations made by the experts in this study, while informed, may still not fully capture the diversity of perspectives within the agricultural sector.

In addition, the agricultural sector within the EU is a large and highly diverse sector, encompassing a wide range of practices, scales of operation, and regional variations. This diversity poses a significant challenge, as it increases the risk of the analysis being overly generic and not adequately capturing the nuances of different contexts. To address this issue, efforts were made to derive specific results and to connect these findings to targeted policy recommendations. This approach aimed to ensure that the insights were applicable to distinct aspects of the sector, culminating in the development of 19 separate policy briefs. By tailoring policy suggestions to specific contexts and issues and providing recommendations on both an EU and member state level, our research sought to mitigate the risk of overly broad generalizations and to provide meaningful guidance for policy development.

3. Results

This section synthesizes the main results relevant to policy recommendations coming out of the activities of AgroFossilFree. More detailed results can be found in the associated AgroFossilFree deliverable on identified policy gaps and policy guidelines [28]. The results presented have been synthesized into 19 policy briefs presented in the discussion section.

3.1. Research on Energy Use in EU Agriculture

AgroFossilFree research that combined results from hundreds of studies on the energy use status of EU agriculture revealed that EU agriculture is a significant energy consumer and that this sector is considerably fossil fuel-dependent. More specifically, agriculture is responsible for 3.2% of total energy consumption in the EU, according to Eurostat. Energy used in EU agriculture is mainly derived directly from crude oil and petroleum products (56%), leaving electricity second in importance (17%), and natural gas third (14%). Renewables and biofuels account only for 9%. Nonetheless, these results do not consider indirect energy use from the production and transportation of fertilizers and pesticides, something that would increase agriculture's proportion of energy use in the EU-27 up to 62% higher overall [29]. Another important finding was that the three main production systems (open-field, livestock, greenhouses) vary significantly in terms of energy use, its concentration, and breakdown.

Considering indirect energy consumption as shown in Figure 2, the production and use of fertilizers plays the most important role in energy consumption in EU open-field agriculture (around 50% of all energy inputs). Diesel fuel for on-farm activities is also significant (around 30%), electricity for irrigation, storage and drying covers 8% of the total energy consumption, and pesticides and seeds consume around 5% each [24]. Regarding livestock production, except for beef farms, animal feed accounts for around 75% of all energy use. A total of 60% of cereal production in the EU is used for animal feed, while high-protein agricultural products (e.g., soybean) are significantly imported. Electricity use in livestock facilities is also important and it is mainly derived from fossil sources; however, it varies considerably depending on the production system [25]. The greenhouse sector is split into two main categories in terms of energy consumption. High-tech facilities are also high-yielding- and high-energy-intensive, using energy mainly for heating and cooling. By contrast, low-tech facilities present lower yields and are less energy-intensive, as they use little to no heating/cooling and their main energy use is more similar to open-field agricultural systems with fertilizers, diesel use for machinery, irrigation, and other activities dominating their energy use profile [26].

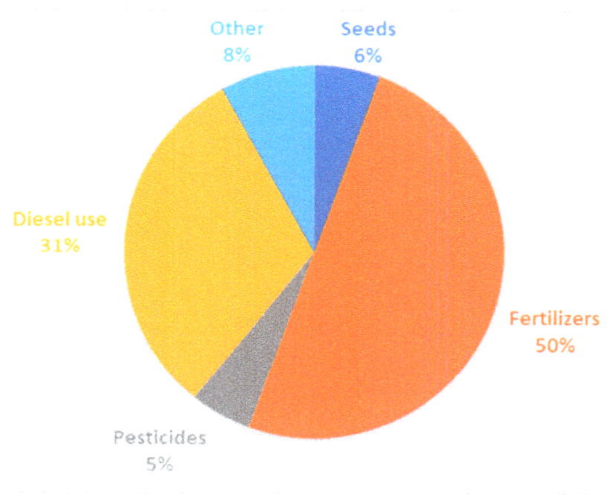

Figure 2. Energy inputs of open-field agriculture for EU-27 (%) [24].

A series of policy-related needs and outputs from this literature review are summarized in Table 1:

Table 1. Policy-related results originating from research on energy use status in EU agriculture.

Policy-Related Results	Description
Immediate increase in RESs powering agricultural facilities	In order to reduce the dependency of EU agriculture on fossil-based energy systems, it is vital that RES integration in farming facilities should become a priority.
Energy efficiency improvement across the agricultural sector	The first step before RESs are applied in agriculture is energy efficiency improvement, in order to reduce the energy requirements of the farms and integrate smaller RESs in terms of energy capacity. This way, the RES integration will be cheaper and the total energy consumption of agriculture will be minimized.
Shifting to non-conventional systems (organic, conservation)	Non-conventional systems (organic, conservation) do not necessarily increase energy use efficiency (energy use per production unit), but they definitely use more sustainable energy sources.
Unification of small holdings	Large farms, due to economies of scale, have a higher possibility of adopting FEFTS of all types, which is shown in the general positive correlation between larger farms and energy input per hectare. Therefore, small farms should either be incorporated from larger holdings or become members of new cooperation models between small farmers to take advantage of this and invest in both energy efficiency and RESs.
Use reduction in chemical fertilizers and imported animal feed	Since indirect energy use dominates agriculture, it is important to reduce chemical fertilizer use (mainly nitrogen) and non-locally produced animal feed. This way, energy and fossil energy inputs in agriculture will be significantly reduced.
Improve energy statistics in agriculture	Given the various limitations of EU agricultural energy statistics, including simplistic and conflicting methodologies and significant data gaps, our comprehension and accuracy of energy use estimates in EU agriculture would improve with the implementation of consistent definitions and methodologies regarding what is measured as energy use in agriculture. Regular and consistent all-farm energy audits could play an important role in this.

3.2. Farmers' Needs, Ideas, and Interests in the Adoption of FEFTS in the EU

In order to identify the barriers and drivers for the adoption of FEFTS, 511 stakeholder were surveyed including 470 direct surveys with farmers, and 41 expert interviews that were carried out across eight EU countries. The results indicate that economic factors are the main motivators, with cost reduction being a key reason for adopting renewable energy sources (RES) and energy-saving technologies/practices. Environmental concerns are also significant. Farmers highlight that financial support and access to customized cost–benefit models tailored to their specific farm operations are crucial incentives for adopting FEFTS. Based on these findings a number of key policy-related needs stood out and are synthesized in Table 2 [18].

Table 2. Key identified policy-related needs from farmer surveys and expert interviews.

Policy-Related Needs	Description
Specific subsidies for FEFTS	Economic considerations are found to be the main driver for the adoption or non-adoption of FEFTS and over two-thirds of adopters of FEFTS indicated that a subsidy was crucial for FEFTS adoption.
Sustainability and environmental concerns	Adoption of FEFTS could be enhanced due to the sustainability and environmental concerns of practitioners, which are the second reason for which FEFTS are adopted.
Reshape agricultural advisory/extensions services	Only around 15% of farmers use advisory services as primary sources of knowledge and awareness on RESs and energy efficient technologies; as such, advisory systems across the EU need to be redesigned and become more supportive of "innovation and knowledge" processes from a technological point of view.
Develop lighthouse demo farms	Demonstrate applied FEFTS in specific farms to act as best practice examples.

Table 2. *Cont.*

Policy-Related Needs	Description
Support small and medium-sized farms	Very small to medium-sized farms require specific policies to adopt FEFTS in their activities.
Support market transparency	The FEFTS industry could possibly mislead farmers on the benefits of the adoption of such technologies and strategies, and clear information could play a significant role in increasing the trust and FEFTS integration of EU farms.
Tune products for farming conditions	The conditions in farms of all types (e.g., humidity, dust, etc.) are very different than residential, industrial, and other sectors, and FEFTS industry should develop optimized and tuned products specifically for farming environment. Regulatory and economic incentives should be given to them for such a transition.
Reduced bureaucracy and steady regulations	State administration should prepare easily interpretable legislation and require reduced paperwork keeping both steady for long periods to provide a transparent situation for farmers to invest in FEFTS.

3.3. Main FEFTS Categories Identified by Stakeholders in the Content of the AgEnergy Platform

One of the main project outputs of AgroFossilFree is the AgEnergy platform [27] where around 2000 publicly available FEFTS have been curated and published by stakeholders, including a decision support toolkit (DST) [30] supporting stakeholders in selecting the most appropriate FEFTS for their farms.

The main FEFTS categories in the AgEnergy Platform were assessed by stakeholders using specific questions (four general, two environmental and three socioeconomic related questions) and a Likert scale (Strongly disagree, Disagree, Neutral, Agree, and Strongly Disagree). Then, this scale was converted into numerical values (−2, −1, 0, 1, 2) and the answers to all nine questions were aggregated to give the final score (max score: 16; min score: −16) of each FEFTS. The FEFTS included in this process were from scientific publications, research projects, and market-ready products, while the related training material and the possible financing tools were not included in this analysis, as they did not show trends in interest about FEFTS, which was the main target of this process. After sorting the FEFTS list according to their score and based on their categories, it was selected to consider specific thresholds for each category (Papers: ≥14; Projects: ≥10; Products: ≥14) above which a FEFTS could be considered as a popular solution for European agricultural stakeholders. The most prevalent FEFTS categories are presented in Table 3. This process was particularly useful in the development of policy recommendations as it provided a list of the most relevant FEFTS in the EU agricultural context.

Table 3. FEFTS categories identified in the content of AgEnergy Platform.

Papers (≥14)	Projects (≥10)	Products (≥14)
Most popular FEFTS categories:		
Solar	Solar	Precision agriculture
Photovoltaics (agrivoltaics)	Photovoltaics (agrivoltaics)	Solar
Bioenergy	Precision agriculture	Conservation agriculture
Biogas/biomethane production	Conservation agriculture	Wind/windmills
Conservation agriculture	Bioenergy	Electricity storage
Nutrient management	Wind/windmills	Efficient buildings
Heat pumps	Heat pumps	Bioenergy

Table 3. *Cont.*

Papers (≥14)	Projects (≥10)	Products (≥14)
Wind/windmills	Hydropower	Biogas/biomethane production
		Liquid biofuels production
		Efficient vehicles
		Heat pumps
		Hydropower
Additional categories found:		
Community energy	Community energy	Insulation
Biochar	Biochar	Building management systems

3.4. Grassroots Workshop Results

Central to AgroFossilFree's methodology was capturing grassroots stakeholders' opinions on FEFTS by bringing together relevant agricultural stakeholders in national and transnational workshops to develop community grassroot ideas around topics that included energy consumption in EU agriculture, the main aspects influencing the integration of innovative FEFTS in EU farms, the most interesting commercial FEFTS, the related current and possible future policies around FEFTS application in EU farming, the future changes in terms of energy efficiency improvements and renewable energy sources applied in agricultural production systems, and the possible novel research paths and collaboration schemes. Regarding the eight national hubs, information exchange was held in three multi-actor workshops per hub (twenty-four workshops), connecting researchers, agricultural and energy advisors, and FEFTS providers and farmers in order to capture their ideas and needs regarding energy use in agriculture and FEFTS integration on a regional level. As for the EU level, three workshops were carried out to bring stakeholders from all eight countries together to interact explaining the advantages and disadvantages of the situation around FEFTS in their countries, expand their knowledge about practices in other states and conclude in common and local strategies to increase FEFTS adoption in the EU. The discussions and findings from these workshops were analyzed and synthesized by relevant experts. The main findings are presented in the following sections.

3.4.1. Generic Results

The most common policy related outputs from all workshops combined are summarized in Table 4. The main discussion issues that arose were (i) farms' energy needs at a technical, financing, and knowledge-sharing level, (ii) obstacles impeding FEFTS adoption, and (iii) motivations to enhance FEFTS adoption. More detailed information is given in the relevant project deliverables [22,23].

Table 4. Key identified policy-related needs from all workshops.

Policy-Related Outputs/Needs	Description
Support RESs/energy efficiency technologies to improve energy effectiveness in farming.	To significantly enhance energy efficiency within the agricultural sector, it is essential to support the integration of RESs and energy efficiency technologies. These technologies can reduce dependency on fossil fuels and lower operational costs for farmers.
Founding agricultural energy communities for optimum use of local energy sources.	Establishing agricultural energy communities can play a crucial role in optimizing the use of local energy resources. These communities would enable farmers to collaboratively invest in and manage renewable energy projects, such as biogas plants, small-scale hydroelectric installations, or shared solar farms. By pooling resources and sharing the benefits of locally produced energy, agricultural energy communities can enhance energy security, reduce costs, and promote a decentralized energy system that supports regional economies and reduces environmental impacts

Table 4. *Cont.*

Policy-Related Outputs/Needs	Description
Farm energy audit services could give a major role to agriculture in EU defossilization.	Implementing farm energy audit services is a vital step towards positioning agriculture as a key player in the EU's defossilization strategy. Energy audits can identify areas where energy consumption can be reduced and efficiencies can be gained, helping farmers to implement targeted improvements. By offering these services, the EU can ensure that agricultural practices are optimized for energy use, contributing to lower greenhouse gas emissions and fostering a transition towards a low-carbon economy. This initiative not only supports environmental objectives but also enhances the economic viability of farms through cost savings and improved energy management.
Joint training, economic assistance, advisory, and networking should be supplied (+CAP knowledge).	To facilitate the transition to renewable energy and energy efficient systems, the EU should provide a comprehensive support package including training, financial aid, extension services, and networking opportunities. This will equip farmers with the necessary knowledge and resources to implement and maintain these technologies effectively, enhancing their capacity to adopt sustainable practices.
Assist the quadruple helix of innovation to grow in agriculture for FEFTS integration.	Support the collaboration between academia, industry, government, and civil society (the quadruple helix) to foster innovation in agriculture. This approach will accelerate the integration of FEFTS by leveraging the strengths of each sector, promoting sustainable development in agriculture.
Creation of "transfer centers/centers of excellence".	Establish transfer centers or centers of excellence to showcase functional FEFTS. These centers will serve as a bridge between farmers and researchers, demonstrating practical applications of advanced technologies and facilitating knowledge transfer, thus encouraging widespread adoption of these solutions.
Put forward ready-to-use FEFTS.	Encourage the development and dissemination of plug-and-play FEFTS that require minimal maintenance and are easy to operate. Simplifying the user experience will make it more feasible for farmers to adopt these technologies, reducing barriers to entry and promoting sustainable practices.
Economic support for farms introducing climate-friendly measures.	Offer financial incentives to farmers who implement climate-friendly measures. This could include subsidies, tax breaks, or grants, making it economically viable for farmers to transition to RES and energy efficient technologies.
Reduce taxes for farm RES-derived energy use (especially in alternative fuels).	Implement tax reductions for agricultural operations using RESs, especially alternative fuels. This policy will lower the cost burden on farmers, encouraging the shift away from fossil fuels and promoting the use of cleaner energy.
Develop grants and ease the application to these grants for FEFTS integration in farming processes.	Expand and streamline the application process for grants aimed at investing in FEFTS. By making these grants more accessible, the EU can support more farmers in adopting sustainable technologies, fostering a broader transition across the agricultural sector.
Regulations should not cause difficulties to FEFTS adoption.	Ensure that regulations do not hinder the adoption of FEFTS. Simplify rules related to agriPV grid connections and the self-sufficiency of biofuels, removing bureaucratic obstacles that can delay or prevent the implementation of these technologies.
Policies promoting measurable/result-based subsidy financing of FEFTS.	Implement subsidy programs that are measurable and result-based, focusing on the tangible benefits of FEFTS adoption. This will ensure that subsidies are effectively promoting the desired outcomes, such as reduced greenhouse gas emissions and increased energy efficiency.
Differentiation of regulations between EU countries (geographical, climate and cultural differences + farm size).	Adapt regulations to account for the geographical, climatic, and cultural differences among EU countries, as well as varying farm sizes. This differentiation will ensure that policies are relevant and effective across diverse agricultural contexts, promoting more equitable and widespread adoption of sustainable practices.
Merge energy efficiency and agri-food policies and simplify bureaucracy for FEFTS application in EU farms.	Integrate energy efficiency and agri-food policies to streamline administrative requirements for FEFTS use in agriculture. Establish shared standards among different governmental sections to simplify processes, reduce bureaucratic hurdles, and enhance the coherence and effectiveness of policy measures aimed at promoting sustainability in agriculture.

3.4.2. Open Field Agriculture

The open field agriculture workshops addressed a wide range of pertinent topics. Some of the key, though non-exhaustive, results include the fact that the growing fuel and electricity prices will adversely impact open-field food production and increase their prices, with fertilization, tillage, and irrigation being the most energy-intensive practices. Alternative production systems, like organic farming, often require more energy as compared to conventional practices but these are seen as generally environmentally acceptable as they use more sustainable and renewable energy sources. Reducing fertilization costs through cover crops, conservation agriculture, and resilient crop varieties can reduce dependence on fossil energy sources. Legislative actions enabling electricity sales and storage could revolutionize large-scale farming as they can support on farm renewable energy production. In addition, policies should prioritize energy efficiency, carbon storage, alternative fuels like biomethane, and the effective dissemination of information on FEFTS. Proposed measures for promoting FEFTS in open-field agriculture include training programs, pilot platforms, subsidy schemes, tax reductions on renewable energy, carbon credit schemes, and EU-level incentives for sustainable practices. Policies should be results-based, linking subsidies to measurable outcomes rather than specific technologies [22,23]. Table 5 below shows the main policies proposed in the workshops about open-field agriculture.

Table 5. Key identified policy-related needs from open-field agriculture workshops.

Policy-Related Outputs/Needs	Description
Subsidies and funding for FEFTS	The most frequently mentioned incentive is the provision of subsidies and funding for the purchase of FEFTS and associated technical support.
Focus on energy efficiency solutions	Policy initiatives should prioritize solutions aimed at enhancing energy efficiency in agriculture. By reducing energy consumption, farmers can lower operational costs and minimize their environmental impact, contributing to the EU's overall sustainability goals.
Support for biogas production systems	Promote biogas production systems that utilize open-field agricultural products as feedstock. This approach will help farmers convert waste into renewable energy, providing a sustainable energy source and reducing greenhouse gas emissions.
Assistance for using nitrification inhibitors	Provide assistance to farmers for the use of nitrification inhibitors, which can significantly reduce CO_2 emissions in farming. Supporting the adoption of these inhibitors will enhance soil health and contribute to climate change mitigation.
Policies for farm-level biofuel production and consumption	Revise policies and regulations to facilitate the transition of agriculture into a prosumer of biofuels. By producing and consuming their own biofuels, farmers can achieve greater energy independence and sustainability.
Support for gradual change in cultivation practices	Encourage the gradual change in cultivation practices, e.g., Conservation Agriculture (CA), through staged implementation and soft transitions. This approach will allow farmers to adapt to new methods while maintaining productivity.
Carbon credits market as a financial incentive	Utilize the carbon credits market as a financial incentive to promote the adoption of new cultivation practices. Farmers can generate additional revenue by participating in carbon offset programs, which reward them for reducing emissions.
CAP support for CA practices	Ensure that the CAP provides support for the adoption of CA practices. CAP funding should be directed towards encouraging sustainable farming methods that enhance soil health and reduce environmental impact.
Subsidies for positive organic matter balance	Subsidize farms that accomplish soil organic matter surplus. This will incentivize practices that improve soil fertility and structure, leading to long-term agricultural sustainability.
Utilization of untapped raw material for RESs	Motivate farmers to utilize untapped raw materials left in fields for the development of RESs in rural areas. By turning agricultural residues into energy, farmers can contribute to rural energy security and sustainability.
Promotion of biofertilizers and alternatives to synthetic fertilizers	Promote the production and use of biofertilizers and alternatives to synthetic fertilizers. This will reduce the reliance on chemical inputs, improving soil health and reducing environmental pollution.
Encouragement of carbon storage solutions	Encourage the adoption of solutions for carbon storage, including Precision Agriculture (PA), CA, and carbon farming. These practices will enhance carbon sequestration in soils, contributing to climate change mitigation and sustainable agriculture.

3.4.3. Livestock

The livestock workshops addressed a wide range of pertinent topics. Some of the key, though non-exhaustive, results include the fact that feed production is identified as the primary energy consumer, followed by transport and building energy use. The impact of rising fuel and electricity costs on livestock production is not significant for the sector, affecting only 5–10% of production costs, thus not driving significant investment. However, rising energy prices incentivize larger farms to invest in RESs, like solar PV and biogas, while smaller farms struggle without subsidies and may cease operations due to unsustainable costs. Participants advocated for analyzing energy consumption across supply chains and implementing energy-saving technologies. Suggested policies include practical and online training, pilot platforms, economic incentives for adopting FEFTS, and policies to stabilize prices and ensure investment returns. Financial support should focus on green technology adoption, with tax policies encouraging shifts from fossil fuels and subsidies based on farm size and RES production levels. Table 6 below shows the main policies proposed in the workshops about livestock production.

Table 6. Key identified policy-related needs from livestock production workshops.

Policy-Related Outputs/Needs	Description
Promotion of integrated heat pumps and PV panels	Promote the use of integrated heat pumps and PV panels in livestock facilities. These combined systems will provide efficient heating and electricity, reducing reliance on fossil fuels and lowering greenhouse gas emissions in livestock production.
Higher prices for low carbon footprint livestock products	Encourage higher market prices for livestock products with a low carbon footprint, similar to the policies supporting organic products. This will incentivize farmers to adopt practices that reduce emissions, contributing to more sustainable agriculture.
Certification of FEFTS in livestock facilities	Develop a certification system for the integration of FEFTS in livestock facilities. Certification will ensure that these technologies are effectively implemented and provide a standard for evaluating energy efficiency and sustainability.
Support for replacing existing feed with alternatives	Provide support for replacing traditional livestock feed with more sustainable alternatives. This will reduce the environmental impact of feed production and promote the use of locally produced feed resources.
Sustainable and self-sufficient farms	Assist farms in becoming sustainable and self-sufficient in terms of feed and energy. Policies should focus on providing the necessary resources and knowledge to enable farmers to produce their own feed and generate their own energy, reducing external dependencies.
Regulatory support for farm-level biogas	Facilitate the growth of biogas production at the farm level by adjusting regulations to support its application in livestock farming. This will help farmers turn waste into valuable energy, promoting a circular economy within the agricultural sector.
Industrial symbiosis for energy flows	Encourage industrial symbiosis for energy-related flows between livestock facilities and other sectors. This approach will optimize resource use, reduce waste, and enhance the overall efficiency and sustainability of energy consumption in agriculture.
CAP support for FEFTS adoption	Leverage the CAP to assist in the adoption of FEFTS, addressing the current scarcity of related subsidies. Redirecting CAP funds towards FEFTS will provide financial incentives for farmers to adopt energy-efficient and RESs.
Stimulation of circular economy through CAP	Stimulate the circular economy (CE) in livestock production through the CAP and encourage increased collaboration among stakeholders. This will promote the recycling of resources and the reduction in waste, contributing to a more sustainable agricultural sector.
Cooperation between livestock actors	Facilitate collaboration among livestock actors to secure higher value for their products and access to FEFTS. Collaborative efforts can lead to shared resources and knowledge, enhancing the economic viability and sustainability of livestock farming.
Combining animal welfare with energy-saving policies	Integrate animal welfare considerations with energy-saving policies to support sustainable livestock production. Ensuring high standards of animal welfare alongside energy efficiency measures will enhance the overall sustainability and ethical standards of the industry.

Table 6. *Cont.*

Policy-Related Outputs/Needs	Description
Specific support for RES integration in livestock	Create specific support categories within policies for the integration of RESs in livestock production. Targeted financial and technical assistance will help farmers adopt RES technologies, reducing their carbon footprint.
labeling for reduced/green energy use products	Implement a well-established labelling system for livestock products that use reduced or green energy, avoiding greenwashing. Clear and credible labelling will help consumers make informed choices, driving demand for sustainably produced livestock products.

3.4.4. Greenhouse

The greenhouse agriculture workshops addressed a wide range of pertinent topics. Some of the key, though non-exhaustive, results include the fact that stakeholders acknowledge the substantial difference in energy use and Energy Use Efficiency (EUE) and agree that rising energy costs will possibly reduce food production in greenhouses. This will drive the shift to energy-efficient technologies and practices, though the pace depends on fossil fuel prices, technological advancements, and EU efficiency goals. The European greenhouse industry should prioritize policies to reduce direct and indirect energy consumption and increase renewable energy production. Proposed incentives include policies, subsidies, communication approaches, information sharing, and training. While some countries have applicable policies, more effective support is needed, particularly policies focusing on returns on investment to promote the adoption of FEFTS. Table 7 below shows the main policies proposed in the workshops about greenhouse production.

Table 7. Key identified policy-related needs from greenhouse production workshops.

Policy-Related Outputs/Needs	Description
Improvements to existing greenhouse production sites	Implement targeted funding, incentives, and support solutions for the modernization of existing greenhouse production sites. Simplifying access to financial resources will enable the adoption of advanced technologies, increasing energy efficiency and productivity while reducing environmental impact.
Financial support for CO_2 injection into greenhouses	Provide financial support and necessary permits for the injection of exhaust gasses (CO_2) into greenhouses. This practice can enhance plant growth and productivity, contributing to more efficient resource use and lower greenhouse gas emissions in the agricultural sector.
Support energy/carbon audits for greenhouses	Introduce and fund energy and carbon audits for greenhouses, particularly those utilizing CO_2 influxes. These audits will identify areas for improvement in energy use and carbon management, helping greenhouse operators to implement more sustainable practices.
Use CO_2 tax (CO_2 reduction rewards) for green production	Leverage CO_2 tax revenues to create rewards for green production in greenhouses. This policy would incentivize growers to adopt practices that reduce CO_2 emissions, promoting a shift towards more sustainable and environmentally friendly greenhouse operations.
Develop support schemes for renewable heat and power	Establish support schemes to promote the use of renewable heat and power in greenhouse production. These schemes should provide financial incentives and technical assistance to encourage the adoption of RES, such as solar thermal or biomass heating.
Coordinated approach for district heating opportunities	Adopt a coordinated approach to integrate greenhouses into district heating systems. This will provide multiple opportunities for farms to access efficient heating solutions, reducing energy costs and emissions while enhancing the sustainability of greenhouse operations.

4. Discussion

In recent years, the EU agriculture sector has been shifting towards FEFTS integration in agriculture. The adoption of RES technologies has accelerated while, additionally, more energy-efficient machinery is becoming available and being developed by the industry [27]. Simultaneously, there is an increased focus on finding appropriate ways to reduce synthetic fertilizer and pesticide use through biological alternatives and the adoption of more sustainable agricultural practices. These trends are expected to continue in the medium term as these technologies develop and policy is supportive of these transitions.

4.1. Policy Briefs

Based on the presented results, policy recommendations were extracted about technical, financial, regulatory, and information-sharing aspects of the agricultural domain in order to give a guideline to national and EU policymakers for developing strategies to defossilize farming by introducing FEFTS in EU farms. The main three categories of policy recommendations derived by this work are as follows:

- Horizontal policies that are suitable for any farm type and any FEFTS and can act as fundamental for other specific FEFTS policies to be followed easier by farmers.
- Policies specific to agricultural production systems that are associated with open-field, livestock, or greenhouse systems and related FEFTS and designate the essential stages for farmers to apply them.
- Generic policies refer to actions that would benefit any innovation towards green growth to grow in EU farming.

These policy recommendations were then synthesized into 19 policy briefs, presented in Table 8 [31]. These policy briefs portray current challenges, policy recommendations on the EU and Member States' levels, and the expected impact of implementing these policies. More details about them are given in the references shown within Table 8 that presents their titles.

Table 8. Titles of developed policy briefs.

Policy Brief No.	Policy Brief Title
Horizontal Policy Recommendations	
1	Enabling the creation and growth of energy communities in rural areas [32]
2	Farm energy audits [33]
3	European Low Energy/Carbon Label of agricultural products [34]
Policy Recommendations Specific to agricultural production systems	
4	Agrivoltaics for open-field agriculture [35]
5	Alternative fuels for agricultural machinery [36]
6	Precision agriculture as energy consumption reduction strategy [37]
7	Carbon farming for carbon removals [38]
8	Conservation agriculture to enhance soil carbon stock and reduce GHG emissions in European agriculture [39]
9	Alternative crop nutrient providers (green fertilizers/biofertilisers, biostimulants/biochar) [40]
10	Building management systems (BMSs) for agricultural constructions [41]
11	Heat pumps for HVAC of agricultural constructions [42]
12	Photovoltaics (PV) and photovoltaic thermal (PVT) collectors and systems for agricultural constructions rooftops [43]
13	Biogas production from agricultural waste and other innovative feedstock/biomethane upgrading for local consumption or grid injection [44]
14	Facilitating the development of energy independent farming in livestock [45]
15	Livestock building energy upgrading/renovation [46]
16	The use of thermochemical fluids for energy saving and storage in agriculture [47]
Generic Policy Recommendations	
17	Financial support to fossil-energy-free technologies and strategies [48]
18	Regulatory support to fossil-energy-free technologies and strategies [49]
19	Technology, knowledge transfer, and awareness-building provisions to support the diffusion of fossil-energy-free technologies and strategies [50]

4.2. Generic Policy Recommendations in the Context of the New CAP

The current CAP has a large focus on environmental sustainability and climate resilience and complements the Green Deal. It also places an increased focus on innovation and knowledge transfer while supporting rural development. The Generic Policy Recommendations in Table 8 align well with the objectives and mechanisms of the new CAP (2023–2027).

Policy Brief 17 "Financial Support to FEFTS" aligns well with some of the mechanisms as part of the CAP including the emphasis on providing financial support for sustainable practices, which can include FEFTS. Through eco-schemes under Pillar I and targeted measures under Pillar II, the CAP can channel funds towards these technologies. Eco-schemes are specifically designed to reward farmers for adopting environmentally friendly practices, such as using RESs and energy-efficient technologies. Similarly, rural development programs can provide grants and subsidies for investments in FEFTS, promoting energy efficiency and RES use on farms, while the focus on fairer distribution of payments includes supporting small and medium-sized farms to adopt new technologies, which can be directed towards FEFTS [51].

Regarding Policy Brief 18 "Regulatory Support to FEFTS", under the CAP, member states have the flexibility to design regulatory frameworks that support the adoption of FEFTS within their national CAP strategic plans, ensuring alignment with EU-wide objectives, while addressing local needs. Similarly, the new CAP's focus on a performance-based approach allows for regulatory measures to be designed in a way that incentivizes compliance and adoption of sustainable technologies, such as FEFTS [52].

Regarding Policy Brief 19 "Technology, Knowledge Transfer, and Awareness Building Provisions to Support FEFTS Diffusion", the new CAP places significant emphasis on research, innovation, and knowledge transfer. Programs like Horizon Europe and CAP-funded advisory services are aimed at promoting technological advancements and sustainable practices. Enhanced advisory services under the new CAP will provide farmers with the necessary knowledge and skills to adopt and effectively use FEFTS. This includes demonstration projects, training sessions, and extension services. In addition, the European Innovation Partnership for Agricultural Productivity and Sustainability (EIP-AGRI), and the EU-CAP network, can play a key role in facilitating the transfer of knowledge and innovation related to FEFTS to the farming community [3].

4.3. Policy Recommendations Specific to Agricultural Production Systems

These policy briefs are unique as they are to a large extent based on inputs from grassroots stakeholders. The 13 policy briefs that are related to specific agricultural production systems are particularly relevant as they promote specific FEFTS. By promoting these, based on evidence-based results, it can be of particular importance for policymakers, especially on a national level, where the relevant context for each of these technologies exists.

In particular, policy briefs 4–9 are related to open-field agriculture while 10–16 are related to agricultural constructions (livestock and greenhouse facilities). Some respond to the energy use make-up of European agriculture. For instance, in open-field agriculture, our research indicates that around 30% of all energy use is associated with the use of diesel for agricultural machinery, while over half of all energy use is associated with the production of chemical fertilizers. As such, policy brief 5 on alternative fuels and policy brief 9 on alternative crop nutrient providers are particularly relevant. For instance, the development of biomethane for tractors in localized livestock production farms could significantly reduce fossil use and dependency for livestock farmers, while development and support for local biofertilizers would significantly reduce the dependence on natural gas for the production of chemical fertilizers. In other cases, the policy briefs respond to the emergence of innovative and new technologies and practices. For instance, carbon farming and rewarding farmers for sequestering carbon is increasingly gaining traction in policy discourse [53] and the relevant policy brief calls for an acceleration in developing a standard for monitoring, reporting and verification of soil carbon stocks in order to make a transparent and trustworthy framework for carbon farming credits application in agriculture. Certain policies that promote and incentivize the adoption of more sustainable agricultural practices, such as precision agriculture in policy brief 6 and conservation agriculture in policy brief 8, could considerably reduce the use of fossil fuels through the adoption of more efficient practices, such as no-tillage, and the more efficient use of agricultural inputs.

4.4. Integrating Policy Guidelines into Policy Processes

It is notable that the sustainability dimensions in the policy briefs promoted by stakeholders are already, in many cases, included in or under discussion by policymakers. Integrating policy guidelines into the policy processes of the EU and member states' agricultural policies presents both significant challenges and opportunities. One major challenge is the heterogeneity of agricultural practices and economic conditions across member states, which complicates the uniform application of policy guidelines. Additionally, the varying degrees of technological readiness and adoption among farmers further exacerbate these disparities [54,55].

However, there are substantial opportunities in harmonizing these guidelines with the CAP to promote sustainable practices and the adoption of FEFTS. The presented policy briefs and their presentation to relevant policymakers is an initial step and following from this there is a pressing need for more in-depth discussions between policymakers and stakeholders to ensure the successful integration of FEFTS promotion in agricultural policy. These discussions should focus on practical implementation strategies, aligning with the CAP's sustainability goals, and fostering a collaborative approach to policy development and execution.

5. Conclusions

In conclusion, the results presented in this paper can be useful for EU and national policymakers due to the fact that they are the result of a direct interaction with European agricultural and energy stakeholders working in all agricultural production systems through their participation in surveys, interviews, and workshops over three years (2020–2023). Based on these results, 19 key policy recommendations have been synthesized, including horizontal policies for any farm and FEFTS type, specific agricultural production system policies that support FEFTS application in farms of certain production systems, and generic policies considered essential for a green transition of EU farming. The proposed policies should be addressed by the future amendments of the CAP, the RED, the Green Deal, and the Farm-to-Fork strategy amendments and future member state legal initiatives.

The independence of EU agriculture from fossil fuels will require decades. This paper presents the increased availability of key FEFTS for farm use with relatively acceptable costs and the clear willingness of farmers to shift from conventional technologies and strategies to more sustainable ones, but simultaneously identifies several obstacles that hinders this transition. The outlined policies should be effectively implemented with a long-term view, supporting farms from a technical and economical perspective for the transition to more sustainable and resilient EU agriculture. Accelerating the adoption of FEFTS in EU agriculture is imperative for achieving climate neutrality and ensuring long-term sustainability. The policy guidelines presented in this paper provide a strategic roadmap for policymakers to drive this transformation.

The implications of these findings extend beyond the agricultural sector. On one hand, shifting away from fossil fuels in agriculture is a crucial component of the EU's broader climate strategy. On the other hand, while other major economic sectors have generally made more progress towards integrating RESs in recent years compared to agriculture, it is particularly necessary to implement policies that support sustainable energy transitions within the agricultural sector. Such policies are not only essential for agriculture but are also likely to generate positive spillover effects across other economic sectors. For instance, an increase in RES supply from agriculture, combined with the rising demand for electrified machinery, will contribute to creating a larger and more stable energy market, thus minimizing distortions and benefiting the broader economy.

Author Contributions: Conceptualization, A.T.B.; methodology, A.T.B.; formal analysis, A.T.B. and B.P.; investigation, A.T.B. and B.P.; data curation, A.T.B.; writing—original draft preparation, A.T.B. and B.P.; writing—review and editing, A.T.B. and B.P.; supervision, A.T.B.; project administration, A.T.B.; funding acquisition, A.T.B. All authors have read and agreed to the published version of the manuscript.

Funding: This research was funded by the European Union's Horizon 2020 research and innovation program under grant agreement ID 101000496.

Data Availability Statement: The original contributions presented in the study are included in the article, further inquiries can be directed to the corresponding author.

Acknowledgments: This study has been developed as part of the Horizon 2020 AgroFossilFree project (www.agrofossilfree.eu). We would like to thank the following partners for their contributions and insight into developing this article: Agricultural University of Athens (AUA), Agricultural & Environmental Solutions (AGENSO), Aarhus Universitet (AU), the Comite Europeen Des Groupements De Constructeurs Du Machinisme Agricole (CEMA), the CenterCentre for Research & Technology Hellas (CERTH), the Confederazione Generale Dell Agricoltura Italiana (CONFAGRICOLTURA), DELPHY European Conservation Agriculture Federation (ECAF), Iniciativas Innovadoras Sal (INI), Innovationscenter for Økologisk Landbrug (ICOEL), the Instytut Uprawy Nawozenia I Gleboznawstwa, Panstwowy Instytut Badawczy (IUNG-PIB), Landbrug & Fodevarer F.M.B.A. (L&F), Lubelski Osrodek Doradztwa Rolniczego W Konskowoli (LODR), RESCOOP EU ASBL, the Agriculture and Food Development Authority (TEAGASC), Trama Tecnoambiental S.L. (TTA), and Wirtschaft Und Infrastruktur Gmbh & Co Planungs Kg (WIP).

Conflicts of Interest: The authors declare no conflicts of interest. The funders had no role in the design of the study; in the collection, analyses, or interpretation of data; in the writing of the manuscript; or in the decision to publish the results.

References

1. European Commission. *Farm to Fork Strategy*; European Commission: Brussels, Belgium, 2020.
2. European Commission. A European Green Deal | European Commission. Available online: https://ec.europa.eu/info/strategy/priorities-2019-2024/european-green-deal_en (accessed on 7 December 2022).
3. European Commission. The New Common Agricultural Policy: 2023-27. Available online: https://ec.europa.eu/info/food-farming-fisheries/key-policies/common-agricultural-policy/new-cap-2023-27_en (accessed on 14 September 2021).
4. Long, T.B.; Blok, V.; Coninx, I. Barriers to the Adoption and Diffusion of Technological Innovations for Climate-Smart Agriculture in Europe: Evidence from the Netherlands, France, Switzerland and Italy. *J. Clean. Prod.* **2015**, *112 Pt 1*, 9–21. [CrossRef]
5. Scarlat, N.; Fahl, F.; Lugato, E.; Monforti-Ferrario, F.; Dallemand, J.F. Integrated and Spatially Explicit Assessment of Sustainable Crop Residues Potential in Europe. *Biomass Bioenergy* **2019**, *122*, 257–269. [CrossRef]
6. Musiał, W.; Zioło, M.; Luty, L.; Musiał, K.; Bianco, V.; Klepacki, B.; Stolarski, M.J. Energy Policy of European Union Member States in the Context of Renewable Energy Sources Development. *Energies* **2021**, *14*, 2864. [CrossRef]
7. The French ministry of Agriculture and Food Le Plan Énergie Méthanisation Autonomie Azote | Ministère de l'Agriculture et de La Souveraineté Alimentaire. Available online: https://agriculture.gouv.fr/le-plan-energie-methanisation-autonomie-azote (accessed on 18 September 2024).
8. Poponi, D.; Basosi, R.; Kurdgelashvili, L. Subsidisation Cost Analysis of Renewable Energy Deployment: A Case Study on the Italian Feed-in Tariff Programme for Photovoltaics. *Energy Policy* **2021**, *154*, 112297. [CrossRef]
9. Netherlands Enterprise Agency. Stimulation of Sustainable Energy Production and Climate Transition (SDE++). Available online: https://english.rvo.nl/subsidies-financiering/sde (accessed on 18 September 2024).
10. Plan Up. PAREER-CRECE Programme [Spain] | PlanUp. Available online: https://www.planup.eu/en/resources/policies/pareer_crece_programme_[spain]/871 (accessed on 18 September 2024).
11. Dział: Aktualności Polskie Towarzystwo Fotowoltaiki » A Rising Star: PV Market Poland. Available online: https://pv-polska.pl/en/a-rising-star-pv-market-poland/ (accessed on 18 September 2024).
12. Land & Forst Federal Energy Efficiency Programme: These Are the Funding Criteria | LAND & FOREST. Available online: https://www.landundforst.de/energie/bundesprogramm-energieeffizienz-foerderkriterien-569931 (accessed on 18 September 2024).
13. Pe'er, G.; Finn, J.A.; Díaz, M.; Birkenstock, M.; Lakner, S.; Röder, N.; Kazakova, Y.; Šumrada, T.; Bezák, P.; Concepción, E.D.; et al. How Can the European Common Agricultural Policy Help Halt Biodiversity Loss? Recommendations by over 300 Experts. *Conserv. Lett.* **2022**, *15*, e12901. [CrossRef]

14. Bojnec, Š.; Fertő, I. Do Different Types of Common Agricultural Policy Subsidies Promote Farm Employment? *Land Use Policy* **2022**, *112*, 105823. [CrossRef]
15. Jenner, S.; Groba, F.; Indvik, J. Assessing the Strength and Effectiveness of Renewable Electricity Feed-in Tariffs in European Union Countries. *Energy Policy* **2013**, *52*, 385–401. [CrossRef]
16. Fieldsend, A.F.; Cronin, E.; Varga, E.; Biró, S.; Rogge, E. 'Sharing the Space' in the Agricultural Knowledge and Innovation System: Multi-Actor Innovation Partnerships with Farmers and Foresters in Europe. *J. Agric. Educ. Ext.* **2021**, *27*, 423–442. [CrossRef]
17. Herrera, B.; Gerster-Bentaya, M.; Tzouramani, I.; Knierim, A. Advisory Services and Farm-Level Sustainability Profiles: An Exploration in Nine European Countries. *J. Agric. Educ. Ext.* **2019**, *25*, 117–137. [CrossRef]
18. Paris, B.; Kanaki, V.; Koutsouris, A.; Balafoutis, A.T.; Papadakis, G. Farmers' Needs, Ideas and Interests on the Adoption of Fossil Energy Free Technologies and Strategies in the EU. *Int. J. Sustain. Energy* **2024**, *43*, 2326454. [CrossRef]
19. Koutsouris, A.; Vassilki, K. *Report on Framework, Methodology and Standards*; AgroFossilFree Consortium: Athens, Greece, 2020.
20. Koutsouris, A.; Kanaki, V. *Report on Farmers' Needs, Innovative Ideas and Interests*; Agricultural University of Athens: Athens, Greece, 2021.
21. Agricultural University of Athens. *Multi-Media Material*; Zenodo: Geneva, Switzerland, 2023. [CrossRef]
22. Aarhus University. *Final Report on AFF Regional Workshops Outcome*; Zenodo: Geneva, Switzerland, 2023. [CrossRef]
23. Aarhus University. *Report from the Transnational Workshops*; Zenodo: Geneva, Switzerland, 2023. [CrossRef]
24. Paris, B.; Vandorou, F.; Balafoutis, A.T.; Vaiopoulos, K.; Kyriakarakos, G.; Manolakos, D.; Papadakis, G. Energy Use in Open-Field Agriculture in the EU: A Critical Review Recommending Energy Efficiency Measures and Renewable Energy Sources Adoption. *Renew. Sustain. Energy Rev.* **2022**, *158*, 112098. [CrossRef]
25. Paris, B.; Vandorou, F.; Tyris, D.; Balafoutis, A.T.; Vaiopoulos, K.; Kyriakarakos, G.; Manolakos, D.; Papadakis, G. Energy Use in the EU Livestock Sector: A Review Recommending Energy Efficiency Measures and Renewable Energy Sources Adoption. *Appl. Sci.* **2022**, *12*, 2142. [CrossRef]
26. Paris, B.; Vandorou, F.; Balafoutis, A.T.; Vaiopoulos, K.; Kyriakarakos, G.; Manolakos, D.; Papadakis, G. Energy Use in Greenhouses in the EU: A Review Recommending Energy Efficiency Measures and Renewable Energy Sources Adoption. *Appl. Sci.* **2022**, *12*, 5150. [CrossRef]
27. AgroFossilFree. What Is the AgEnergy Platform? Available online: https://www.agrofossilfree.eu/about/ (accessed on 14 July 2021).
28. Centre for Research and Technology Hellas. *Report on Identified Policy Gaps and Policy Guidelines*; Zenodo: Geneva, Switzerland, 2023. [CrossRef]
29. Agricultural University of Athens; Centre for Research and Technology Hellas. *Report on Current Energy Use Status in EU Agriculture*; Zenodo: Geneva, Switzerland, 2021. [CrossRef]
30. Kyriakarakos, G.; Balafoutis, A.T.; Vaiopoulos, K.; Abdul, M.; Voskakis, M.; Kaminiaris, M.D.; Tsiropoulos, Z.; Bochtis, D. Design and Implementation of a Decision Support Tool to Assist the Ranking of Fossil-Energy-Free Technologies and Strategies for a given Farm. *Smart Agric. Technol.* **2023**, *4*, 100169. [CrossRef]
31. AgroFossilFree. Consortium AgroFossilFree Policy Briefs. Available online: https://www.agrofossilfree.eu/2023/09/01/policy-briefs/ (accessed on 29 July 2024).
32. AgroFossilFree Consortium. Enabling the Creation and Growth of Energy Communities in Rural Areas. Available online: https://www.agrofossilfree.eu/2023/09/01/policy-briefs/ (accessed on 29 July 2024).
33. AgroFossilFree Consortium. Policy Brief No 2 Farm Energy Audits. Available online: https://www.agrofossilfree.eu/2023/09/01/policy-briefs/ (accessed on 29 July 2024).
34. AgroFossilFree Consortium. Policy Brief No 3 European Low Energy/Carbon Label of Agricultural Products. Available online: https://www.agrofossilfree.eu/2023/09/01/policy-briefs/ (accessed on 29 July 2024).
35. HyPErFarm Consortium. Policy Brief No 4 Agrivoltaics for Open-Field Agriculture. Available online: https://www.agrofossilfree.eu/2023/09/01/policy-briefs/ (accessed on 29 July 2024).
36. AgroFossilFree Consortium. Policy Brief No 5 Alternative Fuels for Agricultural Machinery. Available online: https://www.agrofossilfree.eu/2023/09/01/policy-briefs/ (accessed on 29 July 2024).
37. AgroFossilFree Consortium. Policy Brief No 6 Precision Agriculture as Energy Consumption Reduction Strategy. Available online: https://www.agrofossilfree.eu/2023/09/01/policy-briefs/ (accessed on 29 July 2024).
38. AgroFossilFree Consortium. Policy Brief No 7 Carbon Farming for Carbon Removals. Available online: https://www.agrofossilfree.eu/2023/09/01/policy-briefs/ (accessed on 29 July 2024).
39. AgroFossilFree Consortium. Policy Brief No 8 Conservation Agriculture to Enhance Soil Carbon Stock and Reduce GHG Emissions in European Agriculture. Available online: https://www.agrofossilfree.eu/2023/09/01/policy-briefs/ (accessed on 29 July 2024).
40. AgroFossilFree Consortium. Policy Brief No 9 Alternative Crop Nutrient Providers (Green Fertilisers/Biofertilisers, Biostimulants/Biochar). Available online: https://www.agrofossilfree.eu/2023/09/01/policy-briefs/ (accessed on 29 July 2024).
41. AgroFossilFree Consortium. Policy Brief No 10 Building Management Systems (BMS) for Agricultural Constructions. Available online: https://www.agrofossilfree.eu/2023/09/01/policy-briefs/ (accessed on 29 July 2024).
42. RES4LIVE Consortium. Policy Brief No 11 Heat Pumps for HVAC of Agricultural Constructions. Available online: https://www.agrofossilfree.eu/2023/09/01/policy-briefs/ (accessed on 29 July 2024).

43. AgroFossilFree Consortium. Policy Brief No 12 Photovoltaics (PV) and Photovoltaic Thermal (PVT) Collectors and Systems for Agricultural Constructions Rooftops. Available online: https://www.agrofossilfree.eu/2023/09/01/policy-briefs/ (accessed on 29 July 2024).
44. AgroFossilFree Consortium. Policy Brief No 13 Biogas Production from Agricultural Waste and Other Innovative Feedstock/Biomethane Upgrading for Local Consumption or Grid Injection. Available online: https://www.agrofossilfree.eu/2023/09/01/policy-briefs/ (accessed on 29 July 2024).
45. AgroFossilFree Consortium. Policy Brief No 14 Facilitating the Development of Energy Independent Farming in Livestock. Available online: https://www.agrofossilfree.eu/2023/09/01/policy-briefs/ (accessed on 29 July 2024).
46. AgroFossilFree Consortium. Policy Brief No 15 Energy Upgrading/Renovation of Livestock Buildings. Available online: https://www.agrofossilfree.eu/2023/09/01/policy-briefs/ (accessed on 29 July 2024).
47. The Greefa Consortium. Policy Brief No 16 The Use of Thermochemical Fluids for Energy Saving and Storage in Agriculture. Available online: https://www.agrofossilfree.eu/2023/09/01/policy-briefs/ (accessed on 29 July 2024).
48. AgroFossilFree Consortium. Policy Brief No 17 Financial Support to Fossil Energy Free Technologies and Strategies (FEFTS). Available online: https://www.agrofossilfree.eu/2023/09/01/policy-briefs/ (accessed on 29 July 2024).
49. AgroFossilFree Consortium. Policy Brief No 18 Regulatory Support to Fossil Energy Free Technologies and Strategies (FEFTS). Available online: https://www.agrofossilfree.eu/2023/09/01/policy-briefs/ (accessed on 29 July 2024).
50. AgroFossilFree Consortium. Policy Brief No 19 Technology, Knowledge Transfer, and Awareness Building Provisions to Support Fossil Energy Free Technologies and Strategies (FEFTS) Diffusion. Available online: https://www.agrofossilfree.eu/2023/09/01/policy-briefs/ (accessed on 29 July 2024).
51. European Commission. *Common Agricultural Policy for 2023–2027 28 CAP Strategic Plans at a Glance*; European Commission: Brussels, Belgium, 2022.
52. Pe'er, G.; Zinngrebe, Y.; Moreira, F.; Sirami, C.; Schindler, S.; Müller, R.; Bontzorlos, V.; Clough, D.; Bezák, P.; Bonn, A.; et al. A Greener Path for the EU Common Agricultural Policy. *Science* **2019**, *365*, 449–451. [CrossRef] [PubMed]
53. European Commission. Carbon Removals and Carbon Farming. Available online: https://climate.ec.europa.eu/eu-action/carbon-removals-and-carbon-farming_en (accessed on 31 July 2024).
54. Serebrennikov, D.; Thorne, F.; Kallas, Z.; McCarthy, S.N. Factors Influencing Adoption of Sustainable Farming Practices in Europe: A Systemic Review of Empirical Literature. *Sustainability* **2020**, *12*, 9719. [CrossRef]
55. Barnes, A.P.; Soto, I.; Eory, V.; Beck, B.; Balafoutis, A.; Sánchez, B.; Vangeyte, J.; Fountas, S.; van der Wal, T.; Gómez-Barbero, M. Exploring the Adoption of Precision Agricultural Technologies: A Cross Regional Study of EU Farmers. *Land Use Policy* **2019**, *80*, 163–174. [CrossRef]

energies

MDPI

Article

Who Should Own the Residual Rights over Distributed Resources?

Nilton Bispo Amado [1,2,*], Erick Del Bianco Pelegia [1,2], Ildo Luís Sauer [1,2], Welson Bassi [1,2], Julieta Andrea Puerto Rico [1,2] and Carlos Germán Meza González [2,3]

1 Instituto de Energia e Ambiente, Universidade de São Paulo, São Paulo 05508-010, Brazil; erickpelegia@usp.br (E.D.B.P.); illsauer@iee.usp.br (I.L.S.); welson@iee.usp.br (W.B.); julietapuerto@gmail.com (J.A.P.R.)

2 Centro de Análise, Planejamento e Desenvolvimento de Recursos Energéticos (CPLEN), Universidade de São Paulo, São Paulo 05508-010, Brazil; cgmgcimp@gmail.com

3 Instituto Verde, Universidad Americana (UAM), Managua 14157, Nicaragua

* Correspondence: nilton@iee.usp.br; Tel.: +55-11-3091-8646

Abstract: Residual rights are the right to choose asset uses not specified in the contract. They are essential in situations of uncertainty. The current energy transition presents a greater variety of assets. Moreover, it is impossible to anticipate all relevant states that the assets and the environment can embody, making it impossible to optimize them contractually. Furthermore, there is consensus that the transition must occur quickly, which means high levels of investment in new specific assets. How should we distribute property rights to maximize social benefits in a context with specific and dispersed assets? Because of the complementarity between network and distributed resources, this article questions the premise that deverticalization is invariably beneficial to consumers and argues for the need to revise the concept of network and develop the regulatory implications of such a reconceptualization. We defend the need to evaluate alternative network concepts considering the technological repertoire available to operationalize them. When considering the technological repertoire available today, characterized by the competitiveness of information and communications technology (ICT) and distributed resources, we should recognize the inherently incomplete nature of the contracts signed between network operators and users.

Keywords: distributed resources; energy transition; incomplete contracts; smart grid; unbundling

1. Introduction

According to the media, the ongoing energy transition threatens network operators [1,2]. The expression *utility death* has been in the news to denote this threat for some time. However, it is hard to reconcile this utility death image with any credible energy system scenario.

Traditionally, supply and demand are aligned by dispatchable generators according to changes in load. When demand is higher than supply, the system frequency tends to decrease. If this occurs, dispatchable generators inject more power into the grid to compensate for such a decrease and return the operating frequency at the reference frequency. If demand is lower than supply, the control acts in the opposite direction, reducing power injection into the grid. Therefore, controlled firing determines the dynamic behavior of dispatchable generators, which controls the rate at which primary energy is converted into useful work, permanently maintaining the coupling between generation and load.

However, with the increasing insertion of intermittent renewable sources and other asset classes (such as energy storage and demand control technologies), the grid must actively align between supply and demand, acting on emerging energy and para-energy technologies. For example, grid controls can trigger curtailment of non dispatchable generators (e.g., solar photovoltaics or wind power) or activate flexible loads (e.g., air conditioning) to compensate for up and down ramps from intermittent sources. In this emerging context, the economic value of the grid increases since, in addition to the passive transmission of power flows, it should actively perform the coupling function with dispatchable generators. On the one hand, such a coupling function requires more complex network operations, but on the other hand, it allows the system to operate with lower operating costs (reduced fuel costs). But such a scenario will be possible only if investments in the grid (*The power system can be subdivided into generation, transmission, distribution, and end-users. The grid is composed of transmission and distribution assets. Our focus in this paper is distribution grids. We see the grid as a subset of the power system with specific attributes and functions.*) enable it to process generation and end-users' state information in real-time and redirect flows as necessary. In this new context, energy autonomy would happen without depending on the continuous fossil material processing to obtain energy, but it would require a massive increase in information processing. In other words, the transition from a fossil-based system to a renewable-based one corresponds to replacing matter processing with the capture of energy flows plus information processing. It is not easy to see how the social utility of the grid could diminish in such a context.

Therefore, there is no reason to assume that the potential for power system networks generating value for society will be smaller. Everything indicates that ensuring adequate investments in the network can significantly increase society's capacity to use flow energies (solar, wind, etc.) and reduce dependence on fossil fuels. Then, it is reasonable to claim that the supposed death spiral is not a result of disruptive competition from resources at the customer side of the meter but from the inherent design of rate structures [3]. This article deepens such understanding by proposing a redescription of the grid's role, considering the implication of the emergence of distributed resources (distributed generation, automatic demand control, energy efficiency, energy storage) for the grid concept. In methodological terms, our approach is a development of the premise that the effectiveness of the regulatory apparatus always has to consider the technological repertoire available to agents instead of viewing it as a black box.

A limitation of our network redescription is that, although we recognize that the energy system does not involve only the electrical networks, we will only deal with part of the electrical subsystem, which corresponds to the distribution networks. Thus, while part of our redescription applies to transmission or non-electrical networks (e.g., gas networks), this article focuses on electrical power distribution networks. Our proposal to focus on electricity distribution networks stems from our assessment of the more remarkable emergence of the problem of rethinking networks in the scope of distribution due to the increasing competitiveness of distributed resources.

Our approach emphasizes the functional role of the grid, clearly differentiating it from the generation's role, but also recognizes that part of the generator load-following function could be executed by the grid with increased economic efficiency. Although dispatchable generators have traditionally implemented the supply–demand coupling function, emerging technologies make it possible and even necessary to perform an essential part of this coupling function through the grid. However, such a transformation is impossible without a change in the grid concept, including developing the regulatory implications of this change. We propose moving away from the naive conception of the grid as a passive electricity transporter and demonstrate the need to understand the grid in more general terms, that is, as a provider of flexibility. Section 2 details our understanding of the grid as a flexibility provider that increases the socioeconomic value of energy resources. Section 3 presents a typology of possible structures to organize the grid. Section 4 uses the so-called unbundling problem to demonstrate the naive grid conception's influence on the current power sector's organization. Section 5 explains why the advent of distributed resources increases the importance of incomplete contracts [4–6]. In conclusion, we defend the need to systematically incorporate a comprehensive grid concept, capable of considering the exponential increase in the quantity and variety of the energy technologies repertoire of the 21st century, that is more compatible with the reality of prosumer capitalism.

2. The Role of Grid in Power Systems

Although the role of the energy system, in general, and the power system, in particular, in triggering and maintaining the development trajectories of contemporary civilization is increasingly recognized [7–9], the specific mechanisms by which such systems guarantee agents autonomy in their decisions are still not sufficiently clarified.

For example, it is customary to emphasize the role of dispatchable generation in balancing supply and demand at any given moment. This role is proper and essential, but such a role would be impossible without the permanent connection between energy resource users and resource sources guaranteed by the grid. An essential feature of this coupling between sources and users guaranteed by the joint action of generators and grid operators is that all users are served simultaneously, without queues—all the users are attended at the same time, no matter their location or the moment of the demand. This arrangement ensures that the power system constantly adapts to users' needs, thus giving them ample flexibility to organize their activities. In practice, this process replaces the serial service characteristic of queues with the parallel service characteristic of operation at a reference frequency (50 Hz or 60 Hz, depending on the country). When demand is higher than supply, the frequency tends to decrease. If this occurs, more power is injected into the grid to compensate for such a decrease and return the operating frequency at the reference frequency. If demand is lower than supply, the control acts in the opposite direction, reducing power injection into the grid. It is worth emphasizing that this dynamic process is only possible because everyone is permanently connected to the electric field produced by the set of generators, which must operate synchronized at the reference frequency. The grid physically guarantees this total connection of all agents to the same electric field—without it, the flexibility in using energy resources would be impossible despite the enormous amount of primary sources available. Thus, the grid is essential to the coupling between sources and users.

At the same time, the generation transforms primary resources into electricity and makes temporal allocation of aggregate power; *the grid translates this aggregate capacity into temporal and spatial electrical power for each user*. As a result, the combined action of generation and grid transforms energy into a social object that can be activated whenever the users consider it convenient. That is, generation and grid perform equally necessary but distinct functions.

Another way of describing the process that guarantees energy autonomy to the socioeconomic system is by observing that every user of the power system inevitably depends on two elementary contracts: an energy purchase contract with generation and an options contract with the grid. The generation contract guarantees total work capacity, while the grid contract makes it possible for users to decide when to use the work capacity produced by generation. For most of the power system history, these two elementary contracts with generation and grid appeared together, and there was no need to distinguish them for users.

Nevertheless, the current energy transition context has become increasingly important to understand that generation and grid are distinct functional components of the power system. It is crucial to understand that the service value provided by the grid is derived from the energy value but does not correspond to energy sales. The grid service is a peculiar form of derivative. The first mention we have identified of the grid-users relationship as an options contract comes from the context of a legal dispute:

> "Capacity is not actual electricity. It is a commitment to produce electricity or forgo the consumption of electricity when required. Generation resource owners sell capacity to utilities, which need sufficient capacity to provide electricity to their customers reliably. This creates a kind of options contract. When a utility experiences a high demand for electricity, it can call on the capacity resource to produce that electricity". ([10], p. 3)

This options terminology comes from the derivatives market. In derivatives markets, a long position implies purchasing a right, which corresponds to a short position referring to the sale of such a right. In this context, rights and obligations refer to the exercise of a purchase (long call, short call) or a sale (long put, short put) [11]. The grid user pays both for the right to dispose freely of amounts of power and for the total integrated power in a certain period (total consumed energy).

When signing a contract with distribution, the consumer pays for the right to use work capacity at any time, assuming a peculiar long call position. The distribution, in turn, receives for assuming the obligation to guarantee this right, thus placing itself in a short call position. The short call position assumed by distribution allows users to use variable capacity quantities and decide the best time to use any fraction of the total work capacity produced by generation.

Historically, the power system started as a means to obtain a specific energy service (lighting). Nevertheless, it evolved towards becoming a real-time general energy service provider. More and more agents began to select electricity as the most suitable means to obtain energy services in various forms. This trend involved employing electric motors; communication in telegraphs, radios, and televisions; information processing in microprocessors, etc. This evolution often involved replacing other energy technologies with electrical ones, with the electric motor being a particularly illustrative example (*([12], p. 513): "Shortly after steam power began to yield to electricity, installation of electric motors called attention to the obvious restraints placed upon efficiency by the steam engine. Its systems, practices, and factory organization became almost visibly redundant. Thus, as "unit drive" electric power grew in plant after plant, thorough reorganization of factory layout and design took place. Machines and tools could now be put anywhere efficiency dictated, not where belts and shafts could most easily reach them. To these advantages were simultaneously added to revamped industrial processes, leading to mass-production and batch-processing technique."*). The use of electric motors to provide mechanical power went from 5% of total mechanical horsepower in 1899 to 90% in 1939 [13]. With the advent of electricity, the degree of energy customization to users' needs has staggeringly increased:

> "Electricity's rapid rise to dominance as a source of industrial power, then, was based on a number of compelling advantages. Electricity could be "packaged" in almost any size, whereas steam engines became highly inefficient below a certain size. Thus, "fractionalized" electric power sources of precisely the right capacity for each industrial application meant large energy and capital savings. Recourse to large steam engines that generated large amounts of power in situations that required only small or intermittent doses, could now be avoided. Electricity offered opportunities for "fine tuning" the supply of power to specific needs. Furthermore, the electric motor reduced the requirements for floor space and offered much greater freedom in the organization and layout of the workplace. Electric motors meant that the flow of work in factories did not have to accommodate a clumsy system of overhead belting and shafting in order to transmit power, very wastefully, from a central power source to a large number of machines". ([13], p. 15)

In other words, the advent of electricity represented a more flexible energy source for economic agents. The benefit of the power system is that it delivers the most flexible work capacity in real-time. By combining generation and grid functionalities, it is possible to guarantee real-time work capacity tailored to the needs of any user. The power system guarantees access to flexible capacity work, preserving socioeconomic agents' autonomy and the pricing system's effectiveness by combining generation and grid functions in complementary ways. The century-old history of the power system does not indicate the possibility of reducing one function to another.

3. Types of Power Flexibility

So far, we have argued that the grid does not sell electricity but flexibility to use electrical work capacity. However, how does the grid operationalize this energy option? Today, it is possible to point out the following possibilities:

(i) Grid Physical Capacity. In this case, work capacity produced in a different location from the user, usually through centralized generation, is directed to satisfy the user's needs. The grid must be able to carry centralized generation for any user physically. It corresponds to the traditional way power systems provide flexibility.

(ii) Behind-the-Meter. In this case, flexibility is offered by producing energy at the consumption point. Therefore, the required energy does not need to circulate through the grid. This strategy includes both distributed generation and energy storage. In this arrangement, providing flexibility to use energy is not separate from generation. The effectiveness of this arrangement depends on a high degree of control over the local generation profile or the possibility of exploring correlations between the generation and the demand profiles (e.g., photovoltaic generation versus air conditioning demand). Power system users have increasingly evaluated the opportunity costs involved in this option as lower than in the flexibility provision solution traditionally adopted in the electricity sector. However, this increase in the viability of the behind-the-meter solution results from the possibility of grid users selling the energy they produce to the power system, which indicates that this strategy does not decrease the grid value. On the contrary, the behind-the-meter solution partially explores a new grid valorization frontier.

(iii) Demand-Side Management. Usually, more power generation is injected into the grid whenever there is an increase in demand. Alternatively, it is possible to satisfy the increase in user A's demand by reducing user B's demand, provided that this reduction somehow increases the utility of user B. This alternative satisfies user needs without additional generation resources or inflows through the grid. In this case, flexibility does not rely on additional power injections but on the temporal reallocation of demands actively triggered by the power system. In economic terms, demand-side management proposes to weigh, for the same moment, the power injection marginal cost vis-à-vis the user idleness marginal cost. Therefore, it departs from the traditional power systems postulate that the user idleness cost is always untenable.

Even today, grid physical capacity is often considered the only way to provide flexibility. The exclusive focus on this approach forces grid operators to design the grid in such a way as to support the maximum expected value of users' aggregate demand at all grid points. In this approach, the grid operator investments must always guarantee that the demand for the entire electrical work capacity can circulate inside the network. Therefore, any expansion of services by grid operators invariably involves increasing the physical capacity to support inflows through the grid. Since the physical volume of resources demanded determines the planning and operation of the grid in technical and economic terms, we can call them volumetric grids. In volumetric grids, acquiring equipment such as transformers, cables, feeders, etc. is paramount to ensure safe transit of any level of demandable capacity. That is, the standard solutions of the volumetric grids are wired solutions in opposition to so-called non-wired solutions [14–16], a way of expression that suggests the dependence on physical investments in the grid's transport capacity. Indeed, volumetric grids have been the paradigm for planning and operating electrical networks. The paradigm provides the mental models and the toolbox regulators and engineers use to build solutions for power system problems.

While still the industry standard, the grid physical capacity option has drawbacks. Designing the network so that every observed increase in demand circulates within it produces substantial economic inefficiency in the invested capital. No matter how small the time interval, the network operator observes a demand value—he must invest so that this demand can circulate through the network, which impacts the level of asset utilization.

A graphical representation of the asset utilization rate helps make the impact of organizing the grid parameterized exclusively according to maximum demand more straightforward. In this sense, a representation widely used in energy planning is the load duration curve (LDC). The individual daily peak loads can be arranged in descending order to form a cumulative load model, composing the LDC ([17], pp. 37–40). Figure 1 shows 1993 LDCs at the system generation level and for a typical distribution substation of the Pacific Gas and Electric Company (PGeE) [18]. The figure shows that there are only a few hours in the year when the rated capacity of these assets is fully utilized [18]. Also, the distribution system is not as well utilized as the generation assets [18], indicating that the problem of economic efficiency of investments is even more pronounced in distribution than in generation.

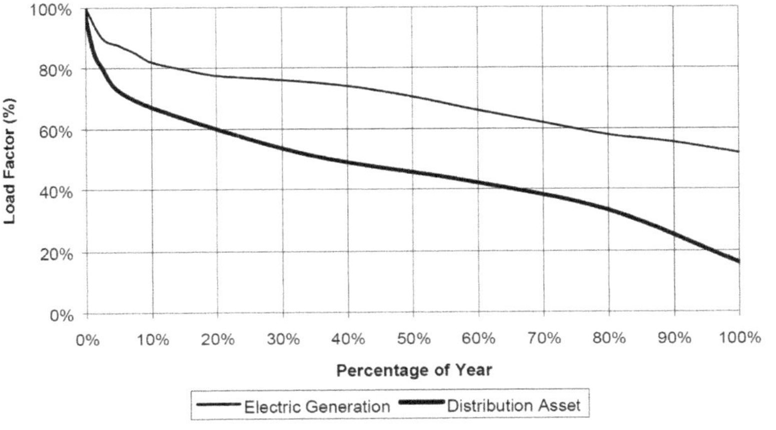

Figure 1. PG&E load duration curves for 1993 illustrate decreased utilization of assets closer to the customer ([18] apud [19]).

We have used measured data to build an LDC for a distribution asset located on the University of São Paulo campus as indicated by Figure 2 below.

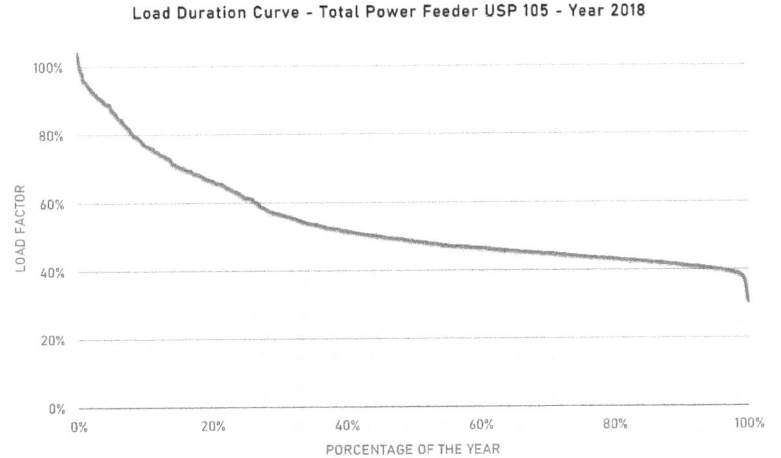

Figure 2. Load duration curve of network distribution asset in the year 2018. Source: Own elaboration by the authors using data from feeder at University of São Paulo.

The ordinate indicates the percentage of the total apparent power used by the asset (feeder), while the abscissa indicates the percentage of the year such use occurs. Only 5% of the time is more than 88% of the feeder capacity used. Therefore, 95% of the time, at least 12% is idle. Moreover, only 20% of the time, more than 65% of the total capacity is used.

This situation indicates idleness in this asset, which causes economic inefficiency and poor resource allocation.

Although the above figures are not encouraging from the point of view of economic efficiency, they are a normal condition in traditional planning. Indeed, high levels of idleness combined with short periods of demand peaks are not unique to the electricity sector; this is a common feature in infrastructure assets. What is new is the possibility that we can start operating with higher asset utilization rates, *if we can find a way to use distributed resources in a coordinated manner*. The emergence of distributed resources in the electricity sector makes it increasingly possible to customize energy services and reduce asset idleness. The potential massive penetration of distributed resources makes relying exclusively on grid physical capacity solutions unnecessary and even counterproductive. The more distributed resources' financial costs decrease, the more opportunity costs associated with grid physical capacity solutions increase. This situation makes it advisable to determine the regulatory implications of the emerging competitiveness of distributed resources (*In 2022, the world installed 45% more solar power capacity than the year before [20]. The data also indicate that compound annual growth of more than 15% is typical among the top 20 markets [20]. If not all, a significant part of this trend comes from the growing technological competitiveness of renewable sources; the average levelized cost of electricity (LCOE) for stand-alone and island 100% renewable energy systems has decreased by 9% annually between 2016 and 2021 from* $\$_{2021}0.54/(\text{kW h})$ *to* $\$_{2021}0.29/(\text{kW h})$ *[21].*). Outside the solar PV market, there are also reasons for optimism. For example, it is estimated that in Europe, distributed energy resources can offer 100 GW of demand response potential [22]. The subsequent sections of this article provide contributions in this regard.

4. Naive Grid Concept and Unbundling Corollary

We defend the need to explicitly develop the regulatory implications of considering the grid as a flexibility provider, not merely an electrical capacity transporter. We must recognize that the grid organization can be structured in multiple ways to see how emerging energy technologies can develop and mature. Here, we use the terms organization and structure in a sense presented by Maturana and Varela:

> "Organization denotes those relations that must exist among the components of a system for it to be a member of a specific class. Structure denotes the components and relations that actually constitute a particular unity and make its organization real. Thus, for instance, in a toilet, the organization of the water-level regulation system consists of the relations between an apparatus capable of detecting the water level and another apparatus capable of stopping the inflow of water. The toilet unit embodies a mixed system of plastic and metal comprising a float and a bypass valve. This specific structure, however, could be modified by replacing the plastic with wood, without changing the fact that there would still be a toilet organization". ([23], p. 47)

The distinction between organization and structure is pertinent when there are different ways to operationalize a given organization and irrelevant when there is a single way to instantiate it. During most of the history of the power sector, structuring the network using grid physical capacity was the only option available to agents. This situation has created difficulty for agents in realizing that the grid function is not to transport capacity but to provide flexibility. Because of this, there is a tendency, which sometimes appears implicitly in the analytical models used, to consider the grid as physical capacity as the only structure available to provide flexibility. After the often implicit consideration of the grid as a capacity transporter, it usually follows the observation that the production function available to operationalize such a function has costs that decrease with increases in the production scale, characteristic of a so-called natural monopoly. Undoubtedly, such an approach was very successful throughout the 20th century. Nevertheless, it becomes problematic as one moves from a factory energy model to a service energy model. In the former, the revenue is tied to the physical volume of sales, focused mainly on economies of scale; in the latter, it is necessary to exploit flow energies systematically, which requires a strong focus on using the information to explore scope economies. This transition requires assessing a much higher number of possibilities to provide energy. Investments are necessary to capture energy flows as much as massive information processing. The differences between the traditional context, which prevailed in most power sector history, and the emerging context justify a more precise characterization.

The traditional context presents ubiquitous dispatchable centralized generation, relatively high information processing costs, and generally high distributed resource costs. In comparison, the emerging context combines dispatchable and intermittent sources, minor information processing costs, and high competitiveness of distributed resources.

It is possible to assess the effect of the change in context on industrial organization by evaluating the impact of information technologies outside the electricity sector. The drastic reduction in computing costs has given rise to business models guided by logic and dynamics that radically differ from those observed in industrial capitalism, which experts have called prosumer capitalism [24–26]. In prosumer capitalism, companies exploit acts of production and consumption in the most synergistic way possible [24–26]. This new form of exploitation raises questions about the power of monopolies, not only in the economy but also in the cultural sphere, while at the same time presenting real opportunities for productivity increases. When consumers use the services of companies such as Amazon, Facebook, or Uber, they provide information that allows these companies to adjust in an increasingly profitable way both the set of assets they own and the provision of the services they contract. One of the ways such a strategy can increase productivity is by mining information regarding the states of assets, including assets of third parties. For example, Uber's service connects idle cars with potential users. Ultimately, it increases the use of car assets through available information and proportionally low investment in physical assets from the company. In practice, these companies use information technologies to systematically coordinate specific acts of production with specific acts of consumption, thus operationalizing synergies and increasing productivity.

Despite the possibilities for reorganizing the transformation of primary sources into electricity, the prevalence of the naïve network concept in the power sector has prevented the development of business models that explore prosumer capitalism's full potential in the energy transition context. Despite the extensive regulatory changes triggered by the liberalization process after 1990 and the growing competitiveness of distributed resources, we continue to organize the network under the logic of the naïve grid concept. The technological repertoire of information technologies and distributed resources has had little impact on conceiving the grid.

With the liberalization of energy markets, an issue that began to worry analysts and agents was the role of network operators in commercial activities. After all, the operator could use its privileged access to grid assets to discriminate against third parties competing with its commercial sector ventures. The most accepted solution for this problem is the vertical separation of power sector activities, also known as unbundling. Network unbundling comes in two forms: ownership and legal unbundling. In its most rigorous form, ownership unbundling prescribes mandatory split ownership of the monopoly network business from the commercial businesses, removing the incentives to discriminate against third parties [27]. The other possibility is legal unbundling, where cross-ownership of the network and commercial businesses are allowed [27]. However, extensive administrative rules try to isolate the network from the commercial business [27].

All unbundling forms assume that the inefficiencies produced by deverticalization are smaller than those produced by verticalization. Nevertheless, evidence is generally not presented to corroborate such claims. In a study assessing the impact of deregulation, Mackay and Mercadal [28] note that the literature on the consequences of deregulation is surprisingly scarce. The authors [28,29] investigated the evolution of electricity prices in the United States (1994–2016) by comparing price evolution in states that remained regulated with states that deregulated, controlling for the influence of relevant variables such as size and energy mix (*Ref.*

([28], p. 17): *"By individually matching utilities based on their size and generation mix prior to the onset of deregulation, we are able to nonparametrically control for changes in macroeconomic factors—such as fuel costs and demand for electricity—when measuring a number of outcome variables."*). Figure 3 below is a graphical summary of their results:

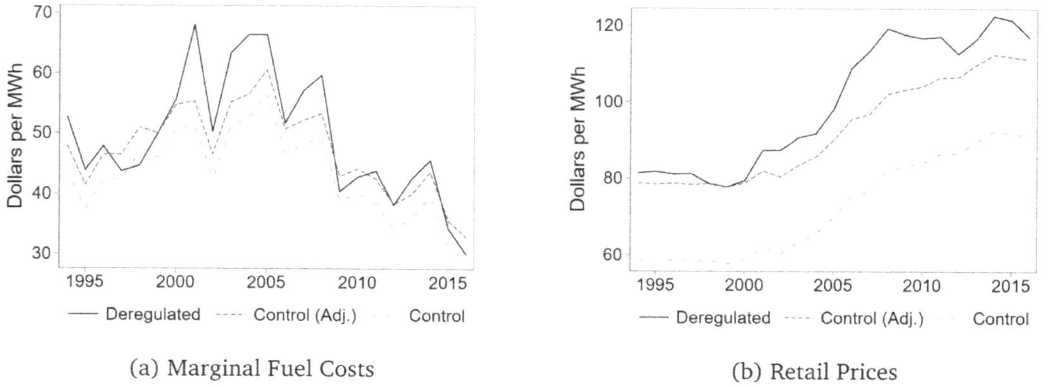

Figure 3. Source: [28]. Panel (**a**) plots the average marginal fuel costs of generation for all generating facilities that in 1994 belonged to utilities in deregulated states (solid black line) and control states (dotted line). Panel (**b**) plots the quantity-weighted retail price for investor-owned utilities in deregulated states and in control states. The dashed line in both panels plots retail prices and fuel costs for control states after adjusting for level differences in 1999.

Panel (a) in Figure 3 indicates that marginal fuel costs have recently fallen in regulated and deregulated markets. However, panel (b) indicates retail prices have risen in both markets. In deregulated markets, the increase has been substantially more significant. Thus, the empirical results do not indicate gains for consumers.

The above results corroborate some experts' criticism that unbundling can negatively affect coordinating activities that produce energy services [30,31]. Yet in 1990, Nobel Prize economists Hart and Tirole considered that the debate about the conditions under which vertical mergers are anticompetitive still needed to be settled [32].

However, despite the theoretical doubts and the few empirical results, the economic regulation of power systems carried out since the 1990s keeps assuming that vertical integration is always harmful to grid users. Why does unbundling enjoy such an acceptance? One reason is that in the traditional context of the 20th century, the grid had no active function, and the power system used almost entirely dispatchable centralized generation, with tiny use of distributed resources (*examples of distributed resources used in the 20th century include diesel generators working as backup and cogeneration plants.*). Because of the high costs of information processing and distributed resources, the technological repertoire is such that there is one single production function to provide flexibility, operated in the following terms:

(i) The grid has a passive function, not actively interfering in the optimal allocation of energy resources. Only central generators should play an active role in meeting end-users' demands, ideally without operator interference. In this context, a characteristic property distinguishes the relationships between the power system and the users: the demand to be met by the network operators is defined in a way that is entirely exogenous to the investment decisions of the network operators. Such a context justifies accepting the naive grid concept and related implications, including the regulatory ones.

(ii) Another crucial characteristic is that the production function available to operate the grid constitutes a so-called natural monopoly. More specifically, the available technology repertoire is exclusively made up of wired solutions with scale-decreasing costs. As it operates in a natural monopoly, the grid can inflict considerable losses on other agents despite its passive operational nature if the regulators do not adequately supervise it.

The risks of granting greater autonomy to grid operators in the above context seem high without clear benefits. However, when we stop seeing the grid as a mere carrier of capacity and start understanding it as a provider of flexibility, the limits of the grid's physical capacity typology and the possibilities opened up by distributed generation and demand control become clearer. In the electrical system, flexibility is the right to have power available where and when we want it. As explained in Section 2, this right is different from accessing quantities of energy. In economic terms, it corresponds to the possibility of modulating one's load curve according to autonomously defined criteria, which, in turn, means starting and stopping transformations in the production and consumption processes in a completely autonomous manner, without having to refer to external processes (for example, decisions by an external entity or the fact that it is day or night). Such flexibility has costs, which depend on the structures used to operationalize it. In practice, ensuring flexibility in the use of energy resources amid high load variability by operating under grid physical capacity forces us to accept operating assets with high levels of idleness and, therefore, with low economic efficiency; for a significant part of the time, nobody will use the investment in physical capacity.

One way to get around the grid physical capacity inefficiencies on the distribution grid is to use distributed resources. However, doing so requires incorporating emerging distributed generation and demand control technologies dispersed in a given region to the preexisting grid, which suggests the relevance of increasing verticalization, at least in the distribution grid. However, the emergence of these technologies has not changed the verticalization debate, which remains stuck in the dominant discussions (and technological repertoire) available in the context of the 1990s, when the deregulation processes effectively unfolded. The regulatory process has kept unbundling as a corollary of the naive grid concept notwithstanding the technological changes since then.

Despite the importance of the accumulated debate on the effects of deverticalization, we must maintain sight of the specificity of the current context. Even if one ignores the theoretical and empirical arguments against deverticalization or considers that deverticalization has been successful, it needs to be evaluated if the justifications for deverticalization in the traditional context are transferable to the emerging moment. The fact that segregating centralized generation from the grid is rational does not mean that segregating distributed resources from the grid should also be an optimal solution. It is at least plausible to speculate that the complementarity of distributed resources is greater than that of centralized generation with the grid. In short, upstream unbundling (separation between centralized generation and grid assets) should not be taken as proof that downstream unbundling (separation between distributed resources and the distribution grid) is rational. However, without a more comprehensive grid concept, conducting systematic simulations and empirical studies that would allow us to evaluate such speculation becomes impossible. A naive grid concept has epistemologically blocked [33] the emergence of new studies and business models.

In the emerging context, the repertoire of energy technologies increases and differentiates, imposing new requirements on the operation of the grid. Although it is easy to use terms such as smart grid to denote the new requirements that agents demand, it is not evident that such requirements are consistent with the naive grid concept. It seems unreasonable to expect that an element that should only provide physical capacity to connect different points comes to have decision-making functions, something implicit in the smart grid concept. The naive grid approach has been an obstacle because it does not adequately represent the economic value of information or distributed resources. In the next section, we depart from such guidance and use incomplete contract theory to explain why the growing competitiveness of distributed resources challenges the assumption adopted since the 1990s that grid deverticalization always benefits its users.

5. The Increasing Importance of Incomplete Contracts

Since the development of the theory of incomplete contracts, the weight of arguments in favor of verticalizing production chains has increased considerably [4,5,34]. However, the theory of incomplete contracts does not defend that the system must invariably be verticalized. Its contribution goes in another direction: it forces us to explicitly represent different ways of organizing the industry instead of tacitly sticking to a form of organization taken as evidently correct and not subject to scrutiny. Today, the regulatory discussion has been conducted as if increased competition, merely understood as a higher number of companies, invariably benefits consumers as long as it does not occur in transmission and distribution. There is a very restricted view of the possible organizational models without reevaluating such models with the emergence of new technologies. To overcome this overly simplistic scenario, it seems appropriate to view the firm as a set of assets whose mobilization depends on property rights that some agent eventually must activate in unanticipated ways, particularly in the context of the energy transition, which requires investments in specific assets whose technical and commercial models are not yet fully known.

The emerging context combines dispatchable and intermittent sources, which tends to increase the complexity of the grid. On the other hand, the drastic reduction in information processing costs and the increase in the competitiveness of zero marginal costs resources make the new context not merely a cost that society will have to bear but also a real opportunity to obtain productivity gains. However, taking advantage of possible benefits requires the development of an institutional apparatus that adequately aligns incentives. Investments with the potential to increase productivity (e.g., by reducing fuel costs) and reduce dependence on fossil fuels (e.g., by increasing solar photovoltaics) will only increase quickly enough if they are associated with high rates of return. No matter how worthy Environmental, Social, and Governance (ESG) practices (*[35]: "The most surprising finding is how thoroughly the "Big 3" investors—BlackRock, Vanguard and State Street—turned against environmental and social shareholder proposals in the past year. In the U.S., BlackRock's support for E&S proposals dropped from 41.3% in the 2020–2021 season, to 23.7% in 2021–2022, to a mere 8.7% in 2022–2023. Vanguard's drop was equally precipitous, from 29.6% to 12% to a tiny 3% in the past year. And State Street went from 43.7% to 28.6% to 21.2%. Clear signs of the backlash."*) are, the change in productive structures under capitalism can only occur if it is possible to organize interests and assets towards the goals valued by society.

Currently, we process material extracted from fossil stocks to generate power in central plants. Such an approach requires a not-so-complex grid but changes the planet's heat balance [26,36], causing increasing environmental problems. Nevertheless, there are alternatives to this traditional approach. In some situations, the grid operator could process information and act on assets that capture energy or end-use equipment. In such cases, operator interference is an action to coordinate the transformation of flow energies into electricity, thus increasing the relative share of renewable energies and reducing energy service costs. Such a possibility is not empty speculation.

A feeder in an urban environment with a high air conditioner load could lighten its load if the grid operator identifies the situation and profits by selling photovoltaic solar energy. In line with the principles of the circular economy [37–39], using waste treatment to generate biogas [40] would make it possible to transform a relevant environmental problem into a dispatchable energy source. Demand control makes building load flexible, especially if it is possible to aggregate the load profile of thousands and even millions of users in real-time, facilitating the insertion of intermittent renewable sources [41]. These and other hypothetical situations are increasingly conceivable, but fitting them into a consistent regulatory model has been difficult.

Nevertheless, it is crucial to recognize that even the correct identification of the above possibilities is not automatic, requiring relevant information processing investments to continue mining data and prospect opportunities. Once a promising opportunity is identified, investments in specific assets are necessary to transform the information into economic value. The need to combine investments in information processing with specific physical assets suggests the convenience of operationalizing business models typical of prosumer capitalism.

Moreover, the urgency of climate change imposes an additional complicating factor: the speed with which the energy transition is expected to happen [42]. Such a rapid change in energy production and use structures requires a high rate of innovation. However, innovations are hitherto unconceived relationships that an agent introduces at a given moment. It is strange to think they are compatible with full contractual relationships.

Therefore, the dependence on flow energies and the required innovation rate indicates the impossibility of fully specifying ex ante all the relationships relevant to the provision of energy services. It is plausible to consider that any conceivable contract specifying the terms of flexibility provision services in the emergent context must leave out a large set of relevant situations.

Having overcome the naive concept grid, recognizing the existing uncertainties and heterogeneity of assets, how should property rights be distributed in order to maximize social benefit and accelerate the energy transition? In other words, how should assets and their uses be combined?

> "Grossman and Hart (1986) [4] used the theory of incomplete contracts to develop answers to the question "What is a firm, and what determines its boundaries?" in their path-breaking paper on the costs and benefits of vertical integration. Perhaps the central issue is that economic actors are only boundedly rational and cannot anticipate all possible contingencies. It might well be that certain states of nature or actions cannot be verified by third parties after they arise, like certain qualities of a good to be traded in the future, and thus cannot be written into an enforceable contract. When contracts are incomplete, and consequently not all uses of an asset can be specified in advance, any contract negotiated in advance must leave some discretion over the use of the assets; and the "owner" of the firm is the party to whom the residual rights of control have been allocated at the contracting stage. Grossman and Hart argue persuasively that the optimal allocation of property rights—or governance structure—is one that minimizes efficiency losses. Thus, in a situation where party A's investment is more important than party B's investment, it is optimal to allocate property rights over the assets to party A, even if this discourages investment by party B. More generally, the incomplete contracts/property rights approach produces a theory of ownership and vertical integration. It also directly addresses the question of what is a firm. Grossman and Hart argue that a firm is a collection of assets over which the owner has residual control rights". ([43], pp. 182–183)

More specifically, if the firm is a collection of assets, how should property rights over distributed resources be allocated to minimize economic losses, given the technological portfolio available to agents today? Should the distribution operator be understood as segregated, without control over distributed resources? Or should it be seen in an integrated way, despite the risks to competition involved?

We saw in Section 4 that starting from the naive grid concept, the unbundling postulate is a natural choice, requiring strong evidence to the contrary before it can be rejected. However, when starting from a comprehensive grid conception, the following factors suggest that deverticalization can be problematic:

(i) The high number of contingent situations and relevant states that cannot be anticipated. According to the theory of incomplete contracts [4,6,32], such situations cause underinvestment if the operations are carried out unbundled.
(ii) No evidence exists that exploring the high number of contingent situations is possible without using the grid. On the contrary, everything we know indicates that it is essential to use the grid to combine resources in such a way as to transform uncertainties into acceptable risks. The theory of incomplete contracts indicates that when assets are strictly complementary, some form of integration is optimal (Hart).

(iii) The high idleness level of the assets of the current distribution network indicates that the combination of decoupled tariffs with the use of distributed resources by the grid operators has a high potential to postpone investments. The growing competitiveness of distributed resources makes it possible to increase economic efficiency by reducing the idleness level of assets and postponing new investments in traditional distribution assets. Therefore, an increase in the rate of profit for the operators and a reduction in costs for the power system users seem possible (empirical studies should be executed to investigate such a hypothesis, drawing comparisons with the mainstream approach). If the distribution grid operators had the autonomy to explore such possibilities, this would open a new frontier of accumulation and increase investments toward energy transition.

As it seems unlikely that (i), (ii), or (iii) are irrelevant considerations, it would be up to the proponents of the segregated solution to demonstrate that, despite appearances to the contrary, the economic losses of the segregated solution are minor. Because of its interpretation of what ownership of an asset is, the theory of incomplete contracts is critical in investigating scenarios in which it is important to transform uncertainty into acceptable risk; more than right over revenue streams, ownership of the asset means having control over the asset. One of the meanings of this control is the right to use the asset in unspecified ways. This ownership perspective clarifies that an opportunity ex post verified can be readily exploited by the asset owner but will have to be negotiated by third parties. Hart (1995) [6] expresses that by saying that the owner of the assets has residual rights over them. It is not obvious that we should always develop distributed resources through third parties who will have to pay to access the grid instead of granting grid operators full autonomy to develop business models with distributed resources ([6], pp. 5–6):

> "Economists have written a great deal about why property rights are important, and in particular, why it matters whether a machine, say, is privately owned or is common property. However, they have been less successful in explaining why it matters who owns a piece of private property. To understand the difficulty, consider a situation where I want to use a machine initially owned by you. One possibility is for me to buy the machine from you; another possibility is for me to rent the machine from you. If contracting costs are zero, we can sign a rental agreement that is as effective as a change in ownership. In particular, the rental contract can specify exactly what I can do with the machine, when I can have access to it, what happens if the machine breaks down, what rights you have to use the machine, and so on. Given this, however, it is unclear why changes in asset ownership ever need to take place. In a world of contracting costs, however, renting and owning are no longer the same. If contracts are incomplete, not all the uses of the machine will be specified in all possible eventualities. The question then arises: Who chooses the unspecified uses? A reasonable view is that the owner of the machine has this right; that is, the owner has residual rights of control over the machine, or residual powers. For example, if the machine breaks down or requires modification and the contract is silent about this, the owner can decide how and when it is to be repaired or modified".

The above considerations imply that the residual rights value increases when the economical use of the required assets involves many contingent and not previously specifiable situations, which is the case when exploring distributed resources. When one recognizes the importance of controlling specific assets for the profitability of the business models in question, the speed with which energy transition must happen, and the variety of contingent situations, it does not seem that the segregated grid can adequately answer the uncertainties involved. A final decision requires model formalization, including empirical data and forecasts of technological evolution. Nevertheless, this empirical analysis can only happen if we have an adequate grid concept and imagine alternative ways of structuring grid services.

Even if upstream unbundling were indisputably successful, this would not prove that downstream unbundling should be adopted. As an empirical test to evaluate possibilities, we should be able to compare grid distribution operator autonomy with the traditional regulation. By grid distribution operator autonomy, we mean the following:

(i) Decoupling between distribution tariffs and volumes of energy delivered by the distribution would guarantee a specific revenue for the provision of flexibility service by the distribution grid, defined within quality criteria that encompass the entire range of consumers. Such tariff decoupling is consistent with grid operators not selling energy but offering flexibility services associated with energy use.
(ii) Grant broad autonomy to the distribution grid operator to combine assets and build business models, including selling distributed generation at a profit. This possibility creates a strong incentive for distribution operators to develop distributed generation as an alternative to centralized generation.

The hypothesis that the autonomy of the distribution grid operator using downstream verticalization will benefit the grid's users must be subjected to rigorous scrutiny. The point here is not to demonstrate its validity but the relevance of explicitly comparing it with traditional regulation. More generally, it is essential to explicitly analyze different ways of organizing the power system, considering both the alignment of incentives and the characteristics of the repertoire of technologies existing at a given historical moment. The traditional approach has been to idealize a particular form of organization rather than comparing the advantages and disadvantages of different forms of organization.

A promising alternative to the traditional approach is to develop institutional arrangements to reconcile verticalization's positive and negative aspects. Proposals to reconcile competition and coordination should be encouraged (see, for instance, [27]). However, the institutional imagination of too many specialists has assumed, without further evidence, that the economic inefficiencies arising from unbundling are minor. Such a premise is understandable when starting from a naive grid concept. Nevertheless, this approach is unconvincing if we see the grid as a flexibility provider in a prosumer capitalist context, not a mere capacity transporter.

As the naive grid concept tends to limit the possible combination of assets, it underestimates the possibility of relying on non-dispatchable sources. On the other hand, conceiving the grid in more general terms as a provider of energy options encourages the integration between energy and information [44], facilitating the approximation between energy technologies and information technologies. Since information technologies are currently the most dynamic sector of the economy, such an approach would accelerate the energy transition. In a context where accelerated technological change is needed, a regulatory framework that opens up space to combine assets seems more promising than creating complex and artificial structures for utilities to accelerate innovation [45]. Giving up the naive grid concept to incorporate a more comprehensive concept makes it possible to integrate the power sector into the reality of prosumer capitalism. Such integration certainly brings problems and contradictions, but we must deal with them directly rather than unilaterally idealizing the energy transition under capitalism. It is only possible to make the smart grid a reality with such integration.

6. Conclusions

The article has demonstrated the need to consider the grid as an energy options provider, not a mere capacity transporter. We have also demonstrated that the unbundling postulate is a consequence of applying a naive grid concept in the power sector liberalization moment. Furthermore, understanding the grid as a flexibility provider allows us to demonstrate that the increased competitiveness of distributed resources imposes the need to recognize the inherently incomplete nature of contractual relationships between power system operators and users, particularly for distribution networks. In the current regulatory situation, simulations and empirical studies comparing different ways of organizing the grid have yet to be conducted because the naive grid concept has functioned as an epistemological obstacle. In possession of an adequate grid concept, such studies comparing the current regulation with downstream verticalization are the next step of the research. Although the complementarity of the distribution grid and distributed resources seems strong, much empirical evaluation is required to confirm it. A first comparison is to simulate the system's evolution within current regulation with its evolution under alternative regulation, in which the distribution operator has broad autonomy to build business models, including the possibility of obtaining profit from operating in distributed generation. Despite the risks to competition, this scenario transforms the current conflict between distributed generation and distribution operator into an alliance. If the economic surplus is sufficiently high, the regulator can offset the economic losses resulting from the loss of competition via lump transfers to grid users. At the moment, a simplistic conception of the grid has prevented us from studying grid organization forms that do justice to the emerging heterogeneity of assets available in the ongoing energy transition.

Finally, although this article has focused on distribution grids for reasons of scope, a conceptual and later empirical reassessment is most likely necessary for other energy systems networks, such as electricity and natural gas transmission networks. After all, both the context in which the deverticalization of networks began to be adopted, and the technological repertoire has changed radically since the 1990s. However, reassessments should always consider the specific characteristics of such networks and the technological repertoire effectively available to network operators and society as a whole.

Author Contributions: Conceptualization and methodology, N.B.A. and I.L.S.; validation, N.B.A., I.L.S. and E.D.B.P.; measurement and data processing: W.B.; writing—original draft preparation, N.B.A.; writing—review and editing, I.L.S., J.A.P.R., C.G.M.G. and E.D.B.P.; visualization, N.B.A. and W.B.; supervision, I.L.S.; project administration, I.L.S.; funding acquisition, I.L.S. All authors have read and agreed to the published version of the manuscript.

Funding: This research was funded by the Institute of Energy and Environment of University of São Paulo (IEE-USP).

Data Availability Statement: The original contributions presented in the study are included in the article, further inquiries can be directed to the corresponding author.

Acknowledgments: The first author gratefully recognizes the discussions with João Marcos Mott Pavanelli about institutional aspects of energy transition, which stimulated the approach adopted in the paper.

Conflicts of Interest: The authors declare no conflicts of interest.

References

1. Denning, L. Lights Flicker for Utilities. *The Wall Street Journal*, 22 December 2013.
2. Clavenna, S. Unraveling the Metaphors of Utility Evolution. *Greentech Media*, 15 August 2013. Available online: https://www.greentechmedia.com/articles/read/unraveling-the-metaphors-of-utility-evolution (accessed on 25 August 2024).
3. Felder, F.A.; Athawale, R. The life and death of the utility death spiral. *Electr. J.* **2014**, *27*, 9–16. [CrossRef]
4. Grossman, S.J.; Hart, O.D. The costs and benefits of ownership: A theory of vertical and lateral integration. *J. Political Econ.* **1986**, *94*, 691–719. [CrossRef]
5. Hart, O.; Moore, J. Property Rights and the Nature of the Firm. *J. Political Econ.* **1990**, *98*, 1119–1158. [CrossRef]
6. Hart, O. *Firms, Contracts, and Financial Structure*; Clarendon Press: Oxford, UK, 1995.
7. Smil, V. *Energy and Civilization: A History*; The MIT Press: Cambridge, MA, USA, 2017.
8. Crosby, A.W. *Children of the Sun—A History of Humanity's Unappeasable Appetite for Energy*; W. W. Norton & Company: New York, NY, USA, 2006.
9. Love, T.; Isenhour, C. Energy and economy: Recognizing high-energy modernity as a historical period. *Econ. Anthropol.* **2016**, *3*, 6–16. [CrossRef]
10. Advanced Energy Management Alliance—Business Management Consultant. Advanced Energy Management Alliance v. FERC, No. 16-1234, United States Court Appeal for the District of Columbia, Legal Case Document. 2017; p. 3. Available online: https://law.justia.com/cases/federal/appellate-courts/cadc/16-1234/16-1234-2017-06-20.html (accessed on 30 April 2024).
11. Hull, J.C. *Options, Futures and Other Derivatives*; Pearson: London, UK, 2018.
12. Du Boff, R.B. The introduction of electric power in American manufacturing. *Econ. Hist. Rev.* **1967**, *20*, 509–518. [CrossRef]

13. Rosenberg, N. The role of electricity in industrial development. *Energy J.* **1998**, *19*, 7–24. [CrossRef]
14. Moghe, R.; Tholomier, D. Grid Edge Technology as a Non-Wires Alternative. In Proceedings of the IEEE Power & Energy Society General Meeting (PESGM), Montreal, QC, Canada, 2–6 August 2020; IEEE: Piscataway, NJ, USA, 2020; pp. 1–5.
15. Hile, S.; Murdock, D.; Robison, M. Procuring Distribution Non-Wires Alternatives: Practical Lessons from the Bleeding Edge. *ICF White Paper*, 2017. Available online: https://www.icf.com/-/media/files/icf/white-paper/2017/icf-procuring-distribution-non-wires-alternatives-july-2017.pdf?rev=a50967f3ef6a4b648dcc12d0c701f4f9 (accessed on 23 March 2020).
16. Cowart, R. Recommendations on Non-Wires Solutions. DOE Electricity Advisory Committee Momorandum to the Honorable Patricia Hoffman. 2012. Available online: https://www.energy.gov/oe/articles/eac-recommendations-doe-action-regarding-non-wires-solutions-october-17-2012 (accessed on 23 March 2023).
17. Billinton, R.; Allan, R.N. *Reliability Evaluation of Power Systems*; Plenum Press: New York, NY, USA; London, UK, 1996.
18. Kannberg, L.D.; Chassin, D.P.; DeSteese, J.G.; Hauser, S.G.; Kintner-Meyer, M.C.; Pratt, R.G.; Schienbein, L.A.; Warwick, W.M. *GridWiseTM: The Benefits of a Transformed Energy System*; Technical Report; Pacific Northwest National Laboratory (PNNL): Richland, WA, USA, 2003.
19. EPRI. *Distributed Utility Valuation Project Monograph. Report TR-102807*; Technical Report; Electric Power Research Institute (EPRI): Palo Alto, CA, USA, 2003.
20. Europe, S. *Global Market Outlook for Solar Power 2023–2027*; Technical Report; SolarPower Europe: Brussels, Belgium, 2023.
21. Weinand, J.M.; Hoffmann, M.; Göpfert, J.; Terlouw, T.; Schönau, J.; Kuckertz, P.; McKenna, R.; Kotzur, L.; Linßen, J.; Stolten, D. Global LCOEs of decentralized off-grid renewable energy systems. *Renew. Sustain. Energy Rev.* **2023**, *183*, 113478. [CrossRef]
22. IRENA. *Market Integration of Distributed Energy Resources-Innovation Landscape Brief*; Technical Report; IRENA: Masdar City, United Arab Emirates, 2019.
23. Maturana, H.R.; Varela, F.J. *The Tree of Knowledge: The Biological Roots of Human Understanding*; Shambhala Publications: Boston, MA, USA; London, UK, 1992.
24. Ritzer, G. Prosumer capitalism. *Sociol. Q.* **2015**, *56*, 413–445. [CrossRef]
25. Ritzer, G. Contemporary Capitalism and the "New" Prosumer. In *The Oxford Handbook of Consumption*; Wherry, F.F., Woodward, I., Eds.; Oxford University Press: New York, NY, USA, 2019; pp. 75–93.
26. Jakimowicz, A. The future of the energy sector and the global economy: Prosumer capitalism and what comes next. *Energies* **2022**, *15*, 9120. [CrossRef]
27. Brandstätt, C.; Brunekreeft, G.; Buchmann, M.; Friedrichsen, N. Balancing between competition and coordination in smart grids—A Common Information Platform (CIP). *Econ. Energy Environ. Policy* **2017**, *6*, 93–110.
28. Mackay, A.; Mercadal, I. *Do Markets Reduce Prices? Evidence from the U.S. Electricity Sector*; Working Paper 21-095; Harvard Business School: Boston, MA, USA, 2024.
29. Penn, I. Why Are Energy Prices So High? Some Experts Blame Deregulation. *The New York Times*, 2023. Available online: https://www.nytimes.com/2023/01/04/business/energy-environment/electricity-deregulation-energy-markets.html (accessed on 24 August 2024).
30. De Nooij, M.; Baarsma, B. Divorce comes at a price: An ex ante welfare analysis of ownership unbundling of the distribution and commercial companies in the Dutch energy sector. *Energy Policy* **2009**, *37*, 5449–5458. [CrossRef]
31. Brunekreeft, G. Network unbundling and flawed coordination: Experience from the electricity sector. *Util. Policy* **2015**, *34*, 11–18. [CrossRef]
32. Hart, O.; Tirole, J.; Carlton, D.W.; Williamson, O.E. Vertical integration and market foreclosure. *Brook. Pap. Econ. Act. Microecon.* **1990**, *1990*, 205–286. [CrossRef]
33. Bachelard, G. *A Formação do Espírito Científico*; Contraponto: Rio de Janeiro, RJ, Brasil, 1996.
34. Sánchez, R. *The Impact of "Incomplete Contracts" on Long-Term Infrastructure Concessions*; United Nations: San Francisco, CA, USA, 2020.
35. Murray, A.; Gordon, N. BlackRock, Vanguard, and State Street Turned Against Environmental and Social Proposals This Year, a Clear Sign of Backlash. *Fortune*, 2023. Available online: https://fortune.com/2023/10/31/blackrock-vanguard-state-street-esg-proposals-voting/ (accessed on 14 April 2024).
36. Schwartzman, D. Solar communism. *Sci. Soc.* **1996**, *60*, 307–331.
37. Borrello, M.; Pascucci, S.; Cembalo, L. Three propositions to unify circular economy research: A review. *Sustainability* **2020**, *12*, 4069. [CrossRef]
38. Laurenti, R.; Singh, J.; Frostell, B.; Sinha, R.; Binder, C.R. The socio-economic embeddedness of the circular economy: An integrative framework. *Sustainability* **2018**, *10*, 2129. [CrossRef]
39. Lewandowski, M. Designing the business models for circular economy—Towards the conceptual framework. *Sustainability* **2016**, *8*, 43. [CrossRef]
40. D'Aquino, C.A.; Santos, S.C.; Sauer, I.L. Biogas as an alternative source of decentralized bioelectricity for large waste producers: An assessment framework at the University of São Paulo. *Energy* **2022**, *239*, 122326. [CrossRef]
41. Short, J.A.; Infield, D.G.; Freris, L.L. Stabilization of grid frequency through dynamic demand control. *IEEE Trans. Power Syst.* **2007**, *22*, 1284–1293. [CrossRef]

42. Masson-Delmotte, V.; Zhai, P.; Pirani, A.; Connors, S.L.; Péan, C.; Berger, S.; Caud, N.; Chen, Y.; Goldfarb, L.; Gomis, M.I.; et al. Policymakers. In *Climate Change 2021: The Physical Science Basis. Contribution of Working Group I to the Sixth Assessment Report of the Intergovernmental Panel on Climate Change;* Cambridge University Press: Cambridge, UK, 2021. Available online: https://www.ipcc.ch/report/ar6/wg1/downloads/report/IPCC_AR6_WGI_SPM.pdf (accessed on 12 April 2024).
43. Aghion, P.; Holden, R. Incomplete contracts and the theory of the firm: What have we learned over the past 25 years? *J. Econ. Perspect.* **2011**, *25*, 181–97. [CrossRef]
44. Spreng, D. Possibilities for substitution between energy, time and information. *Energy Policy* **1993**, *21*, 13–23. [CrossRef]
45. Thomas, S. The RIIO Formula: Old Wine in New Bottles. 7 February 2022. Available online: https://ssrn.com/abstract=4028934 (accessed on 25 August 2024). [CrossRef]

energies

MDPI

Article

Carbon Capture and Storage (CCS) Implementation as a Method of Reducing Emissions from Coal Thermal Power Plants in Poland

Michał Kopacz [1], Dominika Matuszewska [2] and Piotr Olczak [1,*]

[1] Mineral and Energy Economy Research Institute, Polish Academy of Sciences, 7A Wybickiego St., 31-261 Cracow, Poland; kopacz@min-pan.krakow.pl

[2] AGH University of Kraków, Faculty of Energy and Fuels, 30 Mickiewicza Ave., 30-059 Cracow, Poland; dommat@agh.edu.pl

* Correspondence: olczak@min-pan.krakow.pl

Abstract: The Polish economy, and especially the energy sector, is facing an energy transformation. For decades, most electricity in Poland has been generated from hard coal, but in recent years, renewable energy sources have been gaining an increasing share of the market. The aim of the energy transformation is to reduce the carbon footprint in electricity production, which translates into the decarbonization of the economy, including manufactured products. Currently (2024), increasing the share of renewable energy sources raises major challenges in terms of energy storage or other activities and forces cooperation with flexible sources of electricity generation. One of the challenges is to determine what a decarbonized energy mix in Poland could look like in 2050, in which there would be sources (with a smaller share of coal sources in the mix than currently) of electricity generation based on hard coal with CCS technology. In order to do this in an economically efficient manner, there are aspects related to the location of power plants that would remain in operation or repower current generating units. The added value of the study is the simulation approach to the analysis of the problem of assessing the effectiveness of CCS technology implementation together with the transport and storage infrastructure, as well as the multi-aspect scenario analysis, which can determine the limits of CCS technology effectiveness for a given power unit. Positive simulation results (NPV amounted to 147 million Euro) and the knowledge obtained in the scope of the correlated and simultaneous impact of many important cost factors and prices of CO_2 emission allowances make this analysis and its results close to reality. Examples of analyses of the effectiveness of CCS system implementations known from the literature are most often limited to determining linear relationships of single explanatory variables with a specific forecasted variable, even if these are multifactor mathematical models.

Keywords: CO_2 reduction emission; CCS; CCUS; coal thermal power plant; optimization; Poland

1. Introduction

1.1. Energy Transformation

The current energy mix in Poland is dominated by the volume of hard coal and brown coal power plants. The increase in the share of renewable energy sources is related to issues regarding the significant increase in energy production from stable sources (water, biogas, biomass) and weather-dependent fuel-saver sources (photovoltaic and wind energy) [1]. Plans for the energy mix are evolving in Poland, with new updates increasing the future share of low-emission sources [2,3]. In this respect, Wojtaszek et al. [4] conducted a comparative analysis of Poland and Germany for challenges (resignation from electricity generation from coal-fired power plants) and changes in energy policy with particular emphasis on the perspective of 2050. Wierzbowski et al. [5] also analyzed the Polish energy policy from the same time perspective, in which coal-fired power plants currently play the main role. There are plans for the development of small and large nuclear power as well as storage of electrical energy within the framework of the policies [6]; the former is relatively expensive considering the investment outlays, and the latter is not economically viable for now [7], and we are extracting less and less coal and at higher operating costs [8]. Currently (2024), there are plans to eliminate or significantly reduce hard coal production in Poland [9]. Some researchers see a partial analogy to the energy problems (present and past) of Finland, which already has a nuclear power plant [10].

Nyga-Łukaszewska et al. [11] analyzed the cointegration of the hard coal and natural gas markets in the period from 2011 to early 2019. The main conclusion from the analyses was that the Polish power industry is more closely linked to the international coal market than to the natural gas market, while for heating the relationships are reversed. Lorenc et al. [12] used a research method based on statistical analysis, including mainly the analysis of the correlation between the prices of shares of the dominant company (from the coal mining and other areas) and the prices of shares, e.g., a company separated from existing structures. The above aspects are related to new challenges faced by energy companies, such as sustainable development and circular economy.

Pluta et al. [13] analyzed various scenarios of changes in the energy mix from the perspective of 2050, with particular emphasis on the possibility of achieving a 95% reduction in CO_2 emissions while maintaining energy security (mainly as certainty of energy supply). The subject of energy security and its impact on the shaping of generation capacity in Poland was also discussed in the following works [14,15].

Progress in the field of CO_2 capture and utilization methods has been noticeable for many years [16]. Its main barrier to development in the EU was the high level of investment outlays and operating costs compared to the level of CO_2 emission fees, the so-called ETS, or compared to alternative investments in sources with lower emission coefficients calculated per unit of generated electricity [17–21]. Economic barriers are not the only type of factors influencing this technology; issues related to legal requirements for countries and life cycle assessment (LCA) analyses of electricity production are of great importance [22]. Moreover, there are still issues of infrastructure, investment, and the use of electricity in other areas as well [23]. Additionally, some countries see sources of competitive advantages in the economic field in decarbonization issues.

Bui et al. analyzed various industries, particularly in the context of the development of renewable energy technologies. They also examined different CO_2 capture and storage methods [24]. Wang et al. investigated the use of Convolutional Neural Networks to evaluate combustion technologies, including their efficiency [25].

The main novelty lies in the methodology for selecting existing power plants suitable for adding carbon capture and storage (CCS) technology. This approach considers multiple parameters, including Monte Carlo simulations, to assess potential sites. For traditional thermal power plants, key factors include proximity to thermal coal supplies (i.e., transportation distance) and access to the electricity grid. However, in the context of CCS, additional considerations, such as the distance for CO_2 transport and the associated costs, are also crucial.

1.2. Energy Mix in Poland Currently and in 2050

In the coming years, the departure from coal in Poland will involve repowering or new locations for coal thermal power plants. In the context of capacity market requirements, carbon production footprints, and fees for CO_2 emissions into the atmosphere, companies whose operations are associated with a high level of carbon footprint are facing new challenges.

An analysis of the impact of potential changes in the price ratio of domestic and imported coal and their impact on the volume of coal imports to Poland was carried out by Kamiński et al. [26]. In contrast, a mathematical model using thirteen scenarios of the price ratio was developed by [27].

Repowering old coal units to gas or nuclear, but also CCS, provides the possibility of expanding the range of possibilities for coal thermal power plants. In each of these possibilities, due to, among others, the uneven demand for energy in the system (taking into account energy production from renewable energy sources, the so-called Duck curve), a thermal energy storage installation with a capacity of 200–1200 MWh was analyzed by Bartela et al. [28]. Hard coal mining in Poland is not only about economics, business, and energy but also government, climate, and social policy, including adapting to plans [29]. On the other hand, the development of nuclear energy and hydrogen is planned. The idea of switching new electricity generation units to natural gas is questionable due to prices, availability, and issues related to energy security. One solution is to introduce a mixture of natural gas and hydrogen into the gas network, which, thanks to zero-emission hydrogen production, reduces the total carbon footprint [30].

One of the methods of CO_2 management is the production of synthetic methane from CO_2 obtained from exhaust gases using amine scrubbing and hydrogen from renewable sources. This was conducted, among others, in an experimental installation in a power plant located in Łaziska Górne [31]. However, since 2013, experiments were carried out in this area in which the achieved efficiency of CO_2 separation was above 85%, and energy demand was 4.8–7.5 MJ/kg CO_2 [32].

The share of electricity production from hard coal is decreasing year by year (Figure 1). Additionally, the plans for changes in the production structure in the energy sector, as well as in the mining sector, are important because they indicate the disappearance of the share of the above-mentioned carrier in the future energy mix in Poland. The location of the main coal thermal power plant unit in Poland is presented in Figure 2.

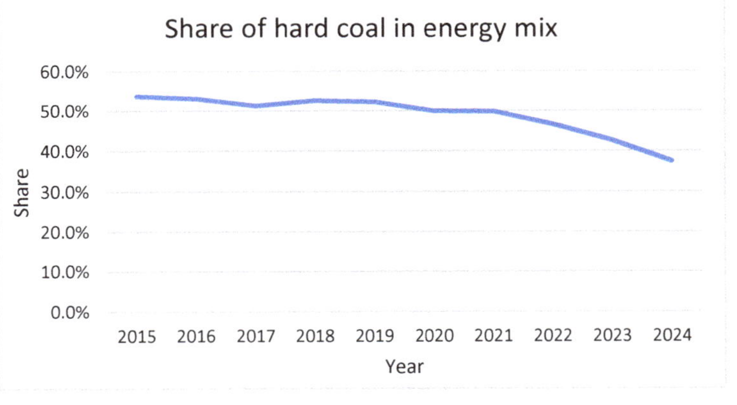

Figure 1. Share of energy from coal thermal power plants in the energy mix in Poland from 2015 to June 2024. Source: own study based on [33].

The scientific value of the publication is the simulation approach to the analysis of the problem of assessing the effectiveness of CCS technology implementation together with the transport and storage infrastructure and the multi-aspect scenario analysis, which determines the limits of CCS technology efficiency for a given power unit capacity. Positive simulation results and the knowledge obtained in the field of correlated and simultaneous impact of many important cost factors and CO_2 emission allowance prices make this analysis valuable and its results reliable. Examples of analyses of the effectiveness of CCS implementations known from the literature are most often limited to determining linear relationships between single decision variables and a specific forecasted output, even if these are multifactorial mathematical models.

Figure 2. Location of coal thermal power plants (only power plants with a capacity higher than 900 MW) in Poland. Source: own study based on [33–35].

2. Calculation Methods

A proprietary methodology for the effectiveness of CCS technology implementation for coal-fired power units was developed. The CCS system assumes the construction of a carbon dioxide capture installation, the transport and storage of CO_2, and the monitoring of the facility for the next 30 years after the end of storage. In order to apply it, the data necessary for the analyses were selected.

2.1. Data for Calculations

Over the past dozen or so years, the prices of hard coal for energy have been lower than the prices of natural gas per 1 MWh of energy. Exceptionally, there were short periods when the prices of hard coal for energy were higher than the prices of natural gas: at the beginning of 2020 and 2023. The detailed price profile is marked in Figure 3a for 2023, and half of 2024 is shown in Figure 3b.

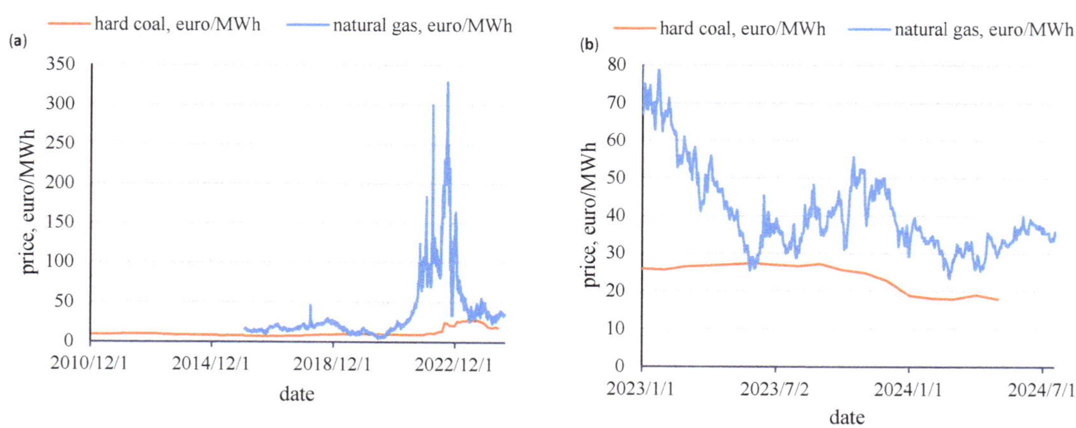

Figure 3. (**a**) Hard coal (2010–2024) and natural gas (2015–2024) prices per 1 MWh. (**b**) Hard coal and natural gas prices per 1 MWh. Values from (**a**) reduced to a range of 2023–2024. Source: own study based on [36].

Historic record prices of natural gas occurred in 2022 (330 Euro/MWh), contributing to the increase in electricity prices and inflation in many economies, including Poland. The increase in electricity prices was caused, among other things, by the fact that gas power plants operated as one of the peak sources (merit order principle [37]) for electricity production. Additionally, much higher price fluctuations are noticeable for natural gas (on world exchanges) than for hard coal. Prices of hard coal also achieved records in 2022.

Since the beginning of 2023, the price of natural gas has been on a downward trend (from 72 to 35 Euro/MWh), except for the period from August to October 2023. The prices of thermal hard coal have also shown a downward trend from around 27 to 19 Euro/MWh. The prices of thermal hard coal against the background of the prices of CO_2 emission allowances are presented in Figure 4.

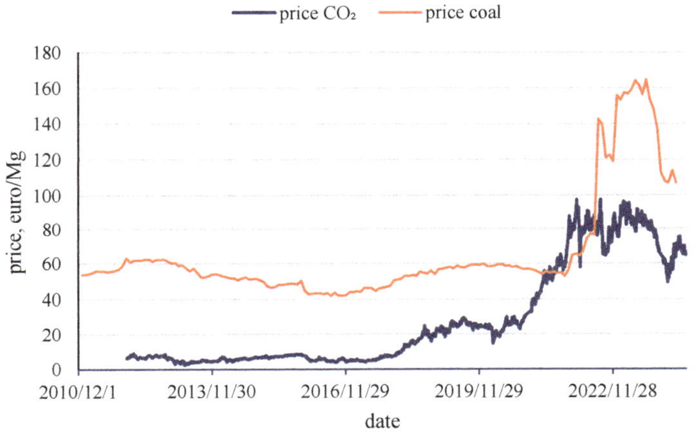

Figure 4. Hard coal and CO_2 emission prices per 1 Mg. Data for the period: 2010–2024. Source: own study based on [38–40].

Electricity prices are presented in Figure 5.

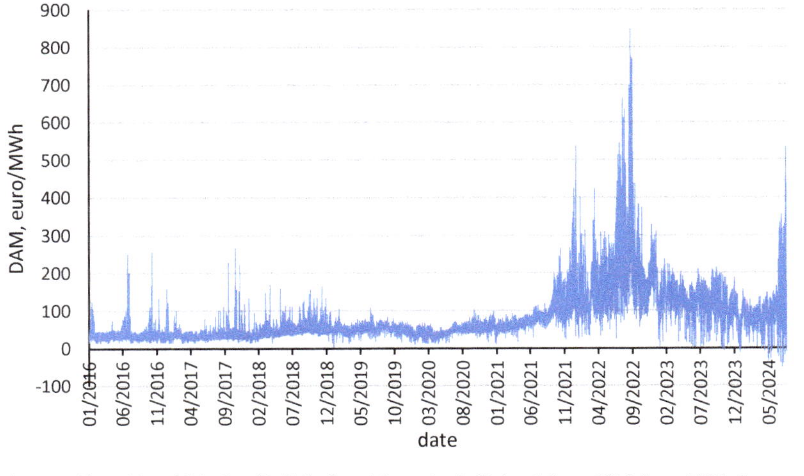

Figure 5. Day-Ahead Market (DAM) electricity price in Poland from 2016–June 2024. Source: own study based on TGE data [41].

In a study prepared by the IISD organization [42,43], the costs for carbon dioxide capture and storage in Canadian conditions were indicated, depending on the concentration of CO_2 in the stream from which capture is to take place. These values range from 27 to over 150 CAD per ton of CO_2 (Figure 6). Additionally, considering the learning curve [44] in this technology, the decreases in cost over the years do not change significantly without clear support; e.g., with subsidies, it is difficult to obtain profitability at least at a minimal level [43].

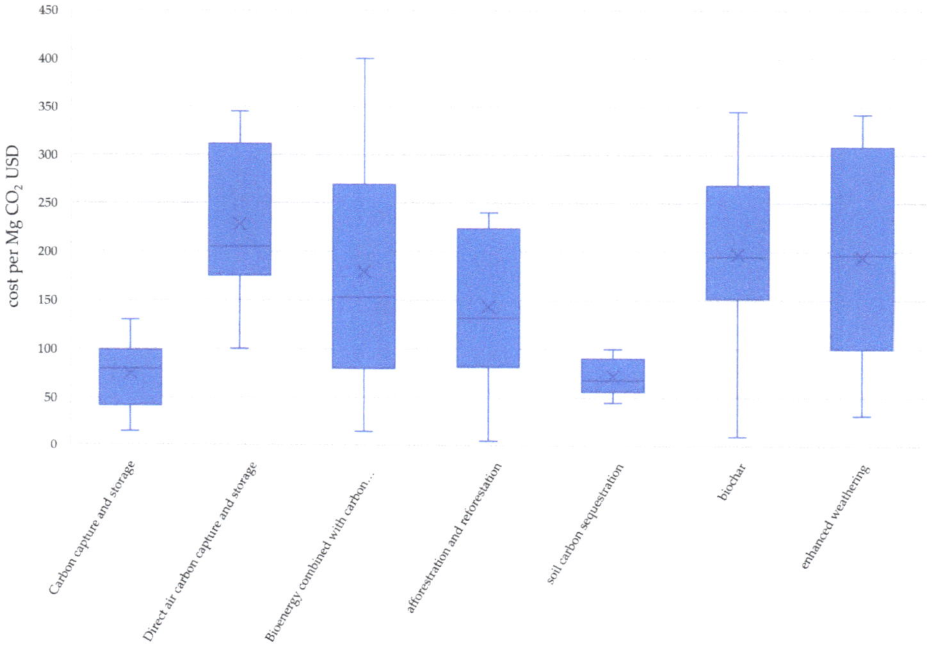

Figure 6. Unit cost values for CO_2 technology components. Source: own study based on [42].

2.2. Methodology

2.2.1. Research (Decision) Problem

The key research problem concerns the resolution of the prospects for the operation of coal-fired power units (plants) in Poland by selecting the optimal strategy for further operation. This strategy assumes an assessment of the justification for the following:

- operation in the current technological and business model, i.e., burning coal until the technical capacity of the units and paying for CO_2 emissions according to market prices, and
- expansion of existing power units with CCS systems.

2.2.2. Approach to Analysis

The research was conducted using a scenario approach.

The adopted decisive criterion is the difference between the differential financial flows for both scenarios of the operation of a model power unit (power plant) with a nominal capacity of approximately 200 MW of electric power. The expansion scenario of the power unit assumes the following:

- the addition of a carbon dioxide capture system,
- transport infrastructure for compressed CO_2,

- construction of an installation for injecting CO_2 into the rock mass (saline structures), and subsequent monitoring during the period of operation and 30 years after the mine is liquidated.

2.2.3. Simulation Model

In the analysis, we assume that all calculations will be performed using an original model of economic efficiency assessment based on Monte Carlo simulation techniques. The model allows for estimating total net cash effects measured as differences in cash flows of Scenario 2 and Scenario 1 in a differential approach regarding variations in production scale and availability of CCS. In particular, the cash flow structure includes costs/expenses related to the following:

- coal transport,
- construction of a CO_2 capture installation (CCS unit) at the power plant,
- processes of carbon dioxide capture, transport, and storage,
- distance for coal deliveries and carbon dioxide injection,
- construction of infrastructure for carbon dioxide transport (pipelines) and injection in salt caverns, and
- cavern decommissioning and storage site monitoring for the next 30 years.
- The idea of the analysis is presented in Figure 7.

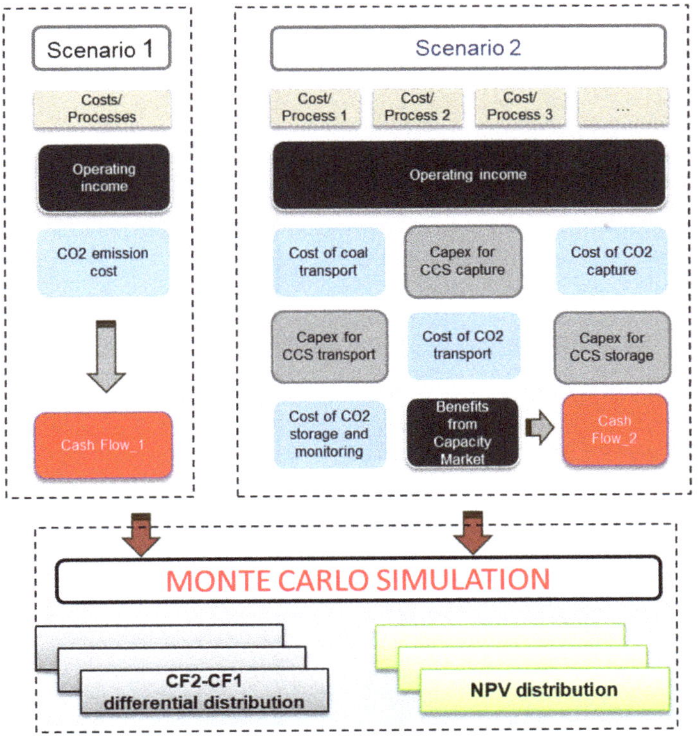

Figure 7. Income and cost components for the analyzed scenarios in the simulation model. Source: own study.

The key performance measures were as follows:

- averaged (annualized) difference in the cash flow of both scenarios: CF_1 and CF_2 for 25 consecutive periods, and
- the net present value (NPV), understood as the sum of 25 annual differences in the cash flow of individual scenarios: $\sum_{ij=0}^{25}(CF1\_i; CF2\_j)$.

These variables were marked as forecasted (outputs) in the developed simulation model. The principle of interpretation of the results was as follows:

- if the average updated difference was positive, the NPV was also positive, and
- an NPV greater than 0 indicated the advantage (higher value) of Scenario 2, i.e., the model of the power plant with a CCS unit and transport infrastructure and storage.

As you can see in the attached diagram (Figure 7), we did not analyze the variables that cancel each other out in both scenarios. As an example, we can give the price of coal and the amount of coal purchased to be burned, or the costs associated with the construction of a power unit. In turn, the key uncertain decision variables in the simulation model were as follows:

- productivity of the power unit with and without CCS infrastructure (P),
- unit cost of coal transport (uCCT),
- unit cost of carbon dioxide capture (uCC_CO2)
- length of the pipeline transporting CO_2 (dP),
- unit cost of electricity (Ep),
- unit cost of CO_2 transport (uCT_CO2),
- market prices of CO_2 emission allowances (Mp_CO2),
- unit costs of CO_2 storage and monitoring (uCS_CO2).

For all the above variables, individual distributions were established as follows:

- P: triangular distribution (min: 2500; mean: 6000; max: 7800) [h/year],
- uCCT: triangular distribution (min: 0.2; mean: 0.3; max: 0.4) [Euro/Mg],
- uCC_CO2: triangular distribution (min: 1; mean: 1.2; max: 2) [Euro/Mg CO_2],
- dP: triangular distribution (min: 50; mean: 250; max: 400) [km],
- Ep: Pearson5 distribution (($\alpha = 6.39$; $\beta = 688.8$) [Euro/MWh],
- uCT_CO2: triangular distribution (min: 0.05; mean: 0.07; max: 0.08) [Euro/Mg CO_2],
- Mp_CO2: GLlogistic distribution: ($\alpha = 90$; $\beta = 9$; $\gamma = 0.85$) [Euro/Mg CO_2],
- uCS_CO2: triangular distribution (min: 6; average: 8; max: 10) [Euro/Mg CO_2].

The selection of triangular distributions was motivated by the literature [45–47]. In particular, when constructing the productivity distribution (P), it was assumed that the power plant with a CCS system would operate with the expected productivity of around 6000 h/y (currently, some coal-fired blocks previously reported to the Capacity Market operate at 25–35% of their nominal power). The availability of a CCS system will enable such an installation to be reported to the auction system in the future (currently, high-emission coal-fired units can no longer be reported to the Capacity Market auctions). We also assumed that the CO_2 pipeline system would be carried out over a maximum distance of 350 km, and electricity needed for CO_2 capture, transport, and storage processes would come from the installation's own production, resulting in reduced sales on the market.

The distribution of market prices of CO_2 emission allowances was considered to be crucial, with a strong impact on the results of the analysis. The following steps were taken in its construction:

- historical prices since 2012 were analyzed,
- a representative period was selected, and
- for CO_2 emission prices above 40 Euro/Mg, the parameters of the best-fit distribution were estimated based on empirical data, then the GLlogistic distribution was selected (Figure 8a) and modified for expected values in the future (Figure 8b).

The idea of the modification was to correct the probability density function of the distribution to achieve the expected (average) value close to 90 Euro/Mg and to have most of the sampled values in the range of 40–130 Euro/Mg, with a probability close to 1. We believed that there is a relatively small probability that CO_2 allowance prices would fall below 40 Euro/Mg or exceed 120 Euro/Mg—the functionality of the ETS system at these price levels seems to meet its objectives.

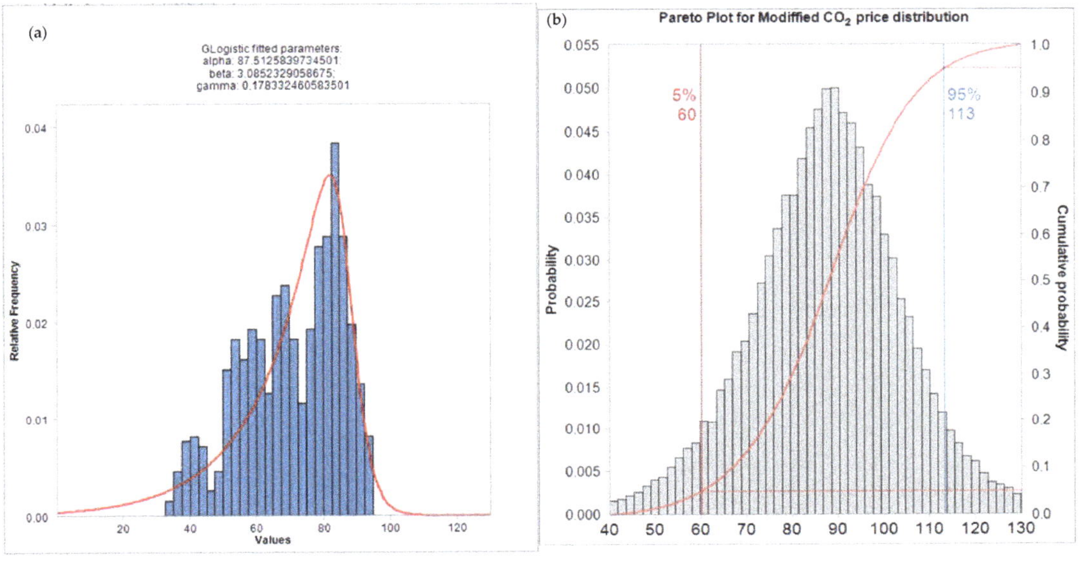

Figure 8. (**a**) Original distribution for the price of CO_2 allowances [Euro/Mg]. (**b**) Modified distribution [Euro/Mg]. Source: own study.

To reflect the differences in energy consumption for a CCS unit, based on [47], we assumed that electricity for self-consumption for a CCS unit would reduce 29% of the total electricity production of the power block. The amount of operating income was a product of market electricity prices and electricity output (production). For modeling electricity prices, we analyzed historical data of electricity from Towarowa Giełda Energii (TGE) in the same period as for CO_2 allowances (2001–July 2024), and then we chose the Pearson5 distribution as the best fit (Figure 9a). We decided not to correct this distribution, allowing for price shocks in the future. The average difference between power units without and with a CCS unit equals 240 MEuro per year with respect to selected electricity distribution.

In order to obtain a relatively correct picture of reality and the relationships between uncertain variables included in the distributions, the following rules were introduced to the sampling procedures:

- with high productivity, the investor will be more willing to pay higher fees for CO_2 transport over longer distances (high probability, strength: 0.5–0.7),
- higher prices of CO_2 emission allowances will result in higher electricity prices (high probability, strength: 0.55),
- higher productivity of the power unit and CCS installation will positively influence the costs of carbon dioxide capture (high probability, strength: 0.7),
- a larger tonnage of delivered coal and received CO_2 will cause the pressure to reduce unit prices in transport (medium and high probability, strength: 0.7–0.8), and
- high prices of CO_2 emission allowances may lead to potentially higher profits from CCS, and thus an increase in pressure on the prices of materials and components, services, and remuneration in this process (low and medium probability, strength: 0.3–0.5).

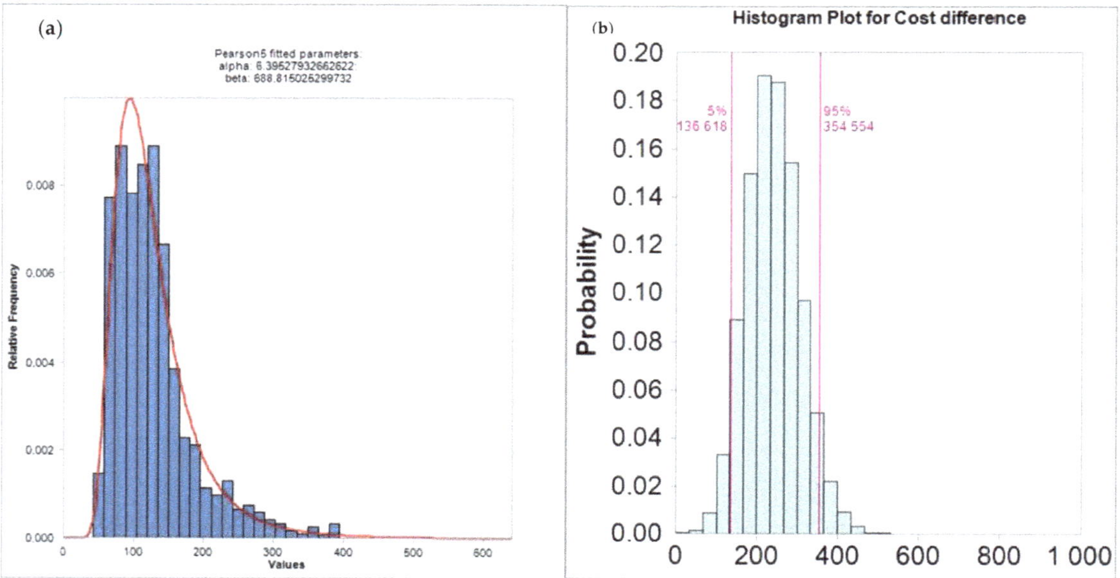

Figure 9. (**a**) Best-fit distribution of electricity price [Euro/MWh]. (**b**) The distribution of income differences between power units without and with a CCS unit [MEuro/year]. Red line in the left chart–fitted distribution curve (Pearson5). Source: own study.

These dependencies in the form of a correlation matrix were entered into the simulation model. The model was recalculated for 10,000 iterations, and then the distributions of the forecasted (output) variables and their descriptive statistics were generated. The calculation model, with the most important items, is presented in Table 1. The values presented in the individual cells are randomly selected in the Monte Carlo simulation.

Table 1. Visualization of the calculation model (spreadsheet)—exemplary numbers. Source: own study.

Years of Analysis	Unit	Sum of 1–25	0	1	2	3	...	23	24	25
Coal consumption	mln Mg/y		1.7	1.7	1.7	1.7	...	1.7	1.7	1.7
Coal LHV (coal calorific value) [48]	MJ/kg		21.0	21.0	21.0	21.0	...	21.0	21.0	21.0
Nominal power	MW		600.0	600.0	600.0	600.0	...	600.0	600.0	600.0
Efficiency	%		40%	40%	40%	40%	...	40%	40%	40%
Productivity	h/y		4381.2	4381.2	4381.2	4381.2	...	4381.2	4381.2	4381.2
Productivity (incl. CCS consumption)	h/y		3285.9	3285.9	3285.9	3285.9	...	3285.9	3285.9	3285.9
Coal transportation distance	km		0.0	0.0	0.0	0.0	...	0.0	0.0	0.0
Unit cost of coal transport [11,40]	Euro/Mg/km		0.3	0.3	0.3	0.3	...	0.3	0.3	0.3
Cost of coal transport	mln Euro		0.0	0.0	0.0	0.0	...	0.0	0.0	0.0
Unit CO_2 emission [33,49]	Mg CO_2/MWh		0.99	0.99	0.99	0.99	...	0.99	0.99	0.99
CO_2 uptake efficiency	%		0.9	0.9	0.9	0.9	...	0.9	0.9	0.9
CO_2 total emission	mln Mg/y	99.9	3.8	3.8	3.8	3.8	...	3.8	3.8	3.8
CO_2 emission without CCS	mln Mg/y	99.9	3.8	3.8	3.8	3.8	...	3.8	3.8	3.8
CO_2 emission with CCS	mln Mg/y	10.0	0.4	0.4	0.4	0.4	...	0.4	0.4	0.4
Unit price of CO_2 [50]	Euro/Mg CO_2	94.7	94.7	94.7	94.7	94.7	...	94.7	94.7	94.7
Cost of CO_2 emission with CCS	mln Euro	966.4	37.2	37.2	37.2	37.2	...	37.2	37.2	37.2
Cost of CO_2 emission without CCS	mln Euro	9663.6	371.7	371.7	371.7	371.7	...	371.7	371.7	371.7
Capex for CCS unit [24,43]	mln Euro	300.0	300.0							
Unit cost of CO_2 capture (without electricity)	Euro/Mg		6.1	6.1	6.1	6.1	...	6.1	6.1	6.1
Cost of CO_2 capture	mln Euro/y	569.3	21.9	21.9	21.9	21.9	...	21.9	21.9	21.9
CO_2 transportation distance	km		0.0	0.0	0.0	0.0	...	0.0	0.0	0.0
Unit cost of CO_2 transport	Euro/Mg/km		0.07	0.07	0.07	0.07	...	0.07	0.07	0.07
Cost of CO_2 transport	mln Euro/y	0.0	0.0	0.0	0.0	0.0	...	0.0	0.0	0.0
Capex for CO_2 storage&monit.	mln Euro	100.0	100.0							
Capex for CO_2 transport pipelines	mln Euro	0.0	0.0							
Unit cost of CO_2 storage&monit.	Euro/Mg		7.1	7.1	7.1	7.1	...	7.1	7.1	7.1
Cost of CO_2 storage&monit.	mln Euro/y	696.9	26.8	26.8	26.8	26.8	...	26.8	26.8	26.8
Unit income from the Power Market	Euro/kW		90.2	90.2	90.2	90.2	...	90.2	90.2	90.2

Table 1. *Cont.*

Years of Analysis	Unit	Sum of 1–25	0	1	2	3	...	23	24	25
Total income from the Power Market	mln Euro/y	1407.5	54.1	54.1	54.1	54.1	...	54.1	54.1	54.1
Cost discount of CO_2 avoidance	mln Euro	0.0	297.8	0.0	0.0	0.0	...	0.0	0.0	0.0
Electricity price	Euro/MWh	127.7	127.7	127.7	127.7	127.7	...	127.7	127.7	127.7
Total operation income with CCS	mln Euro/y	20,269.2	779.6	779.6	779.6	779.6	...	779.6	779.6	779.6
Total operation income without CCS	mln Euro/y	27,025.6	1039.4	1039.4	1039.4	1039.4	...	1039.4	1039.4	1039.4
Income difference	mln Euro/y	6756.4	259.9	259.9	259.9	259.9	...	259.9	259.9	259.9
Total operation cost of DO with CCS	mln Euro/year	1812.3	505.3	52.3	52.3	52.3	...	52.3	52.3	52.3
Total operation cost of DO without CCS	mln Euro/year	9663.6	371.7	371.7	371.7	371.7	...	371.7	371.7	371.7
Cost difference	mln Euro/year	7851.3	−133.6	319.4	319.4	319.4	...	319.4	319.4	319.4
Annual cost difference	mln Euro		38.0				...			
CF (diff.)	mln Euro		−393.5	59.5	59.5	59.5	...	59.5	59.5	59.5
Discount rate	%		10%	10%	10%	10%	...	10%	10%	10%
Discount factor			1.00	0.91	0.83	0.75	...	0.11	0.10	0.09
DCF	mln Euro		−393.5	54.1	49.2	44.7	...	6.6	6.0	5.5
NPV	mln Euro		146.9							

To assess the cash flow and the NPV, the following formulas and conversions are used:

Energy production per year:

$$EPr = CPP * CFPP \tag{1}$$

where:

EPr—energy production, MWh/y;
CPP—power unit, MW;
CFPP—capacity factor of the power plant, h.

Mass of coal consumption (*MC*):

$$MC = \frac{EPr}{nel} \times \frac{3.6}{LHV} \tag{2}$$

where:

LHV—calorific value, which reaches 21 MJ/kg [11,40,48];
nel—efficiency of the coal power plant.

Mass of CO_2 production in the power plant:

$$MCO2 = EPr * CO2f \tag{3}$$

where:

CO2f—emission CO_2 factor, Mg/MWh.

Cost of coal transport (*TCC*):

$$TCC = MC \times uCCT \times CTD \tag{4}$$

where:

uCCT—unit cost of coal transport, euro/km/Mg;
CTD—coal transport distance, km.

Cost of CO_2 transport (*TCCO2*):

$$TCCO2 = MCO2 \times uCT\_CO2 \times dP \tag{5}$$

where:

dP—length of the pipeline transporting CO_2, km;
uCT_CO2—unit cost of CO_2 transport, euro/km/Mg.

Utilization costs of CO_2 storage and monitoring (*UCCO2*):

$$UCCO2 = MCO2C \times uCS\_CO2 \tag{6}$$

where:

MCO2C—mass of CO_2 capture, Mg;
uCS_CO2—unit costs of CO_2 storage and monitoring, euro/Mg CO_2

Emission costs of CO_2 (*ECCO2*):

$$ECCO2 = (MCO2 - MCO2C) \times \text{Mp\_CO2} \quad (7)$$

where:

MCO2—mass of CO_2 produced, Mg;
MCO2C—mass of CO_2 capture, Mg;
Mp_CO2—market prices of CO_2 emission allowances, euro/Mg.

2.3. Power Unit and Supercritical Coal Combustion Process

A unit with a nominal capacity of 600 MW is analyzed. It is powered by crushed and ground coal, which is sent to a boiler, where it is burned with air. Solid waste—ash and slag resulting from coal combustion—is removed from the boiler and then sent to a landfill. The calorific value, *LHV*, of coal reaches 21 MJ/kg. The generated steam at a pressure of about 24 MPa and a temperature of 600 °C is directed through a turbine blade system to a steam turbine coupled to an electricity generator. The working medium then goes to the condenser and then to the boiler. The exhaust gases are dedusted and cleaned, after which they go to the chimney. In the CCS system, after dedusting and cleaning the sulfur, mercury, and nitrogen oxide compounds, exhaust gases are cooled and sent to the CO_2 separation installation, where in the last stage, they are compressed and injected into the pipeline under high pressure. Then, they go to the injection installation and are introduced into the deposit through deep holes. The analyzed system, together with CCS, uses almost 35 MW for its own needs [24].

2.4. Investment Outlays for the CCS Installation

Based on [45–47], it was assumed that the investment outlays for the CO_2 separation system would amount to approximately PLN 450 million (Euro 100 million) for the assumed scale of production (200 MW.

In turn, the investment outlays related to the pipeline construction, in connection with the purchase of land, preparation of technical documentation, exclusions from use, permits, compensations, and permanent infrastructure for a distance of 250 km, were determined at PLN 600 million (approximately Euro 133 million). For the analyzed smaller and larger CO_2 storage distances, these costs were scaled proportionally. This is an acceptable simplification for the purposes of this analysis.

For the purposes of building a salt cavern and injecting and monitoring CO_2, costs of around PLN 300 million (Euro 67 million) were assumed.

3. Results

The results of the analyses were presented for the following values of key technical parameters of the power unit, CCS installation, cost assumptions, and prices of CO_2 emission allowances. Some of these numbers are based on the authors' own data; others are determined based on [47].

- The nominal power of the power plant: 600 MW el.,
- the power plant efficiency: 0.40,
- CO_2 emission factor (CO2f): approx. 1 ton CO_2/MWh,
- the coal calorific value (LHV): 21 MJ/kg,
- detailed forecast period: 25 years,
- installation monitoring period after injection completion: 30 years,
- base values of decision (explanatory) variables, i.e., unit costs of CO_2 management at the level of average values in triangular distributions,

- with a probability of 90%, the value of the CO_2 allowances price is in the range of 60–115 Euro/Mg, with an average of 87 Euro/Mg,
- with a probability of 90%, the value of the electricity price is in the range of 61–240 Euro/Mg, with an average of 130 Euro/Mg,
- the investment outlays for CO_2 capture installation: 300 million Euro; for pipeline construction: 53 million Euro; and for salt cavern and storage installation: 100 million Euro, and
- the reference distance for CO_2 transport: 100 km.

The results of the analyses are illustrated in Figure 10.

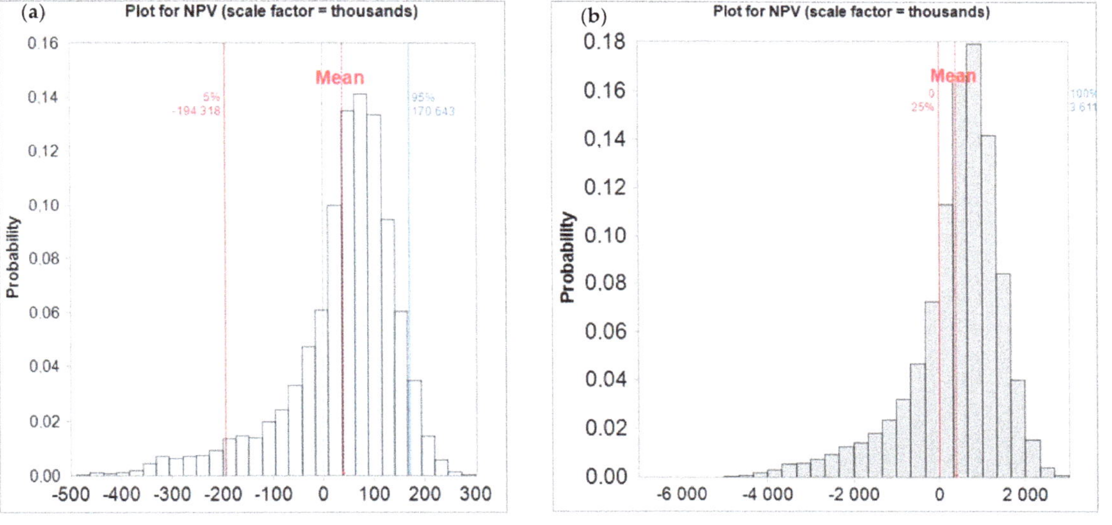

Figure 10. (**a**) Distribution for CF differences between Scenarios 1 and 2. (**b**) Distribution for NPV of Scenarios 1 and 2.

Analyzing Figure 10a,b and Table 2, it can be seen that the expected value of the average annualized cash flow difference (CF_2—CF_1) is approximately 38 million Euro; the annual cash effect of the scenario with CO_2 sequestration is positive and significant. Consequently, the expected value of NPV for the 25-year analysis period amounted to 147 million Euro. It follows that at emission allowance prices of 90 Euro/Mg CO_2 and assumed total outlays of 220 million Euro with respect to coal and CO_2 transport distances, it is economically viable to develop a CO_2 capture unit and transport and storage system. In addition, Table 1 shows that 90% of annualized differential cash flow observations fall within the range of −194 to 170 million Euro. The distributions of the NPV are left-skewed and centrally concentrated. In this case, the probability of receiving an NPV greater than 0 is 75% (Figure 10b) (Table 2).

The tornado plot for annualized cash flow differences (CF2-CF1) attached to the analysis indicates a list of parameters with a distinctive impact on the conditional mean value of CF2-CF1 and their variability. The dominant role is played by the volatility of market prices of CO_2 emission allowances and the productivity of the power unit. The next places are occupied by the unit cost of CO_2 transport, electricity prices, and carbon dioxide capture costs. The influence of the remaining uncertain (explanatory) parameters plays a significantly smaller role in determining the conditional mean of the CF2-CF1 annualized differences (Figure 11).

Table 2. Basic statistics for analyzed distributions in thousands of euros.

Variable Name	CF_2-CF_1 Distribution	NPV Distribution
	Location statistics	
Mean	38,002	147,319
Minimum	−488,156	−5,154,794
Maximum	297,753	2,764,841
	Spread statistics	
St. dev.	109,249	1,100,910
Variance	1×10^{10}	1×10^{12}
Risk ratio	4.29	7.05
CofV	2.87	7.47
	Shape statistics	
Skewness	−1.41	−1.41
Kurtosis	5.53	5.53
	Percentiles	
5%	−194,318	−2,193,777
95%	170,643	1,483,953

Source: own study.

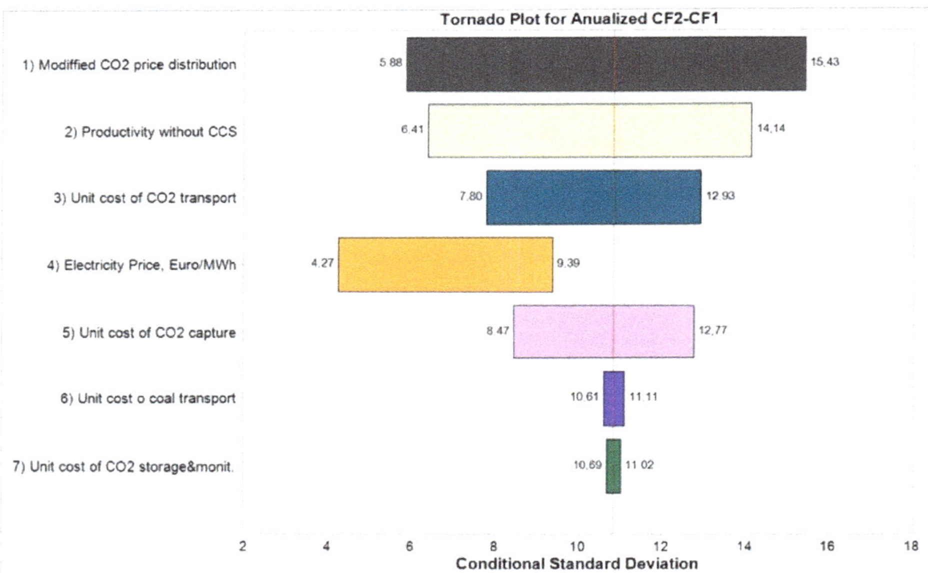

Figure 11. Sensitivity analysis/tornado plot for CF2-CF1 annualized differences under conditional standard deviation. Source: own study.

4. Discussion

In order to deepen the analysis, a scenario analysis was performed. In each scenario, the distances of carbon dioxide storage and the length of coal transport routes to the power plant were changed stepwise. The results are presented in Table 3. Each cell in the table presents the expected NPV value for the established transport distances of coal and CO_2.

In total, the table includes 26 different combinations (scenarios) for the length of the coal and carbon dioxide transport distance. The range from 0 to 350 km corresponds to the realities of Polish coal-fired power plants and covers salt structure locations available in the country, where carbon dioxide can be stored.

Table 3. NPV average (expected) values as a function of the coal distance to the power station and storage unit.

		NPV Average Values [mln Euro]					
	Difference of Transport Distance [km]	CO_2 to Cavern, Distance					
		0	**50**	**100**	**170**	**250**	**350**
Coal to Power Unit	0	388	268	147	−6	−232	−447
	50	164	85	−62		−400	−672
	100	−27	−118	−269		−673	−885
	250	−660	−768	−947		−1242	−1543
	350	−1100	−1202	−1306		−1663	−1941

Source: own study.

Based on the attached table, it can be seen that for the maximum assumed distance in the transport of coal and CO_2, the expected (average) NPV value decreases significantly. At a distance of around 170 km from the storage site, the NPV becomes negative. Simultaneously, a coal transport distance over 100 km causes the NPV to drop below 0. There are only five scenarios (approx. 20% of all analyzed) with a positive NPV. In the developed simulation model, it was assumed that the entire capital expenditures are financed with equity. The decrease in the NPV value is mainly the result of a strong increase in investment outlays on the CO_2 transport pipelines. However, it should be emphasized that these are results obtained with the assumed relatively high price of CO_2 emission allowances.

The importance of the CO_2 emission allowance price model seems to be crucial. Figure 12 presents the simulation results—the distribution of the NPV, defined as highly probable for the range of historical CO_2 allowances price in the period of 2021 (January) to 2024 (August). Based on these data, the average price of CO_2 emission allowances is 70 Euro/Mg, with 90% of all prices falling within the range of 34–91 Euro/Mg CO_2. Based on this distribution, the average value of the differential NPV equals −350 million Euro. A total of 90% of all realizations of the expected results are in the range of −3.083 to 1.184 million Euro. Therefore, with the difference in CO_2 emission allowance prices of around 20 Euro/Mg (other explanatory variables are unchanged), the NPV value is lower by almost 500 million Euro, with a distance for CO_2 injection of 100 km.

An analysis of economic efficiency was also carried out for the power plant (power units) capacity of around 200 MW (3 times decrease). The NPV, with investment outlays of 253 million Euro for the entire CCS system, including pipelines, construction of the cavern, storage, and monitoring infrastructure, also with the same CO_2 allowances prices distribution, dropped significantly, amounting to −1.322 million Euro (Figure 13).

Does underground CO_2 sequestration offer a lifeline for coal technologies, Polish energy security, and the energy sector as a whole? The evidence suggests that it does. However, more comprehensive and detailed analyses are necessary in this area. These analyses should include the expansion of criteria related to energy transmission capacity, specifically addressing the timing and scale of future transmission needs. While this criterion may initially appear marginal due to the repowering of existing generation sites, it becomes significant in the context of evolving energy flow dynamics within the grid. This is particularly relevant given the potential for onshore and offshore wind turbines and power plants to be located closer to Poland's northern regions.

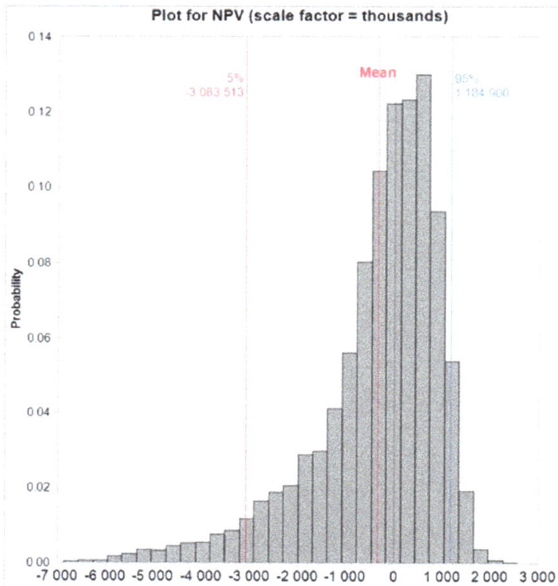

NPV distribution statistics, euro	
Location	
Mean	−350,321
Minimum	−6,879,804
Maximum	2,270,863
Spread	
St. dev.	1,327,991
Variance	2 × 10^12
Risk ratio	−58.27
CofV	−3.79
Shape	
Skewness	−1.44
Kurtosis	5.52
Percentiles	
5%	−3,083,514
95%	1,184,901

Figure 12. The NPV distribution and its basic statistics. Source: own study.

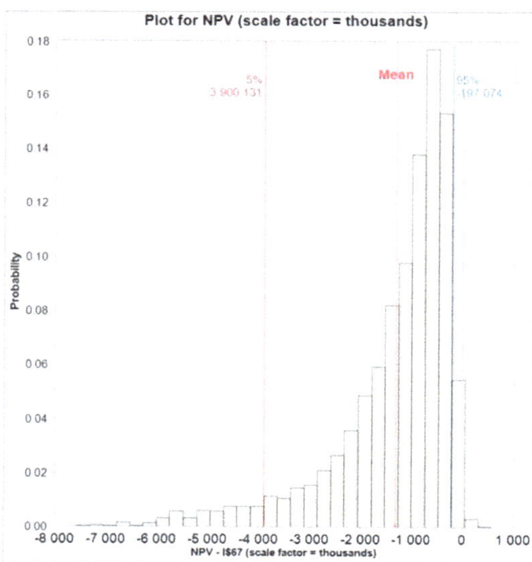

NPV distribution statistics, euro	
Location	
Mean	−1,322,642
Minimum	−7,573,873
Maximum	314,890
Spread	
St. dev.	1,175,550
Variance	2 × 10^12
Risk ratio	−2.74
CofV	−0.89
Shape	
Skewness	−1.81
Kurtosis	6.56
Percentiles	
5%	−3,900,131
95%	−3,217,083

Figure 13. NPV distribution for the energy unit with nominal power at 200 MW el. Source: own study.

The findings indicate that results are highly sensitive to variations in key decision variables. Consequently, effective planning for the development of a CO_2 pipeline network necessitates the establishment of a dedicated network operator and a cost-sharing mechanism among multiple stakeholders through the provision of relevant services.

Future research should prioritize a detailed evaluation of capital investments, operating costs, and potential savings associated with emission reductions. Additionally, incorporating a comparative analysis of alternative technologies will provide a broader understanding of how CCS performs in terms of cost, efficiency, and sustainability. Developing criteria for selecting units for modernization and CCS deployment will be crucial for optimizing implementation strategies. Further exploration of the impacts of environmental changes, along with associated investments and measures in other energy sectors, will enhance the robustness and comprehensiveness of the study.

Some limitations of the study include the following:

- The long-term environmental and geological uncertainties associated with underground CO_2 sequestration.
- The influence of CO_2 pricing and political factors on the adoption and viability of CCS technologies.

5. Conclusions

Poland is facing an energy transformation. In connection with this, the possibilities of a new methodical approach to maintenance (sequence of shutdowns vs. repowering, modernizations), including coal assets (mines) to maintain the generating capacity of coal-fired power plants, were analyzed. In this respect, the analysis covered an example power unit with a capacity of 600 MW el. located close to the mine, for which key technical, energy, and financial values of the appropriate CCS unit were determined. The analysis assumed that waste carbon dioxide could be stored in salt formations (caverns), the potential location of which, as a function of distance from the power plant, was the subject of detailed analyses. The studies assumed that the entire project would be financed by a single entity without sharing the transport and storage infrastructure, which was associated with charging the full cost of CO_2 capture, transport, and storage in the rock mass.

The analysis showed that at the assumed distribution of CO_2 emission allowances price, and assuming the production scale and emissions, transport and storage costs, and necessary investment outlays (453 million Euro), the scenario with CO_2 capture and sequestration is profitable. The resulting NPV amounted to 147 million Euro, with an annual updated value of differential cash flow of 38 million Euro. The implementation of this strategy may constitute an important perspective on stable energy generation for coal-fired power plants operating in Poland. Before their final liquidation, which may take place after 2049 (the expected date of the end of coal mining in Poland), this scenario should be seriously considered, especially keeping in mind that the sharing of transport and storage infrastructure will improve the individual efficiency of CO_2 storage processes for each party. The publication also shows that the limit of effective carbon dioxide storage is a distance of 170 km from the power plant (for the adopted assumptions). It has also been shown that a decrease in the price of CO_2 emission allowances, i.e., a key decision parameter, by almost 20 Euro/Mg CO_2 causes a decrease in the NPV by almost 500 million Euro. This is a significant value. Therefore, selecting the right model for future CO_2 prices is crucial in planning the strategy for CCS with storage in salt caverns.

Author Contributions: Conceptualization, M.K. and P.O.; Methodology, M.K., D.M. and P.O.; Validation, P.O.; Investigation, M.K.; Data curation, D.M.; Writing—original draft, M.K. and D.M.; Writing—review & editing, D.M.; Project administration, P.O.; Funding acquisition, M.K. All authors have read and agreed to the published version of the manuscript.

Funding: This research received no external funding.

Data Availability Statement: The original contributions presented in the study are included in the article, further inquiries can be directed to the corresponding author.

Conflicts of Interest: The authors declare no conflicts of interest.

Nomenclature

CAD	Canadian dollar
CCS	Carbon capture and storage
CF	Cash flow
CO2f	CO_2 emission factor, kg CO_2
DAM	Day-Ahead Market
dP	Length of the pipeline transporting CO_2, kg
ECCO2	Emission costs of CO_2
Ep	Unit cost of electricity, euro/MWh
EPr	Energy production, MWh/y
CFPP	Capacity factor of the power plant, h
CO2f	Emission CO_2 factor, Mg/MWh
CPP	Power unit, MW
CTD	Coal transport distance, km
IISD	International Institute for Sustainable Development
LHV	Calorific value, MJ/kg
MCO2	Mass of CO_2 produced, Mg
MCO2C	Mass of CO_2 capture, Mg
Mp_CO2	Market prices of CO_2 emission allowances, euro/Mg
nel	Efficiency of the coal power plant
NPV	Net present value, Euro
P	Productivity of the power unit with and without a CCS infrastructure, h/year
TCC	Cost of coal transport
TCCO2	Cost of CO_2 transport
TGE	Polish energy market, Towarowa Giełda Energii (in Polish)
UCCO2	Utilization costs of CO_2 storage and monitoring
uCCT	Unit cost of coal transport, euro/Mg
uCC_CO2	Unit cost of carbon dioxide capture, euro/Mg CO_2
uCS_CO2	Unit costs of CO_2 storage and monitoring, euro/Mg CO_2
uCT_CO2	Unit cost of CO_2 transport, euro/Mg CO_2

References

1. Kulpa, J.; Olczak, P.; Stecuła, K.; Sołtysik, M. The Impact of RES Development in Poland on the Change of the Energy Generation Profile and Reduction of CO_2 Emissions. *Appl. Sci.* **2022**, *12*, 11064. [CrossRef]
2. Sobczyk, W.; Sobczyk, E.J. Varying the Energy Mix in the EU-28 and in Poland as a Step towards Sustainable Development. *Energies* **2021**, *14*, 1502. [CrossRef]
3. Szczerbowski, R.; Kornobis, D. The Proposal of an Energy Mix in the Context of Changes in Poland's Energy Policy. *Polityka Energetyczna* **2019**, *22*, 5–18. [CrossRef]
4. Wojtaszek, H.; Miciuła, I.; Modrzejewska, D.; Stecyk, A.; Sikora, M.; Wójcik-Czerniawska, A.; Smolarek, M.; Kowalczyk, A.; Chojnacka, M. Energy Policy until 2050—Comparative Analysis between Poland and Germany. *Energies* **2024**, *17*, 421. [CrossRef]
5. Wierzbowski, M.; Filipiak, I.; Lyzwa, W. Polish Energy Policy 2050—An Instrument to Develop a Diversified and Sustainable Electricity Generation Mix in Coal-Based Energy System. *Renew. Sustain. Energy Rev.* **2017**, *74*, 51–70. [CrossRef]
6. Olczak, P.; Matuszewska, D. Energy Storage Potential Needed at the National Grid Scale (Poland) in Order to Stabilize Daily Electricity Production from Fossil Fuels and Nuclear Power. *Energies* **2023**, *16*, 6054. [CrossRef]
7. Komorowska, A.; Olczak, P. Economic Viability of Li-Ion Batteries Based on the Price Arbitrage in the European Day-Ahead Markets. *Energy* **2024**, *290*, 130009. [CrossRef]
8. Brauers, H.; Oei, P.Y. The Political Economy of Coal in Poland: Drivers and Barriers for a Shift Away from Fossil Fuels. *Energy Policy* **2020**, *144*, 111621. [CrossRef]
9. Pepłowska, M.; Olczak, P. Review of Research on the Impact of Changes Resulting from the Hard Coal Mining Sector in Poland on the GDP Value. *Energies* **2024**, *17*, 1477. [CrossRef]
10. Olkuski, T.; Zubień, P. The Energy Security of Finland after Joining NATO. *Polityka Energetyczna* **2024**, *27*, 49–66. [CrossRef]
11. Nyga-Łukaszewska, H.; Aruga, K.; Stala-Szlugaj, K. Energy Security of Poland and Coal Supply: Price Analysis. *Sustainability* **2020**, *12*, 2541. [CrossRef]
12. Lorenc, S.; Leśniak, T.; Kustra, A.; Sierpińska, M. Evolution of Business Models of Mining and Energy Sector Companies According to Current Market Trends. *Energies* **2023**, *16*, 5212. [CrossRef]
13. Pluta, M.; Wyrwa, A.; Zyśk, J.; Suwała, W.; Raczyński, M. Scenario Analysis of the Development of the Polish Power System towards Achieving Climate Neutrality in 2050. *Energies* **2023**, *16*, 5918. [CrossRef]

14. Rusin, A.; Wojaczek, A. Changes in the Structure of the Polish Energy Mix in the Transition Period to Ensure the Safety and Reliability of Energy Supplies. *Energy* **2023**, *282*, 128831. [CrossRef]
15. Zdonek, I.; Tokarski, S.; Mularczyk, A.; Turek, M. Evaluation of the Program Subsidizing Prosumer Photovoltaic Sources in Poland. *Energies* **2022**, *15*, 846. [CrossRef]
16. Quader, M.A.; Ahmed, S.; Ghazilla, R.A.R.; Ahmed, S.; Dahari, M. A Comprehensive Review on Energy Efficient CO_2 Breakthrough Technologies for Sustainable Green Iron and Steel Manufacturing. *Renew. Sustain. Energy Rev.* **2015**, *50*, 594–614. [CrossRef]
17. European Environmental Agency. EU Emissions Trading System (ETS) Data Viewer. Available online: https://www.eea.europa.eu/en/analysis/maps-and-charts/emissions-trading-viewer-1-dashboards (accessed on 10 September 2024).
18. Kaszyński, P.; Komorowska, A.; Kamiński, J. Regional Distribution of Hard Coal Consumption in the Power Sector under Selected Forecasts of EUA Prices. *Gospod. Surowcami Miner.—Miner. Resour. Manag.* **2019**, *35*, 113–134. [CrossRef]
19. Al-Qadri, A.A.; Ahmed, U.; Jameel, A.G.A.; Ahmad, N.; Zahid, U.; Zein, S.H.; Naqvi, S.R. Process Design and Techno-Economic Analysis of Dual Hydrogen and Methanol Production from Plastics Using Energy Integrated System. *Int. J. Hydrogen Energy* **2023**, *48*, 10797–10811. [CrossRef]
20. Ahmed, U.; Hussain, M.A.; Bilal, M.; Zeb, H.; Zahid, U.; Onaizi, S.A.; Jameel, A.G.A. Utilization of Low-Rank Coals for Producing Syngas to Meet the Future Energy Needs: Technical and Economic Analysis. *Sustainability* **2021**, *13*, 10724. [CrossRef]
21. Ahmed, U.; Hussain, M.A.; Bilal, M.; Zeb, H.; Ahmad, N.; Ahmad, N.; Usman, M. Production of Hydrogen from Low Rank Coal Using Process Integration Framework between Syngas Production Processes: Techno-Economic Analysis. *Chem. Eng. Process.-Process Intensif.* **2021**, *169*, 108639. [CrossRef]
22. Roux, C.; Schalbart, P.; Peuportier, B. Accounting for Temporal Variation of Electricity Production and Consumption in the LCA of an Energy-Efficient House. *J. Clean. Prod.* **2016**, *113*, 532–540. [CrossRef]
23. Markowska, K.; Sękala, A.; Stecuła, K.; Kawka, T.; Sirovitskiy, K.; Pankova, O.; Vnukova, N.; Shulyak, M.; Kharchenko, S.; Shchur, T.; et al. Comparison of the Sustainability and Economic Efficiency of an Electric Car and an Aircraft—A Case Study. *Sustainability* **2023**, *15*, 1238. [CrossRef]
24. Bui, M.; Adjiman, C.S.; Bardow, A.; Anthony, E.J.; Boston, A.; Brown, S.; Fennell, P.S.; Fuss, S.; Galindo, A.; Hackett, L.A.; et al. Carbon Capture and Storage (CCS): The Way Forward. *Energy Environ. Sci.* **2018**, *11*, 1062–1176. [CrossRef]
25. Wang, Z.; Xue, W.; Li, K.; Tang, Z.; Liu, Y.; Zhang, F.; Cao, S.; Peng, X.; Wu, E.Q.; Zhou, H. Dynamic Combustion Optimization of a Pulverized Coal Boiler Considering the Wall Temperature Constraints: A Deep Reinforcement Learning-Based Framework. *Appl. Therm. Eng.* **2025**, *259*, 124923. [CrossRef]
26. Kaszyński, P.; Kamiński, J. Coal Demand and Environmental Regulations: A Case Study of the Polish Power Sector. *Energies* **2020**, *13*, 1521. [CrossRef]
27. Kamiński, J. Domestic Hard Coal Supplies to the Energy Sector: The Impact of Global Coal Prices. *Gospod. Surowcami Miner.—Miner. Resour. Manag.* **2019**, *35*, 141–164. [CrossRef]
28. Bartela, Ł.; Gładysz, P.; Ochmann, J.; Qvist, S.; Sancho, L.M. Repowering a Coal Power Unit with Small Modular Reactors and Thermal Energy Storage. *Energies* **2022**, *15*, 5830. [CrossRef]
29. Kaliski, M.; Gawlik, L.; Szurlej, A.; Sikora, A.P.; Kamiński, J. Hard Coal in the Fuel-Mix of Poland: The Long-Term Perspective. *Arch. Min. Sci.* **2016**, *61*, 335–350. [CrossRef]
30. Hasanbeigi, A.; Arens, M.; Price, L. Alternative Emerging Ironmaking Technologies for Energy-Efficiency and Carbon Dioxide Emissions Reduction: A Technical Review. *Renew. Sustain. Energy Rev.* **2014**, *33*, 645–658. [CrossRef]
31. Chwoła, T.; Spietz, T.; Więcław-Solny, L.; Tatarczuk, A.; Krótki, A.; Dobras, S.; Wilk, A.; Tchórz, J.; Stec, M.; Zdeb, J. Pilot Plant Initial Results for the Methanation Process Using CO_2 from Amine Scrubbing at the Łaziska Power Plant in Poland. *Fuel* **2019**, *263*, 116804. [CrossRef]
32. Więcław-Solny, L.; Tatarczuk, A.; Krótki, A.; Stec, M. The Technological Research Progress of Amine-Based CO_2. *Polityka Energetyczna—Energy Policy J.* **2013**, *16*, 229–241.
33. PSE. Praca KSE. *Generacja Mocy Jednostek Wytwórczych.* Available online: https://raporty.pse.pl/index.html?report=GEN-JW&state=Funkcjonowanie%20KSE,Raporty%20dobowe%20z%20funkcjonowania%20KSE&date=2024-12-12&type=table&search=& (accessed on 12 December 2024).
34. PSE. Dane Systemowe. Available online: https://www.pse.pl/dane-systemowe (accessed on 18 October 2024).
35. PSE. *Plan Rozwoju w Zakresie Zaspokojenia Obecnego i Przyszłego Zapotrzebowania Na Energię Elektryczną Na Lata 2023–2032*; PSE: Konstancin-Jeziorna, Poland, 2022.
36. wnp.pl. Gas Prices. Available online: http://energy.instrat.pl/gas_prices (accessed on 5 April 2021).
37. Figueiredo, N.C.; Silva, P.P. da The "Merit-Order Effect" of Wind and Solar Power: Volatility and Determinants. *Renew. Sustain. Energy Rev.* **2019**, *102*, 54–62. [CrossRef]
38. Quandl EUA Certificate Prices. Available online: http://energy.instrat.pl/co2_prices (accessed on 20 June 2021).
39. Sokołowski, J.; Frankowski, J.; Mazurkiewicz, J.; Lewandowski, P. Hard Coal Phase-out and the Labour Market Transition Pathways: The Case of Poland. *Environ. Innov. Soc. Transit.* **2022**, *43*, 80–98. [CrossRef]
40. Stala-Szlugaj, K. Polish Energy Policy in the Context of the Global Coal Demand Prospects. *Polityka Energetyczna–Energy Policy J.* **2024**, *27*, 211–226. [CrossRef]
41. TGE. SA Polish Day-Ahead Market. Available online: https://tge.pl/ (accessed on 18 October 2024).

42. IPCC; IEA. Estimated Cost Ranges for Carbon Capture and Storage (CSS) and Carbon Dioxide Removal (CDR) Solutions as of 2023, by Approach or Technology. Available online: https://www.statista.com/statistics/1304575/global-carbon-capture-cost-by-technology/ (accessed on 10 August 2024).
43. Cameron, L.; Carter, A.; Sievert, K. *Why the Cost of Carbon Capture and Storage Remains Persistently High*; International Institute for Sustainable Development: Winnipeg, MB, Canada, 2023.
44. Kang, J.-N.; Wei, Y.-M.; Liu, L.; Han, R.; Chen, H.; Li, J.; Wang, J.-W.; Yu, B.-Y. The Prospects of Carbon Capture and Storage in China's Power Sector under the 2 °C Target: A Component-Based Learning Curve Approach. *Int. J. Greenh. Gas Control* **2020**, *101*, 103149. [CrossRef]
45. Uliasz-Misiak, B.; Tarkowski, R. Costs of CO_2 Geological Storage and an Exemplary Analysis for a given Emitter. *Zesz. Nauk. IGSMiE PAN Wydaw. IGSMiE PAN* **2009**, *75*, 21–34.
46. PIG. *Efekt Rzeczowy z Realizacji Przedsięwzięcia z Zakrsu Państwowej Służby Geologicznej—Zadanie „KAPSCO2: Monitorowanie Statusu Projektów CCS (Etap III)*; PIG: Warsaw, Poland, 2022.
47. Kwaśniewski, K.; Kopacz, M.; Grzesiak, P.; Kapłan, R.; Sobczyk, E.J. Coal Gasification—Conditions, Effectiveness and Perspectives for Development. Wydawnictwa AGH: Kraków, Poland, 2015.
48. Szczerbowski, R. The Forecast of Polish Power Production Sector Development by 2050—Coal Scenario. *Polityka Energetyczna—Energy Policy J.* **2016**, *19*, 5–18.
49. Sun, W.Q.; Cai, J.J.; Mao, H.J.; Guan, D.J. Change in Carbon Dioxide (CO_2) Emissions From Energy Use in China's Iron and Steel Industry. *J. Iron Steel Res. Int.* **2011**, *18*, 31–36. [CrossRef]
50. IBISWorld. Recent Trends—Eua Prices. Available online: https://www.ibisworld.com/uk/bed/eua-prices/44112/ (accessed on 6 June 2021).

energies

MDPI

Article

2-Stage Design of E-Moped-Sharing Service for Accessibility, Greenhouse Gas Emissions, and Cost Through Station and Supplier Selections

Seigo Takahashi [1], Yuki Kinoshita [2], Nora Schelte [3], Semih Severengiz [3] and Tetsuo Yamada [1,*]

1 Department of Informatics, The University of Electro-Communications, 1-5-1 Chofugaoka, Chofu 182-8585, Tokyo, Japan; t2330068@edu.cc.uec.ac.jp
2 Department of Informatics, Faculty of Engineering, Kindai University, 1 Takaya Umenobe, Higashi-Hiroshima 739-2116, Hiroshima, Japan; kinoshita@hiro.kindai.ac.jp
3 Sustainable Technologies Laboratory, Department of Electrical Engineering and Computer Sciences, Bochum University of Applied Sciences, Am Hochschulcampus 1, 44801 Bochum, Germany; nora.schelte@hs-bochum.de (N.S.); semih.severengiz@hs-bochum.de (S.S.)
* Correspondence: tyamada@uec.ac.jp; Tel.: +81-(42)-443-5269

Abstract: In recent years, there has been a call for a shift to transportation with lower greenhouse gas (GHG) emissions in order to combat global warming. One of the ecofriendly transportation methods is an electric moped scooter (e-moped)-sharing service that does not emit GHG when it runs. It is necessary to plan the location of charging stations and the material procurement through the manufacturing of e-mopeds in order to reduce the cost and GHG emissions and to improve the accessibility of the service. In this study, a two-stage design on the e-moped-sharing services is proposed to allocate charging stations and select material suppliers for e-mopeds using integer programming. The analysis method to determine the suitable charging station locations and sizes and supplier selection are also presented. Numerical experiments are conducted to illustrate the proposed design and analysis method by assuming Kumpan's 1954 i model installation in Bochum city, Germany. In the numerical experiments, set covering and maximal covering location problems with small coverage radius of charging stations would be better by evaluating accessibility, GHG emissions, and cost comprehensively. Moreover, 11 prioritized demand points were picked out by introducing new indexes such as geographical and demand importance.

Keywords: carbon neutrality; product life cycle; e-moped-sharing service; life cycle assessment (LCA); set covering location problem; maximal covering location problem; integer programming; supplier selection

1. Introduction

To achieve carbon neutrality by 2050 in the EU [1], the EU plans to ban the sale of gasoline and diesel vehicles [2]. The International Energy Agency (IEA) reported that road transportation generated about 15% of the world's total CO_2 emissions [3]. According to NIKKEI, Honda Motor Co., Ltd., which is one of the largest moped manufactures in the world, will cease the production of gasoline mopeds with an engine displacement of 50 cc or less in May 2025 [4].

As a green transportation method, an electric moped scooter (e-moped) with little GHG emissions in use, unlike the gasoline moped, is spreading for short-distance travel [5]. By charging their battery from solar photovoltaic (PV) charging stations, the users can ride their e-moped with little GHG emissions in the usage phase. Moreover, e-moped-sharing services can also save GHG emissions for e-moped production because the volume of e-moped production can be saved by sharing users and potentially also substituting car rides. Thus, the e-moped-sharing service will become an attractive transportation to prevent global warming. UNU by emco reported that the number of scooters used for sharing globally increased by 58% from 66,000 in 2019 to 104,000 in 2020, and that the number of registered users also increased from 5 million to nearly 9 million over the same period despite COVID-19 [6]. There were more than 220 cities where moped sharing was installed in 2022 [7]. Then, 55% of e-mopeds were deployed in the Europe [7].

In addition to GHG reduction through the production and usage phases in e-mopeds, the e-moped-sharing services are useful as an alternative to short-distance transportation, and they help reduce traffic jam and pollution caused by automobiles [8]. For the e-moped-sharing service, accessibility from the demand point to the charging station is a very important factor to obtain users as well as cover demand [9]. As evaluation indexes of accessibility, the average and maximal distance from a demand point to a charging station can be adopted. Moreover, the available number of e-mopeds, which depends on the number and size of charging stations, is also an important accessibility index. If the number of e-mopeds per user is much lower, it leads to lost sales by losing opportunity of riding e-mopeds. In contrast, having many more e-mopeds per user is costly for e-mopeds sharing company. Therefore, locations and sizes of charging stations comprehensively should be determined based on accessibility, the number of e-mopeds, and cost. These types of decisions are called facility location problems [10].

The facility location problem is a mathematical model for finding a desirable location for a facility within a given space [10]. It is used in a wide range of fields such as applied mathematics, operations research, regional science, economic geography, and urban engineering [10]. The problem enables this study to estimate the accessibility, such as the average total distance and max distance. The set covering location problem and maximal covering location problem are major approaches for facility location problems. The set covering location problem minimizes the number of facilities to cover all user demands [11]. Meanwhile, the maximal covering location problem maximizes the cover demand based on the number of facilities and its coverage radius set in advance [11]. That is, the cover radius is static and set in advance in both approaches. The set covering location problem determines the locations and number of facilities to cover all demand, whereas the maximal covering location problem determines the locations of facilities to maximize demand under static number of facilities.

Actually, the e-moped-sharing service can reduce GHG emissions by installing PV charging stations and alleviating traffic jams, but GHG emissions during e-mopeds production cannot be avoided. In particular, material-based GHG emissions can account for a larger percentage against the production process, from material to distribution to users [12]. For example, SHARP reported that GHG emissions in the material manufacturing phase of the 4K LCD TV 2022 model accounted for 23% through its life cycle [12]. Similarly, 28% of GHG emissions come from manufacturing in the base case of the e-moped-sharing service [5]. Additionally, more than 71% of capital costs is from the purchase of the e-moped [13]. To achieve carbon neutrality in transportation, GHG emissions whole life cycle of e-moped needs to be considered. Both material-based GHG emissions and procurement cost differ among supplied countries due to different energy mix and economic situations. Material-based GHG emissions and procurement cost would be saved simultaneously by supplier selection [14].

To save GHG emissions and procurement costs simultaneously, supplier selections can be conducted with two objective functions: minimizing GHG emissions and procurement cost. The suppliers are selected by evaluating both objective functions simultaneously to satisfy the required material demand for e-moped production. Therefore, to install e-moped-sharing services in an environmentally friendly and economical way, two types of decisions, namely charging station location and material supplier selection, are required based on accessibility, GHG emissions, and cost. According to Liu et al. [15], optimizing several competing decision criteria is called a joint optimization problem. The introduction of joint optimization problems in the e-moped-sharing service may provide suggestions on how charging stations should be allocated and how suppliers for e-mopeds' materials are chosen from a different approach of accessibility, GHG emissions, and cost. Its problem will consider the facility location problem in the 1st stage and the supplier selections in the 2nd stage, respectively. However, it is appropriate to apply a two-stage design instead of the joint optimization problem. This is because a two-stage design is not only easier to solve than the joint optimization problem, but also because it is difficult to guarantee that the optimal solution obtained by the joint optimization problem is a better choice in terms of accessibility, GHG emissions, and cost.

Here, research questions (RQs) about the e-moped-sharing service are proposed with as follows:

RQ1. Which charging station scenarios about coverage radius and GHG reduction target should be adopted for the e-moped-sharing service?

RQ2. How should charging stations and material supplier selections be combined for accessibility, GHG emissions, and cost?

RQ3. Which demand points should be prioritized to allocate charging stations through set covering and maximal covering location problems?

RQ4. Which suppliers should be selected for each material?

This study proposes a two-stage design of the e-moped-sharing service for accessibility, GHG emissions, and cost using integer programming. The 1st stage determines charging station locations and numbers based on demand and accessibility using set covering and maximal covering location problems. Because the required number of e-mopeds depends on the number of charging stations, the 1st stage also decides the number of e-mopeds. The 2nd stage determines material suppliers in different countries with different GHG emissions and procurement costs to balance the GHG emissions and cost. To determine the suitable charging station locations and material suppliers through the proposed 1st and 2nd stages, the analysis method is also presented.

The rest of this paper is composed as follows. Section 2 shows the literature review in terms of accessibility, GHG emissions, and cost. Section 3 represents the two-stage design of e-moped-sharing service using two facility location problems and supplier selection. Input data, assumptions of numerical experiments, and scenarios are explained, and results of the numerical experiments are shared in Section 4. Section 5 analyzes the selected locations of charging stations and material suppliers. Section 6 concludes the paper and suggest future studies.

2. Literature Review

Table 1 lists and summarizes related papers in terms of accessibility, GHG emissions, cost, and targets. The ✓ in Table 1 denotes the paper addressing the points. Csiszár et al. [16] developed a two-level charging station-locating method. This method evaluated the potential of electric vehicle use on a macro-level and the p locations of charging stations on a micro-level focused on the land-use using weighted multicriteria methods. This method was applied to a macro assessment of Hungary and a micro assessment of its capital, Budapest. The results showed that charging stations in the park and ride, with their centralized service and high-density areas, provided the city's public charging supply better than gas stations. García et al. [17] investigated the usage and opinions of an e-moped-sharing service based on a web-based survey conducted in Spain. Kruskal–Wallis tests were conducted to find out the district of the urban population that was more likely to use the sharing service, and additional statistical mean differences in specific variables about the service were carried out. They found that age, occupation, income, and environmental awareness appeared to be the main reasons for expanding the service in the future. He et al. [9] developed a location-allocation model which considered the supply-and-demand estimation of EV charging, the government policy, and the spatial constraints in Hong Kong. The goal of their model was to minimize both the penalty for the lack of EV charging demand and the travel time to charging facilities, weighted for demand. Their study showed that the majority of new charging stations should be allocated to new towns and new development areas. Additionally, it is desirable to extend existing charging stations by adding more chargers compared to constructing new charging stations. Polo et al. [18] merged accessibility and location allocation models in geographic information systems as a proposed strategy to improve the spatial planning of public health services. To estimate the spatial accessibility, the two-step floating catchment area (2SFCA) model was conducted based on the Dijkstra's algorithm and the vectorial analysis. Spatial accessibility improvement was estimated by the maximal coverage and the p-median location problems. It was found that the service replacement proposed by the maximal coverage location problem maximized its spatial accessibility more advantageously. Schelte et al. [19] analyzed not only energy supply means such as solar charging stations and battery swapping stations, but also user acceptance and behavior regarding e-moped- and stand-up scooter-sharing services. An online questionnaire was conducted based on the Unified Theory of Acceptance and Use of Technology to identify important factors influencing user acceptance. They clarified that 95% of respondents answered that the sustainability of sharing services was an important factor, and it affected the personal attitude of users. They discussed the user accessibility of the service depending on charging stations. However, these previous papers did not consider the installation cost, especially that of charging stations. It is essential for their owner to reduce its cost and to earn the station usage fee.

Vazifeh et al. [20] proposed a new methodology to conduct the data-driven optimization of EV charging station locations. They solved a discrete optimization problem on a geographical grid to cover the entire demand area and minimize the drivers' total driving distance to charging stations, the related energy cost, and the number of charging stations based on the genetic algorithm. It was found that the algorithm enabled them to reduce the drivers' total driving distance to charging stations, the related energy cost, and the number of charging stations compared to both a locally optimized solution and the current charging station placement in the Boston metro area. However, they did not include the cost of charging facilities and the materials of e-mopeds. Bai et al. [21] proposed a cell-based model which could decide locations, capacity options, and service types of EV charging stations such that all potential charging demand was covered. The bi-objective charging station location problem was conducted to minimize the total cost and to maximize service quality using a hybrid evolutionary algorithm that combined the non-dominated sorting genetic algorithm-II (NSGA-II) with linear programming and neighborhood search. They showed that the hybrid NSGA-II could obtain more Pareto solutions in a much shorter time for large-scale instances with good quality. They discussed the user accessibility and the cost of the service depending on charging stations. However, they did not consider the GHG emissions, especially those of e-mopeds. If charging stations are installed, the environmental analysis should also be evaluated whole the service.

Table 1. Literature review on facility location problems and LCA.

Viewpoint	Accessibility			GHG Emissions			Cost		Research Target
Evaluation Indices	Demand	Distance	Time	Procurement	Usage	Renewable Energy	Station	Vehicle	
Csiszár et al. [16]	✓								charging stations
García et al. [17]	✓								moped-style scooter sharing
He et al. [9]	✓		✓						EV charging facilities
Polo et al. [18]	✓	✓							public health services
Schelte et al. [19]	✓	✓	✓						user acceptance
Vazifeh et al. [20]	✓	✓							public charging stations
Bai et al. [21]			✓				✓		EV charging stations
Chen and Kockelman [22]				✓	✓				car sharing
Schelte et al. [5]				✓	✓	✓			e-moped-sharing service
Schelte et al. [23]				✓	✓	✓			off-grid charging station
Baptista et al. [24]					✓	✓		✓	car sharing
Wortmann et al. [13]					✓	✓		✓	e-moped-sharing system
Takahashi et al. [25]				✓		✓	✓	✓	e-moped-sharing service
Wong et al. [26]							✓		smart charging programs for EVs
Carrese et al. [27]	✓							✓	reposition of e-scooters
Saum et al. [28]	✓	✓						✓	dockless shared e-scooters
This study	✓	✓		✓		✓	✓	✓	charging stations and e-moped-sharing service

Chen and Kockelman [22] investigated the life cycle inventory impacts on energy usage and GHG emissions based on candidate travelers using carsharing in US case. Three different scenarios were built to conduct the sensitivity analysis about the reduction in total life cycle energy and GHG emissions for a candidate. Their results showed that the average individual transportation energy use and GHG emissions could be reduced by approximately 51% if current carsharing members joined a carsharing organization. Schelte et al. [5] compared the e-moped-sharing to alternative transportation in terms of environmental impact and how it could be more reduced. They used the LCA data to estimate how the sharing affected the impact category global warming potential and developed five different usage scenarios of the service as the case study in a German city. Their results showed that the sharing had almost as much environmental impact on global warming potential per person and kilometer as the public transportation to move a lot of people if the e-mopeds have long lifetimes and efficient operation logistics are developed. Schelte et al. [23] indicated the potentials of smart charging stations for urban micromobility. Four different case studies were planned to analyze the usage and environmental influence of the solar charging station using life cycle assessments. It was found that the global warming potential per kilowatt-hour of energy supply would decrease by 73–88% compared to battery swapping with diesel vans if solar charging stations were utilized. They discussed the GHG emissions through the service depending on e-mopeds and renewable energy. However, they did not consider the installation cost, especially that of e-mopeds. It is essential for company to reduce its cost and to expand the service.

Baptista et al. [24] claimed that car sharing contributes to more efficient and rational mobility. The case study was conducted in Lisbon, Portugal, to estimate the reduction of energy and environmental impacts through car sharing. They found that those benefits would be reductions of 35 or 47% in terms of energy consumption and 35 and 65% for CO_2 emissions in case of a shift to hybrid vehicles or to electric vehicles, respectively. Wortmann et al. [13] surveyed the possibility of an e-moped-sharing service if it replaced passenger travels by car and estimated the economic and environmental effects of the service. At first, they modeled three different e-moped scenarios for changing their numbers in Berlin, based on the multi-agent transport simulation framework. Then, the total cost for the service and a life cycle assessment were conducted. Their results showed that the service could replace a large part of all passenger car travel in Berlin. Takahashi et al. [25] evaluated the e-moped-sharing service in terms of GHG emissions and procurement costs during material production, as well as equipment costs for charging stations. They assumed two patterns about the number of e-mopeds; then, supplier selection was carried out to balance costs and GHG emissions for the both patterns. They discussed the GHG emissions and the cost of the service depending on charging stations and e-mopeds. However, they did not consider the user accessibility. A sharing service which ignores user accessibility make it difficult to maintain.

Wong et al. [26] conducted a survey in the USA to estimate willingness for the adoption of the smart charging program. It was determined that both financial and non-financial incentives were effective to participate in its program based on the survey. An incentive of USD 300–400 per year would guarantee that the majority of EV owners/leasers or buyers/leasers interested in EVs would certainly or probably participate in the program considering only the financial incentive. Carrese et al. [27] addressed the problem of optimally managing agents' behavior specially hired by sharing companies to reposition electric scooters and guarantee the urban landscape. Repositioning is primarily intended to ensure a balanced distribution of vehicles within the service area to better meet demand and increase overall profits. The optimization model solution based on mathematical theory was shown to provide a high-value solution for commercial software. Saum et al. [28]

suggested a new data-driven rebalancing framework of dockless shared e-scooters with Monte Carlo sampling to forecast demand. The framework was evaluated on real data from Minneapolis, Minnesota. It was found that overall demand and variance uncertainties were mitigated, and the driving distance and the rebalancing cost were smaller than those of the baseline case. In the above works, they did not address charging stations and supplier selection simultaneously. This study evaluates accessibility, GHG emissions, and cost in the e-moped sharing service by conducting charging station allocation and material supplier selection.

3. Methods

3.1. Overview of the Proposed 2-Stage Design of the E-Moped-Sharing Service

Figure 1 shows the overview of the proposed 2-stage design of e-moped-sharing service for accessibility, GHG emissions, and cost. The yellow and red highlight the decision makings and evaluation indexes in the proposed design method. The 1st stage determined the locations and number of charging stations based on the demand and a covered radius set in advance. The required number of e-mopeds was also determined because it depends on the number of charging stations. Conducting the 1st stage, accessibility such as average and maximal distances from demand points to the charging stations were calculated. The equipment cost of charging stations was also determined at the 1st stage.

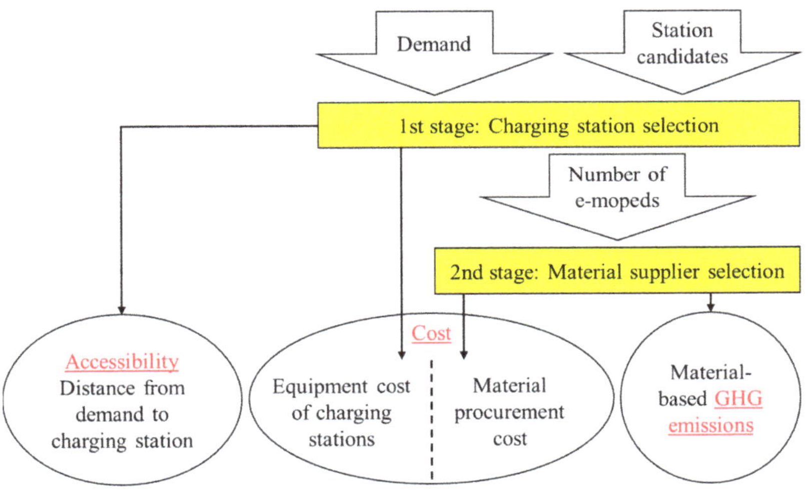

Figure 1. The overview of the proposed 2-stage design of an e-moped-sharing service for accessibility, GHG emissions, and cost.

The 2nd stage conducted the material supplier selection to satisfy material demands for the required number of e-mopeds decided at the 1st stage. Procurement costs and material-based GHG emissions were determined. The sum of equipment cost of charging stations and procurement cost was regarded as the total cost.

Finally, the results of charging station allocation and supplier selection were evaluated from the viewpoints of accessibility, GHG emissions, and cost.

Rigorous pseudocodes detailing the 2-stage approaches are as shown in Appendix A.

3.2. Formulation of the 1st Stage: Charging Station Allocation

The parameters, decision variables, and evaluation variables used in the set covering location problem and maximal covering location problem are as follows:

I, J	Set of demand/charging station candidate points indexed by i, j
A_{ij}	1 if the demand point i can be covered by candidate j, otherwise 0
W_i	User demand at demand point i
$C_{station}$	Cost to install a charging station
P	Installed number of charging stations in maximal covering location problem
x_j	1 if the charging station is installed at demand point j, otherwise 0
z_i	1 if the demand point i is covered, otherwise 0
$N_{station}$	Number of charging stations
CD	Cover demands
EC	Equipment cost of charging stations

3.2.1. Set Covering Location Problem

The set covering location problem aims to minimize the number of facilities as all demand points are covered within a certain distance set as a parameter [11]. Feige [29] pointed out that the greedy algorithm and a linear programming relaxation were investigated as methods for solving the set covering location problem. Equation (1) is the objective function to minimize the number of charging stations $N_{station}$. Equation (2) indicates the equipment cost of charging stations. Equation (3) ensures that all demand points must be covered.

$$N_{station} = \sum_{j \in J} x_j \to min \tag{1}$$

$$EC = C_{station} N_{station} \tag{2}$$

$$\sum_{j \in J} A_{ij} x_j \geq 1 \; \forall i \in I \tag{3}$$

3.2.2. Maximal Covering Location Problem

The maximal covering location problem aims to maximize the cover demand. In contrast to the set covering location problem, the number of stations is static and given as a parameter [11]. Equation (4) is the objective function to maximize the cover demands CD. Equation (5) expresses the equipment cost of charging stations, as well as the set covering location problem. Equation (6) ensures that the installed number of charging stations is equal to P. Equation (7) represents that at least 1 charging stations, which can cover demand point i, must be located if demand point i is covered.

$$CD = \sum_{i \in I} W_i z_i \to max \tag{4}$$

$$EC = C_{station} P \tag{5}$$

$$\sum_{j \in J} x_j = P \tag{6}$$

$$\sum_{j \in J} A_{ij} x_j \geq z_i \; \forall i \in I \tag{7}$$

3.3. *Formulation of the 2nd Stage: Supplier Selection*

The parameters, decision variable, and evaluation variables used in supplier selection are as follows:

K	Set of materials k
L	Set of suppliers l
PC_{lk}	Procurement unit cost of material k in supplier l

E_{lk}	GHG emissions of material k in supplier l
N_k	Number of materials k required to produce one unit
$N_{product}$	Demand
E_{Max}	GHG emissions in the baseline
ε	Target reduction ratio of GHG emissions
f_{lk}	Amount of material k procured from supplier l
MC	Material cost
E	Total GHG emissions

Similar to Takahashi et al. [25], the supplier selection has a bi-objective function: minimizing material procurement cost and minimizing GHG emissions, as shown in Equations (8) and (9). The bi-objective supplier selection is solved using ε-constraint method [14,30]. Equation (10) represents the ε-constraint [30] for the total GHG emissions E. To obtain Pareto-optimal solutions, ε is gradually changed for each GHG-targeted reduction ratio.

Equation (11) shows that the sum of units' amount of material k procured from supplier f_{lk} is equal to the multiplied amount of material k required to produce one unit N_k and the demand quantity $N_{product}$. Demand $N_{product}$ depends on the number of charging stations determined at 1st stage. Thus, the results of charging station allocation at 1st stage affect the input parameter of supplier selection at 2nd stage. Equation (12) is a non-negative constraint.

$$MC = \sum_{k \in K} \sum_{l \in L} PC_{lk} f_{lk} \rightarrow min \tag{8}$$

$$E = \sum_{k \in K} \sum_{l \in L} E_{lk} f_{lk} \tag{9}$$

$$E \leq (1 - \varepsilon) E_{Max} \tag{10}$$

$$\sum_{l \in L} f_{lk} = N_k N_{product} \; \forall k \in K \tag{11}$$

$$f_{lk} \geq 0 \; \forall l \in L, \forall k \in K \tag{12}$$

4. Numerical Experiments and Scenarios

4.1. Input Data and Assumptions of Numerical Experiments and Scenarios

Bochum city in Germany was used as a design example of the two-stage design of an e-moped-sharing service. It was assumed that there were three different sizes of a charging station determining operational e-mopeds per one charging station as shown in Table 2. The first stage was conducted based on the following six scenarios.

Table 2. The 6 scenarios of the 1st stage.

	Small	Middle	Large
Set covering location problem	S1)	S2)	S3)
Maximal covering location problem	M1)	M2)	M3)

Treated facility location problems, namely the set covering and maximal covering location problems, require setting a cover radius in advance. Moreover, the maximal covering location problem also requires setting the number of charging stations. Scenarios small, middle, and large indicate the size of each charging station. The large charging station assumed that more e-mopeds could be installed and that cover radius was wilder, with more costs than small and middle ones.

Regarding the set covering location problems, only the cover radius is required to be set as a parameter to determine locations and number of charging stations. The cover radius was set 2, 4, and 6 (km) in the small, medium, and large stations, respectively. In the case of the maximal covering location problem, the cover radius and the number of charging stations are required to be set as parameters. In this numerical experiment, the cover radius was static, but the number of stations was changed among scenarios. As well as the set covering location problem, as the large station could install more e-mopeds, the number of charging stations was lower than the small and middle station scenarios. The difference between the set covering location problem and the maximal covering location problem in this paper is that the number of charging stations in the set covering location problem is unknown until it is solved, while the number of ones in the maximal covering location problem is static. Table 3 shows the assumptions and input data for each scenario.

Table 3. The assumptions and input data for each scenario.

Scenario	S1)	S2)	S3)	M1)	M2)	M3)
Operational e-mopeds per 1 charging station	169.5	367.0	565.0	169.4	345.1	564.8
Cost to install a charging station $C_{station}$ ($\times 10^4$ USD)	5	12	18	5	11	18
Coverage radius (km)	2	4	6	2	2	2
Installed number of charging stations P	-	-	-	12	9	6

4.2. Assumptions of Supplier Selection

Kumpan's 1954 i model [5] was used to illustrate a design example. The 2nd stage was conducted based on three GHG reduction targets: 26%, 50%, and 70% reductions. As a result, 18 results were obtained through the 1st and 2nd stages. China, Japan, and Korea were assumed as material suppliers. GHG emissions and procurement costs were different among the three countries as shown in Table 4. The GHG emissions were estimated using the life cycle inventory (LCI) database with the Asian International Input-Output Table [31]. The procurement cost was estimated based on a census of manufacture and price level of each country using a decision support tool [32]. All 21 materials in e-mopeds could be supplied from the three countries with the same quality.

Table 4. The BOM of e-mopeds.

No.	Material Name	Weights (g)	Procurement Cost (USD)			GHG Emissions (g-CO_2eq)		
			China	Japan	Korea	China	Japan	Korea
1	Acrylonitrile	10,100	6	14	10	62,658	14,075	21,420
2	Aluminum alloy	18,500	17	40	27	120,333	18,441	23,338
3	Malleable iron pipe fittings (including flange type)	23	0.07	0.16	0.11	637	119	214
4	Other nonferrous metals (from primary smelting and refining)	4	0.01	0.02	0.01	56	9	11
5	Copper wire	938	2	4	3	11,356	1740	2202
6	Copper wire	212	1	3	2	7700	1180	1493
7	Copper wire	41	0.35	0.82	0.56	2482	380	481
8	Plain plate glass	90	2	4	2	11,748	2221	3081

Table 4. *Cont.*

No.	Material Name	Weights (g)	Procurement Cost (USD) China	Japan	Korea	GHG Emissions (g-CO_2eq) China	Japan	Korea
9	Light bulbs	22	7	15	10	30,303	5108	8563
10	Synthetic Fiber Sewing Thread	269	2	5	3	6448	1796	2593
11	Lithium-ion battery	6440	7906	18,386	12,503	19,725,072	5,707,149	8,301,380
12	Polycarbonate	210	0.26	0.61	0.42	1867	471	789
13	Polyethylene	865	0.53	1	0.84	3773	951	1595
14	Methacrylic resin	146	0.17	0.38	0.26	1171	295	495
15	Polypropylene	12,300	7	17	12	51,855	13,072	21,920
16	Fluoropolymer	319	3	8	5	23,413	5902	9897
17	Printed circuit board	4290	2	5	4	6361	2059	2145
18	Synthetic rubber (Synthetic latex included)	8950	9	22	15	97,581	21,919	33,358
19	Cast-iron pipe	1760	2	5	3	18,081	3385	6077
20	Tin free steel	4080	2	5	3	18,162	3400	6104
21	Zinc alloy	15	0.01	0.02	0.01	65	10	13

4.3. Results of Charging Station Allocations

1. Comparison of Accessibility

Table 5 represents the results of the accessibility for each scenario obtained at the 1st stage. Operational e-mopeds were calculated based on the number of charging stations $N_{station}/P$. Cover rate CR represents the ratio of covered potential demands by installed charging stations against all potential demands in Bochum city. The e-mopeds per user were calculated by dividing the value of operational e-mopeds by the potential demands. Average distance AD and maximal distance MD were defined as the average distance and maximal distance between demand points and charging stations. From the perspective of accessibility, it is desirable that e-mopeds per user is greater and approaching one (unit/user). The number of e-mopeds per user becoming lower than one leads a risk of losing users as users cannot ride e-mopeds although they go to the charging stations. In the case of the set covering location problem, as all demands were satisfied, e-mopeds per user depends on only the number of operational e-mopeds. E-mopeds per user decreased as the number of stations and e-mopeds decreased in the set covering location problem. Then, the average distance AD and maximal distance MD increased in the set covering location problem because its cover area expanded.

In the case of the maximal covering location problem, users were changed among scenarios. In contrast to the set covering location problem, the number of stations decreased, and the number of e-mopeds increased in charging station scenarios from M1) and M2) to M3). Moreover, the covered users also decreased. There were trade-offs observed between operational e-mopeds, CR, e-mopeds per user, and AD. For example, operational e-mopeds and e-mopeds per user increased, and CR and AD decreased as charging stations decreased. This change means that users would have more opportunities to use e-mopeds, but more people would not benefit from the service. Additionally, more effort would be required to reach the charging stations. Thus, e-mopeds per user increased. Average distance AD increased in the maximal covering location problem as stations decreased.

The maximal distance MD, increased in the set covering location problem, whereas that in the maximal covering location problem was not changed as charging stations decreased. Additionally, over 75% demands were covered in scenarios M1), M2), and M3). Considering accessibility only, greater e-mopeds per user and shorter AD and MD are desirable. Therefore, it seems that scenario M1) should be adopted owing to the shortest AD and MD.

Table 5. The results of the accessibility for 6 scenarios.

Scenario	S1)	S2)	S3)	M1)	M2)	M3)
$N_{station}/P$	11	4	2	12	9	6
Operational e-mopeds	1864	1468	1130	2033	3106	3389
CR (%)	100	100	100	100	92	76
E-mopeds per user ($\times 10^{-2}$)	0.50	0.39	0.30	0.55	0.90	1.19
AD (km)	0.59	2.04	3.64	0.54	0.64	0.72
MD (km)	1.77	3.89	5.79	1.77	1.77	1.77

2. Comparison of accessibility, GHG emissions, and cost

Figures 2 and 3 represent the 18 results through the 1st and 2nd stages for the maximal distance against total costs per user and GHG emissions per user, respectively. As supplier selection in the 2nd stage was conducted using three GHG reduction ratios for each scenario, each one had three marks in Figures 2 and 3, respectively. Scenario S1) had a lower total cost TC and GHG emissions E than scenarios M1), M2), and M3), as shown in Figures 2 and 3. Moreover, scenario S1) achieved a larger reduction of maximal distance compared to scenarios S2) and S3), with a small increase of the total cost per user and GHG emissions per user, as shown in Figures 2 and 3. The set covering location problem requires the additional installation cost of charging stations to cover the demand points with few users because all demand must be covered. In this study, the extra cost could be reduced by preparing smaller charging station scenarios.

Comparing the three results in scenario S1), the 70% GHG reduction target would be better than that of the 26% and 50% reductions because the total cost TC in the 70% reduction was 4.0% and 7.1%, which was only greater than that of 50% and 26% reductions.

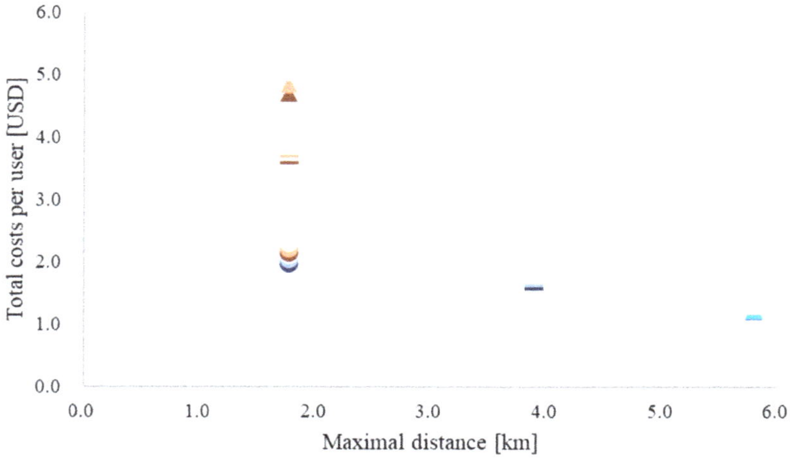

Figure 2. Comparison of accessibility and cost through the 1st and 2nd stages.

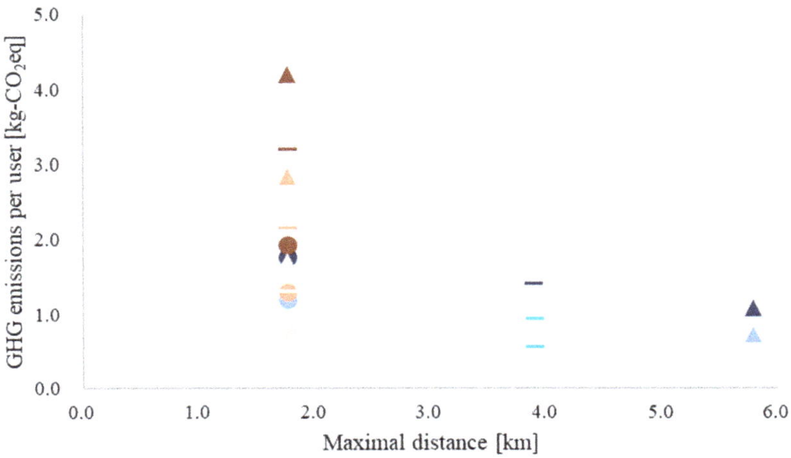

Figure 3. Comparison of accessibility and GHG emissions through the 1st and 2nd stages.

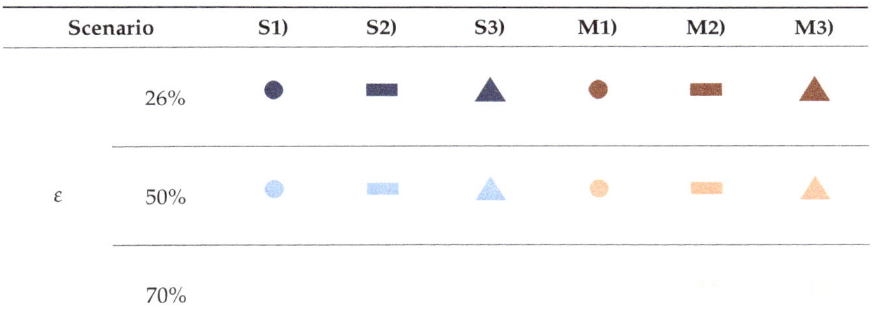

5. Analysis of Charging Stations and Supplier Selections

This section analyzes the selected locations of charging stations and the selected suppliers in detail.

5.1. Analysis of Charging Stations in the First Stage

Figures 4–6 show the results of set covering location problem. Red and blue points represent the demand/candidate for the charging stations. Red points are the demand points, where the charging stations were allocated. Blue points are the demand points, where the charging stations were NOT allocated. Grey circles represent the coverage radius.

In the case of scenario S1), most charging stations covered one demand point. Only one charging station was allocated in the center of Bochum city in spite of the high potential demand, and this may have caused a lost opportunity. Although the number of charging stations was no less than 12 units, there were non-covered areas, where areas were not colored by gray circles due to the small coverage radius. In the case of scenario S2), 4 charging stations were allocated in Bochum city, one in the east, west, north, and south areas, respectively. The western and northern charging stations covered six and seven demand points, whereas the southern and eastern ones covered two and four points, respectively. This was a little unbalanced in terms of where the charging stations were allocated. In the case of scenario S3), two charging stations were allocated by dividing the Bochum city into west and east areas. Covered users on the west side of Bochum city were at risk for lost opportunities because there were many demand points on the west side of Bochum city.

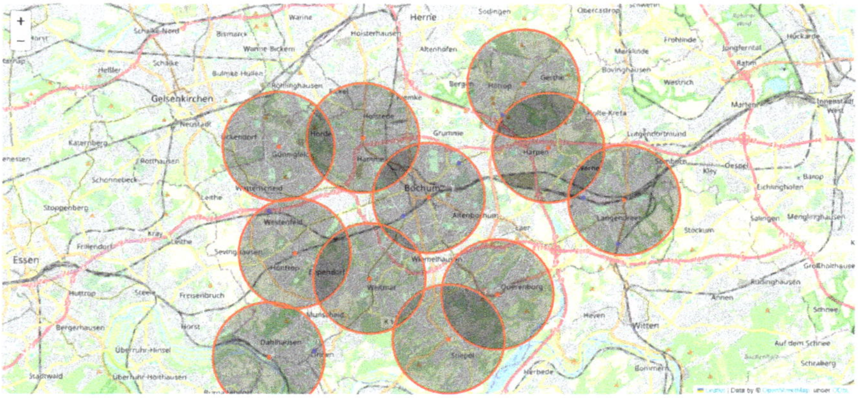

Figure 4. Selected and covered demand in the 1st stage for scenario S1), map source: OpenStreetMap® [33].

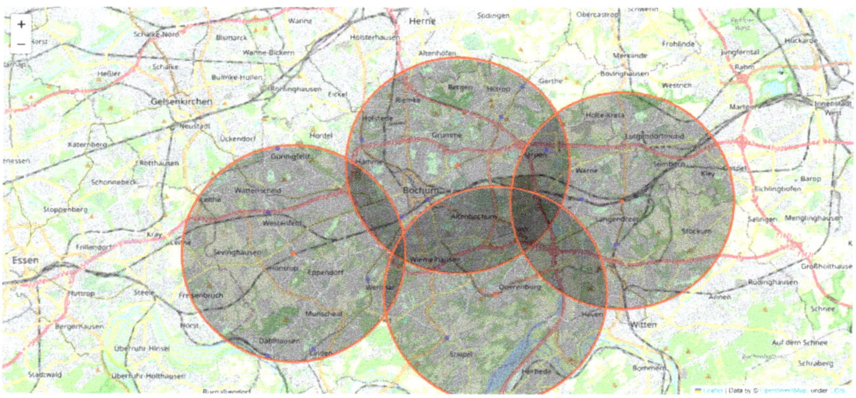

Figure 5. Selected and covered demand in the 1st stage for scenario S2).

Figure 6. Selected and covered demand in the 1st stage for scenario S3).

Figures 7–9 show the allocation results of the charging station for the maximal covering location problem, where the number of them is a given value for each scenario. It is noted that the maximal covering location problem did not cover all demand points. In the case of scenario M1), almost all of Bochum city was covered by the covering area of the charging stations. Moreover, two charging stations were allocated in the central area in Bochum city, even though only one charging station could cover the demand points in the central area as shown in Figure 7. As shown in Figure 7, although the central area had the greatest potential demand, lost opportunity could be avoided by allocating two charging stations in the central area. In the case of scenario M2), 11 out of 12 demand points were covered by nine charging stations as shown in Figure 8. There was an area outside the coverage area in the southeastern region of the city. However, one non-covered demand point was relatively close to the allocated charging station in the southern area. In the case of scenario M3), charging stations were prioritized in the western area of the city, where there were many demand points. There were many non-covered demand points in the eastern area as shown in Figure 9.

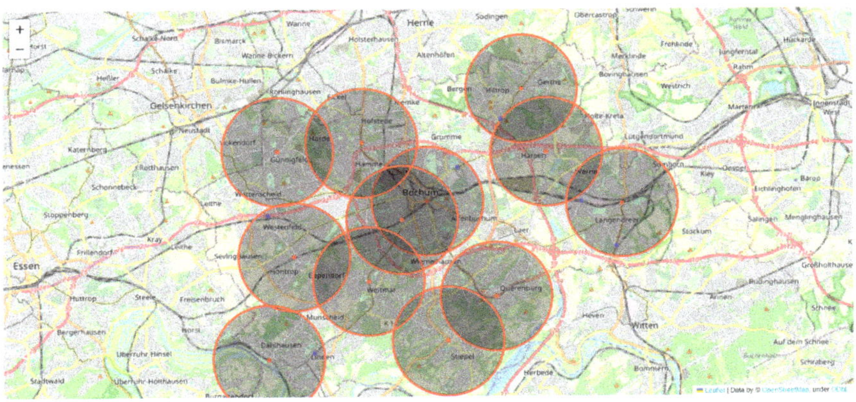

Figure 7. Selected and covered demand in the 1st stage for scenario M1).

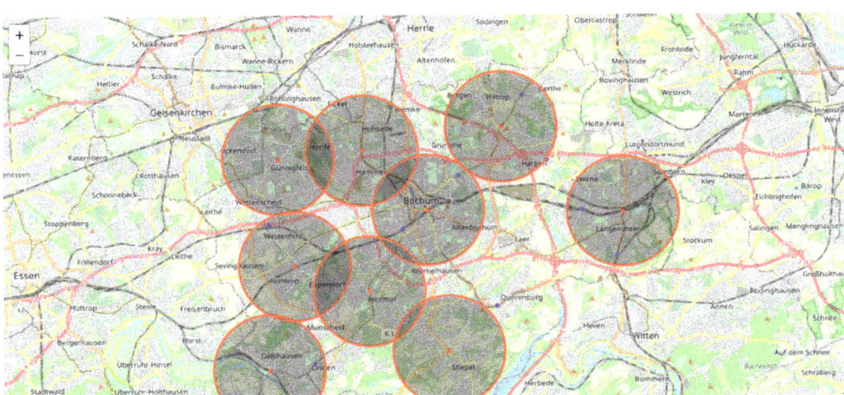

Figure 8. Selected and covered demand in the 1st stage for scenario M2).

Figure 9. Selected and covered demand in the 1st stage for scenario M3).

Table 6 shows allocated charging stations for all scenarios. The ✓ represent the allocated charging stations. To pick out where demand points should be prioritized through the set covering and maximal covering location problems, two new indexes, namely geographical and demand importance, were introduced. Potential demand was assumed to be one-third of the population each district has [34].

Table 6. Allocated charging stations for each charging station scenarios.

City District	Demand Point/ Charging Station Candidate	Potential Demand	Scenario						Geographical Importance	Demand Importance
			S1)	S2)	S3)	M1)	M2)	M3)		
Bochum-Mitte	Bochum Hbf.	34,858	✓			✓	✓	✓		x
	Vonovia Ruhrstadion	34,858		✓					x	
	Bochum Breslauer Str.	34,858	✓			✓	✓	✓		x
Bochum-Wattenscheid	Wattenscheid	24,446								
	Wattenscheid-Höntrop	24,446	✓	✓		✓	✓	✓	x	x
	Bochum Ulrichstr.	24,446	✓			✓	✓	✓		x
Bochum-Nord	Heinrichstraße	11,837	✓			✓				
	Bochum Ruhr Park	11,837	✓			✓				
	Bochum Nordbad	11,837					✓			x
Bochum-Ost	Bochum Langendreer	17,776	✓	✓	✓	✓	✓	✓	x	x
	Bochum-Langendreer West	17,776								
	Auf dem Jäger–Bochum	17,776								
Bochum-Süd	Bochum Ruhr-Universität	17,006	✓	✓		✓			x	
	Bochum-Ehrenfeld	17,006				✓				
	Bochum Haarstr.	17,006	✓			✓	✓			x
Bochum-Südwest	Bochum Südbad	18,362								
	Bochum Dahlhausen	18,362	✓			✓	✓	✓		x
	Bochum Weitmar Mitte	18,362	✓		✓	✓	✓		x	x

The set covering location problem minimized the number of charging stations while satisfying all demand points. Thus, this allocation approach could prioritize the demand point in terms of the geographical viewpoint. Additionally, the selected demand points in scenarios S2) and S3) were considered as geographical importance points as shown in Table 6. On the other hand, as the demand was different, to obtain many users within the limited number of charging stations, the demand importance was introduced. The maximal covering location problem maximized the cover demand under a static number of charging stations. Then, the selected demand points in scenarios M2) and M3) were considered as demand importance points as shown in Table 6. Here, 11 demand points were picked out as geographical or demand importance points as shown in Table 6. The x in Table 6 denotes locations of geographical and demand importances. Comparing six scenarios, scenarios S1), M1), and M2) were better than the other scenarios as all of them allocated 9 out of 11 geographical and demand importance charging stations. The results in Section 4.1 show that scenario M1) was more appropriate than scenario M2) considering only accessibility. Therefore, next Section 5.2 will analyze the results of supplier selection for scenarios S1) and M1).

5.2. Analysis of Supplier Selections in the Second Stage

Figures 10 and 11 show the breakdown of total cost TC and GHG emissions E for scenarios S1) and M1) with three GHG reduction targets. Section 4.3 stated that the 70% GHG reduction target should be adopted because the decrease of GHG emissions E was sharper than the increase of the total cost TC as the GHG reduction target increased. Figure 10 reveals that the equipment cost of charging stations EC, namely the fixed cost, accounted for more than 75% of the total cost TC.

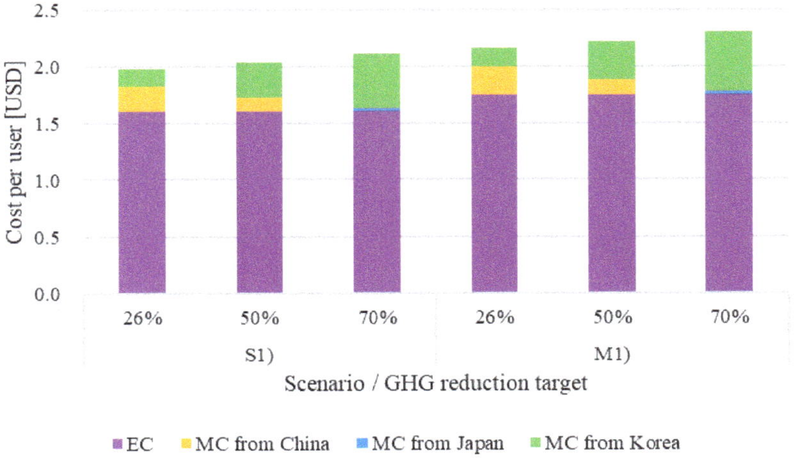

Figure 10. Breakdown of total cost TC for scenarios S1) and M1) with 3 target GHG reduction ratios.

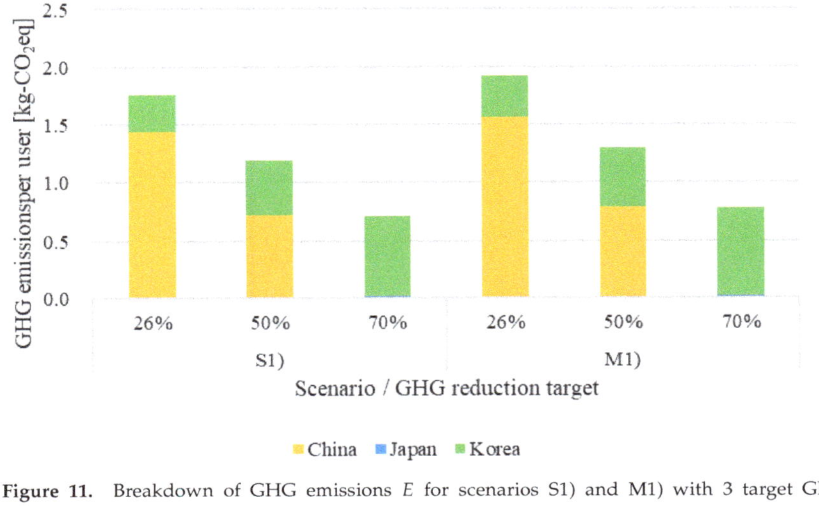

Figure 11. Breakdown of GHG emissions E for scenarios S1) and M1) with 3 target GHG reduction ratios.

In addition, material cost *MC* and GHG emissions *E* generated by supplier selections originated from China and Korea for the 26% and 50% GHG reduction targets. To achieve the 70% GHG reduction target, material suppliers were switched from China to Korea or Japan as shown in Figures 10 and 11. As *MC* from Japan were little more than those from Korea at the 70% GHG reduction target, only little materials were procured. Perhaps, all materials could be supplied from Korea so as to save management efforts about procurement. Moreover, the supplier selections of scenarios S1) and M1) were almost the same behaviors. Therefore, it would be desirable to procure all materials from Korea to save *MC*, *E*, and management efforts for both scenarios S1) and M1).

5.3. Discussion

This section discusses research questions (RQs) in chapter 1 and their answers. The answers found from numerical experiments through 1st and 2nd stages are as follows:

RQ1. Which charging station scenarios about coverage radius and GHG reduction target should be adopted for the e-moped-sharing service?

The set covering and maximal covering location problems were conducted with different coverage radius depending on the charging stations. To compare six scenarios in terms of accessibility, *AD*, *MD* and e-mopeds per user were evaluated. As shown in Table 2, in the numerical experiments for the e-moped-sharing service in Bochum city, scenario M1), the maximal covering location problem with small charging stations should be adopted because the scenario had the lowest *AD* and *MD*. *AD* in scenario M1) became 0.54 (km) so that users could easily go to the charging station. Moreover, the scenario could cover all potential demands.

RQ2. How charging stations and material supplier selections should be combined for accessibility, GHG emissions, and cost?

The 1st stage obtained six results of charging station allocations by setting six scenarios of the coverage radius. The 2nd stage conducted material supplier selections with three different GHG reduction targets for each result obtained in the 1st stage. Then, 18 results through the 1st and 2nd stages were obtained. Total GHG emissions per user and total cost per user were compared as shown in Figures 2 and 3. From Figures 2 and 3, scenario S1) with a 70% GHG reduction target, was the most desirable for accessibility, GHG emissions, and cost. This is because scenario S1) had generally lower *MD*, GHG emissions, and cost per user as shown in Figures 2 and 3. Moreover, the effects of decreasing GHG emissions with the 70% GHG reduction target were greater than those of increasing the total cost compared to the 26% and 50% GHG reduction targets.

RQ3. Which demand points should be prioritized to allocate charging stations through the set covering and maximal covering location problems?

To pick out prioritized demand points for geographical and demand satisfactory, geographical and demand importance were introduced as new indexes so that 11 demand points were prioritized as shown in Table 6. These 11 points, such as Bochum Hbf., Wattenscheid-Höntrop, and Bochum Langendreer, should be prioritized to set charging stations.

RQ4. Which suppliers should be selected for each material?

In the numerical experiments, Korean suppliers should be selected for all materials. By analyzing accessibility, GHG emissions, and cost through the 1st and 2nd stages, the 70% GHG reduction target would be better than the 26% and 50% ones. Although Japanese suppliers were also selected for little materials as shown in Figures 10 and 11, all materials would be procured from Korean suppliers to save management efforts for contracting suppliers in different countries.

6. Conclusions

This study proposed a two-stage design of an e-moped-sharing service for accessibility, GHG emissions, and cost using integer programming. Moreover, analysis methods to determine the suitable demand points for charging stations and suppliers were presented and discussed to evaluate the e-moped-sharing service in the whole product life cycle. To illustrate a design example, installing the e-moped-sharing service in Bochum city was addressed, and Chinese, Japanese, and Korean suppliers were set to balance GHG emissions and procurement cost. One of the key findings is that in scenario M1), the maximal covering location problem with small charging stations should be adopted because the scenario had the lowest average and maximal distance.

The proposed two-stage design was designed for the e-moped-sharing service. The practical implications of this model are as follows:

- One of the advantages is that the proposed method enables us to integrate and treat separated charging station and supplier selection problems together and to evaluate them in terms of accessibility, GHG emissions, and cost. The model can be also adopted to other sharing services such as car sharing and bike sharing. As long as geographic and population data are available, this two-stage planning is applicable in any country. To adjust adopted services, the coverage radius needs to be modified. Moreover, impacts of supplier selections in the 2nd stage will be dependent of adopted service features. For example, in the case of car sharing, the material supplier selections will be more important as cars require many materials. On the other hand, in a case of bike sharing, the importance of supplier selections will be decreased.
- The proposed model also contributes to urban planning. Charging stations should be allocated to crowded areas such as train stations and shopping malls. Users of the e-moped-sharing service will move from a charging station to other stations. Thus, the charging station allocation could partially control flow of people so that traffic jam also could be decreased.

However, there are some limitations as follows:

- All suppliers can provide all materials with the same quantity. Rare metals such as lithium and titanium are limited of their suppliers so that switching suppliers can be difficult. Therefore, additional constrains to limit supplier selections should be added to reflect these situations.
- Only the same size of charging stations can be installed in 1st stage. Based on demand and land cost, it is desirable that their size should be adjusted. Based on the results in the 1st stage, a decisionmaker would be required to change the allocated station sizes manually.
- Updating and maintenance processes were not considered. Both charging stations and e-mopeds are required to be updated and conducted the maintenance. The number of them will affect the GHG emissions and cost of these processes by replacing charging stations and e-mopeds.

Future studies should consider the manufacturing cost of e-mopeds and the land cost of charging stations. GHG emissions of the e-moped manufacturing and charging stations should also be added and evaluated. Furthermore, the facility location problem should be developed to install the different sizes of charging stations simultaneously so as to adjust their size based on demands of each point. In addition, the perspective of the real local traffic situation is lacking and should be investigated. If traffic jams are a regular problem near installed charging stations, this will have a negative impact on user accessibility. Also, tariff costs are not taken into account when procuring materials for e-mopeds. Free trade agreements (FTA) such as the EU and CPTPP will have a significant impact on where they are procured. Finally, the joint optimization problem should be applied to the e-moped-sharing service.

Author Contributions: S.T., Y.K. and T.Y. conceptualized the goals and aims of this study and provided resources. S.T. and T.Y. designed the methodology. T.Y. acquired funds. S.T. and N.S. generated metadata and formulated and operated the numerical experiments. Additionally, S.T. programmed and validated the formulation, visualized the results, and wrote the original draft. Y.K., S.S. and T.Y. managed this project and supervised the overall content and reviewed this paper. Finally, Y.K., N.S. and S.S. revised the paper. All authors have read and agreed to the published version of the manuscript.

Funding: This study was partially supported by the Japan Society for the Promotion of Science (JSPS), KAKENHI, Grant-in-Aid for Scientific Research (B), JP23K26329, from 2023 to 2025.

Data Availability Statement: The most of original contributions presented in the study are included in the article.

Acknowledgments: The authors would like to thank Thomas Spelten in the Bochum University of Applied Sciences for his fruitful comments.

Conflicts of Interest: The authors declare no conflicts of interest.

Appendix A. Rigorous Pseudocodes for the Proposed 2-Stage Design

Algorithm A1 1-A Set covering location problem

```
Input: i, j (demand point/charging station candidate), d_ij (distance from i to j);
Output: x_j (whether the charging station is allocated at j);
let x_j denote the best solution found
R ← 2//coverage radius
while R ≤ 6
    initialize square matrix A_ij with size i × j
    if d_ij ≤ R
            A_ij = 1
    else
            A_ij = 0
    end if
    run the exact search based on Equations (1) and (3) for x_j
    R ← R + 2
end while
return x_j
```

Algorithm A2 1-B Maximal covering location problem

```
Input: i, j (demand point/charging station candidate), d_ij (distance from i to j), W_i
(user demands at i);
Output: CD (cover demands);
let x_j denote the best solution found
P = 6//the number of charging stations
while P ≤ 12
    initialize square matrix A_ij with size i × j
    if d_ij ≤ 2
            A_ij = 1
    else
            A_ij = 0
    end if
    run the exact search based on Equations (4), (6) and (7) for x_j
    P ← P + 3
end while
return CD
```

Algorithm A3 2 Material supplier selection

Input: PC_{lk} (Procurement unit cost of material k in supplier l), E_{lk} (GHG emissions of material k in supplier l), N_k (Number of material k required to produce one unit), $N_{product}$ (Demand), E_{Max} (GHG emissions in the baseline);
Output: MC (material cost), E (total GHG emissions);
1: let f_{lk} denote the best solution found
2: for $\varepsilon = 0.26, 0.5, 0.7$
3: run the exact search based on Equations (8)–(12) for f_{lk}
4: end for
5: return MC, E

References

1. Going Climate-Neutral by 2050—A Strategic Long-Term Vision for a Prosperous, Modern, Competitive and Climate-Neutral EU Economy. Available online: https://op.europa.eu/en/publication-detail/-/publication/92f6d5bc-76bc-11e9-9f05-01aa75ed71a1/language-en (accessed on 21 November 2024).
2. Regulation (EU) 2023/851 of the European Parliament and of the Council of 19 April 2023 Amending Regulation (EU) 2019/631 as Regards Strengthening the CO_2 Emission Performance Standards for New Passenger Cars and New Light Commercial Vehicles in Line with the Union's Increased Climate Ambition (Text with EEA Relevance). Available online: https://eur-lex.europa.eu/eli/reg/2023/851/oj (accessed on 21 November 2024).
3. Cars, Planes, Trains: Where Do CO_2 Emissions from Transport Come from? Available online: https://ourworldindata.org/co2-emissions-from-transport (accessed on 3 February 2023).
4. Honda to End Production of 50cc Mopeds in May 2025 due to Stricter Regulations. Available online: https://www.nikkei.com/article/DGXZQOUC101QS0Q4A610C2000000/?msockid=2b44e3358b9c6a850ac3edc88a766bb9 (accessed on 16 October 2024). (In Japanese).
5. Schelte, N.; Severengiz, S.; Schünemann, J.; Finke, S.; Bauer, O.; Metzen, M. Life cycle assessment on electric moped scooter sharing. *Sustainability* **2021**, *13*, 8297. [CrossRef]
6. UNU Launches Global Moped Sharing Market Report 2020. Available online: https://unumotors.com/en-de/unu-launches-global-moped-sharing-market-report-2020/ (accessed on 28 November 2024).
7. Global Moped Sharing Market Report 2022. Available online: https://go.invers.com/en/resources/global-moped-sharing-market-report-2022?utm_campaign=GMSMR%202022&utm_medium=mopedsharing.com&utm_source=referral (accessed on 21 November 2024).
8. Wang, Y.J.; Kuo, Y.H.; Huang, G.Q.; Gu, W.; Hu, Y. Dynamic demand-driven bike station clustering. *Transp. Res. Part E Logist. Transp. Rev.* **2022**, *160*, 102656.
9. He, S.Y.; Kuo, Y.H.; Sun, K.K. The spatial planning of public electric vehicle charging infrastructure in a high-density city using a contextualised location-allocation model. *Transp. Res. Part A Policy Pract.* **2022**, *160*, 21–44.
10. Owen, S.H.; Daskin, M.S. Strategic facility location: A review. *Eur. J. Oper. Res.* **1998**, *111*, 423–447. [CrossRef]
11. Daskin, M.S. What you should know about location modeling. *Nav. Res. Logist.* **2008**, *55*, 283–294.
12. Sustainability Report 2023. Available online: https://global.sharp/corporate/eco/report/pdf/ssr2023_e.pdf?v=2023 (accessed on 16 October 2024).
13. Wortmann, C.; Syré, A.M.; Grahle, A.; Göhlich, D. Analysis of electric moped scooter sharing in Berlin: A technical, economic and environmental perspective. *World Electr. Veh. J.* **2021**, *12*, 96. [CrossRef]
14. Kondo, R.; Kinoshita, Y.; Yamada, T. Green Procurement Decisions with Carbon Leakage by Global Suppliers and Order Quantities under Different Carbon Tax. *Sustainability* **2019**, *11*, 3710. [CrossRef]
15. Liu, Y.; Ji, Y.; Jiao, R.J. A Stackelberg Solution to Joint Optimization Problems: A Case Study of Green Design. *Procedia Comput. Sci.* **2013**, *16*, 333–342. [CrossRef]
16. Csiszár, C.; Csonka, B.; Földes, D.; Wirth, E.; Lovas, T. Urban public charging station locating method for electric vehicles based on land use approach. *J. Transp. Geogr.* **2019**, *74*, 173–180.
17. García, Á.A.; Gomez, J.; Sobrino, N.; Díaz, J.J.V. Moped scooter sharing: Citizens' perceptions, users' behavior, and implications for urban mobility. *Sustainability* **2021**, *13*, 6886. [CrossRef]
18. Polo, G.; Acosta, C.M.; Ferreira, F.; Dias, R.A. Location-allocation and accessibility models for improving the spatial planning of public health services. *PLoS ONE* **2015**, *10*, e0119190. [CrossRef] [PubMed]
19. Schelte, N.; Severengiz, S.; Finke, S.; Stommel, J. Analysis on user acceptance for light electric vehicles and novel charging infrastructure. In Proceedings of the IEEE European Technology and Engineering Management Summit, Bilbao, Spain, 9–11 March 2022.
20. Vazifeh, M.M.; Zhanga, H.; Santia, P.; Ratti, C. Optimizing the deployment of electric vehicle charging stations using pervasive mobility data. *Transp. Res. Part A Policy Pract.* **2019**, *121*, 75–91. [CrossRef]

21. Bai, X.; Chinc, K.S.; Zhou, Z. A bi-objective model for location planning of electric vehicle charging stations with GPS trajectory data. *Comput. Ind. Eng.* **2019**, *128*, 591–604. [CrossRef]
22. Chen, T.D.; Kockelman, K.M. Carsharing's life-cycle impacts on energy use and greenhouse gas emissions. *Transp. Res. Part D Transp. Environ.* **2016**, *47*, 276–284. [CrossRef]
23. Schelte, N.; Straßberger, H.; Severengiz, S.; Finke, S.; Felmingham, B. Environmental impact of off-grid solar charging stations for urban micromobility services. In Proceedings of the IEEE European Technology and Engineering Management Summit, Dortmund, Germany, 18–20 March 2021.
24. Baptista, P.; Meloa, S.; Rolim, C. Energy, environmental and mobility impacts of car-sharing systems. empirical results from Lisbon, Portugal. *Procedia—Soc. Behav. Sci.* **2014**, *111*, 28–37. [CrossRef]
25. Takahashi, S.; Kinoshita, Y.; Schelte, N.; Spelten, T.; Severengiz, S.; Yamada, T. Comparison of different procurement options and influence on greenhouse gas emissions: Case of Bochum city. In Proceedings of the 7th Asian Conference of Management Science and Applications, Okinawa, Japan, 15–17 December 2023.
26. Wong, S.D.; Shaheen, S.A.; Martin, E.; Uyeki, R. Do incentives make a difference? Understanding smart charging program adoption for electric vehicles. *Transp. Res. Part C Emerg. Technol.* **2023**, *151*, 104123. [CrossRef]
27. Carrese, S.; D'Andreagiovanni, F.; Giacchetti, T.; Nardin, A.; Zamberlan, L. A Beautiful Fleet: Optimal Repositioning in E-scooter Sharing Systems for Urban Decorum. *Transp. Res. Procedia* **2021**, *52*, 581–588. [CrossRef]
28. Saum, N.; Sugiura, S.; Piantanakulchai, M. Optimizing Shared E-Scooter Operations Under Demand Uncertainty: A Framework Integrating Machine Learning and Optimization Techniques. *IEEE Access* **2024**, *12*, 26957–26977. [CrossRef]
29. Feige, U. A Threshold of ln *n* for Approximating Set Cover. *J. ACM* **1998**, *45*, 634–652. [CrossRef]
30. Eskandarpour, M.; Dejax, P.; Miemczyk, J.; Péton, O. Sustainable supply chain network design: An optimization-oriented review. *Omega* **2015**, *54*, 11–32. [CrossRef]
31. Horiguchi, K.; Tsujimoto, M.; Yamaguchi, H.; Itsubo, N. Development of greenhouse gases emission intensity in eastern Asia using Asian international input-output table. In Proceedings of the 7th Meeting of the Institute of Life Cycle Assessment, Chiba, Japan, 24–28 January 2011.
32. Umeta, K.; Kondo, R.; Kinoshita, Y.; Yamada, T.; Yamada, S.; Inoue, M. Development of a decision support tool for GHG emissions and procurement costs with material recommendation. *Jpn. Soc. Des. Eng.* **2020**, *55*, 277–294. (In Japanese)
33. OpenStreetMap®. Available online: https://www.openstreetmap.org/copyright/en (accessed on 27 November 2024).
34. Die Wichtigsten Zahlen zur Bochumer Bevölkerung/Bevölkerung Aktuell. Available online: https://www.bochum.de/Referat-fuer-politische-Gremien-Buergerbeteiligung-und-Kommunikation/Statistik/Die-wichtigsten-Zahlen-zur-Bochumer-Bevoelkerung (accessed on 1 February 2024). (In German).

energies

MDPI

Article

Forecasting Electric Vehicle Charging Demand in Smart Cities Using Hybrid Deep Learning of Regional Spatial Behaviours

Muhammed Cavus [1,*], Huseyin Ayan [2,3], Dilum Dissanayake [4,*], Anurag Sharma [5], Sanchari Deb [2] and Margaret Bell [2]

1 Department of Mathematics, Physics and Electrical Engineering, Northumbria University, Newcastle upon Tyne NE1 8SA, UK
2 School of Engineering, Newcastle University, Newcastle upon Tyne NE1 7RU, UK; h.ayan2@newcastle.ac.uk (H.A.); sanchari.deb@newcastle.ac.uk (S.D.); margaret.bell@newcastle.ac.uk (M.B.)
3 School of Engineering, Istanbul University-Cerrahpaşa, Istanbul 34320, Türkiye
4 School of Geography, Earth and Environmental Sciences, University of Birmingham, Birmingham B15 2TT, UK
5 Faculty of Science, Agriculture & Engineering (SAgE), Newcastle University, Newcastle upon Tyne NE1 7RU, UK; anurag.sharma@newcastle.ac.uk
* Correspondence: muhammed.cavus@northumbria.ac.uk (M.C.); d.dissanayake@bham.ac.uk (D.D.)

Abstract: This study presents a novel predictive framework for estimating electric vehicle (EV) charging demand in smart cities, contributing to the advancement of data-driven infrastructure planning through behavioural and spatial data analysis. Motivated by the accelerating regional demand accompanying EV adoption, this work introduces HCB-Net: a hybrid deep learning model that combines Convolutional Neural Networks (CNNs) for spatial feature extraction with Extreme Gradient Boosting (XGBoost) for robust regression. The framework is trained on user-level survey data from two demographically distinct UK regions, the West Midlands and the North East, incorporating user demographics, commute distance, charging frequency, and home/public charging preferences. HCB-Net achieved superior predictive performance, with a Root Mean Squared Error (RMSE) of 0.1490 and an R^2 score of 0.3996. Compared to the best-performing traditional model (Linear Regression, $R^2 = 0.3520$), HCB-Net improved predictive accuracy by 13.5% in terms of R^2, and outperformed other deep learning models such as LSTM ($R^2 = -0.3756$) and GRU ($R^2 = -0.6276$), which failed to capture spatial patterns effectively. The hybrid model also reduced RMSE by approximately 23% compared to the standalone CNN (RMSE = 0.1666). While the moderate R^2 indicates scope for further refinement, these results demonstrate that meaningful and interpretable demand forecasts can be generated from survey-based behavioural data, even in the absence of high-resolution temporal inputs. The model contributes a lightweight and scalable forecasting tool suitable for early-stage smart city planning in contexts where telemetry data are limited, thereby advancing the practical capabilities of EV infrastructure forecasting.

Keywords: electric vehicles; charging infrastructure; machine learning; deep learning; hybrid models; smart cities; demand forecasting

1. Introduction

The rapid adoption of electric vehicles (EVs) is reshaping urban mobility and necessitating the development of a robust and efficient charging infrastructure, particularly in smart cities. As governments worldwide implement policies to reduce carbon emissions and encourage sustainable transport, EVs are emerging as a key solution to combat climate change and improve air quality. The UK has set ambitious targets to phase out the purchase of new internal combustion engine (ICE) vehicles, with a growing emphasis on enhancing EV accessibility and infrastructure to support this transition by 2035 [1,2]. As EV adoption accelerates, concerns regarding power grid stability and demand fluctuations have emerged. Clusters of simultaneous charging, particularly during peak evening hours, can lead to load imbalances, strain local transformers, and reduce energy efficiency. Studies have demonstrated that uncoordinated charging patterns result in load peaks that challenge distribution network capacity and necessitate costly infrastructure upgrades. For instance, Souza et al. (2025) proposed a valley-filling heuristic to flatten EV charging loads, thereby improving energy efficiency and preventing instability in local grids [3]. Such findings highlight the need for demand forecasting tools that account for user preferences and technical grid constraints, particularly within the evolving context of smart cities and intelligent energy systems.

England's regions are witnessing a notable surge in EV adoption, leading to an increasing demand for reliable and accessible charging points. However, existing charging networks in these areas face significant challenges, including insufficient coverage, disparate geographic distribution, and issues related to charging station reliability. Many EV users express accessibility concerns, long waiting times, and the poor functionality of charging stations, highlighting the urgent need for strategic planning as well as expansion of the infrastructure [4,5]. A well-designed network is essential to alleviate range anxiety, a common concern among EV owners and a barrier to adoption, and to facilitate seamless travel across urban and suburban areas [6].

Smart cities are integrating advanced technologies such as artificial intelligence (AI), machine learning (ML), and the Internet of Things (IoT) to optimise urban mobility solutions. These technologies provide an opportunity to enhance the efficiency of EV charging demand by leveraging predictive analytics based on real-time data monitoring [7,8]. Moreover, integrating EV charging with smart grids can enable bidirectional energy transfer through Vehicle-to-Grid (V2G) and Grid-to-Vehicle (G2V) systems, improving overall energy efficiency and grid stability [9].

Central Research Question

To address current limitations in predictive modelling of EV charging demand, this study poses the following research question:

Can a hybrid deep learning (DL) model, combining Convolutional Neural Networks (CNN) and Extreme Gradient Boosting (XGBoost), achieve higher accuracy in predicting spatial and behavioural patterns of EV charging demand, compared to conventional ML or stand-alone DL models?

Research Hypotheses

- H_0: There is no significant improvement in prediction accuracy using the hybrid HCB-Net model compared to conventional models.
- H_1: The HCB-Net model delivers significantly greater predictive accuracy and improved spatial forecasting of EV charging demand in smart cities.

1.1. Aims

The primary aim of this study is to develop and evaluate a hybrid deep learning model (HCB-Net) that significantly outperforms traditional ML and stand-alone DL techniques in forecasting EV charging demand. This work seeks to advance the field by explicitly integrating user-level behavioural and spatial characteristics into demand forecasting models.

Specific objectives include:

- To benchmark HCB-Net against established models such as Random Forest (RF), XGBoost, SVM, LSTM, and CNN;
- To identify key behavioural and geospatial predictors of EV charging demand through feature engineering;
- To assess regional variations in charging demand using user data from the North East and West Midlands of England;
- To demonstrate that accurate and interpretable demand forecasts can be achieved using survey-derived behavioural data, even where high-resolution sensor data are unavailable;
- To provide actionable recommendations for the strategic planning of EV charging infrastructure in smart cities.

The results of the study confirm that the stated objectives were achieved, with the proposed hybrid model delivering improved predictive accuracy and offering a practical, scalable forecasting tool to support EV infrastructure planning.

1.2. Organisation of This Paper

The remainder of this paper is structured as follows. Section 2 comprehensively reviews existing EV charging demand forecasting models, discussing their methodologies, strengths, and limitations and outlines the proposed HCB-Net framework, detailing the integration of CNN and XGBoost for improved forecasting performance. Section 3 describes the experimental, dataset, collection, preprocessing, and evaluation metrics used to assess the model's performance. Section 4 presents the results and discussions, including a comparative analysis with traditional forecasting methods. Section 5 explores the practical implications, computational efficiency, and potential applications of the proposed model in smart grid infrastructure planning and summarises the key findings, highlights limitations, and suggests future research directions.

2. Literature Review

2.1. Existing Challenges in EV Charging Demand

Transitioning to EVs is essential to reduce carbon emissions and achieve sustainable urban mobility. However, the rapid growth in EV adoption presents significant challenges in developing an adequate charging infrastructure. Several studies have highlighted the primary concerns faced by EV users, including insufficient charging points, unreliable charging networks, and geographical disparities in infrastructure placement [10]. Research indicates that many urban areas lack sufficient public charging stations, leading to long waiting times and range anxiety among EV owners [11]. Furthermore, rural and suburban regions are particularly underserved, limiting the feasibility of longer-distance EV travel [12].

Reliability is another critical issue, with studies reporting frequent malfunctions and connectivity issues in public charging stations [13]. Many EV users encounter technical faults such as broken chargers, slow charging speeds, and poor user interface designs, which reduce overall satisfaction and adoption rates [14]. Additionally, peak-hour congestion at charging stations due to insufficient fast-charging units remains a key concern, emphasising the need for improved planning and investment [15].

In the context of EV–grid interaction, two primary categories of charging control strategies have emerged in the literature: direct control and indirect control. Direct control involves real-time management of EV charging loads by a central authority (e.g., a utility provider or aggregator), using automated scheduling or power-limiting hardware to manage grid balance. Zahedmanesh et al. [16] developed such a system using an in-house control unit to directly modulate the charging behaviour of plug-in EVs based on system constraints. While this method allows for precise control, it often requires significant infrastructure and user compliance.

Conversely, indirect control uses economic incentives, pricing schemes, or tariff structures to influence charging decisions without imposing direct technical constraints. Bordin and Tomasgard [17] examine how behavioural economics and micro-incentive models can shape charging habits to support grid stability, particularly in urban environments. Indirect strategies rely on the assumption that user preferences and routines can be predicted and nudged through well-designed policies or infrastructure placement.

Although several studies have documented the operational limitations of EV infrastructure, including accessibility concerns, hardware faults, and urban–rural disparity [11–14], many focus on technical availability or national-scale summaries. There remains a lack of regionally disaggregated insights into how these challenges manifest across localised smart city contexts like the North East and West Midlands of England.

The present study contributes to the indirect control framework by forecasting individual EV user charging demand based on behavioural, spatial, and demographic data. Rather than enforcing charging patterns, this model anticipates demand trends that can inform incentive design, charger deployment, and urban energy planning in smart cities.

2.2. User Charging Behaviour and Preferences

Several studies have examined EV users' charging behaviours and preferences to determine optimal infrastructure development strategies. Research suggests that many EV users prefer fast-charging stations over standard charging options due to their convenience and reduced charging times [18]. However, fast-charging stations are often limited in availability and may impose higher costs, discouraging frequent use [19].

Studies also indicate that most EV users prefer charging their vehicles at home or workplaces rather than at public charging stations [20]. This trend suggests better integration of residential and workplace charging facilities into future infrastructure plans. Additionally, some users prefer charging from lower battery charge levels and waiting until their battery reaches a critical level before seeking a charging station [21]. This behavioural trend increases demand for charging station data to be available in real time and predictive analytics to optimise the user experience and meet user needs.

Route planning and charging station accessibility are also key considerations for EV users. Many prefer charging stations along major commuting routes and urban centres, highlighting the importance of strategic placement [22]. Integrating EV charging locations with retail centres, office complexes, and leisure facilities has been proposed as an effective way to improve accessibility and encourage widespread adoption [23].

Prior work by Smith et al. (2020) and Rogers (2021) highlighted that older populations are generally slower to adopt EV technology due to both digital barriers and lower tech adoption rates [24,25]. However, these behavioural insights have rarely been integrated into predictive infrastructure planning models. Our study addresses this gap by embedding these demographic variables directly into ML pipelines to quantify their influence on charging demand patterns.

2.3. Role of AI and Deep Learning in EV Charging Demand Forecasting

AI and DL techniques are increasingly used to optimise EV charging demand planning. Recent studies explore the potential of LSTM networks for predicting temporal charging demand trends based on historical usage patterns [26]. LSTMs are well-suited for handling time-series data and can effectively capture fluctuations in charging demand across different times of the day, week, and year [27,28].

A range of ML and DL models have been applied in previous EV demand forecasting studies. Table 1 presents a structured comparison of these models, highlighting their methodological foundations, strengths, and limitations. RF and Support Vector Machines (SVMs) are widely used for structured data problems, providing strong generalisation on small datasets [25]. However, they lack the capability to capture sequential dependencies over time, limiting their applicability for time-sensitive forecasting. Linear Regression (LR) is often employed for baseline comparisons due to its simplicity and interpretability, though it underperforms in the presence of non-linear patterns typical in EV usage data [25]. XGBoost has become prominent in energy-related forecasting tasks due to its scalability, regularisation mechanisms, and superior handling of overfitting in structured datasets [29]. Despite these advantages, it does not natively support temporal modelling.

DL techniques such as Long Short-Term Memory (LSTM) and Gated Recurrent Units (GRUs) have been widely explored for load and EV charging prediction [26,27]. Their strength lies in modelling long-term temporal dependencies. However, they often require large datasets and may suffer from vanishing gradients or overfitting in small-sample contexts. Although originally developed for image processing, CNNs have been adapted to capture spatial patterns in urban infrastructure and EV charger distribution, demonstrating value in location-sensitive forecasting [30]. Yet, on their own, CNNs lack memory of past states and therefore underperform on purely temporal tasks.

Several hybrid models have been proposed to overcome the limitations of individual model types. These include combinations of LSTM with RF [25], SVM [25], or XGBoost [28]. However, such approaches often require complex integration pipelines and lack spatial interpretability. Our proposed HCB-Net model seeks to address these gaps by combining CNN's spatial pattern extraction with XGBoost's regression precision in a unified pipeline, enabling both location awareness and robust forecasting performance.

Table 1. Comparison of forecasting models for EV charging demand.

Model	Captures Time-Series?	Captures Spatial Patterns?	Scalable?	Requires Large Dataset?	Interpretability	Ref.
RF	✗	✗	✓	✗	✓	[25]
SVM	✗	✗	✗	✗	✓	[25]
LR	✗	✗	✓	✗	✓✓	[25]
XGBoost	✗	✗	✓✓	✗	✓	[29]
LSTM	✓✓	✗	✗	✓	✗	[26]
GRU	✓	✗	✓	✓	✗	[27]
CNN	✗	✓✓	✓	✗	✗	[30]
LSTM + XGBoost	✓✓	✗	✗	✓	✗	[28]
HCB-Net (ours)	✗ * (uses XGBoost)	✓✓	✓✓	✗	✓	—

Note: ✓ = moderate; ✓✓ = strong; ✗ = limited or not applicable. HCB-Net uses preprocessed, temporally aggregated inputs rather than sequential inputs.

Table 1 summarises different EV charging forecasting methods and their key attributes, advantages, and limitations. Advanced techniques and XGBoost improve prediction accuracy by reducing overfitting but require careful hyperparameter tuning. DL approaches such as LSTM and GRU excel in handling sequential data, making them suitable for demand forecasting, though they are computationally expensive. CNNs are particularly useful for spatial data analysis but are limited in handling temporal dependencies.

Hybrid models, including LSTM + RF, LSTM + XGBoost, and LSTM + SVM, aim to combine the strengths of DL and ML for enhanced accuracy, albeit at the cost of increased computational complexity. The proposed HCB-Net model has been demonstrated to have superior accuracy, leveraging CNN for spatial feature extraction and XGBoost for regression tasks, making it a promising approach for optimising EV charging demand forecasting (see Table 1).

In addition to temporal forecasting, CNNs have been widely used to analyse spatial patterns in EV infrastructure placement. CNN-based models help identify high-demand regions that require additional charging facilities by processing geospatial data, traffic flow information, and urban development trends [31,32]. This enables urban planners to make informed decisions regarding the optimal distribution of charging stations.

Hybrid DL models that integrate both LSTMs and CNNs have demonstrated superior predictive accuracy in forecasting EV charging needs [30]. These models combine spatial and temporal insights, allowing for a comprehensive understanding of infrastructure requirements. Furthermore, studies incorporating attention mechanisms in DL frameworks have shown promising results in prioritising influential factors such as user demographics, charging habits, and regional energy grid capacities [33].

Recent work continues to expand the understanding of EV charging strategies across both control paradigms. For instance, Lv et al. [34] developed an augmented user equilibrium model to optimise pricing and energy sharing at EV charging stations, effectively accounting for user flexibility and system-level efficiency. Similarly, Altamimi et al. [35] proposed a multi-agent reinforcement learning framework that dynamically adjusts charging strategies using renewable energy sources and storage, aligning with indirect control approaches based on decentralised decision making.

Cavus et al. [36] contributed to the intersection of demand forecasting and optimisation by developing a socio-demographic-aware hybrid DL model for predicting EV charging patterns. This approach bridges predictive modelling with behavioural insight. Complementarily, Cavus et al. [37] introduced a deep-fuzzy logic control architecture for adaptive energy management in grid-connected microgrids, highlighting the integration of learning-based forecasting into real-time operational decisions.

These studies illustrate a broader trend towards hybrid predictive–prescriptive systems. Our present work contributes to this landscape by offering a behaviourally grounded demand forecasting framework that supports data-informed policy, charger deployment, and dynamic energy management strategies in smart cities [38,39].

2.4. *Policy and Planning Implications*

Previous research has highlighted the importance of government policies and regulatory frameworks in shaping EV charging demand expansion. Successful case studies from countries such as Norway and the Netherlands demonstrate how subsidies, tax incentives, and public-private partnerships can accelerate infrastructure development and encourage EV adoption [40]. In the UK, a range of funding schemes and initiatives, such as the On-Street Residential Chargepoint Scheme (ORCS) and Rapid Charging Fund (RCF), aim to support the expansion of charging networks [1]. However, concerns remain regarding these initiatives' long-term scalability and financial sustainability, necessitating continuous evaluation and adaptation of policy measures to align with changing needs [41].

The increasing integration of V2G technology is also gaining attention in the literature because it allows EVs to supply power back to the grid during peak demand periods. This bidirectional energy exchange can enhance grid stability, reduce electricity costs, and contribute to the broader smart city energy ecosystem [29]. However, challenges related to grid compatibility, battery degradation, and regulatory constraints must be addressed to realise the potential of V2G systems fully [42].

While national-level strategies (e.g., ORCS, RCF) aim to expand charging access [1,40], few studies provide actionable forecasting tools aligned with local urban planning needs. Existing literature offers general recommendations [41,43] but lacks granular, user-behaviour-driven predictive frameworks that can assist policymakers in real-time infrastructure placement and demand shaping. This research responds to that need by offering a hybrid AI-driven forecasting tool grounded in user data and regional planning relevance.

Recent work has also highlighted the importance of linking demand forecasting with the commercial viability of charging infrastructure, and of understanding the interplay between public perception and investment decisions. Openshaw et al. (2023), for example, investigated how public attitudes towards dynamic wireless charging technologies may affect their market potential and economic viability [44]. Incorporating such considerations into demand models is critical to ensuring that infrastructure deployment strategies are both financially sustainable and aligned with user expectations. At present, there remains a gap in integrated modelling approaches that connect behavioural demand forecasts with infrastructure economics. This study contributes towards addressing this gap by providing user-informed demand insights that could support more economically robust planning. Future work will seek to integrate explicit economic modelling to strengthen this connection further.

2.5. Research Questions

A critical review of the literature identified research gaps. Five key questions to guide this research were defined as follows:

1. How can hybrid DL models enhance EV charging demand forecasting accuracy compared to traditional ML approaches?
2. What are the key limitations of existing forecasting models, and how can they be addressed using a combination of CNN and XGBoost?
3. How does the proposed HCB-Net, the combined CNN and XGBoost model, perform in real-world EV charging scenarios regarding prediction accuracy and computational efficiency?
4. What role do spatial and temporal dependencies play in EV charging demand forecasting, and how can they be effectively captured in predictive models?
5. How can the findings of this study contribute to optimising EV infrastructure planning in smart cities?

   The answers to these research questions will provide valuable insights into integrating advanced ML techniques in smart grid applications, facilitating efficient EV charging management and infrastructure development.

2.6. Novelty and Key Contributions

This study addresses key limitations in existing EV charging demand forecasting models, particularly the limited incorporation of behavioural factors, lack of regional focus, and architectural complexity. The research makes several original contributions to the field. First, it presents a hybrid DL model (HCB-Net) that combines CNN-based spatial feature extraction with XGBoost regression, specifically designed to model the spatial and behavioural patterns influencing EV charging demand in smart cities. Second, the model explicitly integrates demographic and behavioural predictors, such as user age, charging preferences, commute distance, and battery capacity, which are often neglected in conventional grid-based or sensor-driven approaches. Third, the model is trained on regional, survey-derived data from two distinct urban regions in England (the North East and West Midlands), enabling the capture of localised charging behaviours and spatial disparities in existing infrastructure. Notably, the framework demonstrates that robust and interpretable demand forecasting can be achieved using lightweight survey data, without the need for high-resolution telemetry or time-series inputs.

Finally, the HCB-Net pipeline is designed to be modular and reproducible, making it readily adaptable to other urban contexts as new behavioural or spatial data become available. These contributions advance the methodological toolkit for data-driven smart city planning, offering practical support for early-stage EV infrastructure development and localised grid management.

3. Methodology

This section outlines the methodological framework for predicting EV charging demand. The proposed approach integrates DL and ML techniques into a unified HCB-Net. The methodology consists of data preprocessing, feature engineering, and model training using a combination of CNNs and XGBoost.

3.1. Data Collection

Data for this study were collected via a structured questionnaire administered to EV users in the West Midlands and the North East of England. The survey gathered insights into charging habits, vehicle specifications, and user preferences to inform predictive models of EV charging demand. This data collection effort aimed to explore charging behaviours across varying urban environments, focusing on regional factors such as urbanisation levels, grid capacity, and socio-economic diversity.

In addition to structured, quantitative items (e.g., Likert scales, binary and categorical choices), the survey also included open-ended questions, allowing respondents to elaborate on their charging habits, preferences, concerns, and motivations. These responses constitute the qualitative data collected in this study.

A basic thematic content analysis was performed to categorise recurring themes, such as cost sensitivity, range anxiety, and infrastructure visibility, from these textual responses. While this qualitative input was not directly used in model training, it played a crucial role in guiding feature engineering decisions and interpreting model outputs, particularly in identifying behavioural patterns corresponding to regional demand trends. For instance, frequent mention of limited public chargers in the North East supported the inclusion of region-specific charging availability flags. Similarly, comments on charging time preferences helped frame temporal feature bins used in preprocessing.

To gain comprehensive insights, a structured questionnaire was developed to collect detailed information on various aspects of EV ownership, usage, and charging patterns. The survey captures demographic characteristics, personality traits, and behavioural tendencies of EV users, as well as their driving habits, mobility patterns, and charging preferences. Furthermore, it explores participants' experiences with regional charging infrastructure, including accessibility, reliability, cost, and convenience.

The questionnaire comprises 51 carefully designed questions incorporating multiple formats to ensure a nuanced understanding of user perspectives. These include multiple-choice questions to assess general trends, Likert scale questions to gauge attitudes and satisfaction levels, and open-ended questions to provide space for participants to elaborate on their experiences and opinions. By employing this mixed-question format, the study aims to generate both quantitative and qualitative data, facilitating a holistic analysis of the factors shaping EV charging behaviour in the target regions.

Following ethical approval from the Newcastle University Ethics Committee, the survey was hosted on Prolific, a widely used online research platform known for its high-quality and diverse participant pool. The study specifically targeted EV users in the North

East and West Midlands regions to ensure relevant and region-specific insights. To achieve a representative sample, participants were recruited based on the following criteria:

- Must be a current EV owner or primary user;
- Must reside in either the North East or the West Midlands.

The survey's data collection took place from May 2024 to June 2024 for the North East and from October 2024 to November 2024 for the West Midlands.

A total of 256 EV users participated, with 132 respondents from the West Midlands and 124 from the North East of England. The survey was conducted through online platforms and direct outreach, ensuring a diverse and representative sample of EV users in both urban and suburban settings. Table 2 presents the population distribution of the North East region based on the 2021 Census data [45]. The table categorises the population into broad age groups, allowing for a clearer analysis of demographic trends.

The Under 18 category comprises the largest portion of the population, totalling 379,250 individuals. This is followed by the 45–54 and 25–34 age groups, with populations of 254,985 and 242,915, respectively. The 35-44 category is also significant, with 234,545 individuals.

Table 2. Population distribution by age group in the North East region (Census 2021) [45].

Age Group	Population
Under 18	379,250
18–24	176,709
25–34	242,915
35–44	234,545
45–54	254,985
55–64	229,342
65 and over	315,678

Notably, the population of those aged 55 and above steadily increases, with 229,342 individuals in the 55–64 group and 315,678 in the 65 and over category. This reflects an ageing population trend, which may affect regional healthcare, mobility, and transport infrastructure policies. The lower representation of older individuals in this study's sample aligns with existing literature, which suggests that older populations are less likely to engage with online surveys and adopt EVs. Understanding these demographic differences is essential for policymakers and researchers when planning future infrastructure developments and survey methodologies.

To ensure the validity of the study sample, a Chi-square contingency test was conducted to compare the sample demographics with the population data of the Northeast and West Midlands regions. The test results indicated that, in terms of gender, the sample was representative of the regional population. However, age distribution analysis revealed that participation numbers for individuals over 55 were lower than expected based on the population data of both regions. This discrepancy may be attributed to lower engagement among older individuals in online surveys, as found in studies examining digital survey participation rates across different age groups. Recent studies confirm this age-related trend: Bosehans et al. [46] explored digital and behavioural barriers among older EV users; Rogers [24] explained these patterns using the diffusion of innovation theory; and Smith et al. [25] empirically demonstrated lower EV adoption rates among individuals aged 55 and over.

Figure 1 compares the North East and West Midlands regions in terms of demographics, education, occupation, income, and ethnicity. The main findings draw attention to both disparate geographical spread and similar trends. Each area has marginally more men than women, indicating a fairly balanced gender ratio, with females: 49–51% and males: 49–51% for the West Midlands and North East, respectively. Most people are between the ages of 25 and 34 years (41%) and 35 and 44 years (23%), indicating a sizeable working-age population. This implies that a sizeable fraction of the population in both areas is probably economically active. Only about 13% of the population is in the youngest age group (18–24 years), suggesting that there are fewer young adults, perhaps because they cycle and walk more actively due to migration for work or study. Additionally, the population is primarily concentrated in the middle age range, as evidenced by the lowest representation of older persons aged 55–64 years (6%) and 65+ years (1–2%).

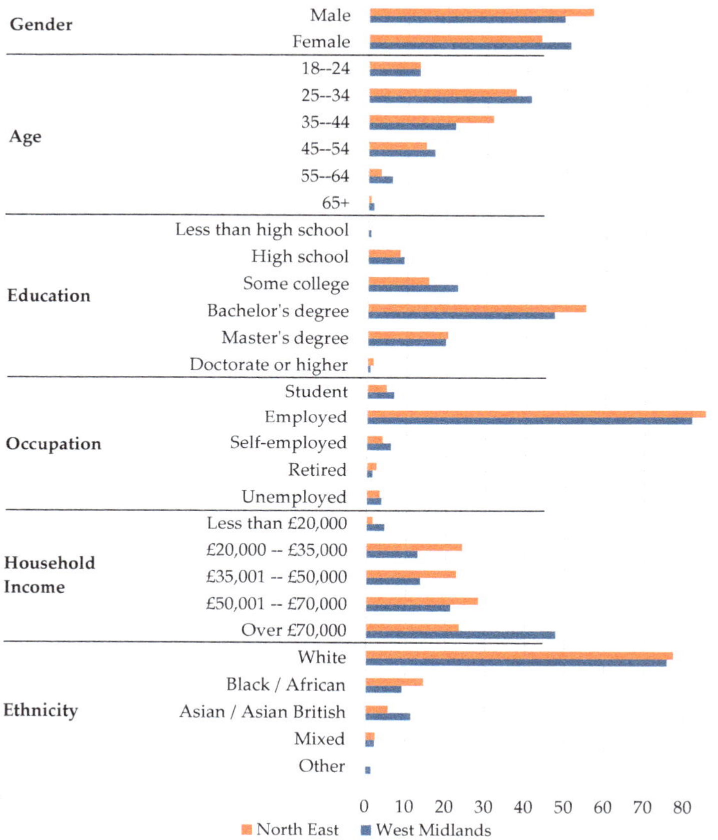

Figure 1. Demographicdistribution of two regions.

Although there are some minor differences in educational attainment between the two areas, bachelor's degrees account for 47% of all qualifications. This suggests that around 50% of people in both regions have earned a university degree. Following this, a considerable percentage of respondents continue their education after their first degree, as evidenced by the prevalence of Master's degrees (20%) and after sixth form some college education (23%). However, less than 1% of people have a doctorate or a lower education than a high school diploma, and only 9% of people have a high school diploma as their highest degree.

In both areas, employment is the most common occupational status, with more than 81% of EV owners being employed. This demonstrates an economic activity in both regions.

Since just 6% of people work for themselves, it is likely that most people choose to or can find traditional employment. The small percentage of students (7%) is consistent with the 18–24 age group's historically lower representation. Relatively low unemployment rates (4%) and retirement (2%) support the notion that most people in both areas are actively working.

The income distribution reveals a significant middle-income section, with the highest percentage of respondents (55%), earning between GBP 20,000 and GBP 70,000. About 16% of people are in the highest income group (above GBP 100,000), with a slightly higher proportion in the West Midlands than in the North East. Similarly, 20% of the population falls into the GBP 70,000–85,000 range, indicating a concentration of people in the upper-middle income range. Conversely, just 4.5% of people are in the lowest income bracket (less than GBP 20,000), indicating that poverty rates are generally low in both areas.

In both areas, the majority of the population (76%) is White, according to the ethnic distribution. Nonetheless, there are statistically significant differences at a 95% confidence level. Compared to the North East, the West Midlands has a somewhat higher percentage of Black and African people (9%) and a larger Asian and Asian British minority (11%). While there is diversity, most of the population are White, as seen by the small minority of Mixed (1.5%) and Other (1.5%) ethnic groupings.

This graphic shows how the North East and West Midlands are mostly similar in terms of education, employment, and income distribution. The population of EV users in these areas is primarily middle-class, highly educated, and employed. Nonetheless, the North East respondents have a marginally higher percentage of employed people and degree holders. At the same time, the West Midlands exhibit marginally higher income earners and a more varied ethnic makeup. The results imply that although social and economic structures are comparable, geographical variations in high-income earners and ethnic variety remain unique.

The data obtained from the regional EV user surveys formed the foundational input for all subsequent modelling tasks. To ensure this dataset could be effectively used for predictive modelling, preprocessing and feature engineering were undertaken to convert raw responses into a machine-readable format.

3.2. Demographic Composition and Generalisability Considerations

The dataset used in this study consists of responses from 256 EV users across two UK regions: the North East and West Midlands. The sample includes participants from diverse demographic backgrounds, with age groups ranging from 18 to 65+, a near-balanced gender distribution (53% male, 47% female), and varied income brackets and educational levels. However, most respondents were early adopters of EV technology, often with higher-than-average environmental awareness and technological literacy. This composition, while reflecting the current EV ownership demographic in the UK, may limit generalisability to broader populations, particularly as EV adoption expands into more diverse socio-economic and geographic groups. Additionally, urban and peri-urban residents are overrepresented in the sample, which could bias demand forecasting outcomes if applied to predominantly rural or underserved areas. Future studies should seek to expand data collection to include more diverse user profiles and geographic contexts to improve model robustness and national scalability.

3.3. Data Preprocessing

Before feeding the dataset into predictive models, cleaning and preparing the data through standardisation and encoding processes was essential. These preprocessing steps served to normalise continuous variables and structure categorical responses, enabling consistent input for ML algorithms. The raw dataset underwent preprocessing to standardise feature values and remove inconsistencies. Let $X = \{x_1, x_2, ..., x_n\}$ represent the feature space and $Y = \{y_1, y_2, ..., y_n\}$ the corresponding charging demand labels. The min–max scaling technique is applied as follows:

$$X' = \frac{X - X_{min}}{X_{max} - X_{min}} \tag{1}$$

where X' represents the normalised dataset and X_{min} and X_{max} are the minimum and maximum values of the feature set, respectively. Additionally, categorical variables are encoded using one-hot encoding:

$$E_{i,j} = \begin{cases} 1, & \text{if category } j \text{ applies to sample } i \\ 0, & \text{otherwise} \end{cases} \tag{2}$$

where $E_{i,j}$ represents the one-hot encoded matrix. Once a clean and consistent dataset was prepared, it was possible to identify and prioritise the most relevant features for model training through a rigorous feature selection and dimensionality reduction process.

3.4. Target Variable Definition

This study aims to predict the average daily charging demand per user, expressed in kilowatt-hours (kWh). This target variable was derived from three survey responses:

1. Charging frequency ("How often do you charge your vehicle?");
2. Charging duration ("On average, how long is each charging session?");
3. Vehicle battery capacity (used as a proxy for energy drawn per charge).

From these, a derived estimate of daily demand was calculated using the following formula:

$$\text{Charging Demand (kWh/day)} = \left(\frac{\text{Battery Capacity (kWh)}}{\text{Full-to-empty cycles}}\right) \times \left(\frac{\text{Charging frequency per week}}{7}\right) \tag{3}$$

Responses were filtered to ensure realism (e.g., capped frequencies, verified charging durations), and outliers were removed. The resulting target variable is a numerical, continuous value representing the estimated average daily charging demand rather than a specific timestamp.

Therefore, this study does not perform time-series forecasting. Instead, it uses behavioural and spatial variables to estimate the expected daily charging load per user, with applications in infrastructure planning, grid load balancing, and spatial demand mapping.

3.5. Feature Engineering

The feature engineering phase focused on identifying the most predictive variables within the dataset to optimise model performance. The feature set was refined for subsequent modelling using XGBoost-based feature importance scoring and Principal Component Analysis (PCA) techniques.

Following preprocessing, the next step was to extract and select the most informative features for training the predictive models. The original dataset included both categorical and continuous variables drawn from 51 questionnaire items, covering:

- User demographics: age group, gender, income bracket, education level;
- Vehicle specifications: EV type, all-electric range, battery capacity;
- Charging behaviour: charging frequency, preferred time of day, charging duration;
- Geospatial indicators: region, residential postcode category, commute distance, urban/suburban classification;

- Behavioural preferences: willingness to travel for charging, home vs. public charging inclination.

These raw variables were initially examined through descriptive analytics and correlation matrices to identify redundancies. Subsequently, the following steps were employed in the feature engineering process:

1. Encoding Categorical Variables: Categorical variables such as region, EV type, and education level were one-hot encoded to preserve their nominal character without imposing ordinal relationships.
2. Scaling Continuous Variables: Numerical inputs such as commute distance, charging frequency, and battery capacity were normalised using min–max scaling to ensure compatibility with neural network layers.
3. Feature Selection Using XGBoost Importance Ranking: An XGBoost model was trained on the full dataset to compute a feature importance score based on information gain (splitting gain), coverage (frequency of use), and weight (impact on model accuracy). Features with negligible contribution to prediction (less than 1% importance score) were excluded from further modelling.
4. Dimensionality Reduction with PCA: Principal Component Analysis (PCA) was applied to the encoded feature space to address potential multicollinearity and reduce noise. Components explaining at least 90% of variance were retained. However, PCA-transformed data was used selectively, primarily for exploratory visualisation and not for feeding into the CNN, where spatial interpretability was needed.
5. Final Feature Set: The refined input for HCB-Net included 24 features: 12 numerical, 8 categorical (expanded to 20 columns post-encoding), and 4 geospatial flags. These were reshaped and structured into a matrix input suitable for convolutional processing.

The proposed HCB-Net is designed to combine the strengths of deep learning and gradient-boosted decision trees for improved forecasting of EV charging demand. The architecture comprises two key components: a CNN for spatial feature extraction and an XGBoost regressor for final prediction. The CNN component is responsible for learning high-level spatial and behavioural representations from the input feature matrix, which includes user demographics, commute patterns, and regional indicators. After convolution and pooling operations, the extracted feature maps are flattened into a one-dimensional vector. This vector serves as the input to the XGBoost model, which performs structured learning and non-linear regression.

The rationale for combining CNN and XGBoost lies in their complementary strengths: CNNs are adept at uncovering spatial relationships and patterns within structured data. At the same time, XGBoost excels at capturing interactions and non-linear dependencies in tabular features. This modular design allows HCB-Net to separate feature representation from regression logic, enabling better generalisation and interoperability.

Although this implementation does not explicitly incorporate a formal attention mechanism, the hybrid design allows for the selective weighting of spatial features via convolution filters and importance-weighted splits in the XGBoost component. Future iterations of the model may explore explicit attention layers to further enhance feature discrimination and interpretability, particularly in multi-regional or temporal behavioural data scenarios.

Feature selection is performed to improve model efficiency and eliminate redundant information. Given a feature set F, the selected features are determined using an importance ranking mechanism based on XGBoost's feature importance score:

$$S_f = \sum_{i=1}^{N} G_i \cdot H_i \tag{4}$$

where S_f represents the feature importance score, G_i is the gain from feature splits, and H_i represents the Hessian matrix, indicating second-order gradient approximations. Principal Component Analysis (PCA) is also employed to reduce dimensionality:

$$Z = W^T X \tag{5}$$

where Z is the transformed feature set, W is the eigenvector matrix, and X is the original feature matrix. The selected features were then used as inputs to train the proposed hybrid DL model, which will be described in detail in the next section.

3.6. Hybrid Deep Learning Framework

The HCB-Net model is designed to integrate the strengths of DL and traditional ML methods for enhanced EV charging demand forecasting. Specifically, CNNs extract spatial and behavioural patterns from structured input data (e.g., user location, demographics, and commute characteristics), while XGBoost handles the final regression task.

Algorithm 1 outlines the operational flow of the proposed HCB-Net model. The pipeline begins with ingesting and reshaping preprocessed data, including spatial, demographic, and behavioural inputs relevant to EV users. These inputs are structured as two-dimensional tensors and passed through convolutional layers to extract spatial and contextual patterns. Pooling operations are then applied to reduce the dimensionality while retaining essential features. The resulting feature map is flattened into a vector representation suitable for structured learning.

Algorithm 1 HCB-Net: Hybrid CNN + XGBoost for EV charging demand forecasting.

Require: Preprocessed dataset D with spatial, demographic, and behavioural features
Ensure: Predicted EV charging demand $\hat{y}$
Load dataset D
Reshape and encode features into a 2D input tensor for CNN
Apply convolutional layers: $F \leftarrow \text{CNN}(D)$
Apply max-pooling to reduce spatial dimensions: $F_{\text{pooled}} \leftarrow \text{Pooling}(F)$
Flatten pooled features: $F_{\text{flat}} \leftarrow \text{Flatten}(F_{\text{pooled}})$
if feature vector F_{flat} is non-empty **then**
    Feed F_{flat} into XGBoost regressor: $\hat{y} \leftarrow \text{XGBoost}(F_{\text{flat}})$
else
    Raise error: "Insufficient features after pooling"
end if
Evaluate model predictions using MAE, MSE, RMSE, and R^2 Score
return $\hat{y}$

If the extracted feature vector is non-empty, it is forwarded to the XGBoost regressor, which performs non-LR to produce the final prediction of daily EV charging demand $\hat{y}$. In cases where feature extraction fails, a condition is raised to ensure model robustness. Finally, the model is evaluated using a suite of performance metrics—MAE, MSE, RMSE, and R^2—to assess both absolute error and goodness-of-fit. This hybrid design allows HCB-Net to capture high-level feature interactions via CNN while benefitting from XGBoost's capability to model structured, tabular relationships in the final prediction layer.

The choice of algorithms in this study was driven by their known strengths in predictive modelling and their relevance to the nature of the data. CNNs were selected for their ability to capture spatial relationships between behavioural and geographic features.

XGBoost was chosen for its strong performance with structured tabular data and robustness against overfitting. LSTM and GRU were tested due to their popularity in sequential modelling, though they underperformed due to the non-temporal nature of the input. Traditional models such as RF, SVM, and LR served as comparative baselines to benchmark the added value of the hybrid approach. LR, in particular, was used as a null model to provide a low-complexity performance reference, as shown in Table 3.

Table 3. Key hyperparameters for all models used in this study.

Model	Key Hyperparameters
HCB-Net	CNN filters = [32, 64], kernel size = 3 × 3, dropout = 0.3, XGBoost max_depth = 6, learning_rate = 0.1, n_estimators = 100
CNN	Filters = [32, 64], kernel size = 3 × 3, activation = ReLU, dropout = 0.3
LSTM	Units = 64, dropout = 0.2, epochs = 50, batch_size = 32
GRU	Units = 64, dropout = 0.2, epochs = 50, batch_size = 32
XGBoost	max_depth = 6, learning_rate = 0.1, subsample = 0.8, n_estimators = 100
RF	n_estimators = 100, max_depth = 8
SVM	Kernel = RBF, C = 1.0, gamma = 'scale'
LR	No Hyperparameters (baseline)

This hybrid structure allows the model to leverage CNN's capability to capture local feature correlations and high-dimensional interactions, essential for interpreting spatial demand trends and usage clustering. Once CNN layers transform and condense the input feature space into informative embeddings, these features are passed into the XGBoost model for prediction. XGBoost's strength in handling structured data, robustness against overfitting, and interpretability make it a suitable complement to CNNs, mainly when the objective is accurate and explainable demand forecasting.

Building on the engineered feature set, the HCB-Net model was designed to integrate spatial insights through CNN and non-LR capabilities via XGBoost. This section outlines the architecture and functioning of the hybrid model for forecasting EV charging demand. The proposed HCB-Net framework integrates CNNs for spatial feature extraction and XGBoost for final regression predictions. The CNN component applies convolutional operations defined as:

$$O_{i,j} = \sum_{m=-k}^{k} \sum_{n=-k}^{k} K_{m,n} \cdot I_{i+m,j+n} \tag{6}$$

where $O_{i,j}$ is the output of the convolution operation at position (i, j), $K_{m,n}$ is the kernel filter, and $I_{i+m,j+n}$ represents the input feature matrix. The CNN layer is followed by a pooling operation to reduce dimensionality:

$$P_{i,j} = \max_{m,n \in R} O_{i+m,j+n} \tag{7}$$

where $P_{i,j}$ is the pooled feature map and R defines the pooling region. The output of the CNN pooling layer forms a feature map that encapsulates dominant spatial and behavioural attributes derived from the dataset. These compressed features are then flattened and fed directly into the XGBoost regressor. By separating spatial pattern extraction (CNN) from the decision-making layer (XGBoost), the model benefits from CNN's generalisation capability in identifying charging hotspots or user routines, while XGBoost interprets these embeddings in the context of the target variable—forecasted charging demand.

The extracted spatial features are then passed into an XGBoost regression model, trained to minimise the objective function:

$$L(\theta) = \sum_{i=1}^{n} \ell(y_i, \hat{y}_i) + \sum_{j=1}^{T} \Omega(f_j) \tag{8}$$

where $\ell(y_i, \hat{y}_i)$ is the loss function measuring prediction error and $\Omega(f_j)$ is the regularisation term controlling model complexity. The final prediction is generated as:

$$\hat{y}_i = \sum_{t=1}^{T} f_t(x_i) \tag{9}$$

where T represents the number of boosting rounds.

While the integration of CNNs and XGBoost has previously been applied in fields such as energy consumption forecasting and traffic flow prediction, this study extends the hybrid approach into the relatively under-explored EV charging demand forecasting domain within the smart city context. The novelty lies not in the reuse of the algorithms themselves, but in:

- The fusion of spatial behaviour data, user demographics, and EV usage patterns within a CNN–XGBoost framework tailored to EV infrastructure planning.
- The modular HCB-Net architecture decouples spatial feature extraction from regression logic, enabling interpretability and flexibility across varied geographic regions.
- Using a region-specific, survey-driven dataset, capturing nuanced behavioural trends from the West Midlands and North East England—two demographically and infrastructurally diverse regions.
- A focus on short-distance travel radius and public-private charging behaviours, which are rarely integrated in previous models.

The contribution is methodological in its context-specific adaptation and empirical validation across regional EV populations, filling a significant gap in urban electrification forecasting literature.

3.7. Model Evaluation Metrics

To evaluate the predictive capabilities of the proposed HCB-Net and benchmark models, four key metrics were employed:

1. Mean Absolute Error (MAE): Measures the average magnitude of prediction errors without considering their direction. It offers a straightforward indication of overall prediction accuracy in real-world units.
2. Mean Squared Error (MSE): Squares the difference between actual and predicted values, penalising larger errors more heavily. Useful for emphasising instances of poor prediction.
3. Root Mean Squared Error (RMSE): The square root of MSE, providing error in the same unit as the output variable. RMSE is particularly suitable for continuous target variables such as charging demand.
4. Coefficient of Determination (R^2 Score): Represents the proportion of variance in the dependent variable that is predictable from the independent variables. Values closer to 1 indicate a better model fit.

These metrics were chosen for their complementary insights into model performance: MAE and RMSE assess absolute accuracy, while R^2 indicates how well the model explains variation in the data.

Since this study involves regression-based forecasting rather than classification, metrics such as accuracy or F1-score were not applicable. Evaluation was conducted using a hold-out test set comprising 30% of the full dataset, with model hyperparameters tuned using five-fold cross-validation on the training set.

This multi-metric approach allows for a robust assessment of each model's practical reliability, generalisability, and consistency when forecasting EV charging demand across diverse regional and user contexts.

To assess the effectiveness of HCB-Net, standard evaluation metrics are employed, including MAE, MSE, RMSE, and R^2. These are computed as follows:

$$MAE = \frac{1}{n}\sum_{i=1}^{n}|y_i - \hat{y}_i| \tag{10}$$

$$MSE = \frac{1}{n}\sum_{i=1}^{n}(y_i - \hat{y}_i)^2 \tag{11}$$

$$RMSE = \sqrt{MSE} \tag{12}$$

$$R^2 = 1 - \frac{\sum_{i=1}^{n}(y_i - \hat{y}_i)^2}{\sum_{i=1}^{n}(y_i - \bar{y})^2} \tag{13}$$

where y_i represents the actual charging demand, $\hat{y}_i$ is the predicted value, and $\bar{y}$ is the mean of observed values.

The methodology integrates CNNs and gradient-boosting techniques to enhance EV charging demand prediction accuracy. Equations (1)–(13) detail the preprocessing, feature extraction, predictive modelling, and evaluation phases, ensuring a comprehensive approach to intelligent infrastructure forecasting. The combination of CNN and XGBoost allows HCB-Net to extract spatial dependencies while leveraging boosting techniques for robust forecasting, making it a suitable solution for predictive analytics in EV infrastructure planning.

3.8. Variable Definitions

Group 1: Equations (3) and (4)—Feature Importance and PCA. S_f denotes the feature importance score for feature f, N represents the total number of boosting rounds (trees) or nodes considered, G_i is the gain (information gain) from splitting on feature f at node i, H_i is the Hessian (second-order derivative) at node i, W is the matrix of eigenvectors obtained from the Principal Component Analysis (PCA), Z denotes the transformed feature set in PCA space, and X is the original feature matrix before dimensionality reduction.

Group 2: Equations (5) and (6)—CNN Layer and Pooling. $O_{i,j}$ refers to the output of the convolution operation at spatial position (i,j), $I_{i+m,j+n}$ represents the input matrix value at offset $(i+m, j+n)$, $K_{m,n}$ is the convolutional kernel (filter) value at position (m,n), k is the kernel size (for example, $k = 1$ corresponds to a 3×3 filter window), the variables m and n are index variables used to iterate over the filter size, $P_{i,j}$ is the output of the pooling operation at position (i,j), and R denotes the region defined for pooling (e.g., 2×2 or 3×3 window).

Group 3: Equations (7) and (8)—XGBoost Objective. $L(\theta)$ represents the overall loss function to be minimised, $\ell(y_i, \hat{y}_i)$ is the prediction loss for sample i, typically using squared error, T is the number of trees in the boosted ensemble, $\Omega(f_j)$ denotes the regularisation term applied to control the complexity of the function f_j, $f_t(x_i)$ refers to the output of the t-th decision tree for the input sample x_i, and $\hat{y}_i$ is the final predicted value for that sample.

Group 4: Equations (9) and (12)—Evaluation Metrics. n indicates the total number of samples in the test set, y_i is the true (ground truth) value for sample i, $\hat{y}_i$ is the predicted value, and $\bar{y}$ represents the mean of the actual values y_i across all test samples.

3.9. Train–Test Splitting and Overfitting Control

To evaluate the performance of the proposed HCB-Net and baseline models, the dataset was divided into training and test sets using a standard 70:30 split. The training set was used to fit model parameters, while the test set evaluated generalisation performance on unseen data.

The data was randomly shuffled before splitting to ensure fair comparison across all models and avoid sample bias. Since the dataset includes both categorical and continuous variables, stratified sampling was applied based on region (North East and West Midlands) to preserve demographic distribution across both sets.

Several strategies were employed to mitigate overfitting during model training, particularly for DL models. First, five-fold cross-validation was applied on the training set for model tuning in the case of RF, XGBoost, LR, and SVM. Second, early stopping based on validation loss was used for LSTM, GRU, CNN, and the HCB-Net hybrid model to prevent over-training. Additionally, dropout layers and regularisation penalties were incorporated in the CNN architecture to enhance generalisation performance. For XGBoost, key hyperparameters such as maximum tree depth, learning rate, and subsample ratios were optimised using a grid search approach.

Model performance metrics (MAE, RMSE, MSE, and R^2) were computed using only the test set, ensuring no information from this subset influenced the training process. Using a held-out test set guarantees a robust and unbiased evaluation of the models' real-world predictive capability.

4. Results and Discussion

The data generated by the questionnaire was used to test each of the models presented in Table 1. The first step in any analysis is determining the statistical significance of the representativeness to regional populations. The performance of the range of ML and DL models for predicting EV charging demand is summarised in Table 4. The table compares models based on four key evaluation metrics: MAE, MSE, RMSE, and R^2 score. Lower values of MAE, MSE, and RMSE indicate higher prediction accuracy, while a higher R^2 score suggests a better fit to the observed data.

While the HCB-Net model achieved the highest performance among all tested methods, its R^2 score of 0.3996 indicates a moderate level of predictive accuracy. This value suggests that the model explains approximately 40% of the variance in user-level EV charging demand based on the available features.

In large-scale deployment scenarios or grid-integrated forecasting systems, R^2 values exceeding 0.5, 0.7, or even 0.8 are typically expected to ensure operational robustness. However, the relatively lower R^2 in this study is influenced by several factors:

- The use of self-reported, survey-based data rather than real-time charging logs or smart meter outputs;
- A limited sample size across just two geographic regions, restricting generalisability;
- The absence of temporal granularity, since the predictions are behavioural-aggregate (e.g., daily demand estimates), not time-series.

Table 4. Performance metrics for various models.

Model	MAE	MSE	RMSE	R^2 Score
RF	0.0565	0.0306	0.1750	0.1723
SVM	0.0814	0.0255	0.1598	0.3094
LR	0.0800	0.0240	0.1548	0.3520
XGBoost	0.0615	0.0266	0.1632	0.2798
LSTM	0.1542	0.0509	0.2255	−0.3756
GRU	0.0786	0.0602	0.2453	−0.6276
CNN	0.0942	0.0277	0.1666	0.2498
LSTM + RF	0.0995	0.0311	0.1762	0.1603
LSTM + XGBoost	0.0987	0.0235	0.1534	0.3635
LSTM + SVM	0.0940	0.0296	0.1722	0.1983
HCB-Net (CNN + XGBoost)	**0.0735**	**0.0222**	**0.1490**	**0.3996**
XGBoost + RF	0.0588	0.0261	0.1615	0.2945

Despite these constraints, the proposed framework demonstrates that meaningful and interpretable demand forecasts can still be extracted from behavioural data, making it valuable for early-stage infrastructure planning, policy simulation, or microgrid load estimation in localised smart city contexts.

Future work will incorporate higher-resolution temporal data, smart meter logs, and vehicle telemetry to enhance predictive power and achieve industry-grade accuracy thresholds.

4.1. Survey Sample Representativeness

Before evaluating model performance, the representativeness of the survey sample was assessed to ensure the validity of subsequent analyses. A Chi-square contingency test was applied to compare the sample demographics (age and gender) with regional census data from the North East and West Midlands of England. The test confirmed that the sample was statistically representative in terms of gender across both regions ($p > 0.05$) but under-represented older age groups, particularly those aged 55 and above. This aligns with findings from prior studies [11,12] that suggested lower digital engagement and slower EV adoption among older populations. Despite this age-related skew, the dataset captures a broad and relevant spectrum of EV user behaviours in urban and suburban contexts.

4.2. Model Performance Comparison

Table 4 summarises the performance of various ML and DL models using four evaluation metrics: MAE, MSE, RMSE, and R^2 score. These metrics allow us to assess both absolute error and the goodness-of-fit between predicted and actual EV charging demand.

The proposed HCB-Net model (CNN + XGBoost) outperformed all other models, achieving the lowest RMSE (0.1490) and the highest R^2 score (0.3996). This result confirms the benefit of combining CNN's spatial feature extraction with XGBoost's robust regression capability. Among traditional ML models, LR and SVMs also performed reasonably well (R^2 = 0.3520 and 0.3094, respectively), with LR surprisingly outperforming CNN and LSTM in terms of R^2, despite its simplicity. This may be due to the linear nature of certain input features or the limited time-series complexity in the dataset. Notably, LSTM and GRU underperformed significantly, both showing negative R^2 scores (−0.3756 and −0.6276, respectively). This suggests that these models failed to capture the temporal dependencies for which they were designed, potentially due to the relatively modest dataset size and low temporal granularity. These findings highlight the importance of data volume and structure when applying recurrent networks.

While a standalone CNN had a moderate performance (R^2 = 0.2498), its predictive capability improved significantly when combined with XGBoost in the HCB-Net architecture.

This reinforces the value of hybrid modelling, where DL extracts high-level patterns and gradient boosting handles refined regression.

Figure 2 visually compares actual and predicted charging demand across test samples for each model. The HCB-Net model's predictions most closely follow the actual demand trend, with minimal deviation. In contrast, models like LSTM and GRU display unstable predictions, with noticeable divergence and volatility across test points. Traditional ML models (RF, XGBoost, LR) show relatively consistent but slightly under-responsive behaviour, smoothing over peaks in demand. While capturing some variation, CNN exhibits irregular patterns, possibly due to its limited ability to model non-spatial relationships on its own. Overall, the figure illustrates that hybrid models offer the most balanced and accurate predictions, capturing both spatial trends and generalised demand levels. In contrast, purely sequential models (LSTM, GRU) appear overfitted or misaligned with the nature of the dataset.

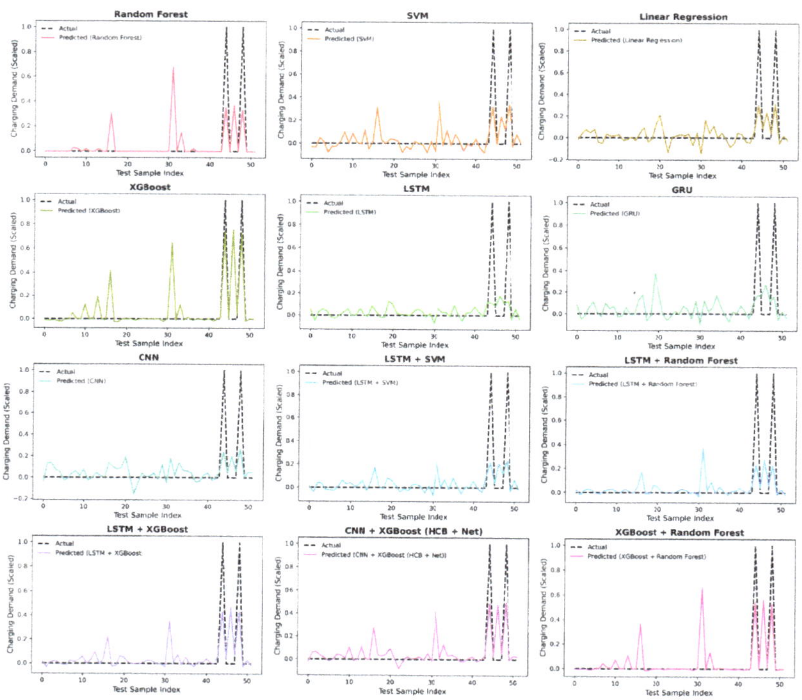

Figure 2. Model performance comparison for EV charging demand prediction.

The figure shows that different models exhibit varying levels of accuracy and robustness in capturing trends in EV charging demand. Traditional ML models, such as RF and SVM, demonstrate moderate performance but fail to capture temporal dependencies effectively. On the other hand, DL models, including LSTM and GRU, perform better in identifying sequential patterns and capturing long-term dependencies in the charging demand data. The proposed HCB-Net (CNN + XGBoost) model significantly outperforms other approaches by leveraging CNN for spatial pattern recognition and XGBoost for enhanced regression performance. The results indicate that HCB-Net reduces prediction errors and improves overall model accuracy compared to stand-alone ML and DL models. This hybrid approach effectively balances computational efficiency and predictive accuracy, making it a promising candidate for real-world EV charging demand forecasting applications.

Furthermore, the evaluation metrics, such as RMSE and R^2, demonstrate that the proposed model performs better than conventional techniques. This improvement highlights the advantage of integrating spatial feature extraction with robust ML regression techniques to enhance forecasting accuracy. Overall, this comparative analysis provides critical insights into the strengths and weaknesses of different forecasting models and validates the effectiveness of the proposed HCB-Net in predicting EV charging demand with high precision.

4.3. Box Plot Analysis of Model Predictions

Figure 3 provides a comparative distribution of model predictions using a box plot. The x-axis represents different models, while the y-axis shows the scaled charging demand. The box plots illustrate the spread, variability, and presence of outliers in each model's predictions. The central tendency and dispersion of the predicted charging demand vary across models, reflecting the extent of their consistency and robustness.

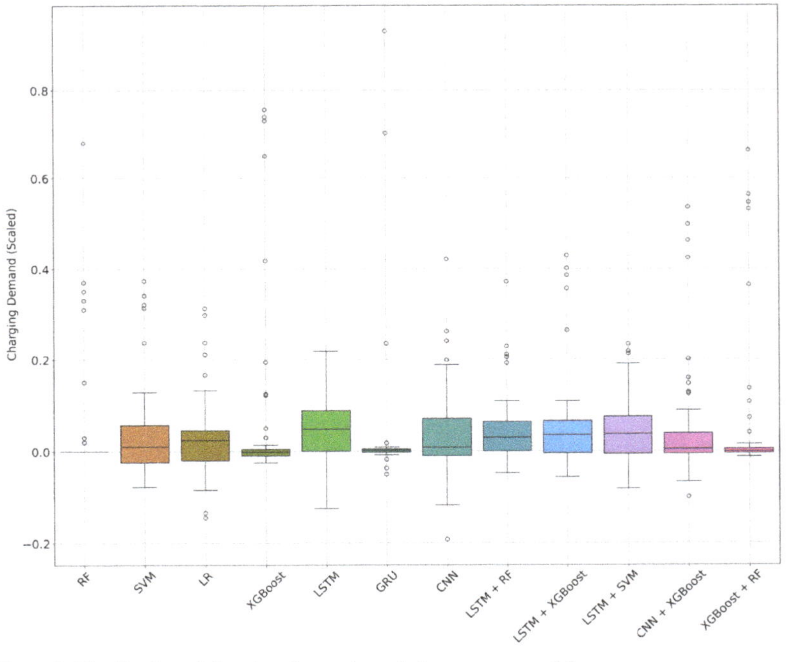

Figure 3. Distribution of charging demand predictions across models.

DL models, such as LSTM and GRU, exhibit a tighter distribution with fewer extreme values, indicating their ability to generalise effectively. In contrast, traditional models with LR and RF show wider interquartile ranges and more pronounced outliers, suggesting challenges in capturing the complex dependencies in the charging data. The proposed HCB-Net model achieves a balanced distribution, reducing outliers while maintaining a narrow spread, highlighting its effectiveness in making reliable charging demand forecasts.

4.4. User Behavioural Trends Analysis

Figure 4 presents a violin plot analysis comparing EV charging behaviour between the West Midlands and North East regions. The plots illustrate the distribution of responses for different charging-related behaviours, such as EV type, all-electric range, public charging frequency, weekly charging habits, charge duration, and willingness to travel a distance for charging. The density variations in the violin plots reveal distinct user preferences in each region.

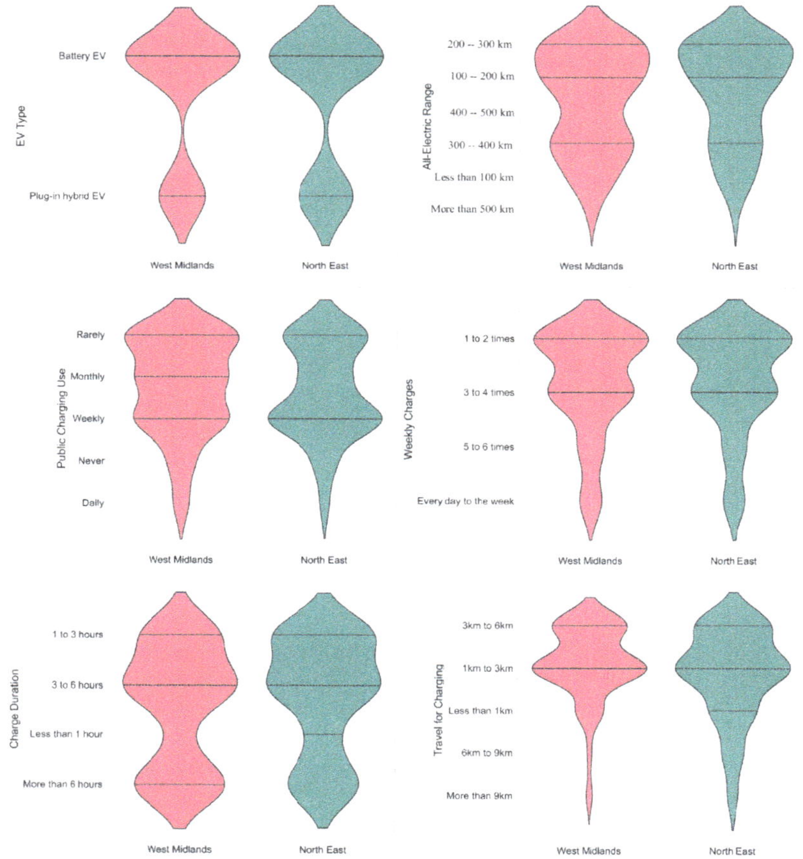

Figure 4. Comparison of user charging behaviour across regions.

Users in the West Midlands exhibit a more diverse distribution of charging habits, with higher variability in public charging frequency and charge duration. This suggests that the charging infrastructure is more developed or widely adopted. In contrast, the North East region shows a more concentrated distribution, indicating a preference for home charging or fewer public charging opportunities. The findings provide insights into regional charging behaviour, which can inform infrastructure planning and policy decisions for optimising EVs for future adoption.

The integration of qualitative insights enriched the interpretation of the model's outputs. User narratives about charging inconvenience, cost constraints, and preferences for workplace charging aligned with quantitative predictors of demand, reinforcing the model's behavioural validity. This mixed-methods approach, although preliminary, points to the value of user-centred forecasting in early-stage smart city planning.

4.5. Charging Demand Forecast

The forecasted demand for EV charging highlights a substantial increase in the need for fast-charging stations, as observed from model predictions in Figure 2. The comparative analysis of ML and DL models indicates that the highest accuracy models, such as HCB-Net (CNN + XGBoost), XGBoost, and RF, demonstrate a rising trend in charging demand across various test samples. This trend suggests that charging facilities must be expanded as EV adoption grows to accommodate increasing usage patterns.

The box plot analysis in Figure 3 further supports this observation by showing that the distribution of predicted charging demand varies across models, with hybrid models providing the most stable and accurate forecasts. The presence of outliers in traditional models such as LSTM and GRU suggests inconsistent demand predictions, which may indicate sudden spikes in charging requirements at specific times. These fluctuations imply the necessity of a reliable fast-charging network that can handle demand surges efficiently.

As illustrated in Figure 4, user behavioural trends show that many EV users prefer integrating charging into their daily commutes and residential areas. The violin plots comparing the West Midlands and North East regions indicate that most users charge their vehicles 1–4 times per week, with a significant portion opting for charging durations of 1–3 h. This preference suggests an increasing reliance on public and home-based fast-charging stations to meet daily commuting needs. Additionally, travel preferences for charging remain concentrated within 1–3 km, implying that new charging points should be installed at convenient locations within urban and suburban areas.

The results from predictive models and user behaviour analysis collectively indicate that EV charging demand must evolve to meet rising demand. Deploying fast-charging stations at strategic locations, such as commercial hubs, residential zones, and major highways, will ensure seamless EV adoption. Furthermore, smart grid integration and dynamic load management can be implemented to effectively balance peak demand periods, thereby preventing overloading and ensuring optimal usage of the charging network.

We acknowledge that traditional behavioural modelling techniques, such as Structural Equation Modelling (SEM) or Confirmatory Factor Analysis (CFA), are well established for analysing survey data and can provide deeper insights into the psychological and socio-demographic drivers of charging behaviour. In contrast, the present study has focused on applying a predictive ML framework to support practical demand forecasting for infrastructure planning. While this approach offers quantitative forecasts of charging demand, it does not fully capture the underlying behavioural mechanisms. A promising direction for future work is to integrate SEM or CFA analyses with the ML pipeline for example, by using latent factors identified through SEM as inputs to forecasting models, thereby combining explanatory richness with predictive power. This would enhance both the interpretability and practical value of the framework for policymakers and transport planners.

4.6. Interpretation of Charging Demand Forecasts and Practical Implications

The charging demand forecasts generated by the HCB-Net model provide valuable insights for practical infrastructure planning in the target regions. The ability of the model to accurately predict average daily charging demand at the user level, based on behavioural and spatial features, highlights the importance of demographic and regional factors in shaping future infrastructure needs. Notably, the model captures regional variation in demand profiles, with users in the West Midlands demonstrating higher variability and greater public charging reliance, whereas North East users exhibit more stable patterns centred on home charging.

These findings suggest that infrastructure investment strategies should be tailored to local contexts: in regions with higher public charging dependency and greater commuting distances, a higher density of fast chargers may be warranted to avoid peak-period congestion and enable flexible travel. Conversely, in areas with strong home-charging preference, investment in grid reinforcement and incentives for smart charging at residential locations could yield greater benefits.

Importantly, the model's ability to provide interpretable forecasts from behavioural survey data, without requiring high-resolution smart meter inputs, offers a practical tool for planners in regions where real-time telemetry is limited. The forecasts can also inform dynamic pricing, load balancing, and phased infrastructure rollout strategies, supporting the transition towards efficient and user-responsive EV charging networks. In future work, further refinement of temporal granularity will enable more detailed peak load management planning.

It is important to acknowledge that the current model is trained on a relatively modest dataset (256 responses), which imposes certain limitations on its predictive capacity. While the use of survey-derived behavioural and spatial data demonstrates promising results, achieving an R^2 of 0.3996, this remains insufficient for highly precise forecasting applications. Moreover, several critical factors known to influence EV charging demand, such as electricity pricing dynamics, weather conditions, seasonal travel patterns, and real-time grid constraints, were not incorporated in the present study due to data availability constraints. The absence of these variables undoubtedly affects model generalisability and limits its immediate applicability for operational decision making. As such, the current framework is best viewed as an early-stage planning tool, capable of informing infrastructure strategy and spatial deployment decisions in data-sparse contexts. Future work will prioritise the integration of dynamic external variables, longitudinal data, and larger-scale behavioural datasets to enhance forecasting robustness and enable real-time applications.

4.7. Policy Implications and Recommendations

The findings from the model performance evaluation, charging demand forecasts, and user behavioural analysis underscore the government intervention's critical role in expanding and enhancing EV charging demand. The demand for EV charging stations is projected to rise significantly, necessitating proactive measures to ensure accessibility, efficiency, and sustainability. From this comparison of data from two contrasting regions, differences in charging behaviours are influenced by the availability of chargers, frequency of use, duration of charging, and the distance travelled. The advantage of AI with ML and DL is that they are applied in real time, with access to data from chargers.

The algorithms derived from this research can be used to perform the following:

1. Short-term forecasting of charging demand can be used to improve the information available to drivers currently using public chargers;
2. Monitoring changing needs of users, allowing evidence-based decision making regarding the provision of rapid DC fast chargers in the long term.

One of the key policy recommendations is investment in expanding and upgrading charging networks. The forecasted demand trends highlight the need for fast-charging stations at strategic locations, such as urban centres, residential areas, commercial districts, and major highways. Governments should implement financial incentives, subsidies, and public–private partnerships (PPPs) to accelerate the deployment of high-speed and ultra-fast charging stations to support this expansion. Additionally, regulatory frameworks should ensure that new developments, such as housing estates and commercial complexes, incorporate EV-ready infrastructure for residents, local travellers, and visitors to the regions. These are necessary to support long-term adoption.

A user-centric design approach is essential to enhance the accessibility and convenience of EV charging. As seen in Figure 4, users typically charge their vehicles 1–4 times per week and prefer charging locations within 1–3 km of their routine commute. To accommodate these preferences, policymakers must encourage installing charging points in high-traffic areas, public parking facilities, and workplaces. Furthermore, integrating smart grid technologies and demand-side management strategies can optimise energy distribution, ensuring the charging infrastructure remains reliable without overloading the power grid.

The data-driven forecasting approach proposed in this study can serve as a foundational tool for implementing both direct and indirect EV charging control strategies. In the context of direct control, predicted regional charging load patterns could inform real-time grid management systems that schedule or throttle charging during peak demand periods, especially when integrated with local distribution network data. Conversely, in the context of indirect control, these behavioural forecasts can support the design of dynamic pricing models, incentive programmes, or infrastructure placement strategies aimed at passively influencing user charging decisions without enforcing strict intervention.

As highlighted by Zahedmanesh et al. [16] and Bordin and Tomasgard [17], successful EV–grid integration depends on a well-calibrated mix of control mechanisms. The proposed model provides a data-informed lens through which such mechanisms can be better targeted, particularly in regions where charging behaviours vary spatially and demographically. Integrating such predictive tools with smart city policy design can enable more agile and adaptive planning responses to increasing EV penetration.

In addition to infrastructure expansion, standardisation of charging protocols and interoperability across networks is crucial. EV users should be able to access charging stations seamlessly, regardless of the operator, through universal payment systems, real-time availability updates, and mobile application integration. By implementing the data-driven decision-making processes, informed by the forecasting algorithms developed in this research, authorities can monitor charging patterns, forecast future needs, and allocate resources efficiently.

Finally, to promote widespread EV adoption, policymakers should launch awareness campaigns and incentive programs, such as reduced charging fees, tax benefits, and rebates for EV users. These initiatives will not only encourage more people to transition to electric mobility but also support a sustainable and efficient EV ecosystem in smart cities.

By implementing these recommendations, governments and stakeholders can ensure the successful expansion of EV charging networks, enhancing both user experience and environmental benefits while supporting the long-term growth of EV adoption.

4.8. Predictive–Prescriptive Integration and Future Directions

While the present study focuses on predictive analytics—specifically estimating user-level EV charging demand using ML—the insights generated by the HCB-Net model also offer strong potential for integration with mathematical optimisation frameworks. This connection forms the basis of prescriptive analytics, wherein predictions guide decision making, such as when, where, and how much EV charging should be encouraged or deferred.

In the context of direct control strategies, the forecasted demand could feed into grid-aware scheduling algorithms that allocate charging loads in real time, optimising for network stability or cost. For indirect control strategies, predictions could inform dynamic pricing mechanisms that influence user behaviour through economic incentives. These themes align with the optimisation-based frameworks presented in prior work by Zahedmanesh et al. [16] and Bordin and Tomasgard [17], who, respectively, explored control hardware for load shaping and behavioural incentive design through microeconomic modelling.

Compared to existing fast-charging infrastructure studies, such as Gnann et al. (2018), which focused on empirical utilisation rates and queuing models calibrated with operational charging data [47], the proposed HCB-Net framework complements this line of work by introducing a predictive tool grounded in user-level behavioural and spatial inputs.

Unlike queuing or capacity-based approaches, HCB-Net enables forecasting of evolving regional demand patterns based on socio-demographic dynamics, offering planners a forward-looking mechanism that can be adapted as EV markets grow and user behaviours shift. Furthermore, the modular design of HCB-Net aligns with advances in vehicle trajectory modelling seen in recent work, such as Ma et al. (2025) [48], where data-fusion and learning-based techniques improve the reconstruction of complex spatiotemporal patterns. The behavioural modelling and spatial feature extraction capabilities of HCB-Net make it highly applicable to related fields, such as demand-responsive transit systems, dynamic traffic assignment, or the integration of electric mobility into connected and autonomous vehicle (CAV) environments. Thus, this study not only advances EV infrastructure forecasting but also contributes a methodological foundation with wider potential across intelligent transport system applications.

Future research could build on this foundation by integrating HCB-Net with multi-objective optimisation models that balance user satisfaction, charging costs, and grid constraints. Such work would enable the development of fully integrated predictive–prescriptive systems for smart city EV infrastructure management, capable of both forecasting demand and recommending optimal charging strategies in real time.

4.9. Generalisability and Contextual Limitations

While the proposed methodology demonstrates promising results in two English regions—the North East and West Midlands—its broader applicability depends on certain contextual similarities. The model is most transferable to urban or peri-urban areas with:

- Moderate to high EV adoption rates;
- A mix of public and private charging infrastructure;
- Varying levels of user charging autonomy (e.g., home vs. workplace charging);
- Diverse demographic and travel profiles.

These features influence user behaviour and infrastructure demand, which are central to the model's feature set and architecture. Applying the model in regions with fundamentally different energy pricing, charger accessibility, or mobility patterns (e.g., rural areas, countries with state-controlled utilities) may require local retraining and contextual adaptation.

Notably, while the numerical outputs of the HCB-Net model may not be directly portable, the framework itself—combining behavioural feature extraction with hybrid prediction—is broadly applicable across smart city planning contexts. Future work should involve external validation across more regions and potentially across countries with different EV policies and urban development stages.

4.10. Limitations in Computational Cost and Scalability

While the HCB-Net model demonstrates improved predictive performance, it also presents certain limitations in terms of computational complexity and scalability. The CNN component requires significant training time and memory allocation due to its convolutional and pooling layers, particularly when working with larger input matrices or higher-dimensional feature encodings. Furthermore, the hybrid architecture necessitates sequential training—first through deep feature extraction, followed by XGBoost regression—making the pipeline more resource-intensive than single-model approaches.

Scalability may also be limited when applied to significantly larger datasets or when extended to national-level EV infrastructure planning. Real-time deployment, for example, would require substantial optimisation or simplification of the current architecture, especially in environments with constrained computational infrastructure. Although the current model is efficient enough for offline planning tasks and regional-scale simulations, future versions may benefit from model pruning, parallelisation strategies, or deployment on edge devices to support broader smart city integration.

We also recognise that large-scale behavioural datasets derived from social media platforms, such as Twitter and Reddit, represent a highly promising complementary source for understanding public attitudes, behavioural trends, and emerging perceptions surrounding EV charging demand. Recent studies by Ruan and Lv (2022) [49] and Qian and Gkritza (2024) [50] have demonstrated that natural language processing (NLP) applied to social media data can capture real-time, spatially resolved public sentiment and topic trends, often at far greater scale and timeliness than traditional surveys. Compared to small-scale questionnaire data, such sources could provide broader insights into public discourse, regional concerns, and emerging behavioural shifts related to EV adoption and infrastructure needs. While the present study focused on structured survey data as a first step towards interpretable demand forecasting, future work will explore integrating social-media-derived insights with ML-based behavioural models. This combined approach would enable more scalable, adaptive forecasting and offer a deeper, dynamic understanding of public attitudes, enhancing the relevance of demand modelling for planners and policymakers.

We also acknowledge that the use of CNNs for this type of structured tabular data introduces certain methodological limitations. CNNs are optimally suited to learning spatially correlated or grid-based data structures, such as images or time-series windows, whereas the present feature set, derived from user survey responses, lacks inherent spatial continuity across features. In this study, CNN layers were used primarily to explore potential latent interactions between behavioural and demographic variables, and to test whether convolutional filters could enhance generalisation in a small-sample regime. However, the comparable performance of a simple linear regression model ($R^2 = 0.3520$) suggests that DL may not be essential for this particular dataset structure. In future work, we will prioritise more methodologically aligned approaches, such as gradient boosting decision trees (e.g., XGBoost, LightGBM), TabNet, or interpretable ensemble models, which are better suited to tabular data and offer greater explainability for practical transport planning applications.

These limitations do not detract from the model's utility in exploratory and planning contexts but should be considered when designing real-time, scalable systems for grid operators or mobility platforms.

5. Conclusions

This study presented HCB-Net, a hybrid deep learning framework integrating Convolutional Neural Networks and XGBoost to predict EV charging demand. Using survey-derived behavioural and spatial data from two UK regions—the West Midlands and the North East—the model achieved a Root Mean Squared Error (RMSE) of 0.1490 and an R^2 score of 0.3996, outperforming a range of traditional machine learning and standalone deep learning models. While the model demonstrated only moderate predictive accuracy, it provided interpretable and regionally grounded insights into user charging behaviours. The study showed that meaningful demand estimates can be extracted even without high-frequency temporal data, making the framework particularly suitable for cities in the early stages of EV infrastructure development. Beyond methodological innovation, the model offers practical value for smart city planners, utilities, and transport authorities. By identifying key behavioural and geospatial predictors of charging demand, HCB-Net can guide the optimal siting of charging stations, prioritise infrastructure investments in high-demand zones, and inform the design of dynamic pricing or demand response strategies. This is especially valuable for urban regions facing high variability in EV usage or preparing for large-scale transitions to electric mobility. Future work will expand the dataset to include more diverse geographic areas, incorporate smart meter or vehicle telemetry data, and extend the model to support real-time forecasting and prescriptive analytics.

Future Work

While this study has provided valuable insights into EV charging demand forecasting, several avenues for future research remain open.

1. Incorporation of Real-time Data: Future work could focus on integrating real-time traffic patterns, weather conditions, and energy grid fluctuations to enhance the accuracy of demand forecasting models.
2. Integration of V2G Systems: As V2G technology becomes more prevalent, future studies should explore how bidirectional energy flow affects charging station demand.
3. Agent-Based Simulations for Charging Infrastructure: Conducting agent-based simulations to model user behaviour, charging station congestion, and dynamic pricing strategies can provide further insights into optimising the charging network and improved information for the public.
4. Geospatial Analysis for Optimal Charging Station Placement: A detailed GIS-based approach could identify optimal charging station locations based on demographics, travel patterns, and urban planning constraints.

Addressing these future directions will allow researchers and policymakers to refine further their EV charging demand predictions. This will ensure a scalable, efficient, and user-friendly charging infrastructure for the widespread adoption of electric mobility.

Author Contributions: Conceptualisation, M.C., H.A., D.D. and M.B.; methodology, M.C., H.A., A.S. and S.D.; software, M.C.; validation, M.C., H.A., D.D. and M.B.; investigation, M.C.; resources, A.S., S.D. and M.C.; writing—original draft preparation, M.C.; writing—review and editing, M.C., D.D. and M.B.; visualisation, M.C. All authors have read and agreed to the published version of the manuscript.

Funding: This research received no external funding.

Data Availability Statement: The data and code are publicly available at https://github.com/cavusmuhammed68/Low_Carbon (accessed on 1 June 2025).

Conflicts of Interest: The authors declare no conflicts of interest.

References

1. Chen, T.; Zhang, X.-P.; Wang, J.; Li, J.; Wu, C.; Hu, M.; Bian, H. A review on electric vehicle charging infrastructure development in the UK. *J. Mod. Power Syst. Clean Energy* **2020**, *8*, 193–205. [CrossRef]
2. Brown, W.; King, M.; Goh, Y.M. UK smart cities present and future: An analysis of British smart cities through current and emerging technologies and practices. *Emerald Open Res.* **2023**, *1*, 5. [CrossRef]
3. Souza, G.G.D.; Santos, R.R.D.; Godoy, R.B. Optimizing power grids: A valley-filling heuristic for energy-efficient electric vehicle charging. *PLoS ONE* **2025**, *20*, e0316677. [CrossRef]
4. Hall, D.; Lutsey, N. *Emerging Best Practices for Electric Vehicle Charging Infrastructure*; The International Council on Clean Transportation (ICCT): Washington, DC, USA, 2017; Volume 54.
5. Suresh, S.; Renukappa, S.; Abdul-Aziz, A.-R.; Paloo, Y.; Jallow, H. Developments in the UK road transport from a smart cities perspective. *Eng. Constr. Archit. Manag.* **2021**, *28*, 845–862. [CrossRef]
6. Heidrich, O.; Hill, G.A.; Neaimeh, M.; Huebner, Y.; Blythe, P.T.; Dawson, R.J. How do cities support electric vehicles and what difference does it make? *Technol. Forecast. Soc. Change* **2017**, *123*, 17–23. [CrossRef]
7. Razmjoo, A.; Nezhad, M.M.; Kaigutha, L.G.; Marzband, M.; Mirjalili, S.; Pazhoohesh, M.; Memon, S.; Ehyaei, M.A.; Piras, G. Investigating smart city development based on green buildings, electrical vehicles and feasible indicators. *Sustainability* **2021**, *13*, 7808. [CrossRef]
8. Apata, O.; Bokoro, P.N.; Sharma, G. The risks and challenges of electric vehicle integration into smart cities. *Energies* **2023**, *16*, 5274. [CrossRef]

9. Torkey, A.; Abdelgawad, H. Framework for planning of EV charging demand: Where should cities start? *Transp. Policy* **2022**, *128*, 193–208. [CrossRef]
10. Bakker, J.J.; Mom, G.P.A.; Schot, J.W. *Contesting Range Anxiety: The Role of Electric Vehicle Charging Infrastructure in the Transportation Transition*; Technische Universiteit Eindhoven: Eindhoven, The Netherlands, 2011.
11. Mashhoodi, B.; van der Blij, N. Drivers' range anxiety and cost of new EV chargers in Amsterdam: A scenario-based optimization approach. *Ann. GIS* **2021**, *27*, 87–98. [CrossRef]
12. Kumar, T. Alleviating range anxiety with 'Enroute': A real-time mobile solution and infrastructure innovation for electric vehicle users. *Procedia CIRP* **2024**, *128*, 37–42.
13. Mohammed, A.; Saif, O.; Abo-Adma, M.; Fahmy, A.; Elazab, R. Strategies and sustainability in fast charging station deployment for electric vehicles. *Sci. Rep.* **2024**, *14*, 283. [CrossRef] [PubMed]
14. Powell, B.; Johnson, C. *Impact of Electric Vehicle Charging Station Reliability, Resilience, and Location on Electric Vehicle Adoption*; National Renewable Energy Laboratory (NREL): Golden, CO, USA, 2024.
15. Yousuf, A.K.M.; Wang, Z.; Paranjape, R.; Tang, Y. An in-depth exploration of electric vehicle charging station infrastructure: A comprehensive review of challenges, mitigation approaches, and optimization strategies. *IEEE Access* **2024**, *12*, 51570–51589. [CrossRef]
16. Zahedmanesh, A.; Muttaqi, K.M.; Sutanto, D. Direct control of plug-in electric vehicle charging load using an in-house developed intermediate control unit. *IEEE Trans. Ind. Appl.* **2019**, *55*, 2208–2218. [CrossRef]
17. Bordin, C.; Tomasgard, A. Behavioural change in green transportation: Micro-economics perspectives and optimization strategies. *Energies* **2021**, *14*, 3728. [CrossRef]
18. Philipsen, R.; Schmidt, T.; Van Heek, J.; Ziefle, M. Fast-charging station here, please! User criteria for electric vehicle fast-charging locations. *Transp. Res. Part Traffic Psychol. Behav.* **2016**, *40*, 119–129. [CrossRef]
19. Sun, X.-H.; Yamamoto, T.; Morikawa, T. Fast-charging station choice behavior among battery electric vehicle users. *Transp. Res. Part Transp. Environ.* **2016**, *46*, 26–39. [CrossRef]
20. Asensio, O.I.; Lawson, M.C.; Apablaza, C.Z. Electric vehicle charging stations in the workplace with high-resolution data from casual and habitual users. *Sci. Data* **2021**, *8*, 168. [CrossRef]
21. Tejaswi, V. Electric Vehicle Charging Behavior: An Analysis of Workplace Charging Heterogeneity to Improve Charging Network Planning. UC San Diego: Climate Science and Policy. 2024. Available online: https://escholarship.org/uc/item/7kp2w0c1 (accessed on 25 June 2025).
22. Zhang, Y.; Li, S.; Blythe, P.; Wardle, J.; Herron, C.; Edwards, S.; Li, D.; Ji, Y.; Namdeo, A. Analysis of electric vehicle charging behaviour in existing regional public and workplace charging infrastructure: A case study in the North-East UK. *Transp. Eng.* **2025**, *19*, 100309. [CrossRef]
23. Hardman, S.; Jenn, A.; Tal, G.; Axsen, J.; Beard, G.; Daina, N.; Figenbaum, E.; Jakobsson, N.; Jochem, P.; Kinnear, N.; et al. A review of consumer preferences of and interactions with electric vehicle charging infrastructure. *Transp. Res. Part Transp. Environ.* **2018**, *62*, 508–523. [CrossRef]
24. Rogers, E.M.; Singhal, A.; Quinlan, M.M. Diffusion of innovations. In *An Integrated Approach to Communication Theory and Research*; Routledge: London, UK, 2014; pp. 432–448.
25. Smith, B.; Olaru, D.; Jabeen, F.; Greaves, S. Electric vehicles adoption: Environmental enthusiast bias in discrete choice models. *Transp. Res. Part D Transp. Environ.* **2017**, *51*, 290–303. [CrossRef]
26. Shanmuganathan, J.; Victoire, A.A.; Balraj, G.; Victoire, A. Deep learning LSTM recurrent neural network model for prediction of electric vehicle charging demand. *Sustainability* **2022**, *14*, 10207. [CrossRef]
27. Koohfar, S.; Woldemariam, W.; Kumar, A. Performance comparison of deep learning approaches in predicting EV charging demand. *Sustainability* **2023**, *15*, 4258. [CrossRef]
28. Cavus, M.; Dissanayake, D.; Bell, M. Deep-Fuzzy Logic Control for Optimal Energy Management: A Predictive and Adaptive Framework for Grid-Connected Microgrids. *Energies* **2025**, *18*, 995. [CrossRef]
29. Tirunagari, S.; Gu, M.; Meegahapola, L. Reaping the benefits of smart electric vehicle charging and vehicle-to-grid technologies: Regulatory, policy and technical aspects. *IEEE Access* **2022**, *10*, 114657–114672. [CrossRef]
30. Tsalikidis, N.; Koukaras, P.; Ioannidis, D.; Tzovaras, D. Hybrid CNN-LSTM forecasting model for electric vehicle charging demand in smart buildings. In Proceedings of the 2024 6th Global Power, Energy and Communication Conference (GPECOM), Budapest, Hungary, 4–7 June 2024; pp. 590–595.
31. Hafeez, A.; Alammari, R.; Iqbal, A. Utilization of EV charging station in demand side management using deep learning method. *IEEE Access* **2023**, *11*, 8747–8760. [CrossRef]
32. Cavus, M.; Dissanayake, D.; Bell, M. Next Generation of Electric Vehicles: AI-Driven Approaches for Predictive Maintenance and Battery Management. *Energies* **2025**, *18*, 1041. [CrossRef]
33. Sasidharan, M.P.; Kinattingal, S.; Simon, S.P. Comparative analysis of deep learning models for electric vehicle charging load forecasting. *J. Inst. Eng. Ser.* **2023**, *104*, 105–113. [CrossRef]

34. Lv, S.; Sun, G.; Wei, Z.; Chen, S.; Chen, Z. Optimal pricing and energy sharing of EV charging stations with an augmented user equilibrium model. *IEEE Trans. Power Syst.* **2023**, *39*, 4336–4349. [CrossRef]
35. Altamimi, A.; Ali, M.B.; Kazmi, S.A.A.; Khan, Z.A. Multi-Agent Reinforcement Learning Optimization Framework for On-Grid Electric Vehicle Charging from Base Transceiver Stations Using Renewable Energy and Storage Systems. *Energies* **2024**, *17*, 3592. [CrossRef]
36. Cavus, M.; Ayan, H.; Oyebamiji, O.K.; Dissanayake, D. Deep Charge-Fusion Model: Advanced Hybrid Modelling for Predicting Electric Vehicle Charging Patterns with Socio-Demographic Considerations. *Int. J. Transp. Sci. Technol.* 2025, *in press*. [CrossRef]
37. Cavus, M.; Allahham, A.; Adhikari, K.; Zangiabadia, M.; Giaouris, D. Control of Microgrids Using an Enhanced Model Predictive Controller. In Proceedings of the 11th International Conference on Power Electronics, Machines and Drives (PEMD 2022), Hybrid Conference, Newcastle, UK, 21–23 June 2022; pp. 660–665.
38. Prajeesh, C.B.; Pillai, A.S.; Khwaja, A.S.; Anpalagan, A. Enhancing Electric Vehicle Charging Infrastructure: A Framework for Efficient Charging Point Management. *e-Prime Adv. Electr. Eng. Electron. Energy* **2025**, *11* , 100926.
39. Cavus, M.; Allahham, A. Spatio-Temporal Attention-Based Deep Learning for Smart Grid Demand Prediction. *Electronics* **2025**, *14*, 2514. [CrossRef]
40. Mastoi, M.S.; Zhuang, S.; Munir, H.M.; Haris, M.; Hassan, M.; Usman, M.; Bukhari, S.S.H.; Ro, J.-S. An in-depth analysis of electric vehicle charging station infrastructure, policy implications, and future trends. *Energy Rep.* **2022**, *8*, 11504–11529. [CrossRef]
41. Qadir, S.A.; Ahmad, F.; Al-Wahedi, A.M.A.B.; Iqbal, A.; Ali, A. Navigating the complex realities of electric vehicle adoption: A comprehensive study of government strategies, policies, and incentives. *Energy Strategy Rev.* **2024**, *53*, 101379. [CrossRef]
42. Dik, A.; Omer, S.; Boukhanouf, R. Electric vehicles: V2G for rapid, safe, and green EV penetration. *Energies* **2022**, *15*, 803. [CrossRef]
43. Wang, S.; Zhuge, C.; Shao, C.; Wang, P.; Yang, X.; Wang, S. Short-term electric vehicle charging demand prediction: A deep learning approach. *Appl. Energy* **2023**, *340*, 121032. [CrossRef]
44. Openshaw, S.; Lu, X.; Pham, H.; Qin, H.; Lukic, S. Investigation of Commercial Viability and Public Perception of Electrified Roadways with Dynamic Wireless Charging. In Proceedings of the IEEE Wireless Power Technology Conference and Expo (WPTCE), San Diego, CA, USA, 4–8 June 2023.
45. Office for National Statistics (ONS). Census 2021, England and Wales. Available online: https://www.ons.gov.uk/census (accessed on 5 April 2025).
46. Bösehans, G.; Bell, M.; Thorpe, N.; Liao, F.; Homem de Almeida Correia, G.; Dissanayake, D. eHUBs—Identifying the potential early and late adopters of shared electric mobility hubs. *Int. J. Sustain. Transp.* **2023**, *17*, 199–218. [CrossRef]
47. Gnann, T.; Funke, S.; Jakobsson, N.; Plötz, P.; Sprei, F.; Bennehag, A. Fast charging infrastructure for electric vehicles: Today's situation and future needs. *Transp. Res. Part Transp. Environ.* **2018**, *62*, 314–329. [CrossRef]
48. Ma, J.; Roncoli, C.; Ren, G.; Yang, Y.; Wang, S.; Yu, J. An integrated framework for fully sampled vehicle trajectory reconstruction using a fused dataset. *J. Intell. Transp. Syst.* **2025** . [CrossRef]
49. Ruan, T.; Lv, Q. Public perception of electric vehicles on Reddit over the past decade. *Commun. Transp. Res.* **2022**, *2*, 100070. [CrossRef]
50. Qian, X.; Gkritza, K. Spatial and temporal variance in public perception of electric vehicles: A comparative analysis of adoption pioneers and laggards using Twitter data. *Transp. Policy* **2024**, *149*, 150–162. [CrossRef]

MDPI AG
Grosspeteranlage 5
4052 Basel
Switzerland
Tel.: +41 61 683 77 34

Energies Editorial Office
E-mail: energies@mdpi.com
www.mdpi.com/journal/energies

www.ingramcontent.com/pod-product-compliance
Lightning Source LLC
Chambersburg PA
CBHW061550220926
51850CB00005B/302